MODERN METHODS IN CARBOHYDRATE SYNTHESIS

Edited by

Shaheer H. Khan
and
Roger A. O'Neill

Perkin Elmer – Applied Biosystems Division
Foster City, CA, USA

harwood academic publishers
Australia • China • France • Germany • India • Japan • Luxembourg
Malaysia • The Netherlands • Russia • Singapore • Switzerland
Thailand • United Kingdom • United States

Emmaplein 5
1075 AW Amsterdam
The Netherlands

British Library Cataloguing in Publication Data

Modern Methods in Carbohydrate Synthesis.
– (Frontiers in Natural Product Research Series; Vol. 1)
  I. Khan, S.H. II. O'Neill, R.A. III. Series
  547.780459

  ISBN 3-7186-5785-6 (hard)
       3-7186-5921-2 (soft)

Front Cover: Solution conformation of Sialyl Lewis X tetrasaccharide as determined by NMR. This molecular modelling projection showing hydrophobic surface was generated by Tripos, Inc., using their Sybyl software for molecular modelling and Molcad software for surface property mapping.

MODERN METHODS IN
CARBOHYDRATE SYNTHESIS

Frontiers in Natural Product Research

A series of books presenting various aspects of the research on differing classes of natural products and providing comprehensive accounts of recent important developments in the field.

Edited by Atta-ur-Rahman, H.E.J. Research Institute of Chemistry, University of Karachi, Pakistan

Volume 1
Modern Methods in Carbohydrate Synthesis
edited by Shaheer H. Khan and Roger A. O'Neill

Other volumes in preparation

Total Synthesis of Indole Alkaloids
edited by Anwer Basha and Atta-ur-Rahman

Manual of Bioassay Techniques
edited by Atta-ur-Rahman and M. Iqbal Choudhary

This book is part of a series. The publisher will accept continuation orders which may be cancelled at any time and which provide for automatic billing and shipping of each title in the series upon publication. Please write for details.

This book is dedicated to the memory of our fathers,

Shabbir H. Khan, L.L.B. (1924–1994)
Gerard K. O'Neill, Ph.D. (1927–1992)

CONTENTS

PREFACE TO THE SERIES

Natural product chemistry is one of the most stimulating branches of organic chemistry. It is therefore not surprising that such giants of organic chemistry as Arigoni, Barton, Robinson, Todd and Woodward, to name but a few, have been closely associated with the development of various facets of this important field.

The enormous structural diversity presented by plant substances offers exciting opportunities to pharmaceutical chemists to discover new lead compounds. The spectacular advances in spectroscopic techniques, as well as isolation technology coupled with the growing understanding at the molecular level about the causes of a number of diseases, have catalysed efforts in new drug design and development. The discoveries in the field of asymmetric synthesis have also greatly facilitated the efforts of the organic chemists to develop chiral syntheses of target molecules.

"Frontiers in Natural Product Research" is a new series of monographs which will comprise focused volumes in various frontier areas of natural product chemistry, written by eminent experts in each field of specialisation. It is hoped that the series will be enthusiastically received by organic chemists, natural product chemists and medicinal chemists.

Atta-ur-Rahman

FOREWORD

We give advice but we cannot give the wisdom to profit by it.
Duc de la Rochefoucauld

This book is a collection of 21 essays by prominent chemists who have played important roles in developing the now well-established field of glycobiology. The goal set by the editors was to produce a one-volume survey of the methodologies published in recent years that are relevant to the synthesis of complex oligosaccharides. It was a timely project since the past decade had seen a veritable frenzy of activity concerned with the availability of complex oligosaccharides – both natural and selected congeners – for possible use in biomedical research. The time was most opportune. Effective funding was available since pioneers in the field had demonstrated that many important diseases result from inappropriate participation of carbohydrate ligands in cellular functions. Furthermore, the instrumentations, techniques and theoretical insights of sufficient sophistication for significant involvement had become well established. The editors are to be congratulated on seizing the opportunity to compile a review that surely will prove to be of as much value to seasoned practitioners as to newcomers to the field.

The incentives for the much heightened interest in carbohydrate synthesis came from the promise that major, even perhaps revolutionary improvements in health care would arise from the rapidly increasing appreciation by molecular biologists of the role of complex carbohydrates in life processes. The "gold rush" began about 20 years ago with the chemical synthesis of several human blood group determinants at the tri- and tetrasaccharide level in forms suitable for use as inhibitors and the preparation of artificial antigens and immunoadsorbents. In this connection, it would be instructive to consult the autobiography that I prepared for the American Chemical Society entitled: "Explorations with Sugars. How Sweet It Was," published in 1990.

The book well documents that the stage is now well set for effective collaboration with biomedical researchers in the general area of glycobiology. The return on the major investment made in the past decade for the development of the powerful arsenal for synthesis awaits challenges to arise from discoveries in biomedical laboratories. Some notable forays have been made, especially in the search for anti-inflammatory mimics of sialyl Le^x. Others will surely come and it will be for the chemists to recognize the opportunities provided. To paraphrase Rochefoucauld, the biologist may give advice but will the chemist have the wisdom to profit by it? My recent editorial, "Toward Enhanced Symbiosis of Chemistry and Biology" (Federation of American Societies for Experimental Biology Journal, March, 1994) addresses this important problem.

The book necessarily rests heavily on the secondary literature which, unavoidably, is always subject to omissions, errors and bias. In fact, a number of misleading statements and even contradictions exist. For example, a serious

deficiency is the reference to the halide-ion catalyzed α-glycopyranosylation reaction as a simple "*in situ* anomerization" since this obscures a thoroughly well
established fundamental property of glycosyl halides, that is of major importance to glycoside synthesis. That α-glycopyranosides can be prepared in high
yield by maintaining the low but effective equilibrium concentration of the reacting β-halide is the key that opened the door to the synthesis of complex oligosaccharides. The basic discovery was that the reactivity of a β-halide on reacting
with an alcohol can be so much greater than that of the α-anomer that the reaction can be guided to α-glycoside formation in high yield simply by employing a
sufficiently high halide concentration to *catalyze* a very rapid β-halide ⇌
α-halide anomerization. It is particularly important to appreciate this fundamental property since many of the modern methods for α-glycosylation surely
proceed by way of an anomeric mixture of highly reactive glycosyl triflate intermediates. These must form at rates much greater than glycoside formation
and undergo anomerization at rates favorable to α-glycoside formation. The
book leaves much food for thought. Nevertheless, the editors' goals were well
met. Overall, the scholarship is first rate and an important milestone in the history of carbohydrate chemistry is well recorded.

Raymond U. Lemieux, CC, FRS
University of Alberta

PREFACE

Carbohydrates have long been recognized as structural and energy storage molecules, examples of which include cellulose, chitin, and glycogen. Recent advances in chemistry and molecular biology have led to the identification of a group of complex oligosaccharides without obvious structural or energy storage roles and which are nearly ubiquitous in living systems, either as constituents of glycoconjugates such as glycoproteins or glycolipids, or as free oligosaccharides. The past decade has seen the elucidation of the roles these molecules play and their usefulness as therapeutic targets. The list of biological functions of oligosaccharides now includes bacterial and viral adhesion to host tissues, leukocyte trafficking and associated inflammatory responses, tumor cell metastasis, clearance of materials from the blood stream, determination of blood group specificity, and regulation of hormone and enzyme activities. The possibility of inhibiting or otherwise regulating carbohydrate-mediated processes such as those listed above has spawned the birth of a small biotechnology industry, and opens up the possibility of the development of oligosaccharide-based or oligosaccharide-mimetic drugs. Synthetic oligosaccharides are currently used as acceptor substrates for glycosyltransferases, the enzymes which synthesize complex oligosaccharides in nature and also as substrates for enzymes that selectively degrade complex carbohydrates. As enzyme substrates, synthetic oligosaccharides are currently used in pharmaceutical development programs where the goal is regulation *in vivo* of the production or modification of specific oligosaccharide structures.

Synthesis of complex carbohydrates and their analogs has long been regarded as among the most challenging areas in synthetic organic chemistry. There are many excellent reviews covering various aspects of synthetic carbohydrate chemistry. This book seeks to present in a concise manner a broad survey of current state-of-the-art synthetic carbohydrate chemistry. The authors have been asked to provide, wherever possible, sufficient synthetic detail to allow the reader to implement the chemistry discussed in their own laboratories, with the emphasis on chemistry which pertains to biologically active molecules.

The most challenging obstacle in synthetic carbohydrate chemistry involves the stereospecific formation of glycosidic bonds. Therefore, this area receives the most extensive coverage, including both natural glycosidic linkage formation, and the attachment of sugars through linkages not occurring in nature. We begin with an historical overview of the most commonly applied strategies for glycosidic linkage formation. A number of chapters then cover a variety of approaches, many of which have been developed recently and represent significant advances in glycosidation techniques. These are followed by chapters detailing the formation of unnatural linkages between monosaccharides, glycoconjugate synthesis, and syntheses involving a variety of naturally occurring and unnatural sugar derivatives. We finish with several chapters which cover one of the most exciting and rapidly evolving areas of glycotechnology: the use of enzymes in oligosaccharide synthesis.

The editors are extremely grateful to Professor Hassan S. El Khadem and Angela R. Karash for their invaluable and thorough critical reading of all of the manuscripts contained in this volume. Their linguistic and chemical insight has no doubt significantly impacted the quality of this book. We are also indebted to Humaira Bano for her editorial assistance. We would like to especially thank the authors for their excellent contributions to this volume.

Shaheer H. Khan, Ph.D.
Roger A. O'Neill, Ph.D.
Perkin Elmer – Applied Biosystems Division
Foster City, California

CONTRIBUTORS

Numbers in parentheses indicate the pages on which the authors' contributions begin.

Frank Barresi (251),
Carbohydrate Research Program,
Alberta Research Council,
250 Karl Clark Road,
Edmonton,
Alberta,
Canada T5K 2E5

Mark Bednarski (316),
Stanford University School of
Medicine,
Lucas MRS Research Center,
Stanford,
CA 94305,
USA

Carolyn R. Bertozzi (316),
Department of Anatomy and Program
in Immunology,
University of California,
San Francisco,
CA 94143-0452,
USA

Mark T. Bilodeau (171),
Laboratory of Bioorganic Chemistry,
Sloan-Kettering Institute for Cancer
Research,
1275 York Avenue,
New York,
NY 10021,
USA

Klaus Bock (352),
Carlsberg Laboratory,
Department of Chemistry,
Gamle Carlsberg Vej 10,
DK-2500 Valby,
Copenhagen,
Denmark

Suzanne C. Crawley (492),
Amgen Inc.,
1840 DeHavilland Drive,
Thousand Oaks,
CA 91320,
USA

Samuel J. Danishefsky (171),
Laboratory of Bioorganic Chemistry,
Sloan-Kettering Institute for Cancer
Research,
1275 York Avenue,
New York,
NY 10021,
USA
and
Department of Chemistry,
Columbia University,
New York,
NY 10027,
USA

Bert Fraser-Reid (155),
Department of Chemistry,
Duke University,
Durham,
NC 27706-0346,
USA

T. Bruce Grindley (225),
Department of Chemistry,
Dalhousie University,
Halifax,
Nova Scotia,
Canada B3H 4J3

Akira Hasegawa (277),
Department of Applied Bioorganic
Chemistry,
Gifu University,
Gifu 501-11,
Japan

Wasimul Haque (403),
Department of Biotechnology,
Alberta Research Council,
250 Karl Clark Road,
Edmonton, Alberta,
Canada T5K 2E5

Ole Hindsgaul (251),
Department of Chemistry,
University of Alberta,
Edmonton,
Alberta,
Canada T6G 2G2

Hiroyuki Iijima (301),
The Institute of Physical and
Chemical Research (RIKEN),
Wako-shi,
Saitama,
351-01,
Japan

Robert M. Ippolito (403),
Department of Biotechnology,
Alberta Research Council,
250 Karl Clark Road,
Edmonton,
Alberta,
Canada T5K 2E5

Pavol Kováč (55),
NIDDK,
LMC,
National Institutes of Health,
Bethesda,
MD 20892,
USA

Jiri J. Krepinsky (194),
Department of Molecular and Medical
Genetics,
Carbohydrate Research Centre and
Protein Engineering Network of
Centers of Excellence (PENCE),
University of Toronto,
Toronto,
Ontario,
Canada M5S 1A8

Robert Madsen (155),
Department of Chemistry,
Stanford University,
Stanford,
CA 94305-5080,
USA

Jean-Maurice Mallet (130),
Département de Chimie,
Ecole Normale Supérieure,
U.R.A. 1686,
24 Rue Lhomond,
75231 Paris,
Cédex 05,
France

Khushi L. Matta (437),
Department of Gynecologic
Oncology,
Roswell Park Cancer Institute,
Elm and Carlton Streets,
Buffalo,
NY 14263,
USA

Seema Mehta (107),
Department of Chemistry,
Simon Fraser University,
Burnaby, British Columbia,
Canada V5A 1S6

Morten Meldal (352),
Carlsberg Laboratory,
Department of Chemistry,
Gamle Carlsberg Vej 10,
DK-2500 Valby,
Copenhagen,
Denmark

Yoshiaki Nakahara (301),
The Institute of Physical and
Chemical Research (RIKEN),
Wako-shi,
Saitama,
351-01,
Japan

Kurt G. I. Nilsson (518),
Glycorex AB,
Sölveg. 41,
S-223 70 Lund,
Sweden

Thomas Norberg (82),
Department of Chemistry,
Swedish University of Agricultural
Sciences,
P.O. Box 7015,
S-250 07 Uppsala,
Sweden

Tomoya Ogawa (301),
The Institute of Physical and
Chemical Research (RIKEN),
Wako-shi,
Saitama,
351-01,
Japan
and
Graduate School for Animal Sciences
and Veterinary Medical Science,
University of Tokyo,
Yayoi,
Bunkyo-Ku,
Tokyo,
113-Japan

Monica M. Palcic (492),
Department of Chemistry,
University of Alberta,
Edmonton,
Alberta,
Canada T6G 2G2

Hans Paulsen (1),
Institute of Organic Chemistry,
University of Hamburg,
Martin-Luther-King-Platz 6,
20146 Hamburg,
Germany

Stefan Peters (352),
Carlsberg Laboratory,
Department of Chemistry,
Gamle Carlsberg Vej 10,
DK-2500 Valby,
Copenhagen,
Denmark

B. Mario Pinto (107),
Department of Chemistry,
Simon Fraser University,
Burnaby, British Columbia,
Canada V5A 1S6

René Roy (378),
Department of Chemistry,
University of Ottawa,
Ottawa,
Ontario,
Canada K1N 6N5

Richard R. Schmidt (20),
Fakultät Chemie,
Universität Konstanz,
Postfach 5560 M 725,
D-7750 Konstanz,
Germany

Pierre Sinaÿ (130),
Département de Chimie,
Ecole Normale Supérieure,
U.R.A. 1686,
24 Rue Lhomond,
75231 Paris,
Cédex 05,
France

Chi-Huey Wong (467),
Department of Chemistry,
The Scripps Research Institute,
10666 North Torrey Pines Road,
La Jolla,
CA 92307,
USA

Chapter 1

TWENTY FIVE YEARS OF CARBOHYDRATE CHEMISTRY; AN OVERVIEW OF OLIGOSACCHARIDE SYNTHESIS

HANS PAULSEN

Institute of Organic Chemistry, University of Hamburg

Martin-Luther-King-Platz 6, 20146 Hamburg, Germany

Abstract   This paper provides a historical survey of the last twenty five years of development and progress in the field of oligosaccharide synthesis. Valuable contributions to the investigation of this field are discussed, and the applicability of these contributions to the syntheses of large complex oligosaccharides is illustrated.

1.   INTRODUCTION

In the past twenty five years carbohydrate chemistry has witnessed an amazing renaissance. During the 1960s all the main aspects of the roles played by carbohydrates in the storage and in the supply of energy in biochemical systems were known. Biosynthesis and biodegradation of carbohydrates became clear, and the role of polysaccharides such as starch as a reserve substance and that of cellulose and chitin as skeleton compounds was under-stood. The isolation of antibiotics containing unusual saccharide structures from microorganisms led to increased activity in the investigation of chemical transformation of saccharides. The development of carbohydrate chemistry reached the present day level with the discovery of the importance of glycoconjugates, especially of glycoproteins and glycolipids. The improvement of analytical methods such as NMR-spectroscopy, mass spectrometry, HPLC-chromatography and electrophoresis enabled the

1

elucidation of the structures of highly complex glycoconjugates in a rapid way. The findings concerning the biological significance of glycoconjugates involved in controlling diverse biochemical processes sparked a renewed interest in carbohydrates. It was found that the carbohydrate part of glycoconjugates may act as antigens or receptors for proteins, and these findings led to the discovery of the essential role of carbohydrates in cell-cell recognition phenomena and cell differentiation.

At the same time the methods of oligosaccharide synthesis underwent a rapid development, in which the oligosaccharide chains of glycoconjugates became leading target structures. New reagents and new methods of synthesis were also developed, and the improved analytical methods mentioned before made it possible to have a close control of all steps of synthesis. To illustrate the progress made consider that in the 1960s the synthesis of two complex disaccharides was regarded as an excellent subject for a Ph.D. thesis, whereas today a Ph.D. thesis may contain the syntheses of twenty or more tri- and tetrasaccharides.

There are several reviews dealing with the synthesis of oligosaccharides,[1,2,3] which have had a positive impact on the work in the field.[1] The most recent overview of oligosaccharide synthesis, and the best, and most comprehensive one (450 pages), was written by Lockhoff.[4] Worth mentioning also are the excellent tables worked out by Liptak et al.[5] which list the synthesized oligosaccharides and describe schematic procedures for all steps of the syntheses. All the oligosaccharides that have so far been isolated from nature as well as all the completely deblocked synthetic oligosaccharides with more than two saccharide residues are stored in the data base CARBBANK.[6] This data base holds more than 30,000 entries, offering a rich source for the documentation of oligosaccharides.

The present paper gives a short overview of the development of oligosaccharide synthesis, but does not lay claim as to completeness. The findings that have led to decisive progress are merely pointed out, and events are described as they appeared. The value of a certain methodological approach in synthesis was recognized when several groups of researchers successfully employed it for different tasks. There are at present three chemical methods for linking saccharide units to oligosaccharides which have proved to be excellent. The first based on the classical Koenigs-Knorr reaction involves the use of glycosyl halides. In its modern form, this method is useful for certain types of reactions,[1] although, there is also a wide range of applications for glycosyl fluorides. The second method is the trichloroacetimidate method.[7,8] This method is very flexible, requires low temperatures, and often needs surprisingly short reaction times. The third important method is the thioglycoside method.[9] Because of their stability, thioglycosides themselves cannot be used directly as glycosyl donors, and their alkylthio group must be activated by electrophiles in order to transform it into a good leaving group.

As a rule, it is difficult to predict which of these three methods will be the best suited for a certain problem. The synthesis of disaccharides involves linking two polyfunctional compounds, and the success of the coupling reaction depends on the reactivity of the donor and acceptor, on the catalyst, on the kind of substituents at both saccharide units and of course, on the preferred selectivity of the reaction towards the α- or the β-anomeric form. The experience and preferences of the person conducting the experiment also play a role. In difficult cases it makes sense to test all three methods and to vary the substitution patterns of the saccharide units, in order to find the best results. This is particularly the case during the syntheses of large oligosaccharides such as those presented by Ogawa[10,11] and Hasegawa.[12,13]

2.   1,2-*trans*-GLYCOSYLATION

In the classical form of the Koenigs-Knorr reaction[14] acetylated glycosyl chlorides or bromides are converted in the presence of insoluble silver catalysts, such as silver carbonate, into glycosides. In the *gluco*- and the *galacto*-series, the β-glycosidically linked oligosaccharides are obtained preferentially. Tipson[15] observed that selectivity depends on the position of the substituent at C-2, because α-glycosidically linked disaccharides will be obtained in the *manno*-series under the same conditions. Tipson summerized these findings in the so called "trans"-rule. Isbell[16] formulated the reaction as a neighboring group assisted procedure, in which the O-acetyl group at C-2 forms an intermediate such as a dioxolanylium cation after the leaving of the halide. The dioxolanylium ion can be opened stereoselectively to the 1,2-*trans*-product (β-D-*gluco* or α-D-*manno*) by a reaction at the anomeric center. If the nucleophilic attack occurs at the dioxolanylium ring, the reaction will yield an orthoester.[16] This principle of directing the stereoselectivity at the anomeric center towards 1,2-*trans*-glycosides has generally been very successful. It is equally effective in nearly all other methods that use different leaving groups at the anomeric center. It must be pointed out, however, that the dioxolanylium intermediate exists as an intimate ion pair with a different structure in each case. As we were able to prove,[17] the production of the naked dioxolanylium cations with $SbCl_6^-$ as an anion will result in intramolecular multiple neighboring group reactions and the classical rearrangement of D-glucose to D-idose will occur.

The Koenigs-Knorr reaction has been continuously improved to become an effective method. Helferich[18] introduced the more efficient mercury catalysts such as mercury cyanide. Later Igarashi[19] used silver perchlorate while Schuerch[20] preferred silver triflate, which proved to be one of the most effective catalysts.[21] As captors of acids tetramethyl urea and different pyridine derivates have been employed. Molecular sieves were also used successfully for capturing the water produced.

For a long time glycosyl fluorides were regarded as too stable for the synthesis of glycosides. In spite of this, Mukaiyama[22] showed that oligosaccharide syntheses are possible when the promoting system tin(II)-chloride-silver perchlorate is used. Following these findings, a range of other Lewis acids such as BF_3, TiF_4, $SnCl_4$, and Trimethylsilyl-trifluoromethansulfonic ester (TMS-Triflate) were employed, successfully.[23] Furthermore, complexes of hafnocenes and zirconocenes, introduced by Suzuki,[24] were found to be very effective. In a neighboring group supported reaction, glycosyl fluorides always yield the 1,2-*trans*-glycosides.

If glycosyl halides with a neighboring group active substituent at C-2 are reacted under basic conditions, the intermediate dioxolanylium ion will produce orthoesters. These can also be converted into 1,2-*trans*-glycosides using Lewis acids as catalysts, but methyl orthoesters are not well suited for oligosaccharide synthesis. The methanol formed leads to large amounts of the methyl glycoside as a by product. A transformation into t-butyl orthoester or a direct formation of the orthoester with the hydroxy group of the saccharide acceptor is very useful. In this case, the rearrangement results in the corresponding disaccharide, as the competitive reaction with methanol is eliminated. Several oligosaccharide syntheses of this kind have been carried out by Kochetkov.[25,26] The rearrangement always leads to 1,2-*trans*-glycosides. It appears that the orthoester procedure is equally suited for polymerization reactions that yield oligosaccharide oligomers with a medium degree of polymerization.[25,26] Orthoesters can also react with the hydroxy groups of polymers, while another variant of the orthoester method is the use of 1,2-*O*-(1-cyanoalkylidene) derivatives,[27] which are activated by triphenyl carbenium salts. The intermediate dioxolanylium ion yields a disaccharide, while triphenylacetonitrile precipates from the solution.

Generally 2-acetamido-2-deoxy-glycosyl bromides reacting under Koenigs-Knorr conditions do not form glycosides but stable 4,5-dehydro-1,3-oxazoles.[28] Only the corresponding chlorides may sometimes be used for glycoside formation. Micheel[29] has found that 4,5-dihydro-1,3-oxazoles can be opened under acidic conditions, so that an amino sugar disaccharide is obtained if a saccharide acceptor is present. However the opening of the 4,5-dihydro-1,2-oxazole neccesitates relatively strong acidic conditions. This is the reason why Lemieux[30] introduced the 2-phthalimido-2-deoxy group for the synthesis of 2-amino-2-deoxy-sugar di-saccharides, which has been applied with great success. It is incapable of forming 4,5-dihydro-1,3-oxazoles and also exerts neighboring group participation, so that 1,2-*trans*-glycosidically linked disaccharides are formed with high stereoselectivity. This reaction behaves equally well when halides or other leaving groups such as trichloroacetimidates and thioglycosides are used. The cleavage of the phthalimido group requires more basic conditions, which may be avoided to some extent by the use of 2-trichloroethoxycarbonylamido-2-deoxy-glycosyl compounds for the synthesis of 2-amino-2-deoxy-sugar oligosaccharides.[31]

3. 1,2-*cis*-GLYCOSYLATION

The 1,2-*cis*-glycosylation, as illustrated by the formation of an α-glycosidic linkage in the *gluco*- and the *galacto*-series, is much more difficult. The first requirement is a substituent at C-2, such as an *O*-alkylether, which can not exert a neighboring group effect, otherwise a reaction with participation of the neighboring groups will lead to the β-glycoside. An S_n2 reaction of a β-pyranosyl halide to give the α-glycoside would seem possible, but it is not practical, because β-pyranosyl halides are very much destabilized by the anomeric effect. Only when highly stable β-chlorides or β-fluorides are used such a reaction may be considered. An important contribution to this problem has been made by Lemieux,[32] who found that pyranosyl halides can easily be anomerized and that the halide can be exchanged nucleophilically.[32,33]

According to Lemieux,[34] α-pyranosyl bromides react in the presence of ammonium bromide with the bromide anion to afford the β-pyranosyl bromide. The β-pyranosyl bromide is highly reactive and reacts much faster than the α-pyranosyl bromide to give the α-glycoside via ionic intermediates. Accordingly, this reaction, which has later been called *in situ* anomerization,[1] yields a product which usually contains a large proportion of α-glycoside. Its proportion will increase with greater reactivity of the donor at the anomeric center and decreased reactivity of the hydroxy group of the acceptor. This is the reason why excellent results can be obtained when using galactose or fucose as donors and less good ones when using glucose, which is less reactive at the anomeric center.[1]

Later it was found that a similar mechanism without addition of halide anions was possible using catalysts such as mercury salts, silver perchlorate, or silver triflate.[1,35,36] This is especially the case in the field of oligosaccharide syntheses of amino sugars. Here, too, the group at C-2 must not have neighboring group activity, which makes acetamido groups unsuitable. Instead the 2-azido-2-deoxy-sugars introduced by us have been employed very successfully,[37,38] because the azido group has no neighboring group activity and can be regarded as a masked amino group. Pyranosyl halides of 2-azido-2-deoxy-sugars will produce high yields of α-glycosides using silver perchlorate or silver triflate as catalysts[39,40] and the azido group can easily be transformed into an acetamido group.

In the case of the reaction of very stable β-pyranosyl chlorides or β-pyranosyl fluorides, the proportion of α-glycosides is often higher. It would seem that apart from the *in situ* anomerization reaction there occur ionic intermediates of the S_n2 type. The reaction of 2-trichloroacetyl-β-pyranosyl halides accomplished by Helferich[41] can be mentioned as an example. Using related methods, Umezawa[42] has described a large number of syntheses of glycosyl antibiotics. In the field of glycosyl fluorides, Mukaiyama[22] tested catalysts other than tin(II)chloride-silver perchlorate. The proportion of the α-product seems to depend to a large degree on the case in question, and different catalysts may have to be tested.[22,23] The Suzuki catalyst, a zirconocene complex, appears to be suited here.[24]

A special problem in 1,2-*cis*-glycosylation is the synthesis of β-manno-glycosides. Independent of whether a neighboring group assisted reaction is involved or not, the glycosylation will only yield α-glycosides, as in the previous case, the *in situ* anomerization is wholly dominant. The direct synthesis of β-*manno*-glycosides was considered impossible for a long time and an indirect way had to be found. Lindberg[44] prepared the β-glucoside first with participation of neighboring groups. Selective oxidation at C-2 to the keto function and selective reduction to the *manno*-configuration resulted in the desired β-mannoside. This method has been used in a large number of cases.[45] Another method is an S_n2 reaction with inversion of the 2-OH group in the *gluco* part of the disaccharide. This nucleophilic substitution can be realized either inter-[46] or intra-molecularly.[47]

After first using silver carbonate for a direct β-*manno* glycosylation, we introduced as a reactive heterogenous catalyst, silver silicate, which is an effective catalyst for β-mannosides.[48,49] This catalyst has to be employed without participation of neighboring groups, in the heterogenous phase, and without any Lewis acids present. It is important, though, that both the α-pyranosyl halide donor and the acceptor must have sufficient reactivity. Silver zeolite, which was introduced later, has a similar reactivity.[50]

Recently, new approaches for β-*manno* glycosylation have been developed. The method introduced by Hindsgaul[51] involves the linking of the saccharide acceptor at the 2-OH group of an alkylthio-α-mannopyranoside via an acetal structure. After activating the sulphur and cleavage of the thio group at the anomeric center, the saccharide acceptor is transferred to the anomeric center of the mannose forming a β-glycosidic linkage. A similar procedure, in which the carbon of the acetal is replaced by silicon has been suggested by Stork.[52] Both methods are not easy to handle. A more elegant method is the one suggested by Ogawa[53] which starts from a mannopyranosyl fluoride with a 4-methoxy-benzyl group at 2-OH. This residue is transformed into a quinoidic mixed acetal which intermediately adds the saccharide acceptor. After cleaving the fluoride, the sac-charide acceptor is introduced and a β-glycosidic linkage is formed. In this form the so called "aglycon delivery" method will probably found further applications.

4.   REACTIVITY AT THE ANOMERIC CENTER

The reactivity at the anomeric center depends on the configuration of the saccharide unit, and to a large degree also on the substitution pattern. The fundamental
work on this problem was published by us in the late 1970s.[54] Quantitative
measurements of the anomerization rate of pyranosyl halides were made to
determine their dependence on the substitution pattern of the saccharide unit. It
was observed that in general acyl groups such as acetyl groups reduce the
reactivity at the anomeric center, while ether groups such as benzyl ethers increase
it. This depends less on the position of the substituents than on their number, and
in these reactions, galactose was about ten times more reactive than glucose.
These findings have been further used in the optimization procedure of different
oligosaccharide syntheses.[1,55] In addition, it is known that deoxy groups increase
the reactivity at the anomeric center even more. It is clear now why in the earlier
research with the α-pyranosyl bromide of benzylated fucose made by Flowers[56] α-
glycosides were obtained almost exclusively by *in situ* anomerization. This
dependency of the reactivity at the anomeric center on the substitution pattern is
valid for all leaving groups.

In his research on pentenyl glycosides, Fraser-Reid[57] later named the
dependence of reactivity on the substitution pattern as "armed" or "disarmed",
which he has successfully employed in larger oligosaccharide syntheses.[58] This
enabled him to connect reactive "armed" saccharides with less reactive ones without losing the possibility of a new activation of a pentenyl residue at the latter
"disarmed" saccharide unit. van Boom[59] was able to prove this effect with
thioglycosides in the same way. Reactive thioglycosides and less reactive thioglycosides can be coupled to disaccharides. In the disaccharide that is obtained the
anomeric thioglycoside group in the "disarmed" part is still intact and can be
activated for a new glycosidation reaction.

5.   TRICHLOROACETIMIDATE METHOD

The history of the trichloroacetimidate method is very interesting. Sinay[60]
observed that in some glycosylation reactions in which 2-acetamido-2-deoxy-
sugars are involved as acceptors the yields were relatively small. He then observed

that under certain conditions a reaction of the pyranosyl halide with the acetamido group of the amino sugar took place. He succeeded in isolating an intermediate product that represents an imidate between the glycosyl donor and the acetamido group of the amino sugar.[60] This can be hydrolyzed very easily and as an unwanted intermediate product it is responsible for the small yield.

Following these findings he used glycosyl halides in a reaction with acetamide and obtained glycosyl acetimidates. He found out that in the presence of an acid catalyst the imidate represents a leaving group as well, and that glycosyl acetimidates are suitable glycosyl donors for disaccharide synthesis.[61] The acetimidate, however, is a leaving group having a low reactivity. It needs a fairly strong acid catalyst and succeeds only with saccharides that are reactive at the anomeric center, because of substituents on the saccharide unit. Thus benzyl activated saccharides can be used for oligosaccharide synthesis but not acetyl ones. According to Fraser-Reid's definition, only the "armed" compounds can be used but not "disarmed" ones.

This problem was convincingly solved by Schmidt.[7,8] He prepared glycosyl trichloroacetimidates, and proved that the trichloroacetimidate group was an excellent leaving group and that such compounds could be used in several ways as glycosyl donors. The preparation of these compounds is relatively simple. By selective deacetylation of peracetylated saccharides with piperidine, 1-OH unsubstituted compounds are obtained which can react directly with trichloroacetonitrile under basic conditions to give glycosyl trichloroacetimidates. If potassium carbonate is used as a base, then the β-compound is formed first and then rearranges to give the α-compound. Use of stronger bases will yield the α-form directly.

The trichloroacetimidate method has found wide use in many areas.[7,8] Catalysts used in this reaction are boron trifluoride or trimethylsilyltriflate. The reaction proceeds rather fast, and often takes only one hour at low temperatures. A reaction under participation of neighboring groups conditions will always yield 1,2-*trans*-products, i.e., the β-glycosides in the *gluco*- and the *galacto*-series and the α-glycosides in the *manno*-series, independent of the use of α- or β-trichloroacetimidates. The β-trichloroacetimidates of benzylated saccharides without neighboring group active substituents are of special interest. These can also be transformed into α-glycosides in the *gluco-* and the *galacto*-series.[62] In this case the reaction conditions have to be optimized by varying the catalyst as well as the solvent. Under such optimized conditions it is also possible to transform α-trichloroacetimidates into β-glycosides without participation of neighboring groups.[63] The trichloroacetimidate method can even make reactions succeed which were not possible by using the halide method. Glycosyl trichloroacetimidates are usually stable enough for storage, and for their activiation they require no metal salts.

6.   THIOGLYCOSIDE METHOD

It is well known that thioglycosides react with halogens such as bromine, to yield glycosyl bromides which can then be employed for the synthesis of glycosides.[64] In this reaction the intermediate bromothionium ion is formed which then cleaves to give alkylsulfenylbromide. The remaining glycosyl cation reacts with bromide anions yielding the glycosyl bromide. Ferrier[65] attempted to activate thioglycosides with mercury salts in order to utilize them for glycoside synthesis. This method, however, resulted in limited success. The decisive contribution which led to the successful development of the thioglycoside method was made by Lönn.[66] He activated thioglycosides by using trifluoromethanesulfonic acid methylester (methyl triflate). The sulphur is methylated to form an intermediate sulfonium ion which can function as a good leaving group in an oligosaccharide synthesis. The neighboring group assisted reaction yields 1,2-*trans*-glycosides with high stereoselectivity i.e. β-glycosides in the *gluco-* and the *galacto*-series. The yields are very good, and the starting material can be regarded as very stable. If neighboring groups do not participate, the reaction may be directed by variation of the solvent to obtain considerable amounts of α-glycosides. This process can be applied in limited cases only, since the β-glycoside which is also formed has to be separated.

As methyl triflate is volatile and extremely toxic, there was a search for other promotors. Meerwein's salt (dimethylmethylthiosulfonium)-boron tetrafluoride, which was proposed by Trost, had no effect. Garegg[67] first employed (dimethylmethylthiosulfonium)-trifluoromethanesulfonate   (DMTST),   which proved to be an excellent thiophilic promotor. This substance has been widely used in glycosylations to form disaccharides and oligosaccharides.[9,68] Under participation of neighboring groups 1,2-*trans*-glycosides are obtained selectively. If the reaction conditions and the solvents are carefully optimized, a reaction without participation of neighboring groups, can yield varying amounts of 1,2-*cis*-products besides the 1,2-*trans*-products.

After the use of the promotor DMTST, a whole range of other promotors were discovered, which were able to activate the thioglycoside group and which are well suited for oligosaccharide synthesis. These include nitrosyl-tetrafluoroborate,[69] phenylselenyl-triflate,[70] alkylsulfenyl-triflate,[71] *N*-iodosuc-cinimide-trifluoroacetic acid,[72] and bis(2,4,6-trimethyl-pyridine)-iodonium-per-chlorate.[73] Furthermore, thioglycosides may be transformed by oxidation into sulfoxides or sulfones, which may then function as leaving groups thereby expanding the usefulness of this class of glycosyl donors. It has been found that in addition to the thioglycoside group *O*-ethyl-*S*-glycosyl-dithiocarbonates can be employed.[74] In these compounds the sulphur can be activated by thiophilic reagents, which can be advantageous in some oligosaccharide syntheses. It is obvious that the thioglycosylation method can be widely used in numerous variations, provided that all the variants are optimized. The choice of the promotor may depend on the experience of the working group and the special problem of the intended oligosaccharide synthesis.

7.   OTHER METHODS OF GLYCOSYLATION

Apart from the three major methods of glycosylation discussed above there are several other methods, which can be employed when one is faced with special problems. It was almost fourty years ago that Lemieux[75] intensively tested the opening of 1,2-epoxides, the so called Brigl-anhydrides, in the work on his classical sucrose synthesis. At that time it was found that alcohols with a high nucleophilicity can lead to 1,2-*trans*-glycosides, while alcohols with low nucleophilicity also yield small amounts of 1,2-*cis*-glycosides.[75,76] When 3,3-di-methyldioxirane became available as a reagent for the epoxidation of enol ethers, Danishefsky[77] investigated this method again. The reagent makes it possible to convert glycals into 1,2-epoxides, thus making the Brigl-anhydride type compounds much more easily accessible. The coupling reaction with a saccharide acceptor to give oligosaccharides takes place at low temperatures with zinc chloride as catalyst and yields 1,2-*trans*-glycosides.[77] The anomeric center contained in the epoxide donor has to be activated by ether substituents at the saccharide unit. The stereochemistry is thus preserved when an "armed" donor in the Fraser-Reid[57] sense is used. This method has also been proposed by Danishefsky[78] for an inverse solid phase oligosaccharide synthesis.

The well-known acid catalyzed addition of alcohols to glycals yields 2-deoxy-sugars and a number of by products. The preparation of 2-deoxy oligosaccharides by the methods described above is often difficult due to the absence of a substituent at C-2 that controls the stereoselectivity of the glycosidic linkage formed. It is more practical to prepare the 2-deoxy-2-iodo compounds first, which can be obtained from glycals by addition reactions. Lemieux[79] has shown that the addition of saccharide acceptors to glycals in the presence of bis(2,4,6-trimethyl-pyridin)-iodonium-perchlorate yields 2-deoxy-2-iodo-gly-cosides, from which the iodine can be removed by reduction.[79,80] The N-iodosuccinimide method developed by Thiem[81,82] works equally efficiently. In this method, glycals react with saccharide acceptors in the presence of N-iodosuccinimide. Usually 2-iodo-α-trans-glycosides are obtained. After cleavage of the iodine by reduction, α-linked 2-deoxy-glycosides are produced. The α-linked 2-deoxy oligosaccharides can easily be prepared by this route. It is much more difficult to produce 2-deoxy-β-glycosides, and it is often necessary to employ special methods.

A new method for glycosylation which is chemically very interesting has been developed by Fraser-Reid.[58,83] He uses 4-pentenyl glycosides as glycosyl donors. These can be activated by bis(2,4,6-trimethylpyridine)-iodonium-perchlorate, N-iodosuccinimide-trifluoromethanesulfonic acid, or N-iodosuccinimide-trimethylsilyltriflate. Addition of a halogen at the pentenyl double bond will produce a halonium ion intermediate. This intermediate then rearranges to a second intermediate containing the leaving group, 2-halomethyltetrahydrofuran. Elimination of this leaving group results in a glycosyl cation, which then reacts to form the glycoside. This method usually produces 1,2-*trans*-glycosides and is suited for syntheses of larger oligosaccharide units.[58]

Recently, Wong[84] and Schmidt[85] working independently found that dialkyl-phosphite groups are suitable leaving groups for oligosaccharide syntheses. In the presence of Lewis acids glycosyl dialkylphosphites can be coupled with saccharide acceptors. This method has been employed successfully for the synthesis of 5-*N*-acetylneuraminic acid glycosides. β-Elimination, which takes place in most cases of glycoside synthesis involving 5-*N*-acetylneuraminic acid, was scarcely observed, and a good yield of α-glycoside was obtained. It appears that the glycosyl phosphite method is well suited for the coupling of ketoses and 2-ulosonic acids, where problems often occur with other methods.

Several attempts have been made to develop a solid phase oligosaccharide synthesis, but with limited success. It seems to be obvious that solid phase oligo-saccharide synthesis will never reach the level of solid phase peptide synthesis or solid phase oligonucleotide synthesis. Obstacles, such as the polyfunctionality of the reactants, the lower yields in the coupling reactions, and the required high stereoselectivity with respect to the proportion of anomers are some causes of this limitation. As mentioned above, Danishefsky[78] described a solid phase oligosac-charide synthesis using the 1,2-epoxide opening reaction which could be useful in certain special cases. Krepinsky,[86] for example, has developed a soluble polyether carrier on which a saccharide can be attached with a linker and oligosaccharide synthesis can be performed. van Boom[87] has used this carrier to synthesize a heptasaccharide. This method involves a series of repetitive steps and could be employed for the synthesis of oligosaccharides containing similar saccharide units, by repetitive steps of 1,2-*trans*-glycosylation reactions.

Parallel to the chemical methods of oligosaccharide syntheses explained above, methods of enzymatic oligosaccharide synthesis have been developed in the last several years.[88] These methods have the advantage that they require no protective groups, thereby shortening the synthesis. The glycoside cleaving enzymes like glycosidases, which are relatively easy to access, have been investigated first. The basis for this was the idea of letting the cleavage reaction which is catalysed by the respective enzyme run in opposite direction as a linking step.[89,90] This can be done even when organic solvents are used. As glycosyl donors, p-nitrophenyl glycosides or glycosyl fluorides can be used. The yields are limited and also the regio selectivity is not absolute in some cases, so that purification steps are necessary.

Linking with glycosyltransferases should be much more selective. Access to these enzymes, however, is more difficult, as they have to be isolated from appropriate substrates whose availability is limited. Nevertheless, if the structure of the enzyme is known, it can be cloned and expressed in larger amounts as a recombinant enzyme. In his pioneering work, Paulson[91] succeeded in obtaining several recombinant sialyltransferases which were capable of transferring 5-N-acetylneuraminic acid to oligosaccharide acceptors. Chemical synthesis of glycosides of 5-N-acetylneuraminic acid is not without problems, though remarkable progress has been made in this field.[92]

The sialyltransferase alone is not sufficient for the glycosylation step. The enzyme also requires an activated sugar, in this case the nucleotide CMP-NeuAc, from which sialyltransferase transfers 5-N-acetylneuraminic acid onto the saccharide acceptor. CMP-NeuAc, though, must be regenerated continuously in a cyclic enzymatic process in order to allow a steady flow of the glycosylation reaction. Such reactions with an integrated cyclic cofactor regeneration have been described for 5-N-acetylneuraminic acid[93,94] and other saccharides, such as galactose.[95] A combination of chemical and enzymatic reactions made it possible to synthesize large amounts of the Sialyl-Lewis[x] oligosaccharide.[93,94] For such complex compounds the combination of chemical and enzymatic methods is a good alternative to chemical synthesis alone.

8. CONCLUSION

In summary, there is a rich arsenal of methods for oligosaccharide synthesis at hand. Apart from the three most important chemical methods there are several methods for special cases, and there are other methods which require further improvements. The method to be used depends on the problem at hand. The number of steps, yield, and selectivities have to be checked carefully before a decision is made. Often an optimization of the linking steps will be necessary. In their synthesis of the highly complex modified pentasaccharide structure of the heparin chain van Boeckel and Petitou[96] have demonstrated how useful careful optimization can be, especially if very large amounts of the oligosaccharide are needed.

REFERENCES

1. H. Paulsen, *Angew. Chem.*, **94**, 184 (1982); *Angew. Chem. Int. Ed. Engl.*, **21**, 155 (1982).
2. A.F. Bochkov and G.E. Zaikov, *Chemistry of the O-Glycosidic Bond*, Pergamon Press, Oxford (1979).
3. K. Toshima and K. Tatsuda, *Chem. Rev.*, **93**, 1503 (1993) and references cited therein.
4. O. Lockhoff, *Houben-Weyl: Methoden der Organischen Chemie*, Vol. E **14a/3**, Georg Thieme Verlag, Stuttgart, p. 621 (1992).
5. (a) A. Lipták, P. Fügedi, Z. Szurmai, and J. Harangi, *Handbook of Oligosac-charides*, Vol. **1**, Disaccharides, CRC Press, Boca Raton (1990); (b) Vol. **2**, Trisaccharides, CRC Press, Boca Raton (1991); (c) Vol. **3**, Higher Oligosac-charides, CRC Press, Boca Raton (1991).
6. R. Stuike-Prill, K. Bock, A. Kleen, H. Paulsen, J.A. van Kuik, J.F.G. Vliegenthart, S. Doubet, D. Smith, and P. Albersheim, *Bioinformatics,* **1**, 12 (1992).
7. R.R. Schmidt, *Angew. Chem.*, **98**, 213 (1986); *Angew. Chem. Int. Ed. Engl.*, **25**, 212 (1986).
8. R.R. Schmidt, *Pure & Appl. Chem.*, **61**, 1257 (1989).
9. P. Fügedi, P.J. Garegg, H. Lönn, and T. Norberg, *Glycoconjugate J.*, **4**, 97 (1987).
10. T. Ogawa, H. Yamamoto, T. Nukada, T. Kitajma, and M. Sugimoto, *Pure & Appl. Chem.*, **56**, 779 (1984).
11. F. Goto and T. Ogawa, *Pure & Appl. Chem.*, **65**, 793 (1984).
12. A. Hasegawa, H. Ohki, T. Nagahama, H. Ishido, and M. Kiso, *Carbohydr. Res.*, **212**, 277 (1991) and references cited therein
13. K. Hotta, H. Ishida, M. Kiso, and A. Hasegawa, *J. Carbohydr. Chem.*, **13**, 175 (1993) and references cited therein.
14. W. Koenigs and E. Knorr, *Ber. Chem. Dtsch. Chem. Ges.*, **34**, 957 (1901).
15. R.S. Tipson, *J. Biol. Chem.*, **130**, 55 (1939).
16. H.S. Isbell, *Am. Rev. Biochem.*, **9**, 65 (1940).
17. H. Paulsen and C.-P. Herold, *Chem. Ber.*, **103**, 2450 (1970).

18. B. Helferich and K.-F. Wedemeyer, *Liebigs Ann. Chem.*, **563**, 139 (1949).

19. K. Igarashi, J. Irisawa, and T. Homna, *Carbohydr. Res.*, **39**, 213 (1975).

20. F.J. Kronzer and C. Schuerch, *Carbohydr. Res.*, **27**, 379 (1973).

21. S. Hanessian and J. Banoub, *Carbohydr. Res.*, **53**, C 13 (1977).

22. T. Mukaiyama, Y. Murai, and S. Shoda, *Chem. Lett.*, **431**, 939 (1981).

23. M. Kreuzer and J. Thiem, *Carbohydr. Res.*, **149**, 347 (1986).

24. (a) T. Matsumoto, H. Maeta, K. Suzuki, and G. Tsuchihashi, *Tetrahedron Lett.*, **29**, 3567 (1988); (b) *ibid*, p. 3575.

25. N.K. Kotchetkov, *Pure & Appl. Chem.*, **42**, 301 (1975).

26. N.K. Kotchetkov, A.F. Bochkov, T.A. Sokolovskaya, and V.J. Synyatkov, *Carbohydr. Res.*, **16**, 17 (1971).

27. A.F. Bochkov and N.K. Kotchetkov, *Carbohydr. Res.*, **39**, 355 (1975).

28. F. Micheel and H. Köchling, *Chem. Ber.*, **90**, 1597 (1957).

29. F. Micheel, F.-P. van de Kamp, and H. Petersen, *Chem. Ber.*, **90**, 521 (1957).

30. R.U. Lemieux, T. Takeda, and B.Y. Chung, *ACS Symp. Ser.*, **39**, 90 (1976).

31. J. Banoub, P. Boullanger, and D. Lafont, *Chem. Rev.*, **92**, 1167 (1992).

32. R.U. Lemieux and J. Hayami, *Can. J. Chem.*, **43**, 2162 (1965).

33. T. Ishika and H.G. Fletcher, Jr., *J. Org. Chem.*, **34**, 563 (1969).

34. R.U. Lemieux, K.B. Hendriks, R.V. Stick, and K. James, *J. Am. Chem. Soc.*, **97**, 4056 (1975).

35. H. Paulsen and O. Lockhoff, *Chem. Ber.*, **114**, 3079 (1981).

36. H. Paulsen and H. Bünsch, *Chem. Ber.*, **114**, 3115 (1981).

37. H. Paulsen and W. Stenzel, *Angew. Chem.*, **87**, 547 (1975); *Angew. Chem. Int. Ed. Engl.*, **14**, 558 (1975).

38. (a) H. Paulsen and W. Stenzel, *Chem. Ber.*, **111**, 2334 (1978); (b) *ibid*, p. 2348.

39. (a) H. Paulsen, C. Kolar, and W. Stenzel, *Chem. Ber.*, **111**, 2358 (1978); (b) *ibid*, p. 2370.

40. H. Paulsen and A. Bünsch, *Carbohydr. Res.*, **100**, 143 (1982).

41. B. Helferich, W.M. Müller, and S. Karbach, *Liebigs Ann. Chem.*, 1514 (1974).

42. S. Umezawa, *Pure & Appl. Chem.*, **50**, 1453 (1978).

43. K.C. Nicolaou, T. Caulfeld, H. Kataota, and T. Kumazawa, *J. Am. Chem. Soc.*, **110**, 7910 (1988).

44. G. Eckborg, B. Lindberg, and J. Lönngren, *Acta Chem. Scand.*, **26**, 287 (1972); (b) *ibid*, p. 3287.

45. (a) M.A.B. Shaban and R.W. Jeanloz, *Carbohydr. Res.*, **52**, 103 (1976); (b) *ibid*, p. 115.

46. J. Alais and S. David, *Carbohydr. Res.*, **82**, 85 (1980).

47. W. Günther and H. Kunz, *Carbohydr. Res.*, **228**, 217 (1992).

48. H. Paulsen and O. Lockhoff, *Chem. Ber.*, **114**, 3102 (1981).

49. H. Paulsen and R. Lebuhn, *Liebigs Ann. Chem.*, 1047 (1983).

50. P. Garegg and P. Ossowski, *Acta Chem. Scand.*, **B 37**, 249 (1983).

51. F. Barresi and O. Hindsgaul, *J. Am. Chem. Soc.*, **113**, 9376 (1991).

52. G. Stork and G. Kim, *J. Am. Chem. Soc.*, **114**, 1087 (1992).

53. Y. Ito and T. Ogawa, *Angew. Chem.*, **106**, 1843 (1994).

54. H. Paulsen, A. Richter, V. Sinnwell, and W. Stenzel, *Carbohydr. Res.*, **64**, 339 (1978).

55. H. Paulsen, *Angew. Chem.*, **102**, 851 (1990); *Angew. Chem. Int. Ed. Engl.*, **29**, 823 (1990).

56. M. Dejter-Juszynski and H.M. Flowers, *Carbohydr. Res.*, **18**, 219 (1971).

57. D.R. Mootoo, P. Konradson, U. Udodong, and B. Fraser-Reid, *J. Am. Chem. Soc.*, **110**, 5583 (1988).

58. B. Fraser-Reid, J.R. Merrit, A.L. Handlon, and C.W. Andrews, *Pure & Appl. Chem.*, **65**, 779 (1993).

59. H.M. Zuurmond, S. van der Laan, G.A. van der Marel, and J.H. van Boom, *Carbohydr. Res.*, **215**, C 1 (1991).

60. J.R. Pougny and P. Sinay, *Carbohydr. Res.*, **47**, 69 (1976).

61. J.R. Pougny, M.A.M. Nassr, N. Naulet, and P. Sinay, *Nouveau J. Chem.*, **2**, 389 (1978).

62. B. Wegmann and R.R. Schmidt, *Carbohydr. Res.*, **184**, 254 (1988).

63. R.R. Schmidt, M. Behrendt and A. Toepfer, *Synlett*, 694 (1990).

64. (a) W.A. Bonner, *J. Am. Chem. Soc.*, **70**, 770 (1948); (b) *ibid*, p. 3491.

65. R.J. Ferrier, R.W. Hay, and N. Vethaviyasar, *Carbohydr. Res.*, **27**, 545 (1973).

66. H. Lönn, *Carbohydr. Res.*, **139**, 115 (1985).

67. P. Fügedi and P.J. Garegg, *Carbohydr. Res.*, **149**, C 9 (1986).

68. F. Anderson, W. Birberg, P. Fügedi, P.J. Garegg, M. Nashed, and A. Pilotti, *ACS Symp. Ser.*, **386**, 117 (1989).

69. V. Pozsgay and H.J. Jennings, *J. Org. Chem.*, **52**, 4635 (1987).

70. Y. Ito and T. Ogawa, *Tetrahedron Lett.*, **29**, 1061 (1988).

71. F. Dasgupta and P.J. Garegg, *Carbohydr. Res.*, **177**, C 13 (1988).

72. G.H. Veeneman, S.H. van Leeuwen, and J.H. van Boom, *Tetrahedron Lett.*, **31**, 1331 (1990).

73. H.M. Zuurmond, G.A. van der Marel, and J.H. van Boom, *Recl. Trav. Pays-Bas*, **112**, 507 (1993).

74. (a) W. Birberg and H. Lönn, *Tetrahedron Lett.*, **33**, 7453 (1991); (b) *ibid*, p. 7457.

75. R.U. Lemieux and G. Huber, *J. Am. Chem. Soc.*, **78**, 4117 (1956).

76. L.J. Sargent, J.G. Buchanan, and J. Baddiley, *J. Chem. Soc.*, 2184 (1962).

77. R.L. Halcomb and S.J. Danishefsky, *J. Am. Chem. Soc.*, **111**, 6661 (1989).

78. S.J. Danishefsky, K.F. McClure, J.T. Randolph, and R.B. Ruggeri, *Science*, **260**, 1307 (1993).

79. R.U. Lemieux and A.R. Morgan, *Can. J. Chem.*, **43**, 2190 (1965).

80. R.W. Friesen and S.J. Danishefsky, *J. Am. Chem. Soc.*, **111**, 6656 (1989).

81. J. Thiem, H. Karl, and J. Schwendtner, *Synthesis*, 696 (1978).

82. J. Thiem and W. Klaffke, *Topics in Current Chemistry*, **154**, 286 (1990).

83. D.R. Mootoo, V. Date, and B. Fraser-Reid, *J. Am. Chem. Soc.*, **110**, 2662 (1988).

84. H. Kondo, S. Aoki, Y. Ichikawa, R.L. Halcomb, H. Ritzen, and C.-H. Wong, *J. Org. Chem.*, **59**, 864 (1994).

85. T.J. Martin and R.R. Schmidt, *Tetrahedron Lett.*, **33**, 6123 (1992).

86. S.P. Douglas, D.M. Whitfield, and J.J. Krepinsky, *J. Am. Chem. Soc.*, **113**, 5095 (1991).

87. R. Verduyn, P.A.M. van der Klein, M. Douwes, G.A. van der Marel, and J.H. van Boom, *Recl. Trav. Chim. Pays-Bas*, **112**, 464 (1993).

88. S. David, C. Augé, and C. Gautheron, *Adv. Carbohydr. Chem. Biochem.*, **49**, 176 (1991).

89. K.G.I. Nilsson, *ACS Symp. Ser.*, **466**, 51 (1991).

90. P. Stangier and J. Thiem, *ACS Symp. Ser.*, **466**, 63 (1991).

91. K.J. Colley, E.U. Lee, B. Adler, J.K. Brown, and J.C. Paulson, *J. Biol. Chem.*, **264**, 17619 (1989).

92. A. Hasegawa, M. Ogawa, Y. Kojma, and M. Kiso, *J. Carbohydr. Chem.*, **11**, 333 (1992).

93. Y. Ito, J.J. Gaudino, and J.C. Paulson, *Pure & Appl. Chem.*, **65**, 753 (1993).

94. C.-H. Wong, Y. Ichikawa, T. Kajimoto, K.K.-C. Liu, G.-J. Shen, C.-H. Lin, Y.-F. Wang, D.P. Dumas, Y.-C. Lin, R. Wang, and G.C. Look, *Pure & Appl. Chem.*, **65**, 803 (1993).

95. J. Thiem and T. Wiemann, *Angew. Chem.*, **103**, 1184 (1991); *Angew. Chem. Int. Ed. Engl.*, **30**, 1163 (1991).

96. A.A. van Boeckel and M. Petitou, *Angew. Chem.*, **105**, 1741 (1993); *Angew. Chem. Int. Ed. Engl.*, **32**, 1671 (1993).

Chapter 2

THE ANOMERIC *O*-ALKYLATION AND THE TRICHLORO-ACETIMIDATE METHOD – VERSATILE STRATEGIES FOR GLYCOSIDE BOND FORMATION

RICHARD R. SCHMIDT

Fakultät Chemie, Universität Konstanz, Postfach 5560 M 725
D-78434 Konstanz, Germany

Abstract In the first part of this review, recent results of the anomeric *O*-alkylation procedure for glycoside bond formation will be discussed, with particular emphasis on α-glycoside bond formation with KDO, anomeric *O*-alkylation of partially or completely unprotected sugars, and on results with alkylating agents derived from secondary hydroxy groups of sugars. The second part is devoted to the trichloroacetimidate method displaying recent applications to glycosphingolipid, glycopeptide, glycopeptidolipid, saponin, and antibiotics syntheses. Additionally, methodological improvements such as the "inverse procedure", the "nitrile effect", and the extension to partially *O*-protected *O*-glycosyl trichloroacetimidates are discussed. In the last section, other anomeric oxygen activation procedures are briefly discussed.

1. INTRODUCTION

Glycoscience has seen a surge in interest in recent years as the biological roles of various carbohydrates have been elucidated. This has led to an increased demand for glycosides, oligosaccharides, and glycoconjugates (glycolipids, glycosphingolipids, glycopeptides, glycoproteins, glycosyl phosphates, etc.) and

20

hence for glycoside bond formation.[1-4] To fulfill this goal, two different basic strategies have been applied.[1] These are:

(i)   *Generation of glycosyl donors by anomeric oxygen exchange reactions,* encompassing the Fischer-Helferich procedure (Lewis acid catalyzed acetalization) and the Koenigs-Knorr procedure (glycosyl halide formation, including fluoride formation, thioglycoside and sulfinyl glycoside formation, etc.)

(ii)   *Activation through retention of the anomeric oxygen atom* which is shown in Scheme 1. Besides the acid catalyzed anomeric oxygen exchange reaction (see i), the simplest form of activation is base promoted generation of an anomeric alkoxide structure of a pyranose or furanose (see Scheme 1) which can be directly *O*-alkylated, thus leading to the anomeric *O*-alkylation procedure (Scheme 1, A). Alternatively, the anomeric alkoxide is used to generate glycosyl donors by addition to appropriate triple bond systems $A \equiv B$ or cumulenes $A=B=C$ or via condensation with $A=B-X$ systems, respectively (Scheme 1, B). These additions require only catalytic amounts of an acid in the glycosylation step.

The most successful methods developed thus far using these types of reactions are the trichloroacetimidate method (Scheme 1, $-A=BH = CCl_3C=NH$) and the phosphate and phosphite formation. The analogous sulfate, sulfonate, and sulfite formations have not been as extensively investigated. All these methods for glycoside bond formation are particularly tempting because Nature has a similar approach for generating glycosyl donors, namely glycosyl phosphate formation.

    In this review recent results which illustrate the scope of the anomeric *O*-alkylation procedure and the trichloroacetimidate method will be summarized. It will also be shown that these methods are nicely supplemented by glycosyl phosphites, which have recently been introduced by us as valuable glycosyl donors.[5,6]

22

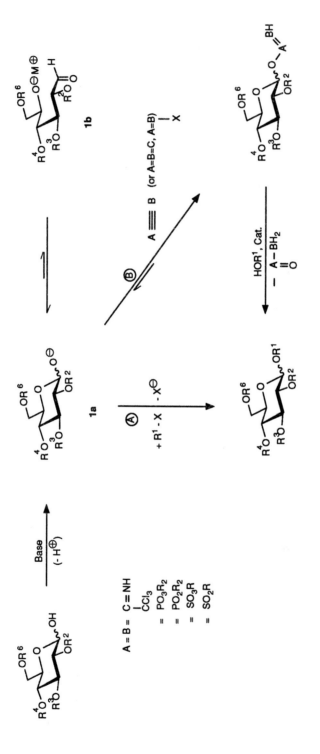

Scheme 1.

2. ANOMERIC O-ALKYLATION METHOD

The base-catalyzed O-alkylation of anomeric hydroxyl groups with various O-triflates (Scheme 1, R'-X = R-OTf) is a very convenient method of glycoside bond formation which we have introduced.[1,7,8] It has been successfully utilized in disaccharide syntheses[1,7-9] and allows the use of a variety of alkylating agents such as methyl iodide and dimethyl sulfate as well as less reactive agents. Potential decomposition reactions, particularly via the acyclic form **1b** or O-alkylation reactions at O-5 of **1b** (Scheme 1), can be generally avoided and conditions which allow high diastereomeric control can often be found.[1] In pyranoses, generally the enhanced nucleophilicity of equatorial oxygen atoms (kinetic anomeric effect)[1,7,8,11] favors the equatorial glycoside (for instance, reaction with **1aβ**); however, thermodynamic reaction control favors the axial glycoside (thermodynamic anomeric effect). Chelation control can also become a dominant factor in determing α/β-selection.

The strong influence of the kinetic anomeric effect and chelation control is exhibited in the successful application of this method to the keto oxygen in KDO (3-deoxy-D-*manno*-2-octulosonic acid) (Scheme 2).[8] Base-catalyzed anomeric O-activation of KDO is problematic due to decreased accessibility of the anomeric oxide atom. The presence of a carboxylate group also lowers the nucleophilicity and favors side reactions in the ring-open form. The application of the anomeric O-alkylation procedure allows these problems to be circumvented. The widely used 4,5,7,8-tetra-O-acetyl protected KDO derivatives did not give the desired α-diastereoseletivity because the equatorial β-oxide is not more reactive than the axial α-oxide. Therefore, we selected a 4,5:7,8-di-O-cyclohexylidene derivative (Scheme 2), which according to the ^{1}H-NMR data, prefers a boat conformation. This allows the carboxamide group and the oxygen atoms at C-5 and C-8 to form a tetradentate chelate ligand which complexes metal ions. This complexation, generating a dianionic species, offered solutions to the problems, especially for nucleophilic reactivity and the desired α-diastereoselectivity. Thus, reaction of this KDO derivative with two equivalents of NaH and subsequent addition of the 6-O-triflate of 2-azidoglucose led to the α-(2-6)-linked disaccharide in good yield

(Scheme 2) and no β-disaccharides were found in the reaction mixture. Subsequent application of the anomeric *O*-alkylation procedure to this molecule

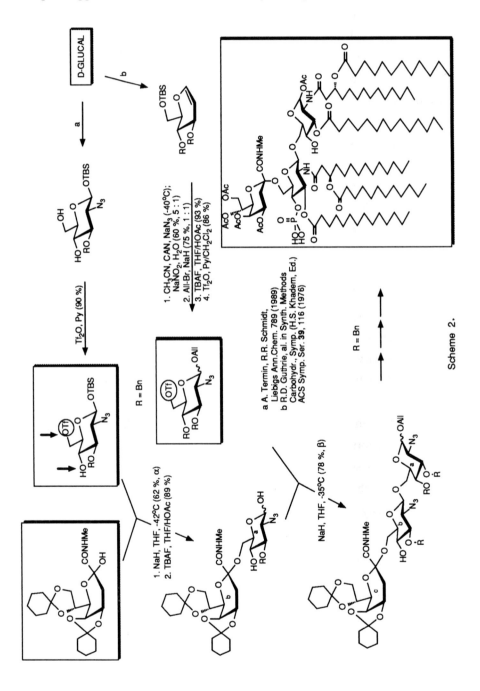

Scheme 2.

led to the corresponding trisaccharide which was subsequently transformed into a lipid A derivative of interest (Scheme 2).

Anomeric O-alkylation has been utilized sucessfully with less reactive alkylating agents such as long chain dialkyl sulfates[10] and O-acylated sugars. [12]

Although the anomeric oxide exhibits an enhanced nucleophilicity, alkylating agents derived from secondary hydroxy groups of sugars could not be employed successfully in this reaction.[1,13] Recent attempts to overcome this limitation are shown in Scheme 3.[14] When the appropriate reaction conditions are used (polar, aprotic solvents and low temperatures) the reactions are very clean and give high product yields. The α/β-selectivities found under these conditions reflect the anomeric composition of the starting materials[1,7,15] because anomerization is slow at these temperatures.

Selective anomeric O-alkylation of free sugars presents a different set of problems. All unprotected hydroxyls can be alkylated, not just the anomeric position and all four possible glycosides (α and β-pyrano- and furanosides) may be formed (Scheme 1). Preliminary studies with partially protected sugars indicate that the lack of solubility of the starting material may be the primary difficulty in direct O-alkylation of the anomeric position with regio- and/or anomeric control playing a secondary role.[10,16] These findings prompted our studies on the direct anomeric O-alkylation of unprotected aldoses in a suitable solvent which is capable of dissolving both the alkylating agent and the free sugar.[10,17] Some results are shown in Table 1. Obviously, anomeric O-alkylation is a convenient method for transforming O-unprotected sugars directly into glycosides, which have found use as surfactants or building blocks in sugar-based natural product syntheses. Although ring-chain equilibration in the starting materials permits the formation of many products, the regiocontrol towards uniform glycoside bond formation is generally very high, a result unknown for any of the other commonly used methods of glycoside formation. Anomer control in pyranoside formation generally favors the equatorial product (see above). Therefore, anomeric O-alkylation of unprotected sugars has the potential to be a very successful methodology in carbohydrate chemistry.

In conclusion, as the recent developments discussed here show, anomeric *O*-alkylation has proven to be a very versatile and competitive method for glycoside bond formation which may gain wide application.

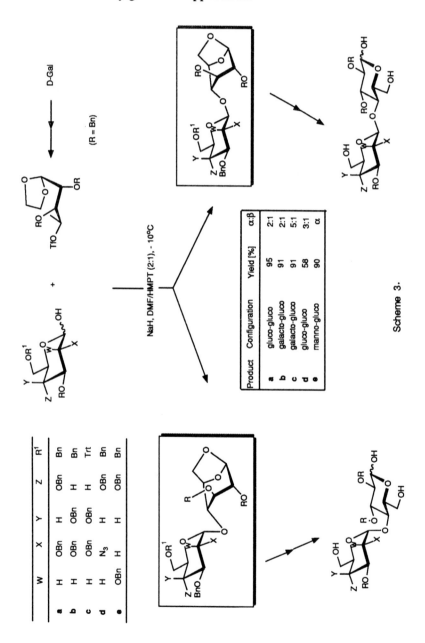

Scheme 3.

Table 1. Anomeric O-alkylation of unprotected sugars.

SUGAR	PRODUCTS AND YIELDS	SUGAR	PRODUCTS AND YIELDS
D-Glc	R = Dec (57 %, α:β = 1:2) R = Bn (64 %, α:β = 1:4) R = All (65 %, α:β = 1:5)	D-GlcNBz	R = Dec (50 %, α:β = 2:1) R = Bn (54 %, α:β = 2:1) R = All (53 %, α:β = 3:1)
D-Gal	R = Dec (62%) R = Bn (63 %) R = All (59 %)	D-Xyl	R = Dec (65 %, α:β = 1:10) R = Bn (58 %, α:β = 1:8) R = All (68 %, α:β = 1:4)
D-Man	R = Dec (48 %, α:β = 1:3) R = Bn (50 %, α:β = 1:3) R = All (52 %, α:β = 1:4)	D-Ara	R = Dec (78 %, 1:3) R = Bn (62 %, 1:3) R = All (61 %, 1:4)

ALKYLATING AGENTS: DIDECYL SULFATE
BENZYL BROMIDE
ALLYL BROMIDE

REACTION CONDITIONS: SOLVENT: DMPU
BASE: NaH
ROOM TEMPERATURE

3. TRICHLOROACETIMIDATE METHOD

As a compliment to the direct O-alkylation at the anomeric oxygen, (Scheme 1, Path A) we have shown that it can also be transformed into a good leaving group (Scheme 1, Path B) with various electron-deficient multiple bond system.[1] However, to achieve a stereocontrolled activation the epimerization of the anomeric oxide ion must be considered (Scheme 1). Thus, in a reversible activation process and with the help of kinetic and thermodynamic reaction-control, both activated anomers are frequently accessible.

Electron deficient nitriles, such as trichloroacetonitrile and trifluoroacetonitrile (= A≡B in Scheme 1), are known to undergo direct and reversible, base-catalyzed addition of alcohols, providing O-alkyl-trihaloacetimidates.[1] A detailed study of the addition of trichloroacetonitrile to 2,3,4,6-tetra-O-benzyl-D-glucose in the presence of base demonstrated this reaction is also applicable for sugars as well.[1,18] Depending on the base, both O-activated anomers can be isolated in pure form and in high yield due to the influence of the kinetic and the thermodynamic anomeric effects.

The significance of the O-glycosyl trichloroacetimidates lies in their ability to act as glycosyl donors under mild acid catalysis. This has been confirmed overwhelmingly in various laboratories and important generalizations derived from these investigations have been summarized in several reviews.[1,4] Some representative examples of recent applications in demanding glycoconjugate syntheses are exhibited in Schemes 4-9.

As the biological importance of glycoconjugates has been revealed, it has sparked a great effort in *glycosphingolipid synthesis*. Of special interest are the compounds which are referred to as tumor-associated antigens.[19] An investigation of their role in tissues may reveal tumorigenesis, support diagnosis, and lead to immunotherapy of cancer.[19,20] An example of a tumor-associated antigen is BGM_1 (Scheme 4), which is an extended GM_1 ganglioside which contains the blood group B determinant in addition to the structural features of the fucosyl-GM_1 (FucGM$_1$).[21] Both compounds, FucGM$_1$ and BGM$_1$ are found in the liver of rats following treatment with fluoroacetamide, which is a carcinogen.[21,22]

The synthesis of the heptasaccharide moiety of BGM₁[23] (Scheme 4) is based on bond disconnections 1 to 5 which leads to suitably protected galactose,

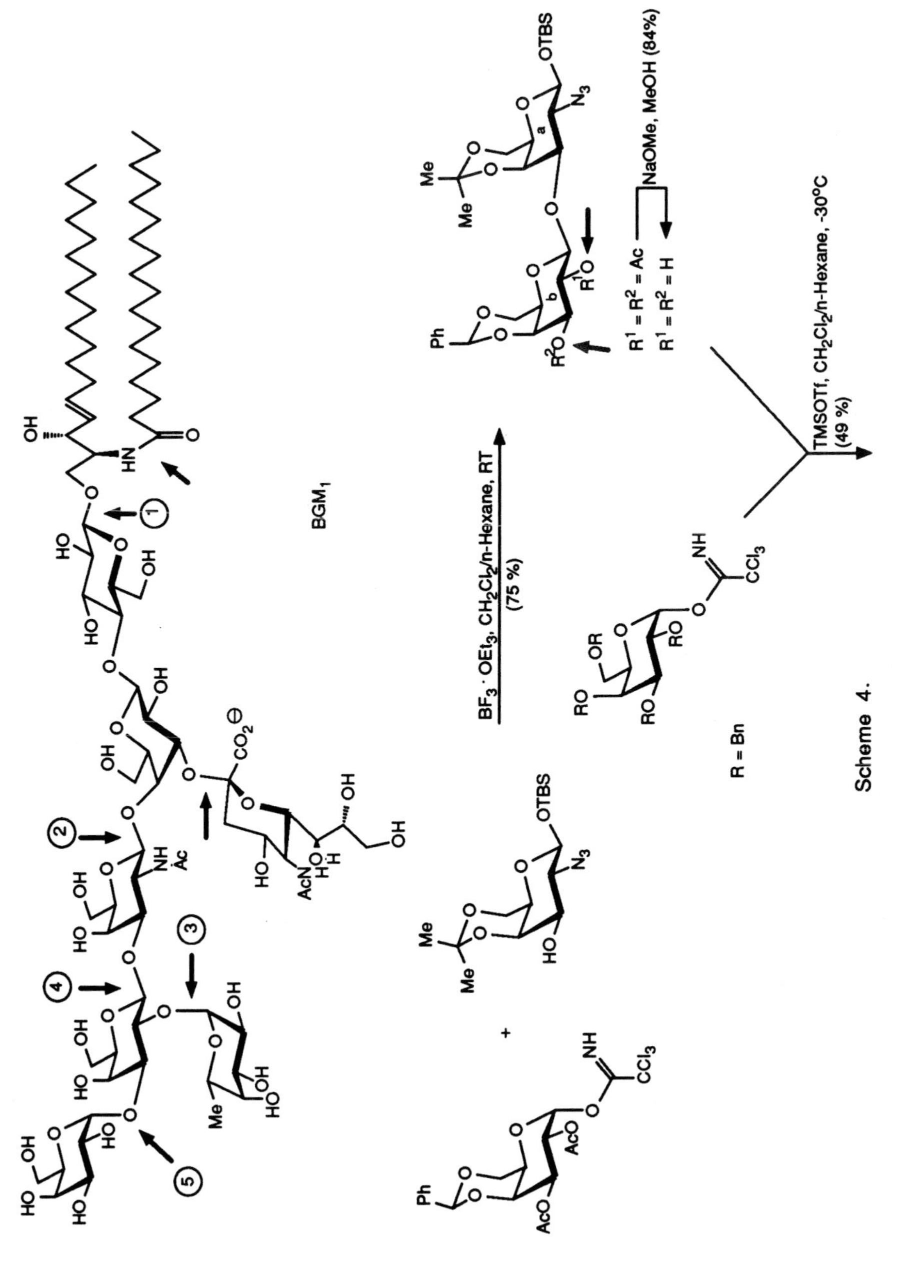

Scheme 4.

30

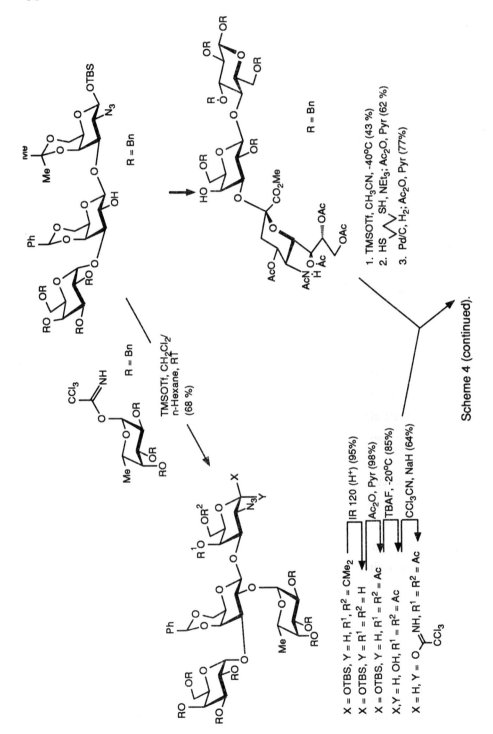

X = OTBS, Y = H, R^1, R^2 = CMe$_2$

X = OTBS, Y = R^1 = R^2 = H

X = OTBS, Y = H, R^1 = R^2 = Ac

X, Y = H, OH, R^1 = R^2 = Ac

X = H, Y = O   NH, R^1 = R^2 = Ac
              CCl$_3$

IR 120 (H$^+$) (95%)

Ac$_2$O, Pyr (98%)

TBAF, -20°C (85%)

CCl$_3$CN, NaH (64%)

1. TMSOTf, CH$_3$CN, -40°C (43 %)

2. HS   SH, NEt$_3$; Ac$_2$O, Pyr (62 %)

3. Pd/C, H$_2$; Ac$_2$O, Pyr (77%)

TMSOTf, CH$_2$Cl$_2$/
n-Hexane, RT
(68 %)

Scheme 4 (continued).

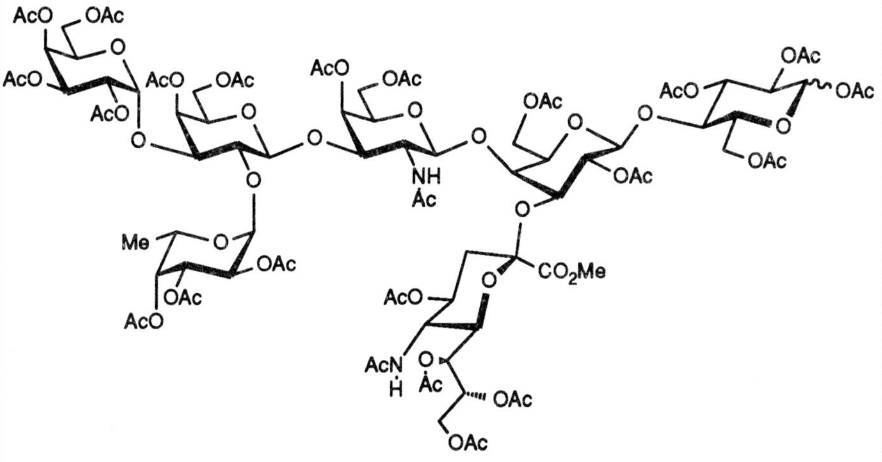

Scheme 4 (continued).

galactosamine and fucose building blocks which are coupled to the required tetrasaccharide moiety. Reaction of the derived donor with the rather unreactive axial 4b-OH group of suitably protected GM_3-trisaccharide[5,24] proved to be very difficult. However, in acetonitrile at low temperature and TMSOTf as catalyst (see below "nitrile effect"[1,25]) an acceptable yield of the desired heptasaccharide was obtained and this could be transformed into the target molecule.

For the completion of the glycosphingolipid synthesis which involves the attachment of the ceramide moiety to the saccharide residue, the azidosphingosine glycosylation procedure which was introduced by us[26] utilizes trichloracetimidates as glycosyl donors. This method has been employed successfully by a number of different groups.[27,29] An example is the GD_{1a} ganglioside synthesis published recently by Hasegawa *et al.*[30] (Scheme 5). The hexaosyl moiety in this case was obtained by thioglycoside activation with an excess of thiophilic promoter. Then the 1-*O*-unprotected hexasaccharide was transformed into the trichloroacetimidate which was reacted successfully with 3-*O*-benzoyl azidosphingosine[26] in the presence of trimethylsilyl trifluoromethanesulfonate (TMSOTf) as the catalyst to afford the β-glycoside in good yield. Subsequent reduction of the azido group, *N*-acylation and removal of *O*-acyl protective groups furnished the unprotected GD_{1a}.

32

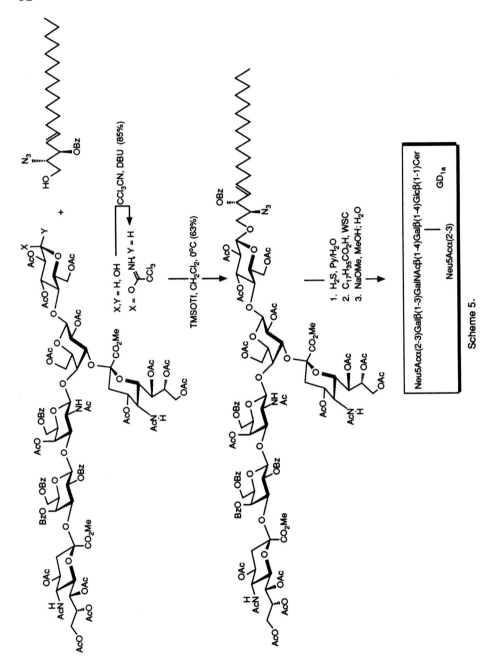

Scheme 5.

Glycopeptides and *glycoproteins* represent a very important area of glycoconjugates for which various efficient synthetic approaches have been reported.[2,3] Glycosyl trichloroacetimidates have been employed successfully for the construction of the required *O*- and *N*-linked oligosaccharide residues[2,31] and for the attachment of sugar residues to amino acids.[32] We have recently extended the trichloracetimidate method to a solid phase synthesis of an *O*-glycosylated hexapeptide of human sialophorin, which is shown in Scheme 6.[33] The disaccharide donor, obtained by the trichloroacetimidate method, can be readily linked to Fmoc-protected serine pentafluorophenylester as the α-isomer, thus providing the building block for the *O*-glycosyl hexapeptide synthesis.

The trichloroacetimidate method has also been employed extensively in the synthesis of *glycopeptidolipidic antigens of mycobacteria*.[34] For instance, a glycotetrapeptide present in the novel structurally variant glycopeptidolipid of Mycobacterium fortuitum, the cause for various mycobacterial infections, has been obtained with the help of this method.[35] Crucial steps of this straightforward synthesis are illustrated in Scheme 7.

Besides glycosphingolipids, other types of glycolipids are found in Nature. Amongst them the *saponins* (steroidal glycosides) play a major role.[2-7,36] They are responsible for the toxicity of sea cucumbers and starfish.[37] The structural variety of marine saponins, both in aglycone and sugar moieties provides a broad spectrum of biological activities.[37,38] Only very recently investigations for the synthesis of saponins have been undertaken in our laboratory.[2-7,36,38,39] Three structural types are found in starfish:[40] sulfated steroidal glycosides (asterosaponins), steroidal cyclic glycosides, and polyhydrated steroidal glycosides. We have recently reported the synthesis of the hexasaccharide moiety of the asterosaponin pectinioside E,[38] which exhibits *in vitro* cytotoxicity against L 1210 and KB cells.[41] The most important aspect of our strategy, outlined in Scheme 8, is the selective glycosylation at 2-OH and 4-OH of a glucose and a 6-deoxyglucose (quinovose) intermediate, thus avoiding extensive protection-deprotection procedures.

Many *antibiotics* contain rare sugar residues which play a critical role in their biological properties. Highlights of the application of the trichloroacetimidate method in this field are the elegant syntheses of enediyne

34

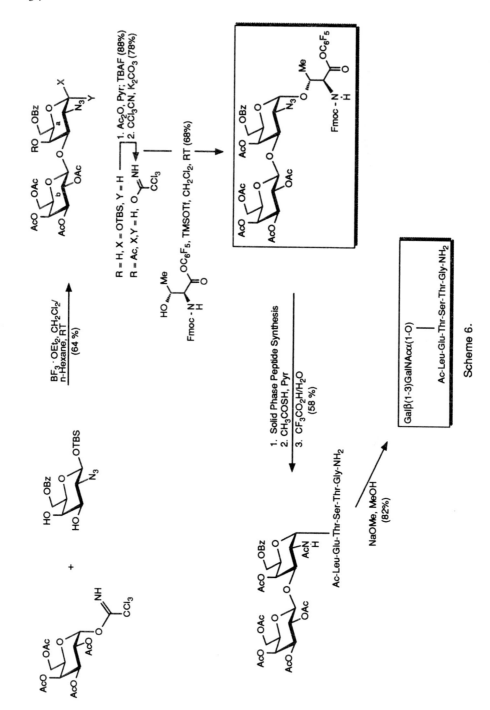

Scheme 6.

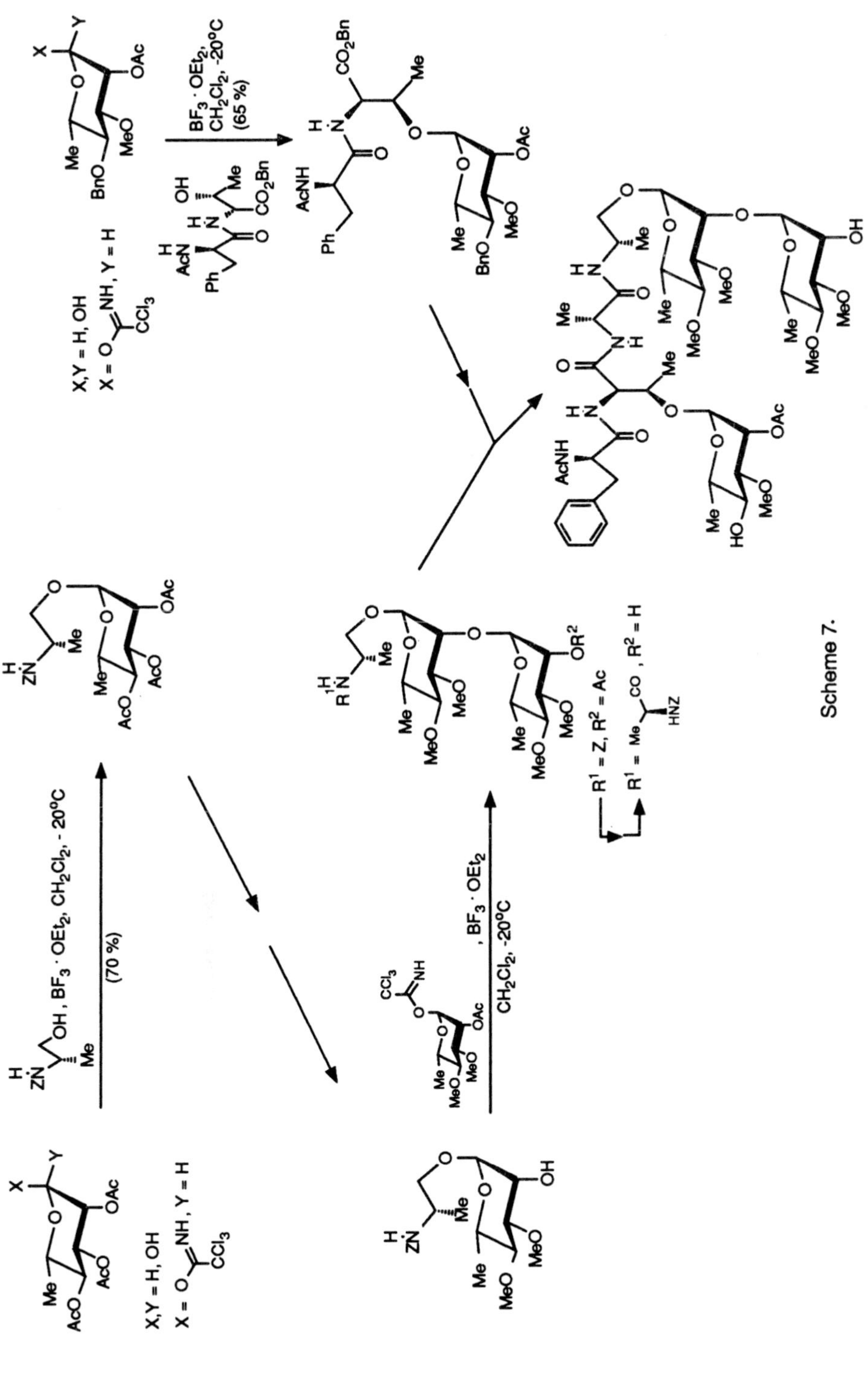

Scheme 7.

36

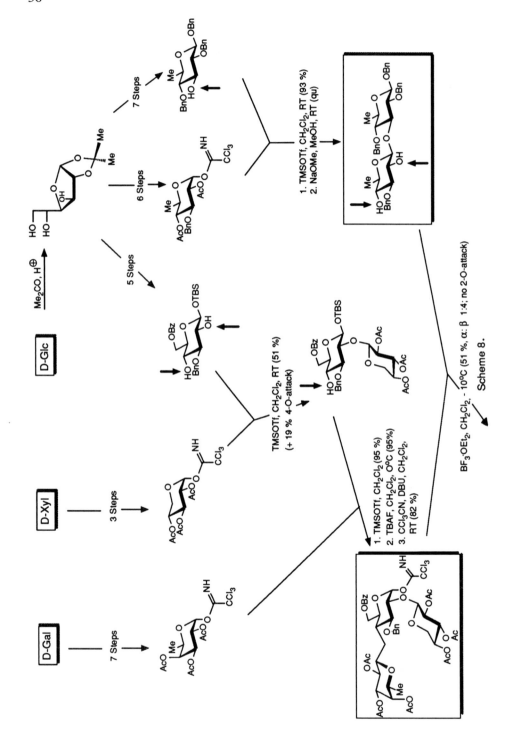

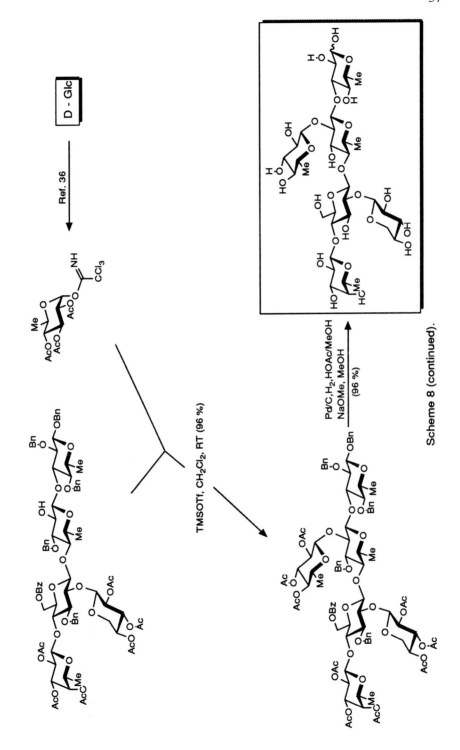

Scheme 8 (continued).

38

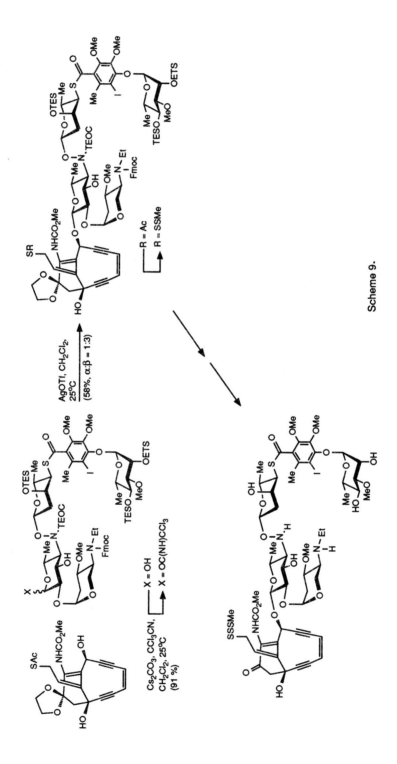

Scheme 9.

antibiotics reported by Nicolaou and collaborators[42] and by Danishefsky and collaborators.[43] The crucial step of the calicheamycin γ^I_1 synthesis of Danishefsky's group[43] is shown in Scheme 9, where silver triflate (AgOTf) is used as catalyst for glycoside bond formation.

Some recent findings have greatly contributed to the scope of the trichloroacetimidate method in terms of improvements in yield and anomeric stereocontrol. Thus for highly reactive glycosyl donors, for instance O-benzyl protected 6-deoxy sugars (fucose, rhamnose, quinovose), the *inverse procedure* (**IP**) has become very useful.[1,44] Glycosylation is generally carried out as a formally termolecular reaction of donor (**D**), acceptor (**A**) and promoter or catalyst (**C**), respectively (depending on the amount required).[1,2] Due to differences in the affinities, the reaction course is expected to be first **DC** interaction and then interaction of the **DC**-complex with **A** (Scheme 10, reaction course I). Obviously, for this sequence of interactions donors and acceptors must have matching reactivities. Therefore, acceptor and donor reactivities are often varied by changing the protective group pattern and, in the case of the donor reactivity by the selection of leaving groups and catalysts.[1,2] However, this strategy is less successful for very reactive glycosyl donors which may decompose in the presence of the catalyst while awaiting reaction with the acceptor. However, if acceptor **A** undergoes complexation with the catalyst **C** prior to interaction with the donor **D** (Scheme 10, reaction course II) this should not be a problem.

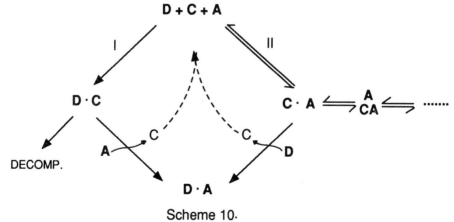

Scheme 10.

The efficiency of this approach is displayed in the synthesis of the Lewis A (Le[a]) antigen pentasaccharide moiety compiled in Scheme 11.[45] Attempts to carry out the fucosylation of the disaccharide acceptor with O-(2,3,4-tri-O-benzyl-L-fucopyranosyl)trichloroacetimidate as the donor (1.5 eq), applying the normal procedure (NP), i.e. adding the catalyst TMSOTf (0.01 eq) to the solution of donor and acceptor in Et$_2$O, led to modest yields of the desired trisaccharide (43%).[44] However, applying the inverse procedure that is first dissolving the acceptor (1 eq) and catalytic amounts of TMSOTf (0.01 eq) and then adding a solution of the donor (1.5 eq), thus enforcing reaction course II (Scheme 10) yielded 78% of the trisaccharide which is an important building block in Le[a] synthesis (outlined in Scheme 11). The efficiency of this principle has been demonstrated with various glycosyl donors.[23]

From these observations it was concluded that under IP-conditions AC-complex (or, due to the requirement of only catalytic amounts of C, AC-cluster formation) takes place first. This complex then reacts with donor D to generate the glycosylating species in the vicinity of acceptor A (Scheme 10, II). Thus, higher product yields are obtained because the competing donor decomposition is not as effective as in reaction course I.

The *influence of solvents on glycosylation reactions* is well known.[1,2,47] For instance, the participation of ethers, due to the reverse anomeric effect[48] results under S_N1-type conditions in the generation of equatorial oxonium ions which favor the formation of the more thermodynamically stable axial products. However, the dramatic effects of nitriles (*nitrile effect*) as participating solvents in O-glycosylation reactions have only recently been observed.[1,25] A highly α-selective glycosylation with the glycosyl halides of D-glucuronic acid as donors could be performed by us when the reaction was carried out in acetonitrile at -15°C by addition of the catalyst (AgClO$_4$) followed by the acceptor.[25] It was found that the reaction is controlled by intermediate formation of β-nitrilium-nitrile conjugates which are generated *in situ* under S_N1 conditions. This principle is shown in Scheme 12 for the trichloroacetimidate of O-benzylated 2-azido-2-deoxyglucose which gave good yields of the β-glycoside[46] when the reaction was conducted in a nitrile solvent at low temperature. This supports the concept of kinetic product control. The explanation for these findings is the following:

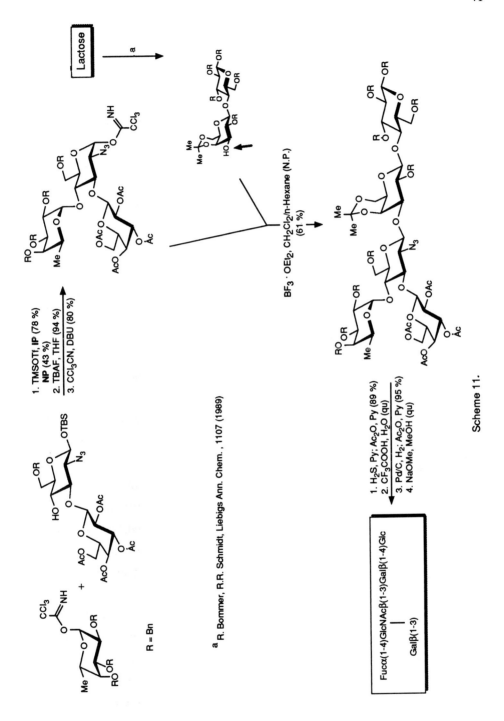

Scheme 11.

a R. Bommer, R.R. Schmidt, Liebigs Ann. Chem., 1107 (1989)

42

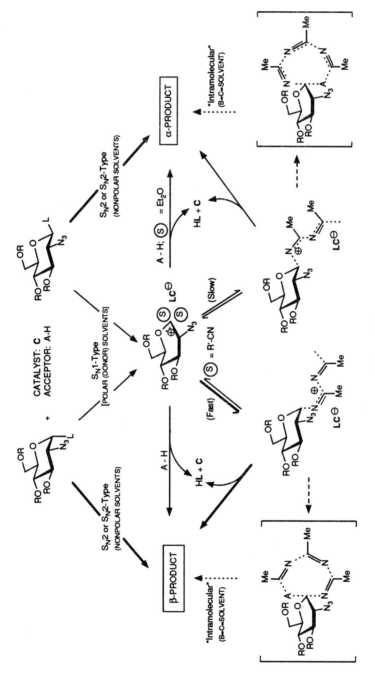

Scheme 12.

provided a good leaving group X, such as the trichloroacetimidoyl group, and low temperatures (-40°C to -80°C) are employed, then the glycosyl donor cleaves off its activating group in the presence of nitriles as solvents and TMSOTf as catalyst. Thus, a highly reactive carbenium ion intermediate is formed, which is then attacked by nitriles preferentially from the axial (α-)face, to afford the kinetically controlled α-nitrilium-nitrile conjugate, leading to the equatorial β-product. On the other hand, the equatorial β-nitrilium-nitrile conjugate is thermodynamically more stable due to the reverse anomeric effect (see the influence of ethers); thus favoring the axial α-product. Therefore, the α/β-ratio in the products is practically independent of the configuration of the initial glycosyl donor and its leaving groups. As long as a good leaving group is used, this leads to the formation of common intermediates for product control. A cyclic eight-membered transition state is proposed to explain the high reactivity and diastereoselectivity of these intermediate nitrilium-nitrile conjugates.[1]

The inverse procedure and the nitrile effect were important developments which facilitated our approach to Lewis X (Le^x) and Lewis Y (Le^y) antigen syntheses.[46] This is demonstrated in Scheme 13 for the synthesis of dimer Le^x, resulting in a very straightforward, efficient, and versatile methodology for the synthesis of glycosphingolipids of the *lactoneo* series and derivatives thereof.

The nitrile effect has also been proven to be quite successful not only for the important azidoglucopyranosyl building block, but also in various other cases.[5,6,24] The most prominent case is α-selective sialylation (equatorial glycoside formation under kinetically controlled conditions), where we demonstrated that phosphites can play an important role as a leaving group[5,24] (see below). For the efficacy of the nitrile effect, the carbenium ion stability and the relative rates of nitrile interception both axially and equatorially obviously plays a decisive role in diastereocontrol; however, these factors cannot be fully rationalzed at this time.

Glycosylations utilizing *O*-protected sugars have become very attractive because protection and deprotection steps can be reduced,[16,17] thus adding greatly to the efficiency of oligosaccharide synthesis. Various examples with highly *regioselective reactions at partially O-protected acceptors* have been reported (see

44

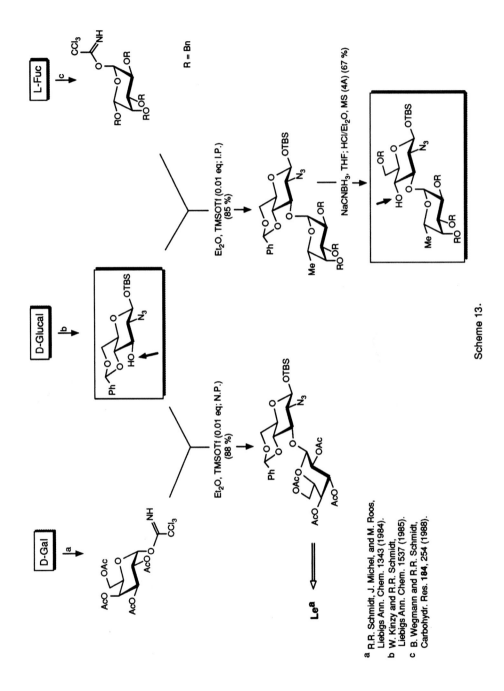

Scheme 13.

a R.R. Schmidt, J. Michel, and M. Roos,
  Liebigs Ann. Chem. 1343 (1984).
b W. Kinzy and R.R. Schmidt,
  Liebigs Ann. Chem. 1537 (1985).
c B. Wegmann and R.R. Schmidt,
  Carbohydr. Res. 184, 254 (1988).

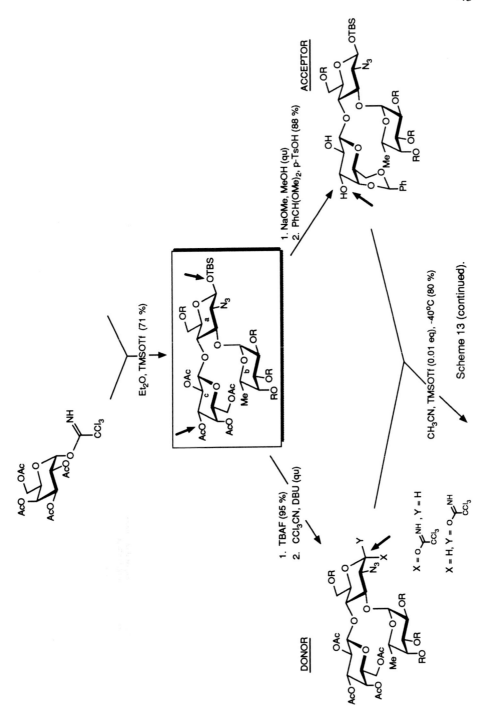

Scheme 13 (continued).

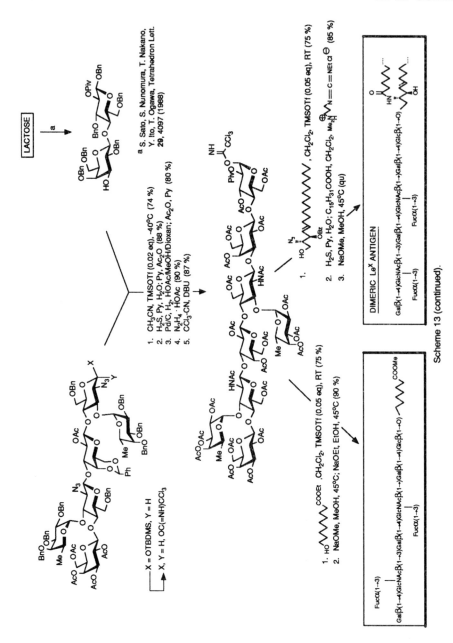

Scheme 13 (continued).

also Scheme 13) which extended the knowledge of relative reactivities of sugar hydroxy groups, especially in the area of sialylation.[5,24] However, partially O-protected glycosyl donors have not been thoroughly investigated. Therefore, we recently undertook preliminary studies to generate *partially O-protected O-glycosyl trichloroacetimidates*.[49,50] Scheme 14 shows two examples where the reactive 3-OH and 2-OH groups are unprotected. These compounds could be obtained with trichloroacetonitrile under mild base catalysis; both compounds seem to exhibit high glycosyl donor properties. However, it was found that for 2-O-unprotected sugars, amide acetal formation is possible in an ensuing base-catalyzed step. Surprisingly, these compounds also exhibit glycosyl donor properties under acid catalysis, thus regenerating the protonated trichloroacetimidate. However, due to lower basicity in the required oxygen protonation, stronger conditions for their activation are required. Obviously, the potential of partially O-protected O-glycosyl trichloroacetimidates as glycosyl donors awaits further exploration.

In conclusion, the trichloroacetimidate method has already become a classical method for glycoside bond formation and it has proven its versatility in complex glycoconjugate syntheses. It continues to gain in scope, due to the contributions from various laboratories.

Besides oxygen, other nucleophiles have also been successfully employed as acceptors. For instance, aryl-C-glycoside have been recently obtained from O-glycosyl trichloroacetimidates and phenol derivatives with TMSOTf as the catalyst in a Fries-type rearrangement.[51]

4. OTHER ANOMERIC OXYGEN ACTIVATION REACTIONS

For anomeric O-activation, base catalyzed addition of glycosyl oxides has also been extended to various electron deficient systems (N-containing triple bond systems, cumulenes, heterocycles, etc.).[1] However, none of the glycosyl donors thus obtained exceeds the O-glycosyl trichloroacetimidates in terms of ease of formation, stability, and reactivity.

One of the most important direct nucleophilic substitutions at activated carbon atoms carried out in Nature is enzymic O- and N-glycosyl bond formation at the anomeric carbon atom.[52] At this activated position, the leaving groups are

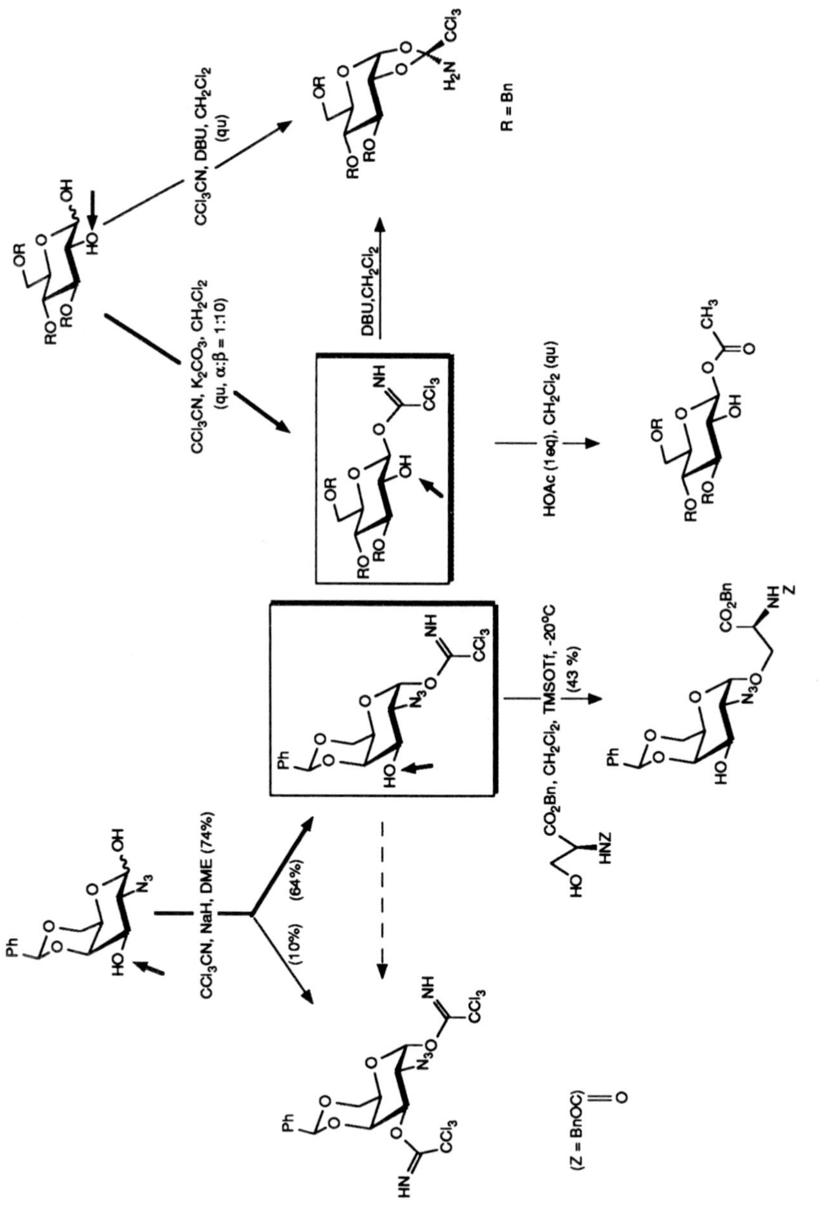

Scheme 14.

phosphates, pyrophosphates, and their nucleoside and lipid ester derivatives which are biosynthesized by anomeric O-phosphorylation reactions. *In vitro,* anomeric O-phosphorylation readily furnishes dialkyl or diaryl glycosyl phosphates.[54] These compounds exhibit good glycosyl donor properties comparable to those of glycosyl fluorides and sulfides, respectively.[1,54] Thus, contrary to a recent statement,[55] not only *in vivo* but also *in vitro* nucleophilic substitution at activated carbon atoms, as exemplified by the anomeric center, can be efficiently performed with glycosyl phosphates by activation of the anomeric oxygen. This was recently demonstrated not only for glycosyl phosphates but also for various glycosyl phosphate derivatives.[56]

For glycosyl donors requiring the highest activation (sugar uronates, aldoses, deoxy sugars) O-glycosyl trichloroacetimidates have been shown by us and others to be very efficient (see above).[1] However, for sugars requiring lower activation to generate glycosyl donor properties (ketoses, 3-deoxy-2-glyculosonates), trichloroacetimidates proved to be less successful.[1,5,24,57] Therefore, for these sugar types a simple leaving group which provided the same advantages as the trichloracetimidate moiety was desired. This leaving group turned out to be the *phosphite moiety* which was recently introduced by us for glycosyl donor generation as a further example of anomeric oxygen activation.[5,24] Meanwhile it could be shown that the phosphite method ideally complements the trichloroacetimidate method in the higher reactivity range as illustrated in Scheme 15, which also points to other important factors in glycosylation reactions.[6]

50

IMPORTANT SUGAR MOIETIES

ALDOPYRANOSYLURONATE	ALDOPYRANOSYL	DEOXYALDOPYRANOSYL	KETOPYRANOSYL	3-DEOXY-2-GLYCULOPYRANOSYL-ONATE
GlcUA	Glc, Gal, Man GlcNAc, GalNAc	Fuc, Rha, Qui 2-Deoxyglc	Fru, Sor	KDO, Neu5Ac

TYPICAL GLYCOSYL DONORS AND THEIR PROPERTIES

TRICHLOROACETIMIDATES ─ ─ ─ ─ ─ PHOSPHITES

STABILITY OF GLYCOSYL DONORS

Increasing ←─────→ Decreasing

"DISARMED" ←── Shift ── O-Acetyl-Protection / O-Benzyl-Protection / No O-Protection ── Shift ──→ "ARMED"

EASE OF ACTIVATION
(ANOMER C$^{\oplus}$ STABILIZATION)

Decreasing ←─────→ Increasing

Scheme 15.

REFERENCES

1. (a) R.R. Schmidt, *Angew. Chem.*, **98**, 213 (1986); *Angew. Chem. Int. Ed. Engl.*, **25**, 212 (1986); (b) *Pure Appl. Chem.*, **61**, 1257 (1989); (c) R.R. Schmidt In *Comprehensive Organic Synthesis* (B.M. Trost, I. Fleming, and E. Winterfeldt, Eds.) Pergamon Press, Oxford, Vol 6, p. 33 (**1991**); (d) R.R. Schmidt and W. Kinzy, *Adv. Carbohydr. Chem. Biochem.*, **50**, 21 (1994).

2. (a) H. Paulsen, *Angew. Chem.*, **94**, 184 (1982); *Angew. Chem. Int. Ed. Engl.*, **21**, 155 (1982); (b) *Angew. Chem.*, **102**, 851 (1990); *Angew. Chem. Int. Ed. Engl.*, **29**, 823 (1990).

3. (a) H. Kunz, *Angew. Chem.*, **26**, 297 (1987); *Angew. Chem. Int. Ed. Engl.*, **26**, 294 (1987); (b) *Pure Appl. Chem.*, **65**, 1223 (1993).

4. K. Toshima and K. Tatsuta, *Chem. Rev.*, **93**, 1503 (1993).

5. T.J. Martin and R.R. Schmidt, *Tetrahedron Lett.*, **33**, 6123 (1992).

6. T. Müller, R. Schneider, and R.R. Schmidt, *Tetrahedron Lett.*, **35**, 4763 (1994).

7. (a) R.R. Schmidt, M. Reichrath, *Angew. Chem.*, **91**, 497 (1979); *Angew. Chem. Int. Ed. Engl.*, **18**, 466 (1979); (b) R.R. Schmidt, M. Reichrath, U. Moering, *Tetrahedron Lett.*, **21**, 3561 (1980); (c) *J. Carbohydr. Chem.*, **3**, 67 (1984); (d) R.R. Schmidt, U. Moering, M. Reichrath, *Tetrahedron Lett.*, **21**, 3565 (1980); (e) *Chem. Ber.*, **115**, 39 (1982).

8. (a) R.R. Schmidt, A. Esswein, *Angew. Chem.*, **100**, 1234 (1988); *Angew. Chem. Int. Ed. Engl.*, **27**, 1178 (1988); (b) A. Esswein, H. Rembold, R.R. Schmidt, *Carbohydr. Res.*, **200**, 287 (1990); (c) H. Rembold and R.R. Schmidt, *Carbohydr. Res.*, **246**, 137 (1993).

9. J.J. Oltvoort, M. Kloosterman, C.A.A. van Boeckel, and H. van Boom, *Carbohydr. Res.*, **130**, 147 (1984).

10. W. Klotz, Ph.D. *Dissertation*, University of Konstanz, Germany (1994).

11. The enhanced nucleophilicity of the anomeric oxygen has been recently discussed in V.G.S. Box, *Heterocycles*, **31**, 1157 (1990).

12. W. Klotz and R.R. Schmidt, *J. Carbohydr. Chem.*, **13**, 1093 (1994).

13.   U. Moering, Unpublished Results, University of Konstanz, Germany (1980).

14.   Y.E. Tsvetkov, W. Klotz, and R.R. Schmidt, *Liebigs Ann. Chem., 371* (1992).

15.   E. Decoster, J.-M. Lacombe, J.-L. Strebler, B. Ferrari, A.A. Pavia, *J. Carbohydr. Chem.,* **2**, 329 (1983).

16.   R.R. Schmidt and W. Klotz, *Synlett,* 168 (1991).

17.   W. Klotz and R.R. Schmidt, *Liebigs Ann. Chem.,* 683 (1993).

18.   R.R. Schmidt and J. Michel, *Tetrahedron Lett.,* **25**, 821 (1984).

19.   (a) S. Hakomori, *Adv. Cancer Res.,* **52**, 257 (1991); (b) *J. Biol. Chem.,* **265**, 18713 (1990) and references cited therein.

20.   Y.A. Hannun and R.M. Bell, *Science,* **243**, 500 (1989) and references cited therein.

21.   E. H. Holmes and S.-I. Hakomori, *J. Biol. Chem.,* **257**, 7698 (1982).

22.   J.F. Bonhours, D. Bonhours, G.C. Hausson, *J. Biol. Chem.,* **262**, 16370 (1987).

23.   U. Greilich, P. Zimmermann, K.-H. Jung, and R.R. Schmidt, *Liebigs Ann. Chem.,* 859 (1993).

24.   T.J. Martin, R. Brescello, A. Toepfer, and R.R. Schmidt, *Glycoconjugate J.,* **10**, 16 (1993).

25.   (a) R.R. Schmidt and E. Rücker, *Tetrahedron Lett.,* **21**, 1421 (1980); (b) R.R. Schmidt and J. Michel, *J. Carbohydr. Chem.,* **4**, 141 (1985); (c) R.R. Schmidt, M. Behrendt, and A. Toepfer, *Synlett,* 694 (1990); (d) Y.D. Vankar, P.S. Vankar, M. Behrendt, and R.R. Schmidt, *Tetrahedron,* **47**, 9985 (1991).

26.   (a) R.R. Schmidt and P. Zimmermann, *Tetrahedron Lett.,* **27**, 481 (1986); (b) *Angew. Chem.,* **98**, 722 (1986); *Angew. Chem. Int. Ed. Engl.,* **25**, 725 (1986); (c) P. Zimmermann, R. Bommer, T. Bär, and R.R. Schmidt, *J. Carbohydr. Chem.,* **7**, 435 (1988).

27.   Y. Ito, M. Numata, M. Sugimoto, and T. Ogawa, *J. Am. Chem. Soc.,* **111**, 8508 (1989).

28.   K. Hotta, S. Komba, H. Ishida, M. Kiso, and A. Hasegawa, *J. Carbohydr. Chem.,* **13**, 665 (1994) and references cited therein.

29. K.C. Nicolaou, T.J. Caulfield, H. Kataoka, *Carbohydr. Res.,* **202**, 177 (1990) and references cited therein.

30. A. Hasegawa, H. Ishida, T. Nagahama, and M. Kiso, *J. Carbohydr, Chem.,* **12**, 703 (1993).

31. C. Unverzagt, *Angew. Chem.,* **106**, 1170 (1994); *Angew. Chem. Int. Ed. Engl.,* **33**, 1102 (1994).

32. (a) W. Kinzy and R.R. Schmidt, *Carbohydr. Res.,* **193**, 33 (1989); (b) A. Toepfer and R.R. Schmidt, *ibid.,* **202**, 193 (1990).

33. (a) J. Rademann, *Diplomarbeit,* University of Konstanz, Germany (1994); (b) J. Rademann and R.R. Schmidt, *Carbohydr. Res.,* Accepted.

34. A. Liptak, A. Borbas, and I. Bajza, *Med. Res. Rev.,* **14**, 307 (1994).

35. M.K. Gurjar and U.K. Saha, *Bioorg. Med. Chem. Lett.,* **3**, 697 (1993).

36. X.-B. Han and R.R. Schmidt, *Liebigs Ann. Chem.* 817 (1992).

37. (a) B.W. Halstead, *Poisons and Venomous Marine Animals of the World,* Vol. I, U.S. Government Printing Office, Washington D.C., p. 537 (1965); (b) B.W. Halstead, *Poisons and Venomous Marine Animals of the World,* (Revised Edition), The Darwin Press, INC., Princeton, New Jersey, p. 145 (1978); (c) D.J. Burnell and J.W. ApSimon, In *Marine Natural Products. Chemical and Biological Perspektives* (P.J. Scheuer, Ed.), Academic Press, New York, Vol. V, p. 287 (1983).

38. Z.-H. Jiang and R.R. Schmidt, *Liebigs Ann. Chem.,* 975 (1992) and references cited therein.

39. (a) X.-B. Han, Z.-H. Jiang, and R.R. Schmidt, *Liebigs Ann. Chem.,* 853 (1993); (b) Z.-H. Jiang, X.-B. Han, and R.R. Schmidt, *ibid.,* 1179 (1993).

40. L. Minale, C. Pizza, R. Riccio, and F. Zollo, *Pure Appl. Chem.,* **54**, 1935 (1982).

41. M.-A. Dubois, R. Higuchi, T. Komori, and T. Sasaki, *Liebigs Ann. Chem.,* 845 (1988).

42. K.C. Nicolaou, H. Saimoto, Y. Mizuno, K.-U. Baldenius, and A.L. Smith, C.W. Hummel, M. Nakada, K. Shibayama, and E.N. Pitsines, *J. Am. Chem. Soc.,* **115**, 7625 (1993) and references cited therein.

43. S.A. Hitchcock, S.H. Boyer, M.Y. Chu-Moyer, S.H. Olson, and S.J. Danishefsky, *Angew. Chem.,* **106**, 928 (1994); *Angew. Chem. Int. Ed. Engl.,* **33**, 858 (1994) and references cited therein.

44. (a) R. Bommer, W. Kinzy, and R.R. Schmidt, *Liebigs Ann. Chem.,* 425 (1991); (b) R.R. Schmidt and A. Toepfer, *Tetrahedron Lett.,* **32**, 3353 (1991).

45. A. Toepfer and R.R. Schmidt, *J. Carbohydr. Chem.,* **12**, 809 (1993).

46. (a) A. Toepfer and R.R. Schmidt, *Tetrahedron Lett.,* **33**, 5161 (1992); (b) A. Toepfer, W. Kinzy, and R.R. Schmidt, *Liebigs Ann. Chem.,* 449 (1994).

47. G. Wulff and G. Röhle, *Angew. Chem.,* **86**, 173 (1974); *Angew. Chem. Int. Ed. Engl.,* **13**, 157 (1974).

48. R.U. Lemieux, *Pure Appl. Chem.,* **25**, 527 (1971).

49. G. Grundler, Ph.D., *Dissertation*, University of Konstanz, Germany (1983).

50. (a) R. Haeckel, C. Troll, H. Fischer, and R.R. Schmidt, *Synlett,* 84 (1994); (b) F. Cinget and R.R. Schmidt, *ibid.,* 168 (1993).

51. J.-A. Mahling and R.R. Schmidt, *Synthesis,* 325 (1993) and references cited therein.

52. M.L. Sinnott, In *Enzyme Mechanisms* (M.I. Page and A. Williams, Eds.), The Royal society of Chemistry, London, p. 259 (1987).

53. M. Inage, H. Chaki, S. Kusumoto, and T. Shiba, *Chem. Lett.,* 1281 (1982) and references cited therein.

54. R.R. Schmidt, H. Gaden, and H. Jatzke, *Tetrahedron Lett.,* **31**, 327 (1990).

55. F.L. Westheimer, *Science*, **235**, 1173 (1987).

56. S. Hashimoto, T. Honda, and S. Ikegami, *Tetrahedron Lett.,* **32**, 1653 (1991) and references cited therein.

57. R. Preuss and R.R. Schmidt, *Synthesis*, 694 (1988).

Chapter 3

SYNTHESIS OF GLYCOSYL HALIDES FOR OLIGOSACCHA-RIDE SYNTHESIS USING DIHALOGENOMETHYL METHYL ETHERS

PAVOL KOVÁČ

NIDDK, LMC, National Institutes of Health, Bethesda, MD 20892, USA

Abstract Dihalogenomethyl methyl ether/Lewis acid reagents are useful for generation of glycosyl halides from various anomeric derivatives of mono and oligosaccharides. When 1,2-*trans*-di-*O*-acyl derivatives of sugars are treated with dihalogenomethyl methyl ethers in the presence of $BF_3 \cdot Et_2O$, the thermodynamically less stable glycosyl halides can be prepared in high yield. The use of other Lewis acids normally yields the thermodynamically more stable glycosyl halides. The reagents are compatible with many blocking groups commonly used in carbohydrate chemistry, including some acid-labile groups. Prolonged treatment of sugars protected with acetal and/or benzyl ether groups may result in their replacement with selectively removable formyl groups, which may be advantageous in designing blocking strategies. A comparison is given of the reactivity of various derivatives from which glycosyl halides can be generated, and the side reactions observed during these conversions are referred to. This is expected to be helpful in a rational choice of reaction conditions for the use of dihalogenomethyl methyl ethers to generate complex glycosyl donors for oligosaccharide synthesis.

1. INTRODUCTION

Two basic reactants are required for the chemical synthesis of oligosaccharides, a glycosyl donor and a glycosyl acceptor. Of the many types of glycosyl donors available to a carbohydrate chemist,[1] glycosyl halides are generally the most useful and are indeed the most frequently used. Historically,[2] the first glycosyl halides used in oligosaccharide synthesis were the *fully acylated* glycosyl halides. In the early days of oligosaccharide synthesis, such compounds adequately fulfilled their role as glycosyl donors because the target substances were either disaccharides or glycoconjugates where only a relatively simple sugar unit was to be attached. During that time, the type of glycosyl halides used were almost exclusively glycosyl bromides. Glycosyl iodides were found too unstable to be practical and, with simple silver or mercury salts as promoters, glycosyl chlorides and fluorides were insufficiently reactive to be of practical value in oligosaccharide syntheses.

When the target oligosaccharides and glycoconjugates became more complex the need arose for more complex glycosyl halides. Synthesis of variously protected sugar derivatives often started with a glycoside of the parent sugar. The chemical manipulations involved the placement of the requisite functional groups at specific positions, followed by acetolysis of the derivatized glycoside. Finally, the formed 1-*O*-acetyl derivative was converted to the corresponding glycosyl halide. Clearly, there was a demand for chemistry allowing a more direct transformation of glycosides to glycosyl halides. At present, an array of reagents which convert glycosides to glycosyl halides are available, *e.g.* ref. 3-10. The present chapter reviews the use of one class of such reagents, namely dihalogenomethyl methyl ethers, for the synthesis of glycosyl halides. Special emphasis is given the use of these substances for the generation of complex glycosyl halides* for oligosaccharide synthesis.

2.   HISTORICAL BACKGROUND AND GENERAL SCOPE OF THE METHOD

The formation of glycosyl halides from anomerically substituted sugars can often be achieved with dihalogenomethyl methyl ethers more conveniently than with any of the reagents described in the papers quoted above. Dihalogenomethyl methyl ethers

---

*Here, the term *complex glycosyl halides* refers to glycosyl halides containing more than one kind of blocking groups, often but not necessarily selectively removable, or glycosyl halides derived from oligosaccharides.

are more generally applicable and the literature shows, *e.g.* ref. 11, that when two methods were compared, the one involving a halogeno ether was superior. The method for the preparation of glycosyl halides using dihalogenomethyl methyl ethers has been refined into a preparatively useful procedure by a group of chemists in Debrecén (Hungary). The original observation in 1959 of Reiche and Gross[12] that treatment of 1,2,3,4,6-penta-*O*-acetyl-β-D-glucopyranose with dichloromethyl methyl ether (DCMME) gives tetra-*O*-acetyl-α-D-glucopyranosyl chloride in good yield was shortly followed by more elaborate treatises. There, the authors described[13,14] the preparation of a number of glycosyl halides from *per-O-acetyl-* *or per-O-benzoyl pento- and hexopyranoses,* the analogous 1-hydroxy derivatives, and *fully acylated disaccharides* (Eq. 1-6).

(1)   X = Cl, 65-70°, 20 min, 82%
             X = Br, 60°, 90 min, 40%

(2)   71%

(3)   61%

(4)   X = Cl, 65-70°, 20 min, 77%
             X = Br, 60°, 90 min, 47%

(5)

$$X = Cl, 50\text{-}60°, 10 \text{ min}, 97\%$$
$$X = Br, 50\text{-}60°, 90 \text{ min}, 73\%$$

(6)   82%

In a similar manner[15] hexa-*O*-acetyl-α-rutinosyl chloride and bromide, and gentiobiosyl chloride[16] were prepared by cleavage of the constituent oligosaccharides from the natural products rutin and amygdalin, respectively.

The cleavage of *glycosides* of mono- and oligo-saccharides with dibromomethyl methyl ether (DBMME) and DCMME was reported[17] from the same laboratory in 1970 (Eq. 7). There, the authors also presented a detailed analysis of the chemistry involved in the conversion. The conclusions most pertinent to the preparation of glycosyl halides can be summarized as follows: *Irrespective of their anomeric configuration*, per-*O*-acylated sugars and their 1-hydroxy analogs, and per-*O*-acyl derivatives of simple alkyl or aryl glycosides (Eq. 7) of mono- or disaccharides, when treated with the dihalogenomethyl methyl ether-$ZnCl_2$ reagents, are converted eventually to the corresponding *thermodynamically more stable* glycosyl halides.

(7)   R = Me, Et, Pr, *i*-Pr, *t*-Bu, Bn, Ph, β-Naphthyl;
             The conversions involving DBMME were carried out in the presence of alcohol-free chloroform

The time needed for the reaction to reach the stage where the thermodynamically more stable product is essentially the only glycosyl halide present depends on the reaction conditions, as nicely demonstrated by a comparative study.[18] Situations [11,18-20] where the β-D-halide is the major end-product (*e.g.* Eq. 8[18] and 9[20]), are rare. The byproducts of the conversion are

(8)   α:β = 88 : 12, 99%

(9)*   92-95%

methyl formate and the respective alkyl or acyl halide. In the case of aryl glycosides, the byproducts are methyl chloride and aryl formate. The stereoselectivity of analogous reactions of substances which have a non participating group at *O*-2 is also dependent on the reaction conditions, as was later

(10)   α : β = 92 :8

---

*The structure given for the end-product was that of the α-anomer, but the reported $J_{1,2}$ 9.5 Hz is indicative of the 1,2-*trans*-configuration, *c.f.*, ref. 20.

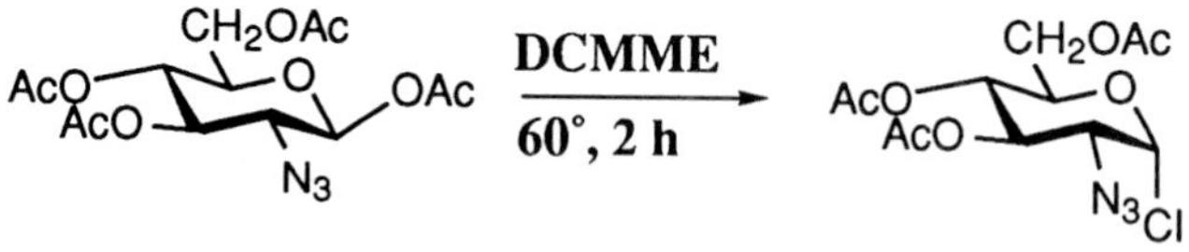

DCMME
60°, 2 h

(11)   Only a minute amount of the β-anomer was
                                              present after a 2 h reaction time

shown by isolation of variable amounts of the thermodynamically less stable products from such conversions (Eq. 10[18] and 11[21]).

Subsequently, the Hungarian group directed by Farkas and Bognár, as well as other groups, studied conversions involving dihalogenomethyl methyl ethers and found that, in addition to 1-*O*-acyl derivatives and simple glycosides, a number of other anomeric derivatives, such as 1,2-ortho esters[22] (Eq. 12), 1,6-anhydro sugars[23], 1-thio-glycosides[24] (Eq. 13, 14), *N*-arylglycopyranosylamines[25] (Eq. 15), and 2-(trimethylsilyl)ethyl glycosides[18,20] (Eq. 8-10) can be conveniently transformed to glycosyl halides in high yields.

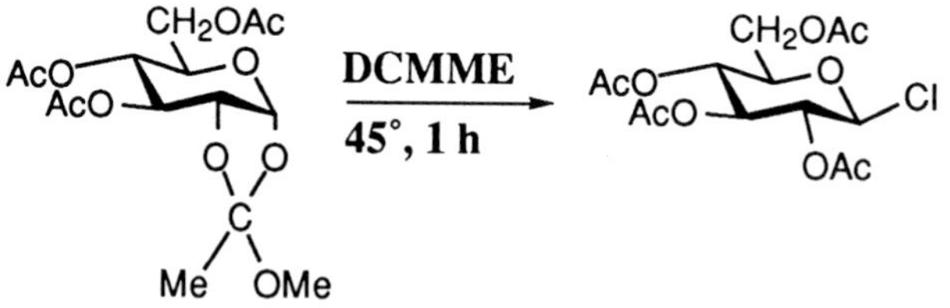

DCMME
45°, 1 h

(12)   60%

The transformation of orthoesters is particularly interesting since it allows preparation of the thermodynamically less stable β-glucosyl chloride, when carried out in the absence of a Lewis acid catalyst (Eq. 12). In the presence of ZnCl$_2$, the reaction yields tetra-*O*-acetyl-α-D-glucopyranosyl chloride in 72% yield within 10 minutes, as a result of anomerization of the initially formed β-chloride.

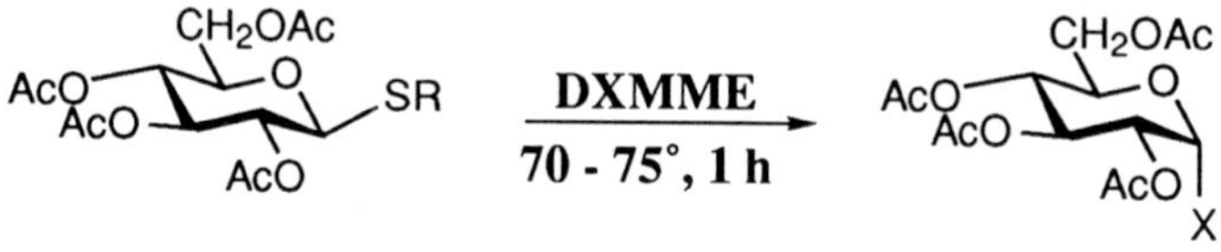

DXMME
70 - 75°, 1 h

(13)   R = Me, Et, *i*-Pr, Bu, *i*-Bu, Bn, Ph   42-66%

(14)   R = Me, Et, i-Pr, Bu, i-Bu, Bn   61-68%

Per-O-acetylated phenyl β-glycosides of cellobiose, lactose and maltose were similarly treated,[24] to give the corresponding α-glycosyl halides in 82, 57, and 39% yield, respectively. These yields, as well as yields from some other conversions, could probably be improved by using chromatography (*vide infra*) for the isolation of the desired products, rather then crystallization of the major product from the crude reaction mixtures.

(15)   R = Ph, Tol, p-BrPh, p-ClPh, p-NO$_2$Ph   41-57%

In addition to the rare examples quoted above (Eq. 8, 9, 12), dihalogenomethyl methyl ethers can be used to prepare the *thermodynamically less stable* glycosyl halides in a more general way. Compounds of this class, the 1,2-*trans*-glycosyl halides, have been often referred to[26] as "glycosyl halides of the unstable series." Of the many reagents available for the preparation of glycosyl halides few can be used for making the 1,2-*trans*-glycosyl halides. Of these, the method developed by Farkas *et al.*[27] involving the use of dihalogenomethyl methyl ether-BF$_3$·Et$_2$O reagent (Eq. 17-19, 21, 25) is unquestionably superior to the procedures developed earlier. The anticipated *stereospecificity* of the reaction was confirmed[28] when the ^{1}H NMR spectrum of a crude mixture of the conversion showed the absence of the 1,2-*cis*-isomer. The conversion involves an acyloxonium ion intermediate and, thus, only 1,2-*trans*-glycopyranoses bearing at O-2 a group capable of anchimeric assistance can be used in such reactions. This is demonstrated by the unreactivity of 1,2,3,4,6-penta-O-acetyl-β-D-mannopyranose under the standard conditions of the conversion. The method is based on the observation originally made by Bock *et al.*,[29] who used boron trifluoride etherate as a Lewis acid catalyst at ambient temperature (Eq. 16). The β-glycosyl chlorides can

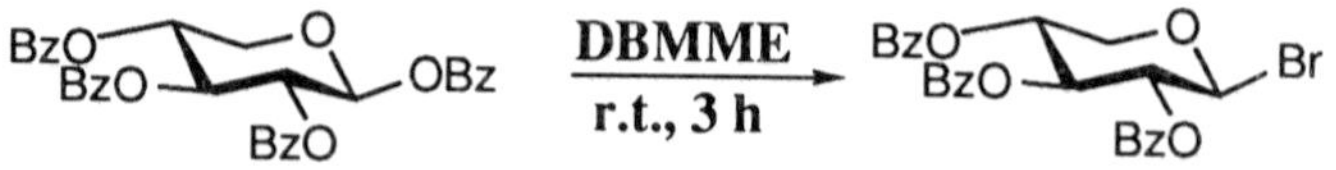

(16)

74%

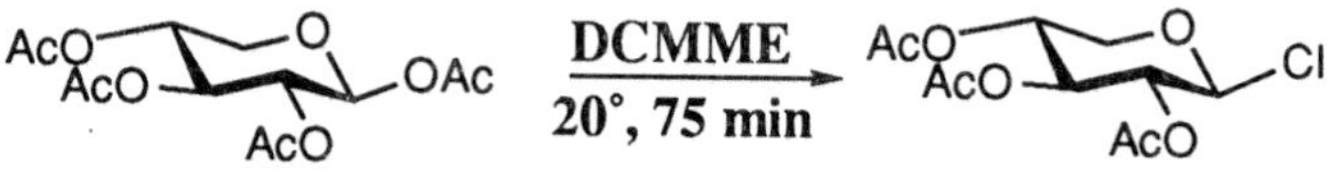

(17)

83%

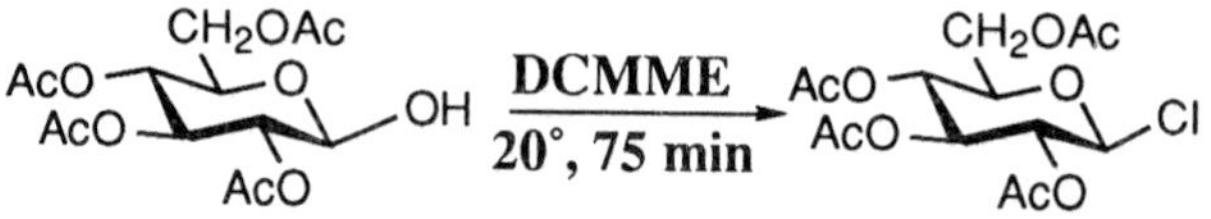

(18)

89%

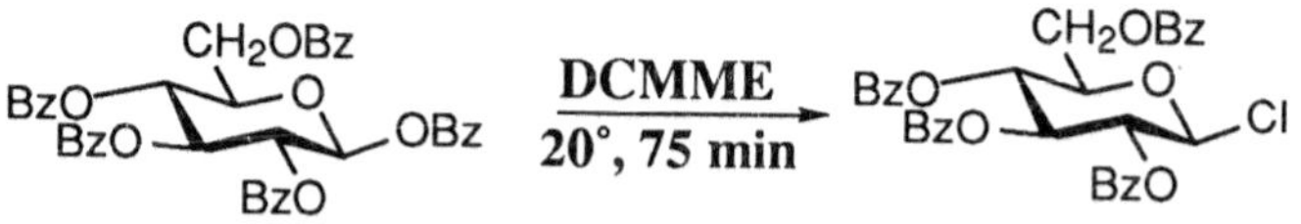

(19)

91%

be obtained under the same conditions from per-*O*-acetylated β-D-galactose and β-maltose in 77 and 81% yields, respectively.

Glycosyl halides can also be conveniently obtained from uronic acids[28] (Eq. 20-25). The rules established with neutral sugars apply here as well: per-*O*-acyl derivatives of esterified uronic acids, or the corresponding glycosides, can be converted to glycosyl halides with dihalogenomethyl methyl ethers. The stereochemical outcome of the reaction depends, as with neutral sugars, on the type of sugar derivative, the kind of Lewis acid catalyst used, and the reaction conditions applied.

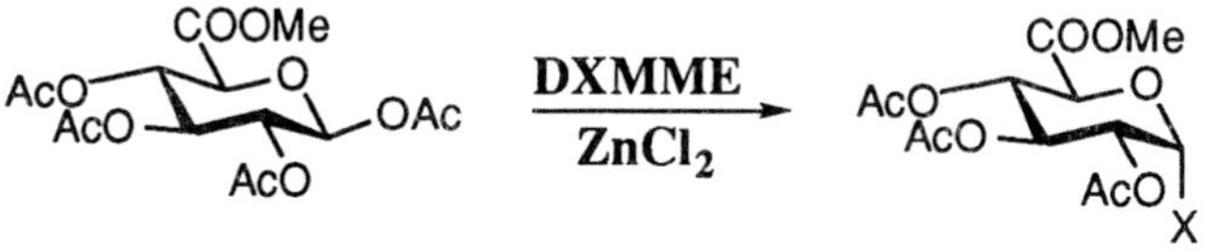

(20)

$$X = Cl, 65\text{-}70°, 0.5\text{ h} \qquad 91\%$$

$$X = Br, 40\text{-}50°, 0.5\text{ h} \qquad 76\%$$

When the treatment with DCMME starts with the α-acetyl derivative, the α- glycosyl chloride is obtained in 85% yield after a reaction time of 1.5 h.

(21)

$$X = Cl, \text{ r.t., } 16\text{ h or } 40° \text{ 8 h} \qquad 93\%$$

$$X = Br, \text{ r.t., } 8\text{ h} \qquad 71\%$$

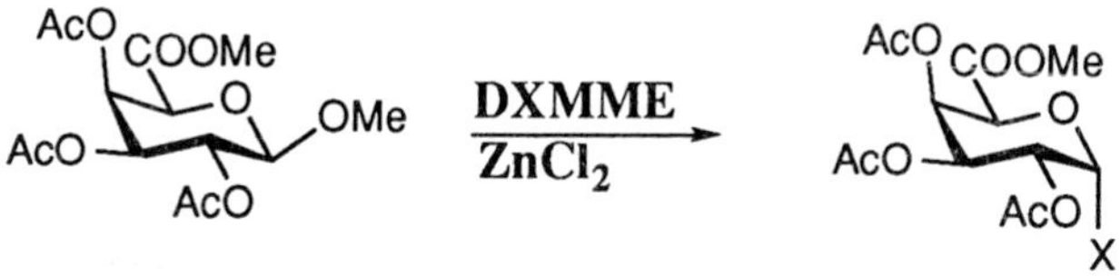

(22)

$$X = Cl, 65\text{-}70°, 2.5\text{ h} \qquad 69\%$$

$$X = Br, 40\text{-}50°, 0.5\text{ h} \qquad 74\%$$

The work in the uronic acid series constitutes[28,30] the first successful use of dihalogenomethyl ethers to prepare glycosyl halides from sugars bearing alkyl ether groups (Eq. 23-25).

(23)

$$X = Cl, 35\text{-}40°, 5.5\text{ h} \qquad 99\%$$

$$X = Br, 40\text{-}45°, 0.5\text{ h} \qquad 74\%$$

The fully methylated glycosyl bromide prepared in this way (Eq. 23) contained ~20% of the free sugar formed during the aqueous work-up by hydrolysis of the very reactive halide.

(24)   86%

(25)   82%

Another valuable feature of the preparation of glycosyl halides by the dihalogenomethyl methyl ether method is that glycosyl halides bearing acid labile substituents, such as *e.g.* cyclohexylidene[31] (Eq. 26) or isopropylidene[32] (Eq. 27) acetal groups (see also ref. 33 for a more recent example), can be prepared in this way. This is particularly important during preparation of complex glycosyl halides used in the synthesis of oligosaccharides as it widens the choice of protecting groups in blocking strategies.

(26)   95%

(27)   ~100%

The preparation and the early use of DCMME in organic chemistry has been reviewed.[34]

3. PREPARATION OF COMPLEX GLYCOSYL HALIDES FOR THE SYNTHESIS OF OLIGOSACCHARIDES

The use of dihalogenomethyl methyl ethers, and particularly the commercially available DCMME, to generate glycosyl halides for oligosaccharide syntheses was duly recognized after silver perchlorate and, particularly, silver trifluoromethanesulfonate were found to be exceptionally useful for highly stereoselective synthesis of 1,2-*cis*-[35,36] and 1,2-*trans*-[37,38] glycosides. Before that, the commercial unavailability, and burdensome preparation of DBMME hampered its widespread use for making glycosyl bromides. On the other hand, the commercially available DCMME was not a very attractive reagent in the synthesis of oligosaccharides: the much less reactive glycosyl chlorides were considered more of a chemical curiosity than useful reagents for practical oligosaccharide synthesis.

In the last two decades, solid scientific evidence confirming the involvement of oligosaccharides in a variety of vital biological processes has resulted in increased interest in synthetic oligosaccharides. Many new useful reagents have been added to the repertoire of the carbohydrate chemists and, at the same time, old classes of derivatives and reagents have been rediscovered and put to new uses. This applies, *inter alia*, to two very useful reagents in the oligosaccharide field, thioglycosides and dihalogenomethyl methyl ethers. At the time they were first prepared, neither of them was thought useful for the synthesis of oligosaccharides. Each of these has its advantages and drawbacks. Thioglycosides are extremely useful reagents because they can withstand many chemical manipulations, and can themselves be used as glycosyl donors, or can be transformed into glycosyl halides (*c.f.* ref. 39). Also, by careful choice of protecting groups their reactivity can be fine-tuned,[40,41] but the odorous nature of the reagents most commonly used for the preparation of thioglycosides may discourage their use. As for the dihalogenomethyl methyl ethers, their ability to generate glycosyl donors from a variety of derivatives, their compatibility with a great number of protecting groups, including some acid labile groups, the simplicity of their use, together with the commercial availability of the inexpensive DCMME, makes this class of reagents very useful. The only drawback seems to be the short shelf-life of DBMME, which makes it unlikely that it will become commercially available.

Reagents that can be conveniently used to generate glycosyl halides from oligosaccharides have been sought when the need arose to synthesize longer oligosaccharides. In such cases, the blockwise synthesis is the strategy of choice. It often requires making glycosyl halides from variously protected oligosaccharides. A few examples of preparations of compounds which belong to the category of complex glycosyl halides have already been given. Below are examples of the use of dihalogenomethyl methyl ethers for transformation of less common mono- or oligo-saccharides, yielding glycosyl donors which are useful in syntheses of higher oligosaccharides.

The reactions shown in Eq. 28-30[42] and 31[43] produced glycosyl halides which were key intermediates in a stepwise and a blockwise synthesis of $(1\rightarrow 6)$-linked galacto-oligosaccharides.

(28)   84%   6%

(29)   85%

(30)   76%

(31)   PhBz = p-phenylbenzoyl   74%

The glycosyl halides shown in Eq. 32 and 33, prepared in a similar way, were used[44] in syntheses of $(1\rightarrow 4)$-galacto-oligosaccharides. They can, however, be of general use as they allow the extension of an oligosaccharide chain at position 4 of D-galactose.

(32)   80%

(33)   90%

The readily obtainable, 6-fluorinated glucosyl chloride, prepared as shown in Eq. 34, was used as the glycosyl donor in the syntheses of methyl α-glycosides of 6'-deoxy-6'-fluoro sophorose and laminaribiose.[45] It may be interesting to note that the yield of the desired fluorinated glycosyl halide, starting from the methyl α-glycoside was 75%. Since milder reaction conditions could be applied when either α- or β-benzoyl derivative was used as the starting material the yield of the same glycosyl chloride was 85-90%.

(34)

The $(1 \rightarrow 3)$-linked di- and tetra-saccharide[46] glycosyl chlorides (Eq. 35, 36), and an analogous fully benzoylated trisaccharide[47] glycosyl donor, made it possible to synthesize, in the blockwise fashion, two heptasaccharides in the series of galacto-oligosaccharides.

(35)   86%

(36)   82%

Glycosyl chlorides prepared under the conditions shown in Eq. 37,[48] 38,[49] 39,[48] 40,[48] and 41[49] were used in the synthesis of oligosaccharides related to the O-specific polysaccharide of *Shigella dysenteriae* type 1.

(37)   73.5%

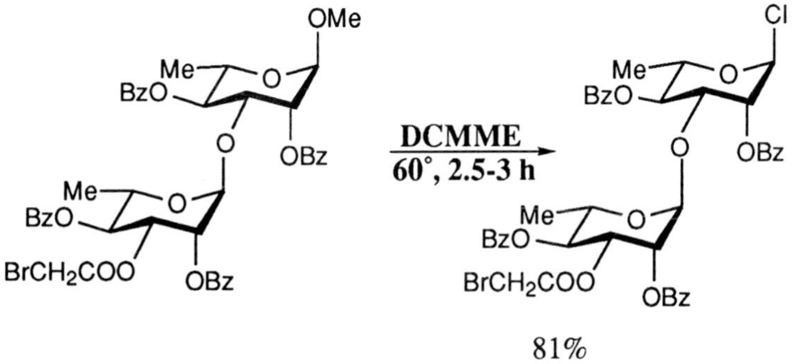
OMe
Me
BzO
O
OBz
BrCH2COO
DCMME
60°, 2.5-3 h
Cl
81%

(38)

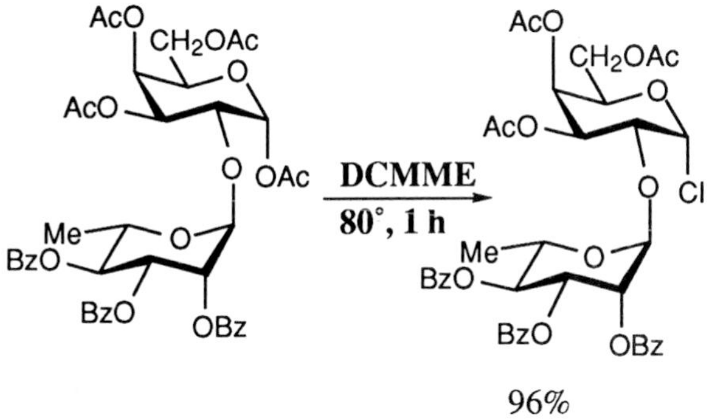
AcO
CH2OAc
OAc
Me
BzO
OBz
DCMME
80°, 1 h
Cl
96%

(39)

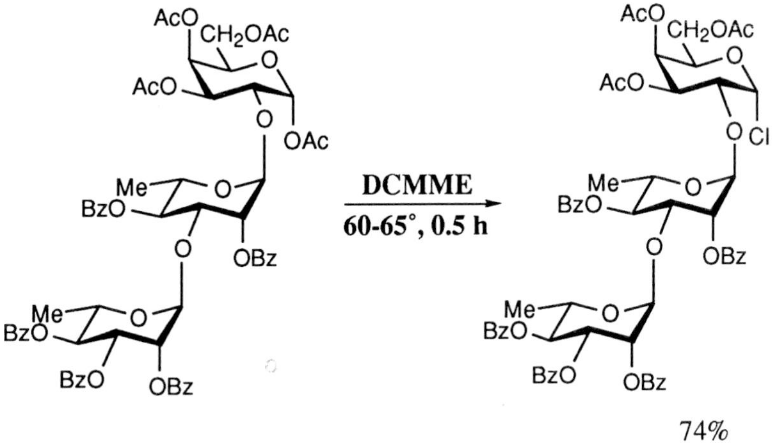
AcO
CH2OAc
OAc
Me
BzO
OBz
DCMME
60-65°, 0.5 h
Cl
74%

(40)

(41)   85%

3.A. Practical Considerations

With the exception of the conversion of orthoesters (Eq. 12), the preparation of glycosyl halides by the use of dihalogenomethyl methyl ethers requires the presence of at least a catalytic amount of a Lewis acid. No reaction was observed in the absence of the catalyst.[13] While other Lewis acid catalysts have been used,[20,29,50] the most useful catalyst to promote *conversions to the thermodynamically more stable glycosyl halides* are anhydrous zinc chloride and zinc bromide. A small amount of the freshly fused catalyst is conveniently prepared in a small test tube by flame-melting a pinch of the salt, and holding the test tube in the flame until the effervescence ceases and a clear melt forms. The melt covering the inner walls of the test tube is disintegrated into a powder by rubbing the walls with a stainless steel spatula. The requisite amount is weighed directly into the reaction vessel that already contains all other components. Efficient mixing can be conveniently provided by magnetic stirring with a teflon-coated stirring bar. Since the hydrogen halide liberated during the conversion has to be allowed to escape, stoppered or screw-capped reaction vials are not to be used. Moisture can be excluded from conventional flasks in the usual way, with the aid of a guard tube filled with an efficient drying agent. The presence of moisture, or a larger amount of impurities often cause the powdered zinc chloride to change into a sticky mass. The addition of more freshly fused catalyst usually corrects the problem without any detriment to the outcome of the reaction. The use of the commercially available solutions of zinc chloride in an inert solvent, instead of freshly fused zinc chloride, has been described.[51] This can be convenient, if *good quality* of such material is available, particularly when the exact amount of zinc chloride catalyst used is critical (for example, in the conversion of a derivative of a higher oligosaccharide to the corresponding glycosyl halide, Eq. 36), since an exact, small amount of the catalyst is easier to add in this way.

Depending on the reactivity of the starting material (*vide supra*) and stability of the various substituents present in the molecule, the reaction can be performed with the excess of DCMME or DBMME serving as both the solvent and the reagent, or with only a slight excess (stoichiometrically) of the halogeno ether. In the latter case, various halogenated hydrocarbons have been used as co-solvents. This set-up makes the reaction conditions milder. Since halogen scrambling during similar conversions has been observed,[52] it is advisable to keep such reaction mixtures halogen-homogeneous.

Some conversions of the type described are accompanied by strong discoloration of the reaction mixture. Unlike darkening during many other reactions in carbohydrate chemistry, this seldom indicates serious decomposition of the desired product or the starting material. Accordingly, examination of such mixtures by thin-layer chromatography (TLC), more often than not, shows no reason for panic. In view of the reactivity of glycosyl halides, hydroxylated solvents should not be used as components of solvent systems when reaction mixtures containing glycosyl halides are analyzed by TLC. Also, one has to remember that the formed glycosyl halides, especially in the presence of the hydrogen halide liberated during the reaction, are likely to hydrolyze partially during the TLC analysis. Thus, the presence of several ghost spots, or streaking during TLC may simply reflect decomposition during TLC, and not in the reaction flask.

Heating and stirring for a longer period of time of reaction mixtures containing DCMME often causes some, or most of the reagent to escape. This can happen even when the reaction flask is equipped with an efficient tap-water condenser. We have been able to largely overcome the need to add fresh reagent to replace that lost due to its volatility by the use of a small dry-ice condenser.

A very useful alternative to the traditional method of isolation of the generated glycosyl chlorides, a partition of the crude products between a cold, aqueous solution of sodium bicarbonate and an organic solvent,[13-15] is column chromatography. For this purpose, fully activated silica gel has been routinely used in our laboratory (see, for example, ref. 48). To prevent decomposition of very reactive halides during chromatography, their stability can be increased by addition of organic, non-nucleophilic base to the elution system.[53]

3.B. Reactivity of Starting Materials

Enough experimental material has accumulated in the literature to serve as a guide for the reaction rates of the conversion of various derivatives to glycosyl halides utilizing dihalogenomethyl methyl ethers. Generally, 1-O-acyl derivatives of sugars are converted to the corresponding glycosyl halides more readily with these reagents than the corresponding methyl glycosides. Also, 1-O-acetates react faster than their 1-O-benzoylated counterparts. Generally, electron-withdrawing groups present as O-substituents around the ring in a carbohydrate decrease the reactivity at the anomeric center. Accordingly, a per-O-acylated methyl glycoside of a monosaccharide will be converted to the corresponding glycosyl halide more slowly than its analog having one or more acyl groups replaced with an alkyl, $e.g.$ benzyl group. In the latter case, to avoid side reactions ($vide\ infra$), the use of an additional solvent is strongly recommended. Such reactions can often be carried out at room temperature.[32,54] Similarly, derivatives of deoxysugars[55,56] are more reactive for this type of reaction, regardless of the position of deoxygenation. The glycosyl halides derived from them are also more reactive than their non-deoxygenated counterparts. Per-O-acylated glycosides bearing a haloacetyl group as a selectively removable protecting group require harsher reaction conditions to be converted to the corresponding glycosyl halides with one of the dihalogenomethyl methyl ether-$ZnCl_2$ reagents than synthons not bearing such groups.[57,58] This is in agreement with the decrease in the reactivity at the anomeric center which is seen when electron-withdrawing substituents are present. The decrease of reactivity at the anomeric position due to the presence of multiple haloacetyl groups in the molecule is so dramatic that fully chloroacetylated methyl α-D-manno- and β-D-galactopyranosides could not be converted to the corresponding glycosyl chlorides at all[59] by treatment with the DCMME-$ZnCl_2$ reagent. The presence of an electron-withdrawing group in the molecule located further away from the anomeric position where the substitution is taking place, as for example in the case of an oligosaccharide, has little effect upon the reactivity at that position. Minor differences in the reaction conditions required for the successful conversion reflect, more likely, the amount of the Lewis acid catalyst used than the change in the reactivity due to the presence of the electron-withdrawing group in the second sugar unit (*c.f.* Eq. 37 and 38).

The stereochemistry at the anomeric position where the exchange for halogen is to take place affects the reactivity as well. This is apparent from a comparison of the conversion of anomeric pairs of compounds (*c.f.* Eq. 28 and 29 as an example). The conversion of the α-benzoyl derivative (Eq. 29) required somewhat harsher conditions, as compared to its β counterpart. The cleavage of the interglycosidic linkage was more extensive under these conditions, resulting in somewhat lower yield of the desired glycosyl halide, than that obtained from the β-anomer. The higher reactivity of a 1-*O*-acetyl compared to a 1-*O*-benzoyl derivative is evident from Eq. 28 and 33: The acetyl group is cleaved from the anomeric center under milder conditions, the rupture of the interglycosidic linkage is less extensive under these conditions and, consequently, the desired glycosyl halide is obtained in a higher yield.

3.C. **Side Reactions**

In one of their early works on the possibility of cleaving various anomeric derivatives of sugars with dihalogenomethyl methyl ethers, Bognár *et al.*[23] studied reactions of DCMME and DBMME with fully acetylated laevoglucosan. Briefly summarized, they found that cleavage of the 1,6-anhydro ring and formation of a glycosyl halide was accompanied by alkylation (dichloromethylation) of the primary hydroxyl group generated at position 6. The dichloromethyl derivative hydrolyzed readily, to give the corresponding 6-*O*-formyl and 6-hydroxy derivatives. The same types of derivatives were formed during our treatment[60] of methyl 2,4,6-tri-*O*-benzoyl-3-*O*-benzyl-β-D-galactopyranoside with DCMME-ZnCl$_2$. Following the course of the reaction by thin-layer chromatography made it possible to determine the order of formation of individual products. This, together with the spectral analysis of the products isolated, allowed us to summarize the events occurring during the treatment of methyl 2,4,6-tri-*O*-benzoyl-3-*O*-benzyl-β-D-galactopyranoside with DCMME-ZnCl$_2$ (Scheme 1): Brief (1 h) treatment of the starting material results in fast conversion of the β-glycoside into the corresponding glycosyl halide, together with the slow anomerization, to give the corresponding α-

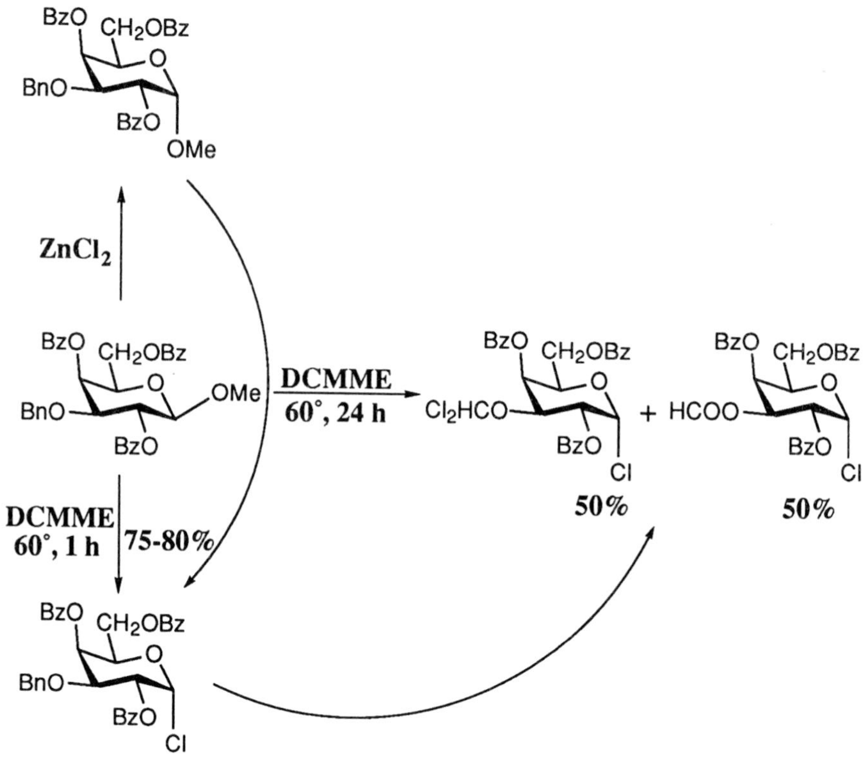

Scheme 1.

glycoside and its conversion to the same glycosyl halide (for another example of the anomerization of the starting glycoside during similar reaction, see Eq. 28). These processes are followed by much slower debenzylation of the substrate and the conversion of the putative 3-hydroxy derivative thus formed into the 3-*O*-formyl and 3-*O*-dichloromethyl derivatives. After 24 h, these two compounds, present in equal amounts, are the only carbohydrate components of the reaction mixture.

Our observations regarding the side reactions during treatment of various carbohydrate derivatives with dihalogenomethyl methyl ethers are in accord with those of others.[3,11,18,32] The debenzylation followed by formylation that may occur in some situations does not necessarily have to be undesirable since a new selectively removable protecting group can be introduced into the substrate in this way, (*e.g.* Eq. 42[18] and 44).

(42)   52%

Even though acetal protecting groups are generally stable under the conditions of the transformations described herein, and this is one of the valued features of this method, prolonged treatment will remove such groups. The hydroxyl groups generated in this way will be modified[32] (Eq. 43).

(43)   89%

If an acetal and a benzyl group are both present, they can be sequentially cleaved, and such conversions[61] (Eq. 44) may be of synthetic utility.

(44)

The examples given show the diversity of transformations which can be effected with dihalogenomethyl methyl ethers when applied to carbohydrate derivatives. Consequently, when applied with a little chemical intuition, these reagents can be very useful not only for the generation of glycosyl donors for oligosaccharide synthesis but they also expand the choice of blocking strategies available.

The possible cleavage of interglycosidic linkages in oligosaccharides during treatment with dihalogenomethyl methyl ethers has already been mentioned. The probability of such unwanted side reactions increases with the length of the oligosaccharide chain, rendering the amount and type of Lewis acid catalyst, as well as fine-tuning of the reaction conditions critical.

A synthetically useful transformation, which can also be looked upon as a side reaction during the preparation of glycosyl halides with DBMME, was observed by Bock et al.[29] They found that by prolonged treatment with DBMME and ZnBr$_2$, glycosyl bromides formed from acylated methyl pentopyranosides (vide supra, Eq. 16) are converted to bromodeoxy pentopyranosyl bromides in high yields (Eq. 45). In the case shown, the two pure anomeric glycosyl bromides could be isolated by fractional crystallization in only moderate yields. However, hydrolysis of the glycosyl bromides present in the crude product gave a high yield of the corresponding free sugar. Thus, this sequence constitutes a useful procedure for the preparation of 2-bromo-2-deoxy derivatives of D-xylose. The observed

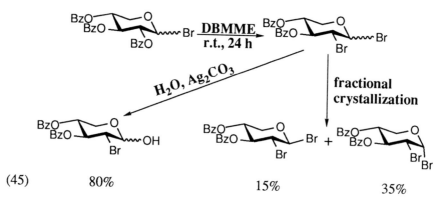

(45)   80%   15%   35%

formation of the deoxybromo pentosyl bromides (Eq. 45) was later used to prepare[62] some 2-bromo-2-deoxy-pentofuranosyl bromides and derivatives by a treatment of methyl per-O-acyl-D-pentofuranosides with DBMME-ZnBr$_2$.

3.D. Safety Considerations

One might surmise, merely by looking at the structure of dihalogenomethyl ethers, that these compounds must be, *a priori*, carcinogenic or mutagenic, or both. Indeed, halo-ethers have been suspect carcinogens, and some compounds in this class have been tested for these undesirable properties. In a study by the National Cancer Institute, DCMME was not found[63] carcinogenic. Nevertheless, DCMME and its bromo analog constitute a potential hazard, and all operations involving these chemicals should be performed in a well ventilated hood.

3.E. Typical Experimental Procedures

The selected procedures given below show preparations of glycosyl halides which differ in structural complexity, and the means used for purification of the desired products.

3.E.i. Hepta-O-acetyl-α-lactosyl bromide[14] (Eq. 5)

A mixture of octa-O-acetyl-β-lactose (1 g) , DBMME (1 mL) and a catalytic amount of freshly fused ZnCl$_2$ in alcohol-free chloroform (5 mL) is heated at 50-60° for 90 min. The mixture is cooled to room temperature, diluted with chloroform (10 mL), and washed successively with cold (0°) water, aqueous sodium hydrogen carbonate, and water. After drying, the cold solution is treated with a little decolorizing charcoal and, after filtration, the filtrate is concentrated at reduced pressure (<50°). Crystallization from chloroform-ether yields the title compound, m.p. 141-142°.

3.E.ii. O-(3,4,6-Tri-O-acetyl-2-azido-2-deoxy-α-D-glucopyranosyl)-(1→3)-2,4-di-O-benzoyl-α-L-rhamnopyranosyl chloride[49] (Eq. 41)

A mixture of methyl O-(3,4,6-tri-O-acetyl-2-azido-2-deoxy-α-D-glucopyranosyl)-(1→3)-2,4-di-O-benzoyl-α-L-rhamnopyranoside[49,64] (2.77 g), freshly fused ZnCl$_2$ (0.1 g) and DCMME (8 mL) in alcohol-free chloroform (8 mL) is heated at 50-55° C for 6-7 h. Thin-layer chromatography (2.5:1, CCl$_4$-EtOAc) then shows complete conversion of the starting material into a faster moving product. Dry toluene (10 mL) is added, the mixture is filtered and, after concentration, the residue is chromatographed, to give the pure, desired halide (2.37 g, 85%), m.p. 129-130° (from isopropyl ether).

3.E.iii.   4-*O*-Benzoyl-2,3-*O*-cyclohexylidene-α-L-rhamnopyranosyl bromide[31] (Eq. 26)

A mixture of methyl 4-*O*-Benzoyl-2,3-*O*-cyclohexylidene-α-L-rhamnopyranoside[31] (0.5 g), DBMME (0.3 mL) and zinc bromide (~30 mg) in alcohol-free chloroform was stirred at 50° C for 1 h. The mixture was filtered through glass wool, the filtrate was concentrated, and dried under reduced pressure. The material obtained (0.55 g, ~100%) was shown by NMR spectroscopy to contain at least 95 % of the desired glycosyl bromide, and it was immediately used for a subsequent glycosylation.

REFERENCES

1. R. R. Schmidt, in *Comprehensive Organic Synthesis* ; B.M. Trost, Ed. Pergamon Press, p. 33, (1991), and references cited therein.
2. L. J. Haynes and F. H. Newth, *Advan. Carbohydr. Chem.*, **10**, 207 (1955).
3. K. Bock and M. Meldal, *Acta Chem. Scand.,* **38B**, 255 (1984).
4. J. Thiem and B. Meyer, *Chem. Ber.,* **113**, 3075 (1980).
5. B. Classon, P. J. Garegg, and T. Norberg, *Acta Chem. Scand.,* **38 B**, 195 (1984).
6. G. R. Perdomo and J. Krepinsky, *Tetrahedron Lett.,* **28**, 5595 (1987).
7. Y. Guindon, C. Yoakim, and H. E. Morton, *J. Org. Chem.,* **49**, 3912 (1984).
8. G. Grynkiewicz and M. Konopka, *Polish J. Chem.,* **61**, 149 (1987).
9. P. M. Collins, P. Premarante, A. Manro, and A. Hussain, *Tetrahedron Lett.,* **30**, 4721 (1989).
10. K. Higashi, K. Nakayama, E. Shioya, and T. Kusama, *Chem. Pharm. Bull.,* **39**, 2502 (1991).
11. H. P. Wessel, T. Iversen, and D. R. Bundle, *Carbohydr. Res.,* **130**, 5 (1984).
12. A. Rieche and H. Gross, *Chem. Ber.,* **92**, 83 (1959).
13. H. Gross and I. Farkas, *Chem. Ber.,* **93**, 95 (1960).

14. I. Farkas, M. Menyhárt, R. Bognár, and H. Gross, *Chem. Ber.,* **98**, 1419 (1965).

15. R. Bognár, I. Farkas-Szabó, I. Farkas, and H. Gross, *Carbohydr. Res.,* **5**, 241 (1967).

16. I. Farkas, I. F. Szabó, R. Bognár, G. Czira, and J. Tamás, *Justus Liebigs Ann. Chem.,* **440** (1976).

17. I. F. Szabó, I. Farkas, R. Bognár, and H. Gross, *Acta Chim. Sci. Hung.,* **64**, 67 (1970).

18. K. Jansson, G. Noori, and G. Magnusson, *J. Org. Chem.,* **55**, 3181 (1990).

19. U. Nilsson, A. K. Ray, and G. Magnusson, *Carbohydr. Res.,* **208**, 260 (1990).

20. K. P. R. Kartha and H. J. Jennings, *Tetrahedron Lett.,* **31**, 2537 (1990).

21. V. Pavliak and P. Kováč, *Carbohydr. Res.,* **210**, 333 (1991).

22. I. Farkas, Z. Dinya, I. F. Szabó, and R. Bognár, *Carbohydr. Res.,* **21**, 331 (1972).

23. R. Bognár, I. Farkas, M. Menyhárt, H. Gross, and H. Paulsen,*Carbohydr. Res.,* **6**, 404 (1968).

24. I. Farkas, R. Bognár, M. M. Menyhárt, A. K. Tarnai, A. Bihari, and J. Tamás, *Acta Chim. Sci. Hung.,* **84**, 325 (1975).

25. M. M. Menyhárt, R. Bognár, and I. Farkas, *Carbohydr. Res.,* **48**, 139 (1976).

26. W. Korytnik and J. A. Mills, *J. Chem. Soc.,* 636 (1959).

27. I. Farkas, I. F. Szabó, R. Bognár, and D. Anderle, *Carbohydr. Res.,* **48**, 136 (1976).

28. P. Kováč, I. Farkas, V. Mihálov, R. Palovčík, and R. Bognár, *J. Carbohydrates, Nucleosides, Nucleosides,* **3**, 57 (1976).

29. K. Bock, C. Pedersen, and P. Rasmussen, *J. Chem. Soc., Perkin Trans.,* **1**, 1456 (1973).

30. P. Kováč and R. Palovčík, *Carbohydr. Res.,* **56**, 399 (1977).

31. T. Iversen and D. R. Bundle, *J. Org. Chem.,* **46**, 5389 (1981).

32. J. Thiem, M. Gerken, and K. Bock, *Justus Liebigs Ann. Chem.,* 462 (1983).

33. W. Klaffke, C. D. Warren, and R. W. Jeanloz, *Carbohydr. Res.*, **244**, 171 (1993).

34. H. Gross, I. Farkas, and R. Bognár, *Z. Chem.*, **18**, 201 (1978).

35. F. J. Kronzer and C. Schuerch, *Carbohydr. Res.*, **27**, 379 (1973).

36. K. Igarashi, J. Irisawa, and T. Honma, *Carbohydr. Res.*, **39**, 213 (1975).

37. S. Hanessian and J. Banoub, *Carbohydr. Res.*, **53**, C13 (1977).

38. P. J. Garegg and T. Norberg, *Acta Chem. Scand.*, **B33**, 116 (1979).

39. P. Fügedi, P. J. Garegg, H. Lonn, and T. Norberg, *Glycoconjugate J.*, **4**, 97 (1987).

40. G. H. Veeneman, S. H. van Leeuven, and J. H. van Boom, *Tetrahedron Lett.*, **31**, 1331 (1990).

41. G. H. Veeneman and J. H. van Boom, *Tetrahedron Lett.*, **31**, 275 (1990).

42. P. Kováč, *Carbohydr. Res.*, **153**, 237 (1986).

43. E. M. Nashed and C. P. J. Glaudemans, *J. Org. Chem.*, **52**, 5255 (1987).

44. P. Kováč and R. B. Taylor, *Carbohydr. Res.*, **167**, 153 (1987).

45. P. Kováč and C. P. J. Glaudemans, *J. Carbohydr. Chem.*, **7**, 317 (1988).

46. P. Kováč, R. B. Taylor, and C. P. J. Glaudemans, *J. Org. Chem.*, **50**, 5323 (1985).

47. T. Ziegler, P. Kováč, and C. P. J. Glaudemans, *Carbohydr. Res.*, **203**, 253 (1990).

48. P. Kováč and K. J. Edgar, *J. Org. Chem.*, **57**, 2455 (1992).

49. P. Kováč, *Carbohydr. Res.*, **245**, 219 (1993).

50. R. F. Helm, J. Ralph, and L. Anderson, *J. Org. Chem.*, **56**, 7015 (1991).

51. V. Pozsgay, B. Coxon, and H. Yeh, *Bioorg. Med. Chem.*, **1**, 237 (1993).

52. T. Lin and R. E. Harmon, *J. Carbohydr. Nucleosides, Nucleotides*, **1**, 109 (1974).

53. J. Niggemann and J. Thiem, *Liebigs Ann. Chem.*, 535 (1992).

54. T. Iversen and D. R. Bundle, *Can. J. Chem.*, **60**, 299 (1982).

55. T. Ziegler, V. Pavliak, T.-H. Lin, P. Kováč, and C. P. J. Glaudemans, *Carbohydr. Res.*, **204**, 167 (1990).

56. K. Jansson, S. Ahlfors, T. Frejd, J. Kihlberg, G. Magnusson, and J. Dahmen, G. Noori, and K. Stenwall, *J. Org. Chem.*, **53**, 5629 (1988).

57. T. Ziegler, B. Adams, P. Kováč, and C. P. J. Glaudemans, *J. Carbohydr.*

Chem., **9**, 135 (1990).

58. T. Ziegler, P. Kováč, and C. P. J. Glaudemans, *Carbohydr. Res.,* **194**, 185 (1989).

59. T. Ziegler, *Liebigs Ann. Chem.,* 1125 (1990).

60. P. Kováč, N. F. Whittaker, and C. P. J. Glaudemans, *J. Carbohydr. Chem.,* **4**, 243 (1985).

61. J. Thiem and M. Gerken, *J. Carbohydr. Chem.,* **1**, 229 (1982).

62. K. Bock, C. Pedersen, and P. Rasmussen, *Acta Chem. Scand.,* **29 B**, 185 (1975).

63. B. L. van Duuren, A. Sivak, B. M. Goldschmidt, C. Katz, and S. Melchionne, *J. Natl. Cancer Inst.,* **43**, 481 (1969).

64. V. Pavliak, P. Kováč, and C. P. J. Glaudemans, *Carbohydr. Res.,* **229**, 103 (1991).

Chapter 4

GLYCOSYLATION PROPERTIES AND REACTIVITY OF THIO-GLYCOSIDES, SULFOXIDES, AND OTHER S-GLYCOSIDES: CURRENT SCOPE AND FUTURE PROSPECTS

THOMAS NORBERG

Department of Chemistry, Swedish University of Agricultural Sciences,
P.O. Box 7015, S-750 07 Uppsala, Sweden.

Abstract   The use of thioglycosides, sulfoxides, and other S-glycosides in carbohydrate chemistry is reviewed, with special focus on their use as glycosylating agents in oligosaccharide synthesis. A survey of available methods for synthesis of thioglycosides is presented, followed by a discussion of available methods of activation of S-glycosides. Both two-step activation (via conversion of a thioglycoside to the corresponding bromide) and direct, one-step activation with thiophilic reagents are considered. Future prospects in the field are also discussed.

1. INTRODUCTION

In the design of a multi-step oligosaccharide synthesis, the choice of the glycosylation method and the protecting group strategies are the most important factors for a chemist to consider. Over the years, knowledge on the subject has accumulated, and several chapters of this book have summarized, under various headings, the more recent findings. This chapter focuses on the protection and activation of anomeric carbons with groups having sulfur in place of the exocyclic hemiacetal oxygen. The best known example of this type of protection/activation group is the alkyl or arylthio group (thioglycosides, see Figure 1), and most of this chapter will be devoted to the discussion of thioglycoside chemistry. Oxidized forms of thioglycosides, such as sulfoxides and other anomeric S-derivatives will also be considered. Anomeric S-xanthates are treated in a separate chapter, and will only be discussed briefly. The intention of the writer is to review developments and trends in the field during the period after 1987, when the most recent review appeared.[1] Earlier reviews[2-5] can be consulted for background.

The versatility of thioglycosides in carbohydrate chemistry stems from the fact that the sulfur atom in a thioglycoside is a soft nucleophile, and is therefore able to react selectively with soft electrophiles, such as heavy metal cations, halogens, and alkylating or acylating reagents. On the other hand, the hydroxyl and ring oxygen atoms of carbohydrates are hard nucleophiles, which can be functionalized with "hard" reagents, without affecting an alkylthio or arylthio function. The sulfur-bearing anomeric center can then be selectively activated with a soft electrophile, to form a reactive glycosylating species that can be used in the creation af a new glycosidic bond (Figure 1).

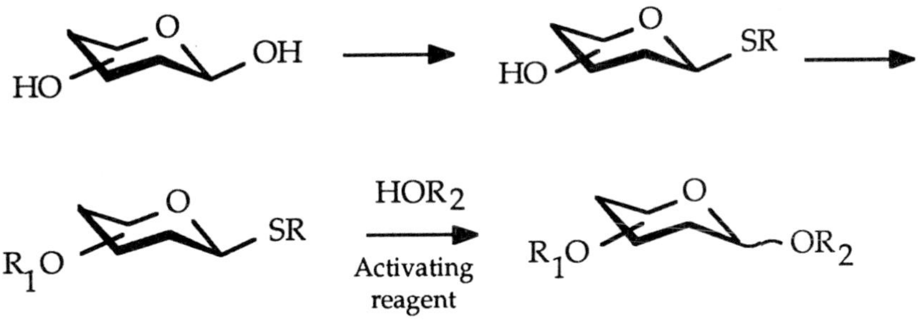

Figure 1.   Protection/activation at the anomeric carbon by use of a thioglycoside.

Although this possibility has been known for a considerable time, it was only in recent years that it was extensively explored. In order to give a perspective on the matter, a brief history of the use of thioglycosides as glycosylating agents in carbohydrate synthesis is given below. Later, the preparation methods used for obtaining thioglycosides are discussed. This is then followed by a discussion of glycosylation reactions of thioglycosides and other S-glycosides. Finally, future prospects will be considered.

2.   USE OF THIOGLYCOSIDES AS GLYCOSYLATING AGENTS

The first conversion of a thioglycoside into a glycosylating agent was reported in 1948 by Bonner,[6] who used bromine in acetic acid to convert O-acetylated monosaccharide thioglycosides into glycosyl 1-acetates, via an initially formed glycosyl bromide. Weygand[7] studied the conversion of thioglycosides to glycosyl bromides further, and, later, the analogous conversion to glycosyl chlorides by treatment of thioglycosides with chlorine was demonstrated.[8] In 1973, Ferrier[9] used mercury(II)sulfate to accomplish the first direct glycosylation with a protected thioglycoside to form a disaccharide derivative. This work triggered an interest in using thioglycosides as glycosylating agents, and several other mercury-(II) and other metal reagents were investigated (for a review, see reference 1). Also, the treatment of thioglycosides with bromine was extended[10] to complex oligosaccharide thioglycosides, which gave oligosaccharide bromides in good yields. Such bromides were difficult to obtain by other routes, because of the acid-sensitivity of the interglycosidic bond(s). The glycosyl bromides obtained were used in heavy metal-salt promoted glycosylations. In 1984, Lönn[11] reported the use of methyl triflate, the first efficient general promoter for direct glycosidation with thioglycosides. This triggered a widespread use of thioglycosides in oligosaccharide synthesis, both with methyl triflate ("one- step activation") or the combination bromine-heavy metal salt ("two-step activation"). The discovery of other general glycosylation promoters (e. g., dimethyl-methylthiosulfonium triflate,[12] N-iodosuccinimide-triflic acid[13]) soon followed, and an ever increasing number of reports on the use of thioglycosides in oligosaccharide synthesis have been published, too many to be fully covered in this review (for representative examples, see references 14-20). Hasegawa[21] demonstrated the usefulness of sialyl thioglycosides in the synthesis of sialic acid-containing oligosaccharides (see another chapter in this book). Other S-glycosides, such as S-xanthates, were also found to be useful in glycosylations[22-24] with sialic acid derivatives. Thioglycoside derivatives of all of the common monosaccharides have been prepared and evaluated as glycosyl donors (see Table 1).

Recently, Kahne[25] showed that glycosyl sulfoxides, prepared from thioglycosides, can be activated and used as glycosyl donors, also in the glycosylation of sugars attached to a solid phase.[26]

In conclusion: the use of thioglycosides and other S-glycosides in carbohydrate chemistry is today wide-spread and well established, and thioglycoside intermediates are seriously considered when planning oligosaccharide syntheses. To aid in such planning, the following sections will summarize some features of thioglycoside chemistry.

3. PREPARATION OF THIOGLYCOSIDES

The first problem that faces a chemist who wants to use thioglycosides is how to prepare the starting monosaccharide derivatives cheaply on a multigram scale, if they are not commercially available. The literature describes a variety of methods for the preparation of simple monosaccharide thioglycosides, most of them give predominantly 1,2-*trans* derivatives. The preparation methods can be grouped in the following categories:

3.A.   Synthesis From a Sugar Derivative, a Thiol and a Lewis Acid (Acid-Promoted Displacement at the Anomeric Center).

This method of preparation, although sometimes applicable to free sugars, works best with fully acylated derivatives. An example is shown in Figure 2, where 1,2,3,4,6-penta-O-acetyl-ß-D-glucopyranose is reacted with thiophenol in the presence of boron trofluoride etherate as promoter to give phenyl 2,3,4,6-tetra-O-acetyl-1-thio-ß-D-glucopyranoside in 71% crystalline yield.[3] Subsequent deacetylation with a catalytic amount of sodium methoxide in methanol gives the crystalline phenyl 1-thio-β-D-glucopyranoside.

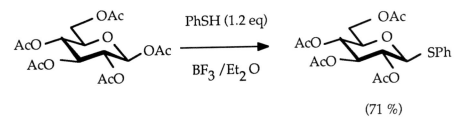

(71 %)

Figure 2.   Synthesis of phenyl β-D-glucopyranoside by Lewis acid promoted displacement at the anomeric center.

The above is a very general method for preparation of 1,2-*trans* thioglycosides from derivatives of glucose, galactose, mannose, rhamnose, fucose, glucosamine, sialic acid, glucuronic acid, and other sugars using either boron trifluoride, zinc chloride[27], or stannic chloride[28] (see table 1, method A1). The reactions are not entirely stereospecific, and minor amounts of 1,2-*cis* thioglycosides are formed concurrently. This does not offer problems when the monosaccharide thioglycosides can be purified by crystallization. Preferrably, one starts with a 1,2-*trans* acetate, which reacts faster with Lewis acids than the 1,2-*cis* isomers, and produces less side-products. Recently, other Lewis acids, such as zirconium(IV)chloride,[29] ferric(III)chloride,[30] and titanium tetrachloride[31] have been evaluated for use with peracylated sugars (see Table 1, method A1). A very popular variation (method A2) uses a tributylstannyl[32] or trimethylsilyl[33] derivative of the thiol, and a Lewis acid. Higher yields and less odor problems are claimed.[32] However, it should be mentioned that the problem of smell with thiols is not unsurmountable in a modern laboratory. Reactions can be performed in enclosed systems in well-ventilated hoods, and the equipment, residual solutions and exhaust gases can be treated with an oxidizing solution (aqueous permanganate or hypochlorite) to eliminate the odor. Once purified, the thioglycosides are odorless.

3.B.   Synthesis by *S*-nucleophilic Displacement at the Anomeric Centre (Base-promoted Displacement at the Anomeric Center).

The example[34] in Figure 3 shows the reaction of tetra-*O*-acetyl-ß-D-glucopyranosyl bromide with a thiophenolate anion to give phenyl 2,3,4,6-tetra-*O*-acetyl-1-thio-ß-D-glucopyranoside (81% yield), which is then *O*-deacetylated to give phenyl 1-thio-ß-D-glucopyranoside. It is an example of the oldest method for thioglycoside preparation, which was first described by Fisher and Delbrück[34] in 1909, and is generally applicable. It has been used for the preparation of a wide variety of 1,2-*trans* thioglycosides, including derivatives of glucose, galactose, glucuronic acid, mannose, xylose, and sialic acid (see Table 1). The best results are obtained with arylthio anions, since they are less basic, which renders the concurrent de-*O*-acetylation a less important side-reaction (with alkylthio anions, a reacetylation step is often added before isolating the product). Recent modifications of the original method include the use of polar, aprotic solvents,[35] or of phase-transfer conditions.[36]

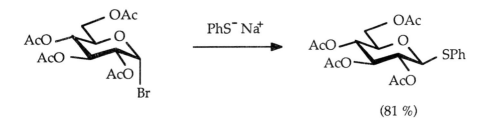

(81 %)

Figure 3. Synthesis of a thioglycoside by reaction of an acetylated glycosyl bromide with thiophenolate anion.

3.C. Synthesis by Preparation of a 1-Thioglycoside Followed by S-Alkylation.

This is a general method for the preparation of primary or secondary alkyl thioglycosides. There are several ways to prepare the first intermediate, namely the 1-thiosugar. Early preparations involved treatment of an acylated glycosyl halide with: (1) Thioacetate anion followed by selective S-deacetylation of the formed anomeric S-acetate[37] (2) Thiourea followed by hydrolysis with weak aqueous base of the formed pseudothiouronium salt[37,38] (3) Xanthate anion, followed by treatment with cold methanolic sodium methoxide.[37,39] Recent variations of the preparation of 1-thiosugar are: Treatment of an acetylated glycosyl chloride with trityl thiolate anion, followed by cleavage of the trityl thioglycoside with mercury salts,[40] treatment of a glycosyl 1,2-*trans* 1-acetate with thioacetic acid and zirconium (IV)chloride[41] or aluminium chloride[42] to give the 1,2-*trans* 1-S-acetate, followed by base-catalyzed hydrolysis; and radical-induced addition of thioacetic acid to glycals to give the 1,2-*cis* 1-S-acetate, followed by hydrolysis.[43]

Once prepared, the 1-thioglycoside is alkylated with an alkyl halide, often *in situ*.[44] Although the total number of steps is higher, the intermediates are often crystalline, the reagents are cheap, and the yields are high throughout. Perhaps more important, this route does not involve use of reagents with strong odor when thiourea or xanthate is used in the first step. The sequence acetylated bromide - pseudothiouronium salt -1-thiosugar - alkyl thioglycoside has been used extensively in the author's laboratory, an   example is shown in Figure 4, for   typical preparations, see references 37 and 44. Using this route, derivatives of glucose, galactose, rhamnose, sialic acid and other sugars have been prepared (see Table 1).

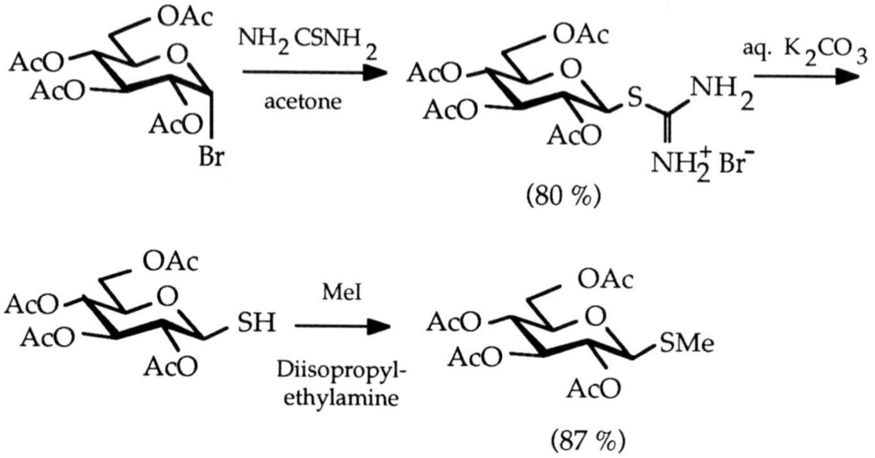

(80 %)

(87 %)

Figure 4. Preparation of a thioglycoside by alkylation of a 1-thiosugar

3.D. Other Methods

Partial hydrolysis of dithioacetals has been used for the preparation of thioglycosides, especially of α–pyranosidic and furanosidic derivatives.[2,45] Aryl thioglycosides have been prepared[46] by treatment of 1-thiosugars with aryl diazonium salts, followed by decomposition of the diazo compounds formed. Glycosyl xanthates, which are easily prepared from the corresponding acetylated glycosyl bromides, can be decomposed to give excellent yields of thioglycosides.[47] Several other, more specialized, routes that lead to thioglycosides have been described recently. They include the radical-induced alkylation of 1-thiosugars with alkenes,[48] and the photochemical addition of thiols to acetylated glycals.[49] Treatment of O-glycosides with Lewis acids and trimethylsilyl derivatives of thiols[50] gives predominantly α-1,2-cis thioglycosides, which can also be prepared by Lewis acid catalyzed anomerization[51] of the corresponding 1,2-trans thioglycosides, as well as by the partial hydrolysis of dithioacetals[45] mentioned above.

Table 1. Preparation of thioglycosides of naturally occuring monosaccharides.

Peracetylated thioglycoside	Method	Reference (Yield)
Methyl ß-D-Glcp	A2	32 (85), 52 (90)
Methyl ß-D-Glcp	B	53 (40)
Methyl ß-D-Glcp	C	44 (85)
Methyl ß-D-Glcp	D	47 (93)
Ethyl ß-D-Glcp	A1	29 (74), 30 (68), 27 (71)
Ethyl ß-D-Glcp	B	53 (19)
Ethyl ß-D-Glcp	C	44 (82)
Ethyl ß-D-Glcp	D	47 (84)
Phenyl ß-D-Glcp	A1	3, 54(71), 30 (75)
Phenyl ß-D-Glcp	B	34 (84), 55 (70)
Phenyl ß-D-Glcp	D	46 (95)
Methyl ß-D-Galp	A2	32 (91), 33 (84), 56 (80), 57 (89)
Methyl ß-D-Galp	B	58 (59), 59 (70), 2 (59)
Methyl ß-D-Galp	C	38 (75)
Methyl ß-D-Galp	D	47 (73)
Ethyl ß-D-Galp	A1	29 (81), 27 (76)
Ethyl ß-D-Galp	C	38 (75)
Ethyl ß-D-Galp	D	47 (70)
Phenyl ß-D-Galp	A1	54 (69)
Phenyl ß-D-Galp	B	60 (54)
Methyl ß-D-GalpN3	C	61
p-Methylphenyl ß-D-GalpN3	B	62 (57)
Methyl α–D-Manp	A2	32 (72)
Ethyl α-D-Manp	A1	29 (43), 28 (68)
Phenyl ß-D-Manp	B	63 (96)
Methyl ß-D-GlcN(Phth)p	A2	33 (93), 57 (93)
Ethyl ß-D-GlcN(Phth)p	A1	30 (85), 16 (72)
Ethyl ß-D-GlcNAcp	A1	30 (62)
Methyl α-L-Rhap	A1	64 (77), 65 (66)
Methyl α-L-Rhap	A2	33 (67), 64 (67)
Methyl α-L-Rhap	C	66 (46)
Ethyl α-L-Rhap	A1	67 (61), 68 (88)
Phenyl α-L-Rhap	A1	64 (66), 69 (93)

Table 1 (continued). Preparation of thioglycosides of naturally occuring monosaccharides.

Peracetylated thioglycoside	Method	Reference (Yield)
Methyl ß-L-Fucp	A2	70 (51), 71 (56)
Ethyl ß-L-Fucp	A1	16
Methyl α-L-Ara	B	58 (43)
Ethyl α-L-Ara	B	72 (56)
Methyl ß-D-Xylp	D	47 (61)
Methyl ß-D-Xylp	B	72 (62)
Ethyl ß-D-Xylp	A1	29 (79)
Ethyl ß-D-Xylp	B	72 (65)
Ethyl ß-D-Xylp	D	47 (90)
Phenyl ß-D-Xylp	A1	54 (56)
Phenyl ß-D-Xylp	B	55 (80)
Methyl ß-D-GlcpA, Me ester	A2	73
Methyl ß-D-GlcpA, Me ester	B	74 (67)
Methyl ß-D-GlcpA, Me ester	D	47 (70)
Ethyl ß-D-GlcpA, Me ester	B	74 (60)
Ethyl ß-D-GlcpA, Me ester	D	47 (69)
Phenyl ß-D-GlcpA, Me ester	B	74 (79)
Ethyl α-KDO, methyl ester	D	75
Methyl α/ß-NeuAc, methyl ester	A2	76 (96)
Methyl α-NeuAc, methyl ester	C	21
Methyl ß-NeuAc, methyl ester	C	21
Ethyl ß-NeuAc, methyl ester	A1	77 (72)
Phenyl α-NeuAc, methyl ester	B	77 (68)
Phenyl ß-NeuAc, methyl ester	A1	77 (81)

4. PREPARATION OF SULFOXIDES AND OTHER S-GLYCOSIDES

Sulfoxides are prepared from the corresponding thioglycosides by oxidation with such reagents as m-chloroperbenzoic acid.[78] Oxidating reagents such as ruthenium tetroxide gives the sulfone derivative[77,79] in high yield. Other S-glycoside derivatives: S-xanthate derivatives can be prepared by treatment of a glycosyl halide with xanthate anion (see separate capter). *S*-Glycosyl-*N,N*-dialkyldithiocarbamates are prepared[80] by displacement of a glycosyl halide with a *N,N*-dialkyldithiocarbamate anion.

5.   GLYCOSYLATION REACTIONS OF THIOGLYCOSIDES

5.A.   Thioglycoside Activation by Conversion to a Glycosyl Halide

Thioglycosides can be used as glycosidation agents in a two-step procedure, in which the thioglycoside is first converted to a glycosyl halide (Figure 5), and then subjected to a conventional metal salt promoted glycosylation either with[10,17] or without[81,82] isolation of the halide.

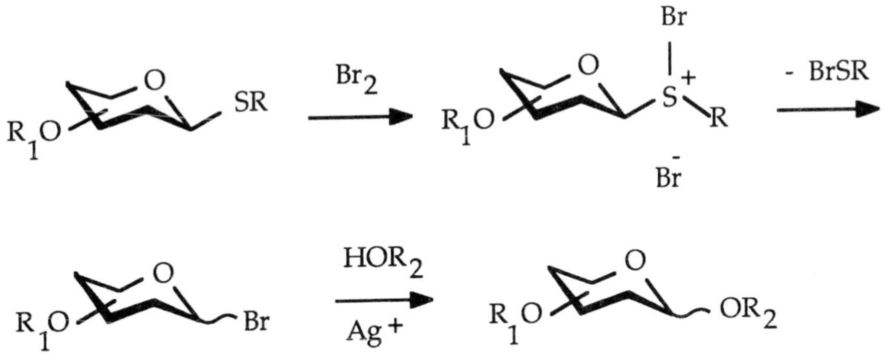

Figure 5.   Activation of a thioglycoside by conversion to a glycosyl bromide, followed by metal ion-promoted glycosylation (the stereochemistry of the bromide and glycoside depends on the protecting group in the 2-position).

The earliest example of this strategy in *oligosaccharide* synthesis was reported by Koto,[10] who prepared malto-oligosaccharides from thioglycoside precursors by a block synthesis approach. The advantage of the two-step approach is, that thioglycoside glycosyl *acceptors* as well as donors can be used. No special precautions have to be taken to avoid activation of the acceptor thioglycoside function, as is the case with the direct activation described below. This is because the acceptor is not present in the first step, namely the treatment with bromine. In the second step (conventional heavy metal salt promoted glycosylation with the bromide formed, at temperatures below 0°C), the acceptor thioglycoside functionality is unaffected. However, in certain cases where "difficult" thioglycoside acceptors have been used, transfer of the alkyl or arylthio function from the glycosyl acceptor to the glycosyl donor has been observed[83] ("transglycosylation," Figure 6) as a side-reaction.

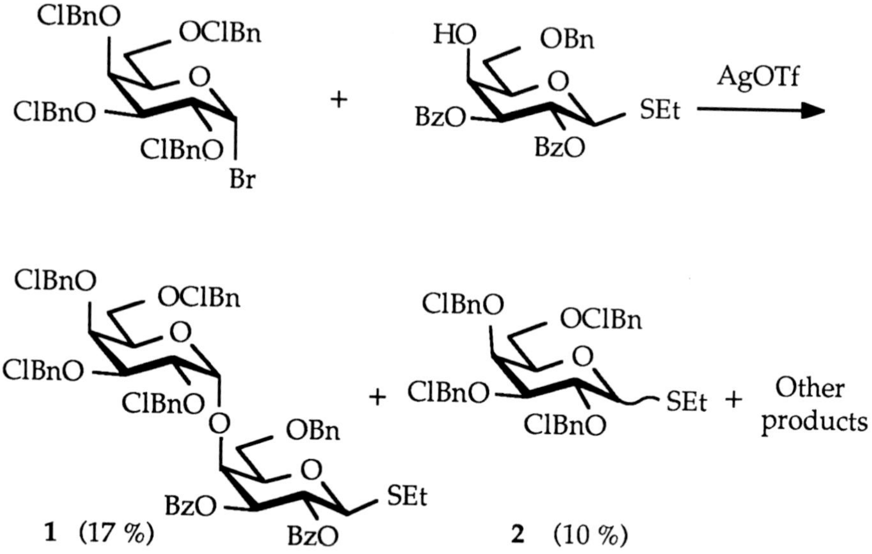

Figure 6. A "difficult" disaccharide coupling, giving a low yield of desired product (1) together with transglycosylation products (2).

This reaction is most likely an electrophilic attack of the glycosylating species on the sulfur atom of the acceptor, followed by cleavage of the sulfonium ion formed to release the thioglycosylated donor and an acceptor glycosyl cation, which in turn can engage in further glycosylations, to produce second generation products.[83]

Although, in the published examples of the above strategy, glycosylation takes place via a glycosyl bromide, the use of glycosyl chlorides is in principle possible, and should be applicable in cases where the bromide is too unstable. Recently, a simple preparation of glycosyl fluorides from thioglycosides was described,[84] suggesting that glycosyl fluorides should also be considered as alternatives in the above strategy.

5.B. Direct Activation by the Use of "Thiophilic" Electrophiles

The thioglycoside functionality, being a soft nucleophile, can be directly converted into a reactive glycosylating species by a variety of soft electrophiles, without the intermediacy of any glycosyl halide (Figure 7). This approach has opened the way for facile, high-yield glycosylations with oligosaccharide blocks as well as with simple monosaccharides. The electrophiles can be of different ionic types, such as carbonium, sulfenium, halonium, or metal ion, which are discussed below.

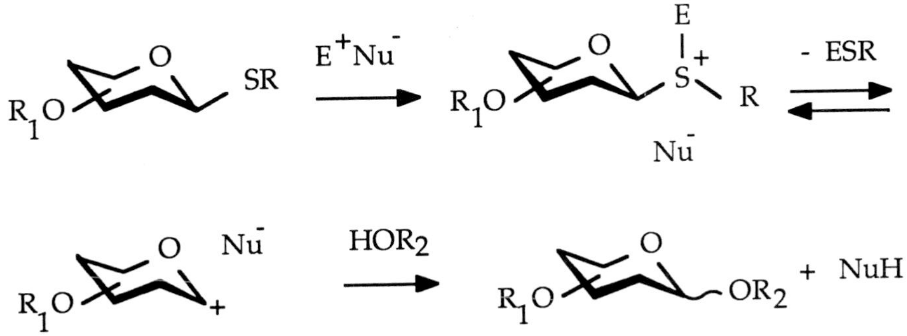

Figure 7. Activation of thioglycosides with a thiophilic reagent (E^+Nu^-).

5.B.i. Carbonium ion-like electrophiles

An example of a reagent of this type is methyl triflate,[11,16,31] which was the first efficient non-metal ion electrophile to be used for thioglycoside glycosylations. Methyl triflate preferentially S-alkylates the thioglycoside function, to give an anomeric sulfonium triflate. This compound, or the derived glycosyl triflate, is the glycosylating species. Methyl triflate is a moderatly reactive promoter, which requires room or slightly higher temperatures, and hours or days of reaction time. The time and temperature depends mainly on the reactivity of the thioglycoside, O-acylated derivatives being less reactive than O-alkylated ones. To neutralize the liberated triflic acid, powdered molecular sieves are added to the reaction mixture. Methyl triflate is now one of the most thoroughly investigated glycosylation promoters, and it has been used sucessfully in a large nuber of cases (examples of typical reaction conditions are found in references 16, 83, 85). Both 1,2-*cis* (with O-benzylated glycosyl donors) and 1,2-*trans* glycosylations (with O-acylated donors) can be produced. The stereochemical outcome in each case follows the same general rules as with other glycosyl donor/promoter combinations. An expected side-reaction with methyl triflate is O-alkylation of the acceptor hydroxyl. However, this has been observed only in special cases, such as with the combination of an unreactive thioglycoside with a reactive glycosyl acceptor.[49]

The reactivity of methyl triflate implies that other alkylating agents could also act as good glycosylation promoters. Indeed, trimethyloxonium tetrafluoroborate was used to activate thioglycosides,[49] but was found to present few advantages over methyl triflate.

Triphenylmethyl perchlorate[86] was found to activate thioglycosides towards O-trityl ether acceptors. Methyl iodide, in combination with 2-pyridyl thioglycosides[87,88] also promotes glycosylations, and this combination has been used in a number of cases.[89,90] The detection of N-methyl-2-thiopyridone and its salt as reaction products indicates that activation of the 2-pyridyl thioglycoside is initiated by N-methylation.

5.B.ii. Sulfenium ion-like electrophiles

The possibility of undesired acceptor O-alkylation occuring with methyl triflate, in addition to safety considerations (methyl triflate is a toxic liquid) have led to a search for more specific S-electrophiles. Dimethyl(methylthio)sulfonium triflate (DMTST) and dimethyl(methylthio)sulfonium tetrafluoroborate (DMTSB) were found to activate both O-acylated and O-benzylated thioglycosides selectively and efficiently.[12] These reagents are "sulfenium ion-like", and presumably alkylsulfenylate the thioglycoside to form a reactive intermediate (see Figure 8). Such intermediates, or their decomosition products, are the glycosylating species. In the case of DMTSB, a glycosyl fluoride is often formed as an intermediate.[84,91] Presently, DMTST, with its more inert triflate counterion, is the most extensively used sulfenium ion-like reagent.

Glycosylations with DMTST are fast (minutes to hours) and proceed at room temperature or below (for examples of typical reaction conditions, see references 17 and 92). The stereochemical outcome of glycosylations follow the the same general trends as with other glycosyl donor/promoter combinations. Interesting results have been reported[21] using alkylthioglycosides of N-acetylneuraminic acid and DMTST, as seen in another chapter. A side-reaction has been reported,[93] in which the nitrogen in an acetamido sugar underwent methylsulfenylation by DMTST. A drawback with the DMTST reagent is that it is hygroscopic and has to be stored under dry and cold conditions, and that it is not commercially available. It can, however, be easily prepared from methyl triflate and dimethyl disulfide by simple mixing in dichloromethane, cooling and filtering of the crystalline product, which is washed with dry ether and then kept under dry conditions in a freezer. Despite its disadvantages, DMTST should be one of the first choices for thioglycoside activations, because of its generality and selectivity.

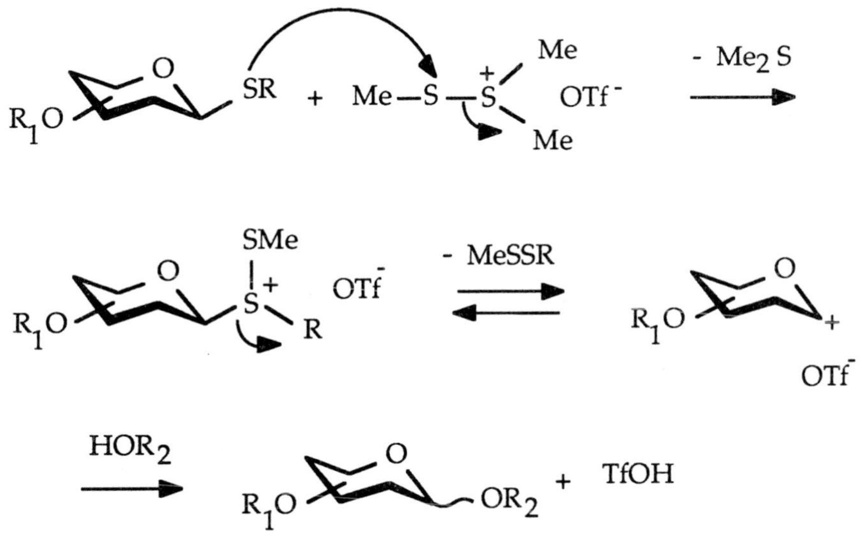

Figure 8. Glycosylation promoted by DMTST.

Reagents with similar modes of action as DMTST are methyl sulfenyl bromide and methyl sulfenyl triflate.[94,95] Neither of these reagents is commercially available, but both are easily prepared. Mixing of bromine and dimethyl disulfide gives methyl sulfenyl bromide, and treatment with silver triflate gives methyl sulfenyl triflate. Glycosidation with these reagents is completed in minutes or hours at room temperature or below, methyl sulfenyl bromide being slower, since it often produces the glycosyl bromide as an intermediate. Slight but interesting differences in stereochemical outcome with O-benzyl thioglycosides have been noted, when the two reagents were compared.[94] For example, methyl sulfenyl bromide gave higher amounts of 1,2-cis-glycosides. A trisaccharide was synthesized with the aid of both reagents,[94] and several reports have confirmed their usefulness.

Phenylselenyl triflate is another glycosylating reagent[96,97] similar in structure, but more reactive (glycosylations within minutes at sub-zero temperatures). Promising results were obtained with sialic acid thioglycosides. Further examples of the use of this reagent should establish its general usefulness.

5.B.iii.   Halonium ion-like electrophiles

The simplest example of reagents with halonium ion character are the halogens themselves, which can be formally regarded as halonium halides. Because of the nucleophilic halide counterion, activation of thioglycosides with bromine or chlorine give the glycosyl halide, as was mentioned above. However, halonium electrophiles exist with less nucleophilic counterions. The first such electrophile used for thioglycoside activation was N-bromosuccinimide.[104] Sulfuryl chloride/triflic acid[98] was the first halonium-type reagent with a triflate counterion. Blood group oligosaccharides were synthesized in good yields using this reagent.[99] The glycosidations were fast, and proceeded at 0°C or below. Presumably, the reagent converts the thioglycoside, via the chlorosulfonium glycoside, to a glycosyl triflate, the byproducts being sulfur dioxide and hydrogen chloride.

More recent examples of halonium ion electrophiles are N-iodosuccinimide-triflic acid (NIS-TfOH)[13,100,101] and iodonium dicollidine perchlorate (IDCP[102]). The former reagent is the more reactive of the two. Glycosylations with NIS-TfOH proceed fast (minutes to hours) at room temperature or below, the reagents are commercially available and cheap (only 0.1 equivalents of triflic acid is necessary), and they are reasonably stable. Because few side-reactions have been reported, this reagent has become the most popular halonium-type glycosylation reagent in recent years (for typical experimental conditions, see references 20 and 100). Because of the moderating influence of the collidine ligands, IDPC is a less reactive reagent than NIS-TfOH, and requires a reactive thioglycoside (one carrying O-alkyl protecting groups). This fact has been used in the armed-disarmed strategy originally introduced[103] for pentenyl glycoside glycosidations to thioglycosides. An O-benzylated (armed) thioglycoside donor was used[102] to glycosylate an O-acylated (disarmed) thioglycoside acceptor, using IDPC as promoter. The reagent did not affect the acceptor thioglycoside function. This strategy is an alternative to the two-step glycosylation discussed previously, if thioglycoside *acceptors* are to be glycosylated with thioglycoside donors. The disadvantage of the approach is that if

1,2-*trans* glycosylations are needed, they have to be formed from 2-*O*-alkylated thioglycosides, which often give low stereocontrol. A word of caution about the use of organic perchlorates such as IDPC: Perchlorates are notoriously unreliable (explosion risk) and should not be used at all on a large scale. Use of iodonium dicollidine triflate as a replacement for IDPC in glycosylations with thioglycosides works equally well[20] and should be a safer alternative.

Several other thioglycoside-activating reagents of the halonium type have been reported, such as *N*-bromosuccinimide/triflic acid[105a] *N*-bromosuccinimide/tetrabutylammonium triflate,[105b] and trifluoroacetic anhydride-iodosobenzene.[106]

5.B.iv.   Metal ion electrophiles

The most common promoters of this type are mercury(II)salts, like the sulfate, nitrate, chloride, benzoate, or reagents like phenylmercury triflate or palladium perchlorate (see reference 1). These salts give good results in some cases. However, in other cases they have shown little or low reactivity, for reasons not fully understood. Copper(II)bromide, in combination with tetrabutylammonium bromide,[107] has also been used, in this case for conversion of thioglycosides into bromides.

5.B.v.   Other reagents or methods

A variety of reagents other than those mentioned above have been used for the activation of thioglycosides. These include nitrosyl tetrafluoroborate,[57,64,108] triflic anhydride[109] and radical cations.[110]

Activation of thioglycosides can also be affected by anodic oxidation[111,112,113] of aryl thioglycosides ("electrochemical glycosylation"). The mechanism involves formation of a radical cation (Figure 9), that cleaves into a a glycosyl cation and an arylthiyl radical (which dimerizes to form a diaryl disulfide). The glycosyl cation, in the presence of an alcohol, reacts to form a glycoside or, under certain conditions, the glycosyl fluoride (presumably by way of reaction between the glycosyl cation and the tetrafluoroborate counterion). *O*-Acylated, *O*-benzylated and unprotected thioglycosides can be used in this electrochemical glycosylation.

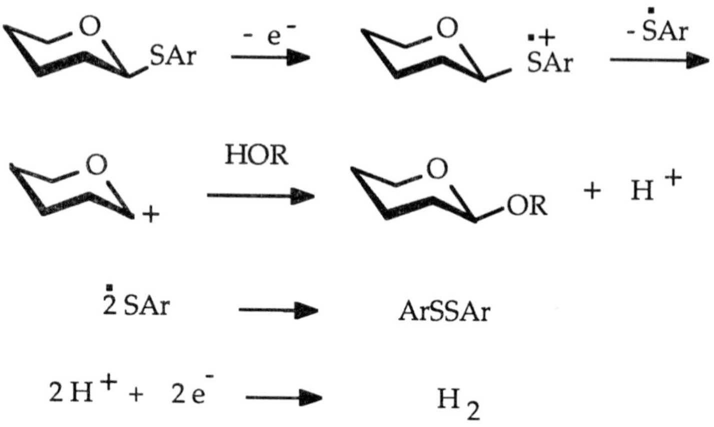

Figure 9. Electrochemical glycosylation using phenyl thioglycosides. The counterion is provided by the supporting electrolyte (lithium or tetrabutylammonium tetrafluoroborate). The acid produced in the reaction is converted to hydrogen at the cathode, if an undivided electrolysis cell is used.

6. GLYCOSYLATION REACTIONS OF SULFOXIDES AND OTHER S-GLYCOSIDES

As shown by Kahne,[25] glycosyl sulfoxides (prepared from the corresponding thioglycosides) can be converted to reactive glycosylating agents, using triflic anhydride as promoter. Several oligosaccharides have been synthesized[25,78,114] by this method, which has also been used[115] in intramolecular glycosylations, and, more recently,[26] in solid phase synthesis of oligosaccharides.

Preliminary experiments with the less reactive glycosyl sulfones have also been performed,[79] using magnesium bromide etherate as promoter. S-Glycosyl-N,N-dialkyldithiocarbamates, which are highly crystalline, stable derivatives, have been shown[80] to function as glycosyl donors with a variety of promoters, such as DMTST, methyl triflate, and silver triflate. The reactions of glycosyl S-xanthates, which are also good glycosyl donors, are covered in another chapter.

7.   FUTURE OUTLOOK

During the last 5 years, several new aspects of the use of thioglycosides in carbohydrate synthesis have been described in the literature. These will undoubtly be further explored in the future. Some of the new ideas are discussed below.

7.A.   Modulation of Thioglycoside Reactivity

Several workers have exploited ways to enhance the reactivity of the thioglycoside donor, and/or to reduce the reactivity of the thioglycoside acceptor, so that direct glycosylation of a thioglycoside acceptor can be performed with a thioglycoside donor. One way is to vary O-protecting groups on the thioglycoside, the 2-position being especially important. Acyl groups give less reactive ("disarmed") thioglycosides, compared to alkyl groups ("armed" thioglycosides). This phenomenon, which is caused by the higher electron-withdrawing properties of acyl groups, is not restricted to thioglycosides. Analogous reactivity differences can be seen with glycosyl halides, pentenyl[103] or other glycosides. The phenomenon has been used[101,116,117] in several oligosaccharide syntheses.

Another attempt[118] to modulate thioglycoside reactivity has been to vary the thioglycoside aglycon. Aryl thioglycosides carrying electron-withdrawing groups were less reactive ("latent donors") towards activating agents than those carrying electron donating groups ("active donors"). It can be expected that future work will shed further light on the usefulness of the above strategies.

7.B.   Stereoselectivity

Although most glycosylations with thioglycoside donors show similar stereochemical outcomes as other glycosylations, there are interesting differences. The high reactivity and solubility of some thioglycoside activating agents allows glycosylations to be carried out at very low temperatures, which drastically alters the α/ß ratios, especially with sialic acid derivatives in nitril solvents.[24] More investigations on the influence on a/ß ratios of factors such as temperature, solvent, acceptor, and counterion are to be expected in the future.

A different approach to the problem of stereocontrol is that presented by Hindsgaul[119,120] and Stork.[115] A temporary linkage between the donor and acceptor is first created, which orients them in a position that sterically favours one particular isomer when the intramolecular glycosylation is carried out (for details and figures, see the chapter by Barresi and Hindsgaul in this book). Several syntheses of ß-mannosides, which are notoriously difficult to obtain by other methods, were performed, with excellent selectivities. This approach has also been used[121] for creating other types of linkages, in all cases thioglycosides or sulfoxides were used as the glycosylating agents.

7.C.   Mechanistic Studies

Available thioglycoside activating reagents have increased in number over the last decade, but careful studies of the mechanism by which these reagents react have only rarely been performed. This is an interesting field for future investigations, especially since the reactions occur in homogenous phase (as opposed to glycosylations with bromides), and should therefore be well suited for study by spectroscopic methods. The results will almost certainly lead to a better understanding of glycosylation chemistry, and to more accurate prediction of the stereochemical outcome of glycosylations.

7.D.   Solid-phase Synthesis of Oligosaccharides

The solid-phase synthesis technique is well established in peptide and oligonucleotide synthesis, but very few reports have appeared in the literature on its application to the synthesis of oligosaccharides. Possible reasons for this are that glycosylations are stereochemically more complicated, but also the simple fact that most glycosylation methods, before the advent of thioglycosides, were associated with either reagents or reaction products that were insoluble, and therefore were not practically suited for solid phase synthesis. Most thioglycoside glycosylations, however, are homogenous, and should give reasonable results. A preliminary report has appeared[26] on the use of sulfoxides (activated by triflic anhydride) in solid-phase synthesis, and the results are encouraging. In the future, more investigations along these lines are to be expected.

7.E.   Chemoenzymatic synthesis

Enzymes can be used as reagents for particulary "difficult" steps in a synthetic sequence, if they can be isolated or produced in preparative amounts. In oligosaccharide synthesis, glycosyltransferases or glycosidases have been used for glycosylations. An interesting combination of enzymatic and chemical methods is the construction, using enzymatic glycosylation, of oligosaccharide "blocks", which are then protected and used in chemical glycosylations of various alcohols.[122,123] Thioglycoside oligosaccharide blocks can be prepared[123] in this way, and this possibility will certainly be further investigated in the future.

ACKNOWLEDGEMENT

I thank Hans Lönn (Symbicom AB, Lund, Sweden) and Per Garegg (University of Stockholm, Sweden) for valuable help during the preparation of this manuscript.

REFERENCES

1. P. Fügedi, P. J. Garegg, H. Lönn, and T. Norberg, *Glycoconjugate J.* , **4**, 97 (1987).

2. D. Horton, *Methods Carbohydr. Chem.*, **2**, 368 (1963).

3. R. Ferrier and R. Furneaux, *Methods Carbohydr.Chem.*, **8**, 251 (1980).

4. D. Horton and D. H. Hutson, *Adv. Carbohydr. Chem. Biochem.*, **18**, 123 (1963).

5. D. Horton and J. D. Wander in *The Carbohydrates*. (W. Pigman, D. Horton and J. D. Wander, Eds.) Academic Press, New York, p. 799 (1980).

6. W. A. Bonner, *J. Am. Chem. Soc.*, **70**, 3491 (1948).

7. F. Weygand and H. Ziemann, *Liebigs Ann. Chem.*, **657**, 179 (1962).

8. M. Wolfrom and W. Groebke, *J. Org. Chem.*, **28**, 2986 (1963).

9. R. Ferrier, R. Hay, and N. Vethaviyasar, *Carbohydr. Res.*, **27**, 55 (1973).

10. S. Koto, T. Uchida, and S. Zen, *Bull. Chem. Soc. Jpn,*, **46**, 2520 (1973).

11. H. Lönn, *Chem. Comm. Univ. Stockholm*, 1 (1984).

12. P. Fügedi and P. J. Garegg, *Carbohydr. Res.*, **149**, C9 (1986).

13. P. Konradsson, D. R. Mootoo, R. E. Mc Devitt, and B. Fraser-Reid, *J.Chem.Soc., Chem. Commun.*, 270 (1990).

14. P. J. Garegg and C. Hällgren, *J. Carbohydr.Chem.*, **11**, 425 (1992).

15. A. Hasegawa, T. Ando, A. Kameyama, and M. Kiso, *J. Carbohydr. Chem.*, **11**, 645 (1992).

16. H. Lönn, *Carbohydr. Res.,* **139**, 105 (1985).

17. M. Nilsson and T. Norberg, *Carbohydr. Res.,* **183**, 71 (1988).

18. S. Nilsson, H. Lönn, and T. Norberg, *Glycoconjugate J.*, **6**, 21 (1989).

19. T. Peters and D. R. Bundle, *Can. J. Chem.*, **67**, 497 (1989).

20. H. Veeneman, S. Van Leeuwen, H. Zuurmond, and J. van Boom, *J. Carbohydr. Chem.*, **9**, 783 (1990).

21. O. Kanie, M. Kiso, and A. Hasegawa, *J. Carbohyd. Chem.*, **7**, 501 (1988).

22. W. Birberg and H. Lönn, *Tetrahedron Lett.*, **32**, 7453 (1991).

23. W. Birberg and H. Lönn, *Tetrahedron Lett.*, **32**, 7457 (1991).

24. H. Lönn and K. Stenvall, *Tetrahedron Lett.*, **33**, 115 (1992).

25. D. Kahne, S. Walker, Y. Chen, and D. Van Engen, *J. Am. Chem. Soc.*, **111**, 6881 (1989).

26. L Yan, CM Taylor, R Goodnow Jr, and D Kahne, *J. Am. Chem. Soc.* **116**, 6953 (1994).

27. R. U. Lemieux, *Can. J. Chem.*, **29**, 1079 (1951).

28. R. U. Lemieux and C. Brice, *Can. J. Chem.*, **33**, 109 (1955).

29. M.-O. Contour, J. Defaye, M. Little, and E. Wong, *Carbohydr. Res.*, **193**, 283 (1989).

30. F. Dasgupta and P. J. Garegg, *Acta Chem. Scand.*, **43**, 471 (1989).

31. H. Lönn, *J. Carbohydr. Chem.*, **6**, 301 (1987).

32. T. Ogawa and M. Matsui, *Carbohydr. Res.*, **54**, C17 (1977).

33. V. Pozsgay and H. J. Jennings, *Tetrahedron Lett.*, **28**, 1375 (1987).

34. E. Fischer and K. Delbrück, *Chem. Ber.*, **42**, 1476 (1909).

35. M. Apparu, M. Blanc-Muesser, J. Defaye, and H. Driguez, *Can. J. Chem.*, **59**, 314 (1981).

36. F. Tropper, F. Andersson, C. Grandmaitre, and, R. Roy, *Synthesis*, 734 (1991).

37. D. Horton, *Methods Carbohydr. Chem.*, **2**, 433 (1963).

38. M. Cerny, J. Stanek, and J. Pacak, *Monatsh.*, **94**, 290 (1963).

39. W. Schneider, R. Gille, and K. Eisfeld, *Chem. Ber.*, **61**, 1244 (1928).

40. M. Blanc-Muesser, L. Vigne, and H. Driguez, *Tetrahedron Lett.*, **31**, 3869 (1990).

41. J. Defaye, H. Driguez, E. Ohleyer, C. Orgeret, and C. Viet, *Carbohydr. Res.*, **130**, 317 (1984).

42. B. Rajanikanth and R. Seshadri, *Tetrahedron Lett.*, **28**, 2295 (1987).

43. A. Gadelle, J. Defaye, and C. Pedersen, *Carbohydr. Res.*, **200**, 497 (1990).

44. M. Cerny and J. Pacak, *Chem. Listy*, **52**, 2090 (1958).

45. E. Ziss, A. Clingmann, and N. K. Richtmyer, *Carbohydr. Res.*, **2**, 411 (1966).

46. M. Cerny, D. Zachystalova, and J. Pacak, *Coll. Czech Chem. Commun.*, **26**, 2206 (1961).

47. M. Sakata, M. Haga, and S. Tejima, *Carbohydr. Res.*, **13**, 379 (1970).

48. J. M. Lacombe, N. Rakotomanomana, and A. A. Pavia, *Tetrahedron Lett.*, **29**, 4293 (1988).

49. P. Fügedi and A. Fentie, *Upublished Results (1986)*.

50. S. Hanessian and Y. Guindon, *Carbohydr. Res.*, **86**, C3 (1980).

51. B. Erbing and B. Lindberg, *Acta Chem. Scand.*, **B30**, 611 (1976).

52. S. Sugawara, H. Nakayama, G. A. Strobel, and T. Ogawa, *Agric. Biol. Chem.*, **50**, 2251 (1986).

53. W. Schneider, J. Sepp, and O. Stiehler, *Chem. Ber.*, **51**, 220 (1918).

54. R. Ferrier and R. Furneau, *Carbohydr. Res.*, **52**, 63 (1976).

55. C. B. Purves, *J. Am Chem. Soc.*, **51**, 3619 (1929).

56. K. Koike, M. Sugimoto, S. Sato, Y. Ito, Y. Nakahara, and T. Ogawa, *Carbohydr. Res.*, **163**, 189 (1987).

57. V. Pozsgay and H. J. Jennings, *Carbohydr. Res.*, **179**, 61 (1988).

58. B. Helferich, H. Grünewald, and F. Langenhoff, *Chem. Ber.*, **86**, 873 (1953).

59. B. Helferich and D. Türk, *Chem. Ber.*, **89**, 2215 (1956).

60. M. Yde and C. K. De Bruyne, *Carbohydr. Res.*, **26**, 227 (1973).

61. M. Haraldsson, H. Lönn, and T. Norberg, *Glycoconjugate J.*, **4**, 225 (1987).

62. B. Lüning, T. Norberg, and J. Tejbrant, *Glycoconjugate J.*, **6**, 5 (1989).

63. V. Pedretti, A. Veyrieres, and P. Sinaÿ, *Tetrahedron*, **46**, 77 (1990).

64. V. Pozsgay and H. J. Jennings, *J. Org. Chem.*, **53**, 4042 (1988).

65. F.-I. Auzanneau and D. R. Bundle, *Carbohydr. Res.*, **212**, 13 (1991).

66. A. Lipták, L. Szabó, and J. Harangi, *J. Carbohydr. Chem.*, **7**, 687 (1988).

67. A. Borbas and A. Liptak, *Carbohydr. Res.*, **241**, 99 (1993).

68. G. H. Veeneman, L. J. F. Gomes, and J. H. Van Boom, *Tetrahedron*, **45**, 7433 (1989).

69. R. D. Groneberg, T. Miyazaki, N. A. Stylianides, T. J. Schulze, W. Stahl, E. P. Schreiner, T. Suzuki, Y. Iwbuchi, A. L. Smith, K. C. Nicolaou, *J. Am. Chem. Soc.*, **115**, 7593 (1993).

70. A. Kameyama, H. Ishida, M. Kiso, and A. Hasegawa, *J. Carbohydr. Chem.*, **10**, 549 (1991).

71. S. Sato, Y. Ito, T. Nukada, Y. Nakahara, and T. Ogawa, *Carbohydr. Res.*, **167**, 197 (1987).

72. H. Zinner, A. Koine, and H. Nimz, *Chem. Ber.*, **93**, 2705 (1960).

73. T. Nakano, Y. Ito, and T. Ogawa, *Tetrahedron Lett.*, **31**, 1597 (1990).

74. B. Helferich, D. Türk, and F. Stoeber, *Chem. Ber.*, **89**, 2221 (1956).

75. G. J. P. H. Boons, F. L. Vandelft, P. A. M. Vanderklein, G. A. Vandermarel, and J. H. van Boom, *Tetrahedron*, **48**, 885 (1992).

76. A. Hasegawa, H. Ohki, T. Nagahama, H. Ishida, and M. Kiso, *Carbohydr. Res.*, **212**, 277 (1991).

77. A. Marra and P. Sinaÿ, *Carbohydr. Res.*, **187**, 35 (1989).

78. S. H. Kim, D. Augeri, D. Yang, and D. Kahne, *J. Am. Chem. Soc.*, **116**, 1766 (1994).

79. D. S. Brown, S. V. Ley, and S. Vile, *Tetrahedron Lett.*, **29**, 4873 (1988).

80. P. Fügedi, P. J. Garegg, S. Oscarson, G. Rosen, and B. A. Silwanis, *Carbohydr. Res.*, **211**, 157 (1991).

81. T. Norberg, M. Walding, and E. Westman, *J. Carbohydr. Chem.*, **7**, 283 (1988).

82. J. O. Kihlberg, D. A. Leigh, and D. R. Bundle, *J. Org. Chem.*, **55**, 2860(1990).

83. S. Nilsson, H. Lönn, and T. Norberg, *Glycoconjugate J.*, **8**, 9 (1991).

84. L. Blomberg and T. Norberg, *J. Carbohydr. Chem.*, **11**, 751 (1992).

85. S. Nilsson, H. Lönn, and T. Norberg, *J. Carbohydr. Chem.*, **10**, 1023 (1991).

86. N. E. Nifantev, L. V. Backinowsky, and N. K. Kochetkov, *Bioorg. Chim.*, **16**, 1402 (1990).

87. G. V. Reddy, V. R. Kulkarni, and H. B. Mereyala, *Tetrahedron Lett.*, **30**, 4283 (1989).

88. H. B.Mereyala and G.V. Reddy, *Tetrahedron*, **47**, 6435 (1991).

89. H. B. Mereyala, V. R. Kulkarni, D. Ravi, G. V. M. Sharma, B. V. Rao, and G. V. Reddy, *Tetrahedron*, **48**, 545 (1992).

90. H. B. Mereyala and V. R. Gurijala, *Carbohydr. Res.*, **242**, 277 (1993).

91. P.-M. Åberg, L. Blomberg, H. Lönn, and T. Norberg, *J. Carbohydr. Chem.*, **13**, 141 (1994).

92. M. Nilsson and T. Norberg, *J. Carbohydr. Chem.*, **8**, 613 (1989).

93. A.-K. Tiden, *Chem. Comm. Univ. Stockholm*, 7 (1991).

94. F. Dasgupta and P. J. Garegg, *Carbohydr. Res.*, **202**, 225 (1990).

95. F. Dasgupta and P. J. Garegg, *Carbohydr. Res.*, **177**, C13 (1988).

96. Y. Ito and T. Ogawa, *Tetrahedron Lett.*, **29**, 1061 (1988).

97. Y. Ito, T. Ogawa, M. Numata, and M. Sugimoto, *Carbohydr. Res.*, **202**, 165 (1990).

98. H. Lönn, *Glycoconjugate J.*, **4**, 117 (1987).

99. E. Kallin, H. Lönn, and T. Norberg, *Glycoconjugate J.*, **5**, 3 (1988).

100. P. Konradsson, U. E. Udodong, and B. Fraser-Reid, *Tetrahedron Lett.*, **31**, 4313 (1990).

101. G. H. Veeneman, S. H. van Leeuwen, and J. H. van Boom, *Tetrahedron Lett.*, **31**, 1331 (1990).

102. G. H. Veeneman and J. H. van Boom, *Tetrahedron Lett.*, **31**, 275 (1990).

103. D. R. Mootoo, P. Konradsson, U. Udodong, and B. Fraser-Reid, *J. Am. Chem. Soc.*, **53** , 5583 (1988).

104. K.C. Nicolaou, S. Seitz, and D. Papahatjis, *J. Am. Chem. Soc.*, **105**, 2430 (1983).

105. (a) M. Sasaki and K. Tachibana, *Tetrahedron Lett.,* **32**, 6873 (1991); (b) K. Fukase, A. Hasuoka, and S. Kusumoto, *Tetrahedron Lett.*, **34**, 2187 (1993).

106. K. Fukase, A. Hasuoka, I. Kinoshita, and S. Kusumoto, *Tetrahedron Lett.*, **33**, 7165 (1992).

107. S. Sato, M. Mori, Y. Ito, and T. Ogawa, *Carbohydr. Res.*, **155**, C6 (1986).

108. V. Pozsgay and H. J. Jennings, *J. Org. Chem.*, **52**, 4635 (1987).

109. H. P. Wessel, *Tetrahedron Lett.*, **31** , 6863 (1990).

110. Y. M. Zhang, J. M. Mallet, and P. Sinaÿ, *Carbohydr. Res.*, **236**, 73 (1992).

111. C. Amatore, A. Jutand, J.-M. Mallet, G. Meyer, and P. Sinaÿ, *J. Chem. Soc., Chem.Commun*, 718 (1990).

112. G. Balavoine, A. Gref, J.-C. Fischer, and A. Lubineau, *Tetrahedron Lett.*, **40**, 5761 (1990).

113. J. M. Mallet, G. Meyer, F. Yvelin, A. Jutand, C. Amatore, and P. Sinaÿ, *Carbohydr. Res.*, **244**, 237 (1993).

114. A. K. Sarkar and K. L. Matta, *Carbohydr. Res.*, **233**, 245 (1992).

115. G. Stork and G. Kim, *J. Am. Chem. Soc.*, **114**, 1087 (1992).

116. H. M. Zuurmond, S. Van der Laan, G. A. Van der Marel, and J. H. van Boom, *Carbohydr. Res.*, **215**, C1 (1991).

117. P. J. Garegg and C. Hällgren, *J. Carbohydr. Chem.*, **11**, 445 (1992).

118. R. Roy, F. O. Andersson, and M. Letellier, *Tetrahedron Lett.*, **33**, 6053 (1992).

119. F. Barresi and O. Hindsgaul, *J. Am. Chem. Soc.*, **113**, 9376 (1991).

120. F. Barresi and O. Hindsgaul, *Synlett.*, 759 (1992).

121. M. Bols, *Tetrahedron*, **49**, 10049 (1993).

122. Y. Ito and J. C. Paulson, *J. Am. Chem. Soc.*, **115**, 1603 (1993).

123. L. Hedbys, L. Johansson, K. Mosbach, P.-O. Larsson, A. Gunnarsson, S. Svensson, and H. Lönn, *Glycoconjugate J.*, **6**, 161 (1989).

Chapter 5

PHENYL SELENOGLYCOSIDES AS VERSATILE
GLYCOSYLATING AGENTS IN OLIGOSACCHARIDE
SYNTHESIS AND THE CHEMICAL SYNTHESIS OF
DISACCHARIDES CONTAINING SULFUR AND SELENIUM

SEEMA MEHTA AND B. MARIO PINTO

Department of Chemistry, Simon Fraser University, Burnaby,
British Columbia, Canada V5A 1S6

Abstract   The use of phenyl selenoglycosides as glycosyl donors and
acceptors in glycosylation reactions is described. The versatility of these
compounds is illustrated by various selective activation strategies that have
been developed. Heteroanalogues of methyl and allyl kojibiosides, methyl
maltoside and methyl isomaltoside, in which the endocyclic and/or the
interglycosidic oxygen atoms have been replaced by sulfur and/or selenium,
are synthesized for evaluation as glycosidase inhibitors.

1. PHENYL SELENOGLYCOSIDES AS VERSATILE GLYCO- SYLATING AGENTS IN OLIGOSACCHARIDE SYNTHESIS

1.A. Introduction

In the last decade there has been a strong impetus for the development of new
and improved methods of oligosaccharide synthesis that require fewer synthetic
manipulations, result in higher yields and increased stereoselectivity, and enable
selective activation.[1]

One of the recent contributions includes the strategy of "armed" and "disarmed" glycosyl units.[2] This method is based on the differential reactivities conferred upon reacting partners by the nature of the protecting groups. An alternative strategy derives from the availability of two reacting units, the glycosyl-X and the glycosyl-Y unit, such that one of these units can be selectively activated under conditions where the other remains latent. This type of selectivity in glycosyl activation was initially described by Silwanis et al.[3] for the case of p-substituted phenyl thioglycosides and was later extended by Roy et al.[4] and Sliedregt et al.[5] We describe here our investigations of phenyl selenoglycosides in glycosylation reactions and the development of various selective activation strategies of the latter type, for the synthesis of oligosaccharides.

1.B. Synthesis of Selenoglycosides
A panel of phenyl selenoglycosides **1-10**, shown in Figure 1, was synthesized in a manner analogous to the method of Ferrier and Furneaux[6] for the synthesis of thioglycosides.[7] All these compounds are crystalline (except **9**), odourless and stable at room temperature.

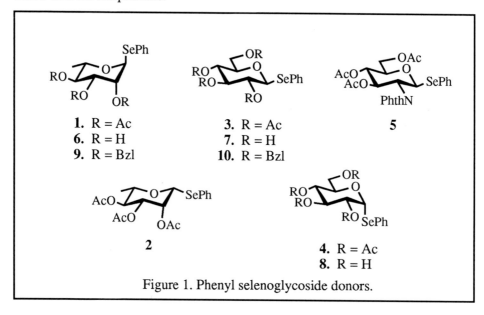

Figure 1. Phenyl selenoglycoside donors.

1.C. Glycosylation of Methyl Glycoside Acceptors with Phenyl Selenoglycosides

Initial activation of the phenyl selenoglycosides was attempted with conventional promoters. They were amenable to activation by thiophilic promoters such as methyl triflate,[8] phenylselenenyl triflate,[9] a mixture of cupric bromide-tetrabutyl ammonium bromide-silver triflate,[10] nitrosyl tetrafluoroborate[11] and mercuric chloride.[12]   However, the most promising conditions for selenoglycoside activation were realized when silver triflate was used as the promoter, in the presence of potassium carbonate. Glycosylation of the methyl glycoside acceptor 11 with selenoglycosides 1 and 5 in the presence of silver triflate and potassium carbonate afforded the 1,2-trans linked disaccharides 12 and 13 in 85% and 84% yield, respectively (Table 1, entries 1 and 2). The reactions proceeded in short reaction times under very mild conditions and proved to be stereoselective and high yielding.   Similarily, secondary glycosyl acceptors 14 and 15 were glycosylated with the selenoglycoside 1 to give disaccharides 16 (70%) and 17 (62%), respectively (Table 1, entries 3 and 4).

1.D. Selective Activation of Phenyl Selenoglycosides over Ethyl Thioglycosides

Experiments were performed to ascertain whether thioglycosides would remain inactive under the conditions established for selenoglycoside activation. A mixture of the phenyl selenoglycoside 1 and the corresponding ethyl thioglycoside 18 was allowed to compete for the glycosyl acceptor 11 under silver triflate promotion in the presence of K_2CO_3. Remarkably, the ethyl thioglycoside 18 remained inactive and was recovered from the reaction mixture in 91% yield. On the other hand, the phenyl selenoglycoside reacted completely to form the disaccharide 12 in a yield of 82% (Table 2, entry 1). In another experiment, the thioglycoside 18 was reacted with the acceptor 11, under the aforementioned conditions; no reaction was observed even after 24 hours of reaction time (Table 2, entry 2).

Encouraged by this observed selectivity, and being cognizant of the potential of the intrinsic greater reactivity of phenyl selenoglycosides, we reasoned that the activation of phenyl selenoglycoside donors in the presence of thioglycoside acceptors might be possible.   The inertness of "armed"

Table 1. Glycosylation of methyl glycoside acceptors with selenoglycoside donors under AgOTf promotion.[a]

Entry	Donor	Acceptor	Base	Molar Ratio[b]	Time (h)	Product	Yield (%)
1	1	11	K_2CO_3	1 : 1.5 : 3.5	1	12	85
2	5	11	K_2CO_3	1 : 1.2 : 6	1	13	84
3	1	14	K_2CO_3	1 : 1.2 : 3.5	1	16	70
4	1	15	K_2CO_3	1 : 1.2 : 2	5	17	62

[a]Reactions in the presence of 5 equivalents of K_2CO_3 with respect to AgOTf. [b]donor : acceptor : AgOTf

Table 2. Control experiments performed.

Entry	Donor	Acceptor	Molar ratio[a,b]	Time (h)	Product	Yield (%)
1	(SePh donor **1**) + (SEt donor **18**)	(OH acceptor **11**)	1 : 1 : 2 : 4	1.5	(**12**) + (SEt compound **18**)	82 / 91
2	(SEt donor **18**)	(OH acceptor **11**)	1 : 1 : 2	24	No reaction	
3	(SEt, OH donor **19**)	(SEt, OH acceptor **19**)	1 : 4.5	24	No cross-coupling	

[a]Reactions in the presence of 5 equivalents of K_2CO_3 with respect to AgOTf. [b]donor : acceptor : AgOTf

thioglycoside acceptors in the presence of AgOTf/ K_2CO_3 was confirmed by a control experiment in which no cross-coupling of the thioglycoside acceptor **19** was observed under these conditions (Table 2, entry 3).

The selective activation of phenyl selenoglycosides over ethyl thioglycosides was effected under silver triflate promotion, in the presence of potassium carbonate. The "armed" thioglycoside acceptor **19** was glycosylated with the phenyl selenoglycoside donor **5** to afford stereoselectively the β-disaccharide **22** in 84% yield (Table 3, entry 1). Similarly, the reaction of the thioglycosides **20** and **21** with selenoglycoside donor **1** produced the corresponding disaccharides **23** and **24** in yields of 80% and 78%, respectively (Table 3, entries 2 and 4). Glycosylations performed with Ag_2CO_3 as the base instead of K_2CO_3 produced analogous results (eg. Table 3, entry 3).

To establish the scope of this novel selective activation method, the reaction of the more reactive perbenzylated selenoglycoside **10**, an "armed" sugar, with an "armed" thioglycoside **19**, was considered next. This glycosylation proceeded in an excellent 90% yield and afforded a mixture of the α- and β- disaccharides **25** and **26** in a ratio of 2.5:1 (Table 3, entry 5).

Thus, the preferential activation of both "armed" and "disarmed" phenyl selenoglycosides over "armed" ethyl thioglycosides is demonstrated and it augers well for the intrinsic higher reactivity of the former class of compounds.

1.E. Selective Activation of a Glycosyl Halide over Phenyl Selenoglycosides

It was observed that the phenyl selenoglycosides were rendered unreactive in the presence of conventional proton acceptors such as 1,1,3,3-tetramethylurea and collidine. Since these are the conditions under which conventional glycosyl halide donors are activated,[13] it followed that there was a potential for the selective activation of glycosyl halide donors over selenoglycosides.

Silver triflate-mediated glycosylations of the selenoglycoside donors **27** and **28** with the glycosyl bromide **29** were performed in the presence of collidine (Figure 2). The desired selectivity was again observed and the corresponding disaccharides **30** and **31** were isolated in yields of 60%.

It was suspected that the preferential coordination of silver triflate to the organic base rendered it unavailable for coordination to, and activation of the selenium atom. This assumption was corroborated by the following experiment.

Table 3. Glycosylation of thioglycoside acceptors with selenoglycoside donors.[a]

Entry	Donor	Acceptor	Base	Molar ratio[b]	Time (h)	Product	Yield (%)
1	5	19	Ag_2CO_3	2 : 1 : 6	1	22	84
2	1	20	K_2CO_3	1 : 1 : 3	1.5	23	80
3	1	20	Ag_2CO_3	1 : 1 : 3	1	23	80
4	1	21	Ag_2CO_3	2 : 1 : 6	1	24	78
5	10	19	K_2CO_3	1.2 : 1 : 6	1.5	25,26 $\alpha{:}\beta = 2.5{:}1$	90

[a]Reactions in the presence of 5 equivalents of K_2CO_3 or 3 equivalents of Ag_2CO_3 with respect to AgOTf.
[b]donor : acceptor : AgOTf

Silver triflate-mediated glycosylation of the methyl glycoside acceptor **11** with the selenoglycoside **1** was performed in the presence of a hindered organic base, 2,6-di-*tert*-butyl-4-methylpyridine. Complexation of the silver triflate with the base was presumably disfavoured and the selenoglycoside was activated under these conditions to afford a 7:1 mixture of the orthoester **32** and the disaccharide **12** (Figure 3).

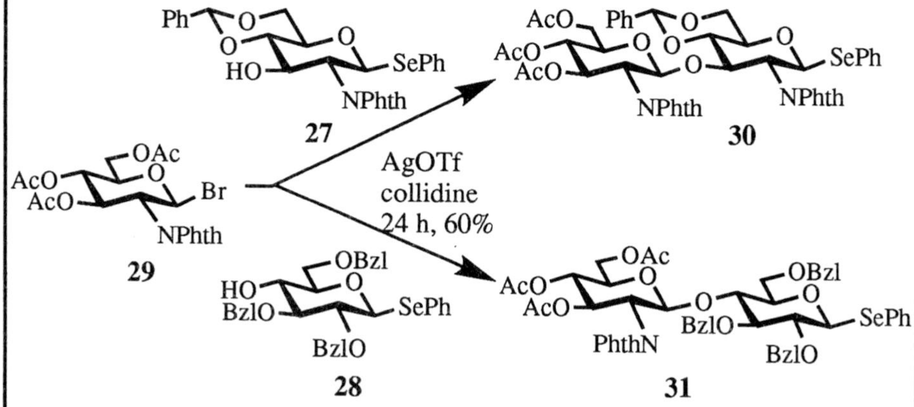

Figure 2. Selective activation of a glycosyl bromide over selenoglycosides.

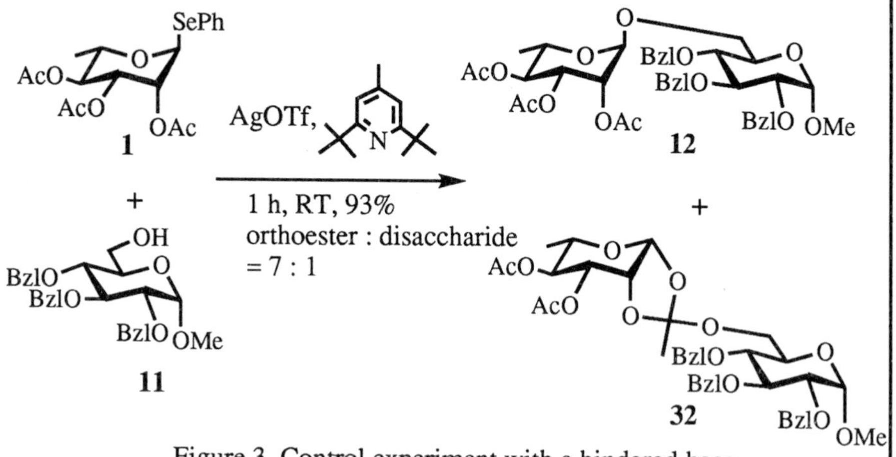

Figure 3. Control experiment with a hindered base.

1.F.   Selective Activation of a Glycosyl Trichloroacetimidate over Phenyl Selenoglycosides

It was envisaged that the selective activation of glycosyl trichloroacetimidates over selenoglycosides would provide even greater versatility in oligosaccharide synthesis. The selective activation of the glycosyl trichloroacetimidate **33** over the selenoglycoside acceptors **27** and **28** was effected by a catalytic amount of triethylsilyl triflate (Figure 4), to give the disaccharides **30** and **31**, respectively, as their selenoglycosides.

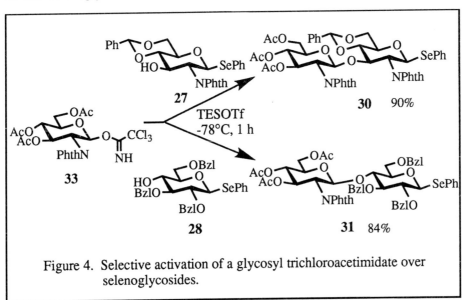

Figure 4. Selective activation of a glycosyl trichloroacetimidate over
selenoglycosides.

1.G.   The Impact of Selective Activation

The convergent block synthesis of oligosaccharides is more desirable and more efficient than the linear, stepwise approach as it minimizes the protecting group manipulation of complex oligosaccharide fragments.   A limitation of this approach is the necessity to functionalize the anomeric center of the oligosaccharide block. The selective activation strategies thus far described are essential in overcoming these types of difficulties.   The ability to achieve selectivity in activation can provide blocks that can function both as glycosyl acceptors or glycosyl donors depending on the choice of the reaction conditions.

Linear, stepwise syntheses are also more effectively accomplished by the sequential, selective activation of different glycosyl donors.

This type of versatility is accomplished with selenoglycosides and is illustrated by the synthesis of the trisaccharide **34** (Figure 5). The selective activation of glycosyl trichloroacetimidates over selenoglycosides enables the glycosylation of selenoglycoside acceptor **27** with the donor **33**. The resulting disaccharide **30** is obtained as a selenoglycoside, which is used directly as a glycosyl donor to afford the trisaccharide **34**. This trisaccharide is obtained as a thioglycoside and is amenable to activation by thiophilic promoters without any further manipulations.

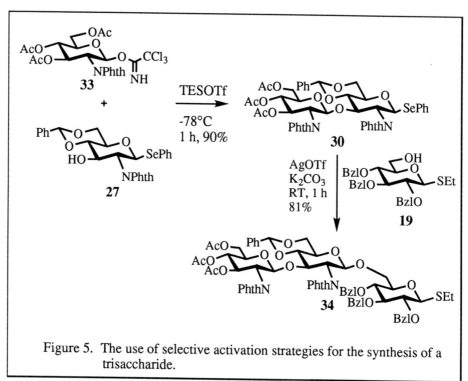

Figure 5. The use of selective activation strategies for the synthesis of a trisaccharide.

Since our initial disclosure,[7] Zuurmond et al.[14] have obtained complementary results with iodonium ion-mediated activation of phenyl selenoglycosides.

at which this reaction is quenched is of great significance in controlling its outcome. At a temperature lower than -50°C, compounds **42** and **43** reacted to afford predominantly the orthoester **48** in 83% yield, in addition to the α-disaccharide **44** (5%) (Table 4, entry 2).

In order to determine the effect of an increase in the reactivity of the reacting partners on the stereochemical outcome of the reaction, we examined the glycosylation of a more reactive benzylated acceptor **49**. In this case, a loss in stereoselectivity was observed, as expected with the more reactive acceptor, and a 1:1 mixture of the α- and β-disaccharides **50** and **51** was obtained (Table 4, entry 3).

The synthesis of the dithiomaltose derivative **37** in which the interglycosidic oxygen atom and the ring oxygen atom of the nonreducing sugar have been replaced by sulfur was examined next. Glycosylation of the selectively protected 4-thioglucopyranoside **52**[35] with the glycosyl donor **42** in the presence of triethylsilyl triflate afforded predominantly the α-disaccharide **53** in 53% yield and a minor amount of the β-isomer **54** (1.5%) (Table 4, entry 4).

A similar approach was followed for the synthesis of the 4-seleno-5'-thiomaltopyranoside **38**. The selectively protected 4-selenoglucopyranoside acceptor **55** was first required. This was synthesized by the initial displacement of the 4-trifluoromethanesulfonate of methyl 2,3,6-tri-*O*-benzoyl-α-D-galacto-pyranoside[35] by potassium selenocyanate, followed by reduction of the seleno-cyanate with sodium borohydride.[36] Glycosylation of the 4-selenol **55** with **42** under triethylsilyl triflate catalysis yielded a mixture of the α- and β-disaccharides **56** and **57** in yields of 46% and 11%, respectively (Table 4, entry 5). The synthesis of these disaccharides represents the first synthesis of reducing disaccharides containing selenium in the interglycosidic linkage.

The disaccharides **44**, **53** and **56** were deprotected by treatment with 0.2N sodium methoxide in methanol to afford the target compounds **36**, **37** and **38**, respectively in 75-90% yields.

2.C. Synthesis of Kojibiose Derivatives Containing Sulfur

A suitably protected glucosyl acceptor **58**, with a free hydroxyl group at the C-2 position, was glycosylated with the trichloroacetimidate donor **42**, by the sequential addition of two aliquots of 0.1 equivalent of triethylsilyl triflate

(TESOTf) to afford the α- and β- disaccharides **59** and **60** in a 3.5:1 ratio (Table 5, entry 1). A minor amount of the orthoester **61** was also isolated. In another experiment with 2 aliquots of 0.7 equivalent of TESOTf, a greater proportion of the orthoester **61** was formed (40%) (Table 5, entry 2) and the α to β ratio of the disaccharides also obtained was 10:1 (40%).

The synthesis of the kojibiose derivatives was also performed with the allyl glycoside acceptor **62**. This block possessed a benzoyl substituent at C-3 rather than a benzyl group, whose removal would be more easily achieved. Moreover, the acceptor was synthesized as its allyl glycoside, which had the potential of being removed to provide the hemiacetal, which could then be functionalized as the trichloroacetimidate of the disaccharide. This block would be useful for further elaboration to the trisaccharide. Compound **62** was glycosylated with the trichloroacetimidate **42** in the presence of 0.1 equivalent of TESOTf to afford an anomeric mixture of the disaccharides **63** and **64** in a ratio of 9:1 (Table 5, entry 3). The improved α- selectivity of this reaction may be due to the lower reactivity of acceptor **62** as compared to **58** due to the replacement of the activating benzyl substituent with a benzoate group. At lower temperatures, the orthoester **65** was obtained (Table 5, entry 4). The disaccharide **63** was deprotected by the initial hydrolysis of the benzylidene acetal with 80% CH_3COOH followed by deesterification under Zemplen conditions to afford the target disaccharide **40**.

2.D. Synthesis of Isomaltose Derivatives Containing Sulfur

Triethylsilyl triflate was employed as the promoter for the glycosylation of the 6-hydroxy acceptor **11** with the glycosyl trichloroacetimidate **42**. The α- and β-disaccharides **66, 67** were obtained in a 1.5:1 ratio (Figure 7).

2.E. Preferential Formation of α-Disaccharides

The exceptional behaviour of 5-thiohexopyranosyl donors to afford preferentially the α-product despite the presence of a participating acetate at the 2-position of the glycosyl donor has been described in the preceding sections. Our observations are in accord with those of Hashimoto and Isumi[29] who have reported a similar preference with 5-thio-L-fucopyranosyl donors.

Table 5. Synthesis of kojibiose derivatives containing sulfur.

Entry	Donor	Acceptor	Ratio[a]	Reaction Conditions	Product	Yield (%)
1	42	58	1 : 1.3 : 0.1 +0.1	-78°C for 4 h	59, 60	90% α:β = 3.5:1
					61	3%
2	42	58	1 : 0.8 : 0.07 +0.07	-78°C for 2 h	59, 60	40% α:β = 10:1
					61	40%

[a]donor : acceptor : promoter.

Table 5. (Continued) Synthesis of kojibiose derivatives containing sulfur.

Entry	Donor	Acceptor	Ratio[a]	Reaction Conditions	Product	Yield (%)
3	**42**	**62**	1 : 2 : 0.1	-78°C for 1 h RT for 1 h	**63, 64**	80% α:β = 9:1
4	**42**	**62**	1 : 2 : 0.1 +0.1	-78°C for 3 h	**63** **65**	16% 54%

[a]donor : acceptor : promoter.

In our opinion, the stereoselectivity may be accounted for by the thermodynamic stability of the axially-oriented aglycon. Whereas D-glucose exists as the α-anomer to the extent of 36% in aqueous solution,[37] 5-thioglucose

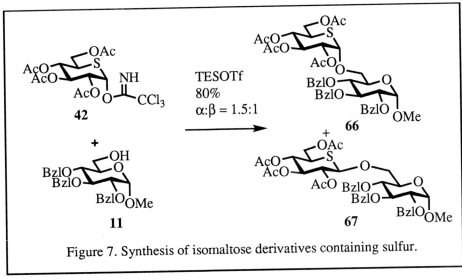

Figure 7. Synthesis of isomaltose derivatives containing sulfur.

exists as the α-anomer to the extent of 80%.[38] This likely results from a greater anomeric effect as well as a lesser steric effect in the latter case. Evidence to suggest this has been published by Pinto and Leung[39] who have studied the anomeric effect of X-C-Y systems. The experimentally observed ΔG values for 2-methoxyoxacyclohexane and 2-methoxythiacyclohexane indicate a greater α-preference for the thiacyclohexane derivative. This has been accounted for in terms of a dominance of the orbital interaction component. Our conclusions are also in agreement with those of Jagannadhan et al.[40] who report a greater thermodynamic stability of the α-S-carbocation vs the α-O-carbocation.

2.F. Orthoester Rearrangement

We speculated that the α-disaccharides might arise from the rearrangement of the orthoesters. To verify this postulate the orthoesters **48** and **65** that were isolated at temperatures below -50°C were reintroduced into the initial reaction conditions but the reaction mixtures were then warmed to room temperature. The orthoester **48** rearranged to afford only the α-disaccharide **44** in a 40% yield (Figure 8).

Also isolated were the acceptor **43**, and the glycals **45** and **46**. The orthoester **65** gave predominantly a mixture of the α- and β-disaccharides **63, 64** in a ratio of 7:1. Thus, the results of the rearrangement of these orthoesters are reflective of results obtained in reactions that were conducted without their isolation. We suggest that the preferential α-disaccharide formation is preceded by orthoester formation. The formation of an α-product from an orthoester rearrangement has some precedent.[41]

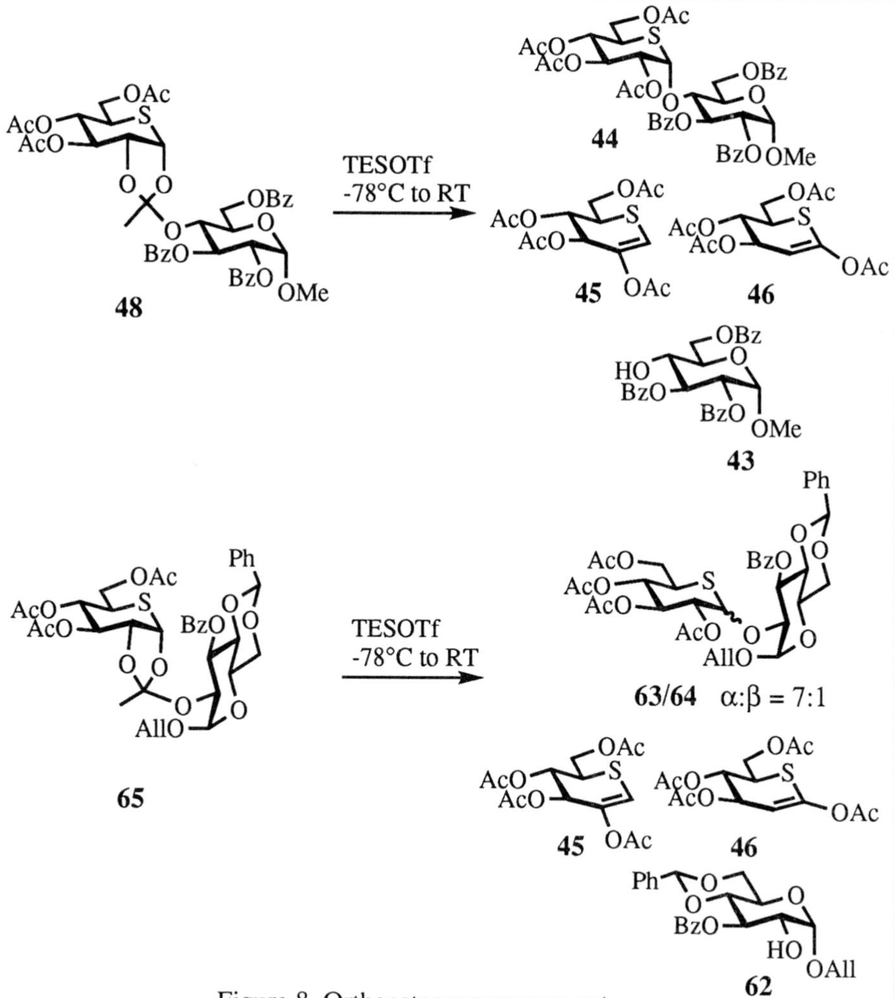

Figure 8. Orthoester rearrangement.

The synthetic methodology and the novel heteroanalogues of disaccharides described in this chapter will provide new tools in the fields of glycochemistry and glycobiology.

REFERENCES

1.   K. Toshima and K. Tatsuta, *Chem. Rev.*, **93**, 1503 (1993).
2.   B. Fraser-Reid, J. R. Merritt, A. L. Handlon and C. Webster, *Pure Appl. Chem.*, **65**, 779 (1993).
3.   B. A. Silwanis, F. Dasgupta and P. J. Garegg, *Abstracts of the Third Chemical Congress of North America,* Toronto, Canada, CARB 92 (1988).
4.   R. Roy, F. O. Andersson and M. Letellier, *Tetrahedron Lett.*, **33**, 6053 (1992).
5.   L. A. J. M. Sliedregt, K. Zegelaar-Jaarsveld, G. A. van der Marel and J. H. van Boom, *Synlett*, **5**, 335 (1993).
6.   R. J. Ferrier and R. H. Furneaux, *Methods in Carbohydrate Chemistry,* Vol. VIII, p. 251 (1980).
7.   (a) S. Mehta and B. M. Pinto, *Tetrahedron. Lett.*, **32**, 4435 (1991); (b) S. Mehta and B. M. Pinto, *J. Org. Chem.*, **58**, 3269 (1993).
8.   H. Lonn, *Carbohydr. Res.*, **139**, 105, 115 (1985).
9.   (a) Y. Ito and T. Ogawa, *Tetrahedron Lett.*, **29**, 1061 (1988); (b) Y. Ito and T. Ogawa, *Carbohydr. Res.*, **202**, 165 (1990).
10. S. Sato, M. Mori, Y. Ito and T. Ogawa, *Carbohydr. Res.*, **155**, c6 (1986).
11. (a) V. Pozgay and H. J. Jennings, *J. Org. Chem.*, **52**, 4635 (1987); (b) *ibid*, **53**, 4042 (1988).
12. R. J. Ferrier, R. W. Hay and N. Vethaviyasar, *Carbohydr. Res.*, **27**, 55 (1973).
13. H. Paulsen, *Angew. Chem. Int. Ed. Engl.*, **21**, 155 (1982).
14. H. M. Zuurmond, P. H. van der Meer, P. A. M. van der Klein, G. A. van der Marel and J. H. van Boom, *J. Carbohydr. Chem.*, **12**, 419 (1993).
15. For example (a) U. Spohr, M. Bach and R. G. Spiro, *Can. J. Chem.*, **71**, 1919 (1993); (b) U. Spohr, M. Bach and R. G. Spiro, *Can. J. Chem.*, **71**, 1928 (1993); (c) U. Spohr and M. Bach, *Can. J. Chem.*, **71**, 1943

(1993); (d) C. Schou, G. Rasmussen, M. Schulein, B. Henrissat and H. Driguez, *J. Carbohydr. Chem.*, **12**, 743 (1993).

16.   T. Feizi and M. Larkin, *Glycobiology*, **1**, 17 (1990).

17.   S. V. Evans, L. E. Fellows, K. T. M. Shing and G. W. J. Fleet, *Phytochemistry*, **24**, 1953 (1985).

18.   M. J. Humphries, K. Matsumota, S. L. White and K. Olden, *Cancer Res.*, **46**, 5215 (1986).

19.   E. Truscheit, W. Frommer, B. Junge, L. Muller, D. D. Schmidt and W. Wingender, *Angew. Chem. Int. Ed. Engl.*, **20**, 744 (1981) and references cited therein.

20.   L. G. A. M. van den Broek, D. J. Vermaas, B. M. Heskamp, C. A. A. van Boeckel, M. C. A. A. Tan, J. G. M. Bolsher, H. L. Ploegh, F. J. van Kemenade, R. E. Y. de Goede and F. Miedema, *Recl. Trav. Chim. Pays-Bas*, **112**, 982 (1993).

21.   K. Bock and B. W. Sigurskjold, In *Studies in Natural Products Chemistry* (Attur-ur-Rahman, Ed.) Vol. 7, p. 29 (1990).

22.   H. Yuasa, O. Hindsgaul and M. M. Palcic, *J. Am. Chem Soc.*, **114**, 5891 (1992).

23.   M. Blanc-Muesser, J. Defaye and H. Driguez, *Chem. Soc. Perkin Trans. 1*, 1885 (1984).

24.   D. H. Hutson, *J. Chem. Soc. (C).*, 442 (1967).

25.   J. Defaye and J.-M. Guillot, *Carbohydr. Res.*, **253**, 185 (1994).

26.   J. S. Andrews and B. M. Pinto, *Carbohydr. Res.*, In Press (1995).

27.   C.-H. Wong, T. Krach, C. Guatheron-Le Narvor, Y. Ichikawa, G. C. Look, F. Gaeta, D. Thompson and K. C. Nicolaou, *Tetrahedron Lett.*, **32**, 4867 (1991).

28.   H. Hashimoto and M. Kawanishi, *Abstracts of the XVIth Int. Carbohydr. Symp.*, Paris, France, p. 167 (1992).

29.   H. Hashimoto and M. Izumi, *Tetrahedron. Lett.*, **34**, 4949 (1993).

30.   W. Schneider and F. Wrede, *Ber*, **50**, 793 (1917).

31.   (a) S. Mehta and B. M. Pinto, *Tetrahedron. Lett.*, **33**, 7675 (1992); (b) S. Mehta, J. S. Andrews, B. D. Johnston and B. M. Pinto, *J. Am. Chem. Soc.*, **116**, 1569 (1994).

32. R. R. Schmidt, *Angew. Chem. Int. Ed. Engl.*, **25**, 212 (1986).

33. G. Excoffier, D. Gagnaire and J.-P. Utille, *Carbohydr. Res.* **39**, 368 (1975).

34. S. Mehta and B. M. Pinto, Unpublished results.

35. L. A. Reed and L. Goodman, *Carbohydr. Res.*, **94**, 91 (1981).

36. B. M. Pinto, J. Sandoval-Ramirez and R. D. Sharma, *Synth. Commun.*, **16**, 553 (1986).

37. (a) R. U. Lemieux and J. D. Stevens, *Can. J. Chem.*, **44**, 249 (1966); (b)
R.   Rudrum and D. F. Shaw, *J. Chem. Soc.*, **52** (1965), (c) S. J. Angyal and V. A. Pickles, *Aust. J. Chem.*, **25**, 1695 (1972).

38. J. B. Lambert and S. M. Wharry, *J. Org. Chem.*, **46**, 3193 (1981).

39. B. M. Pinto and R. Y. N. Leung, *ACS Symp. Ser.*, **539**, 128 (1993).

40. V. Jagannadham, T. L. Amyes and J. P. Richard, *J. Am. Chem. Soc.*, **115**, 8465 (1993).

41. P. J. Garegg, P. Konradsson, I. Kvarnstrom, T. Norberg, S. C. T. Svensson and B. Wigilius, *Acta Chemica Scand. B*, **39**, 655 (1985).

Chapter 6

SYNTHESIS AND USE OF *S*-XANTHATES, CARBOHYDRATE ENOL ETHERS AND RELATED DERIVATIVES IN THE FIELD OF GLYCOSYLATION

PIERRE SINAŸ AND JEAN-MAURICE MALLET

Département de Chimie, Ecole Normale Supérieure, U.R.A.1686,
24 Rue Lhomond, 75231 Paris Cédex 05, France

Abstract : Isopropenyl glycosides have been easily obtained in high yield by reacting the corresponding anomeric acetates with the Tebbe reagent. These compounds are efficient glycosyl donors in the presence of trimethylsilyl triflate or boron trifluoride etherate, probably via a mixed acetal glycoside intermediate. On the basis of this mechanism, glycosylation of monosaccharide hemiacetal donors with acceptors bearing an isopropenyl ether function has been achieved. Glycosylating properties of isopropenyl glycosyl carbonate has also been investigated. Anomeric *S*-xanthates were conveniently prepared by a two step azidonitration-xanthation sequence from a series of galactals. They constitute a novel class of glycosyl donors for the stereoselective preparation of either α- or β- protected precursors of biologically important galactosamine-containing oligosaccharides. Substituted pyranoid glycals were in turn easily derived from glycosyl phenyl sulfones acetylated at C-2 by a reductive samariation-elimination sequence.

1. INTRODUCTION

The efficient and selective synthesis of oligosaccharides is a central problem in carbohydrate chemistry.[1] The use of glycosyl halides as glycosyl donors, first introduced by Koenigs and Knorr,[2] has been the main glycosylation procedure for a very long period of time. The introduction by the Russian School of the orthoester procedure[3] was probably the first significant attempt to discover a substitute for the Koenigs Knorr method. This was followed by the development of the halide catalyzed glycosylation reaction[4] which has been used for the chemical synthesis of blood group antigenic determinants. The discovery of the imidate procedure[5] has probably contributed to our moving further and further away from the use of glycosyl halides. Indeed, as a result of a modification introduced by Schmidt,[6] the trichloroacetimidate glycosylation procedure is nowadays the most frequently used method for the synthesis of complex oligosaccharides. Thioglycosides are also attracting considerable attention in this respect.[7] The purpose of this chapter is to analyze two glycosylation reactions developed in the authors' laboratory: the first is the so-called exoenol ether procedure and its variations,[8] and the second is the xanthate procedure.[9]

2. EXO ANOMERIC ENOL ETHERS AND THE CORRESPONDING CARBONATES AS NOVEL GLYCOSYL DONORS

2.A.   An Introduction to the Theme: the Behavior of a Prop-1-enyl Glucoside

It may be anticipated that electrophilic activation of the double bond of appropriate alkenyl glycosides would generate anomeric carbenium ions and form the basis for glycosylation. Indeed, pent-4-enyl glycosides are currently used as glycosyl donors,[10] on the basis of a 5-exo -trig cyclization reaction (Scheme 1).

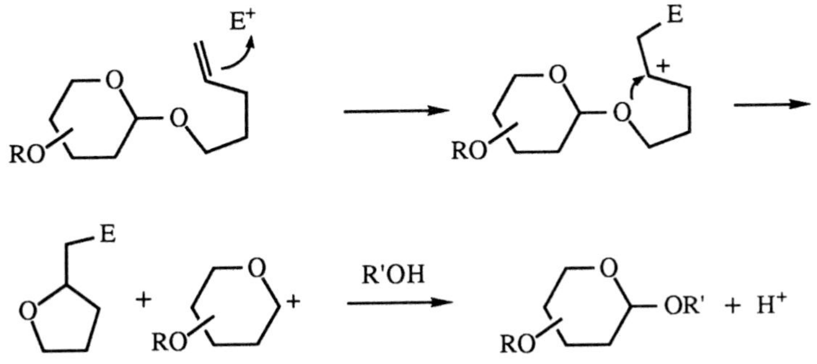

Scheme 1.

Much less studied is the behavior of alk-1-enyl glycosides, molecules which, in principle, should be good candidates for the generation of anomeric oxy carbenium ions (Scheme 2).

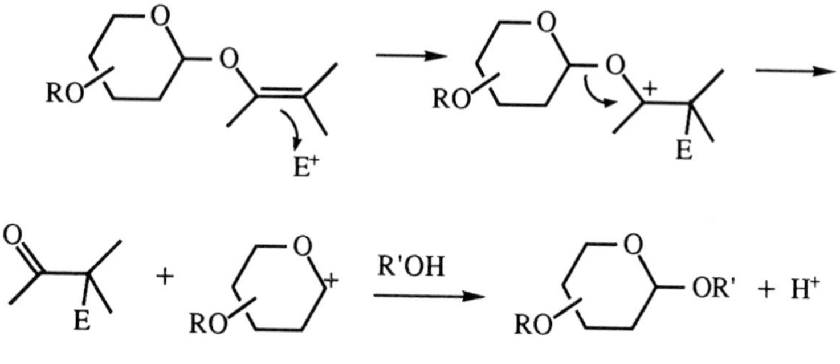

Scheme 2.

Although vinyl glycopyranosides are known derivatives,[11] the prop-1-enyl glycosides are easily prepared and are established members of the alk-1-enyl family. The classical use of the allyl ether as a temporary protecting hydroxyl group in carbohydrate chemistry originates from its easy conversion[12] into a prop-1-enyl ether which, in the presence of an electrophile,[12] regenerates the hydroxyl group (Scheme 3).

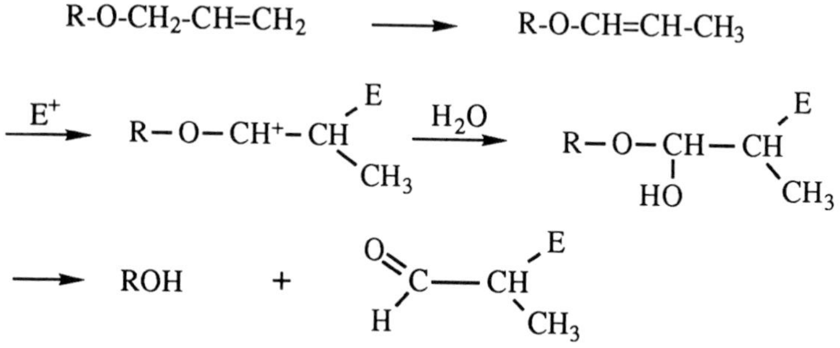

Scheme 3.

It has been shown that mercury (II) chloride induced cyclization of prop-1-enyl glycosides provides oxazoline derivatives[13] (Scheme 4).

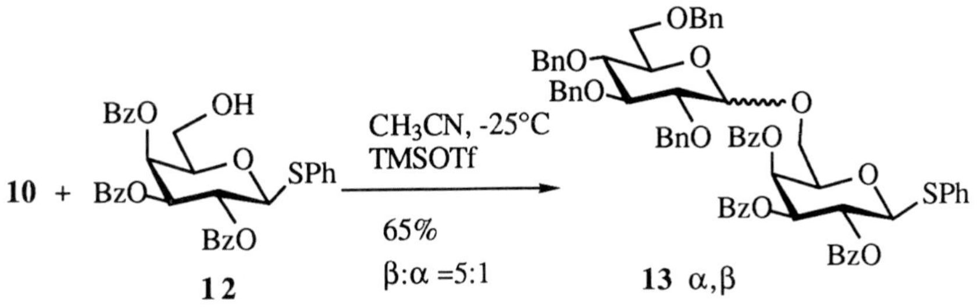

Scheme 9.

The exo-enol ether procedure can thus be used for the synthesis of thiophenyl disaccharides, which are useful building blocks for the preparation of complex oligosaccharides.

We explored next the galactosylation reaction. The isopropenyl galactoside **15** (α,β) was prepared from the known[32] acetate **14** by Tebbe's methylenation in 88% yield. In sharp contrast with the previously reported behavior of the corresponding gluco derivative **10**, galactosylation of **6** with **15** in dichloromethane and in the presence of TMSOTf as promoter gave **16** (α,β) in good yield (70%) (Scheme 10). Conversely, the β-selectivity in acetonitrile was poorer. Whether this already reported [33] behavior is due to a steric or electronic influence of the axially oriented O-benzyl group at C-4 remains to be determined.

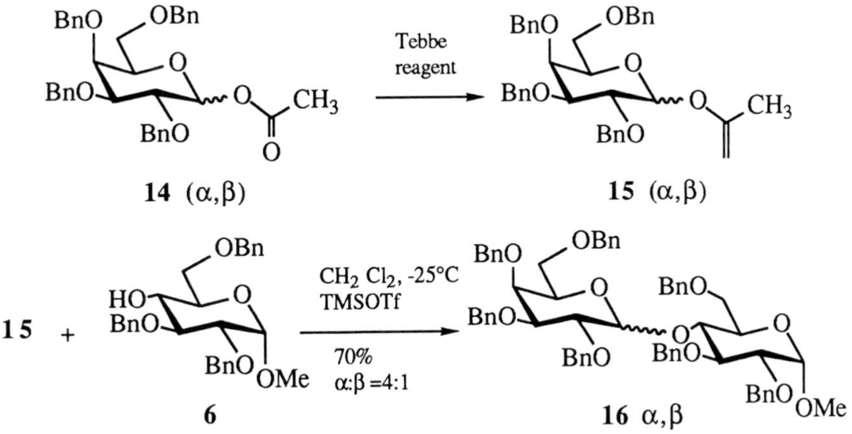

Scheme 10.

At this stage one may speculate on the mechanism of this glycosylation reaction. TMSOTf is known[34] to be a powerful silylating reagent of alcohols in solvents such as dichloromethane or nitromethane. It thus seems reasonable to assume (Scheme 11) that TMSOTf reacts almost instantaneously with the glycosyl acceptor and that the generated triflic acid protonates the enol ether group of A to create the species B. B may eject acetone to provide the oxycarbenium ion C, leading to the glycosylation step, with regeneration of TMSOTf. Alternatively, B may be converted into D, after a kinetic attack by the silylated glycosyl acceptor, followed by a 1,3 O→O silyl migration. D may collapse to give ion H and yield G. The latter reacts with C to provide the observed trehaloses 11 (α,α) and 11 (α,β).

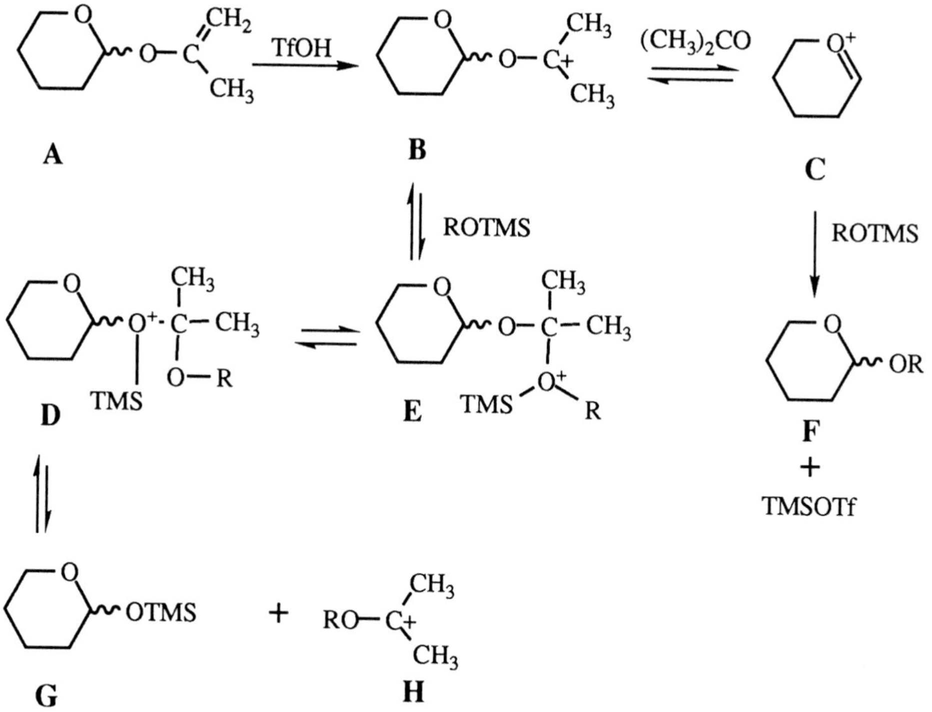

Scheme 11.

It is interesting to note that the recent work reported by Boons[35] confirms our ideas (Scheme 12).

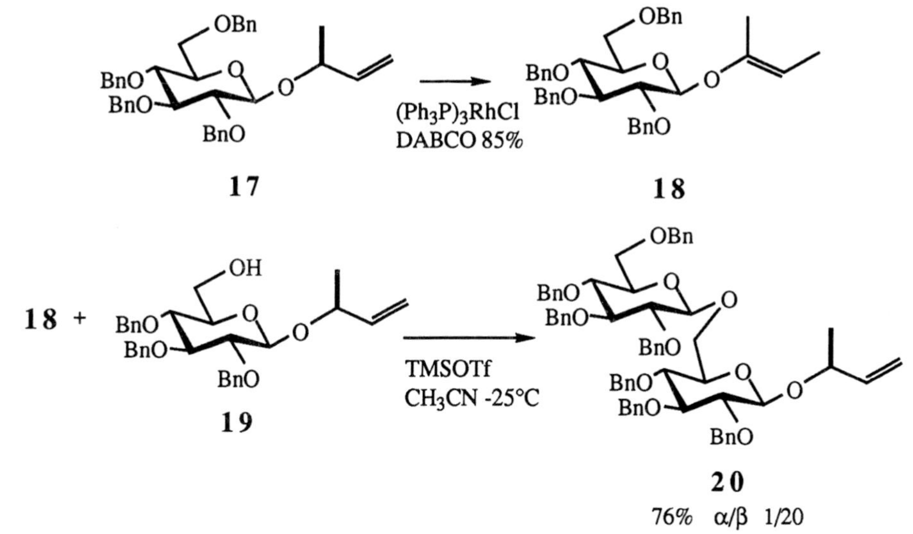

Scheme 12.

2.C. Variation One : The Reverse Isopropenyl Approach

According to the proposed mechanism (Scheme 11) for the isopropenyl glycosylation reaction catalyzed by TMSOTf, reaction of G with H should result in the same reaction pattern, through a mixed acetal glycoside (E) as a key intermediate. Indeed, acid-sensitive mixed acetal glycosides have been prepared under either kinetic[36] or thermodynamic[37] conditions but their potential for glycosylation has not been studied.

The "disarmed" acceptor was "armed" by a two-step sequence (acetylation, Tebbe methylenation) (Scheme 13).

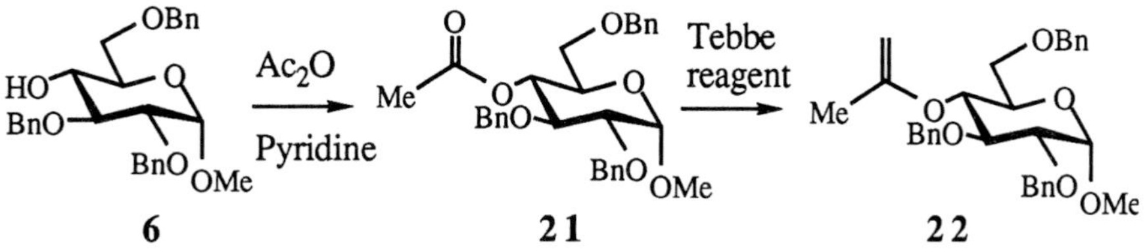

Scheme 13.

As anticipated from the proposed mechanism (Scheme 11), reaction of 22 with 2-azido-3,4,6-tri-O-benzyl-2-deoxy-D-galactopyranose[38] (23) in acetonitrile at -25°C

in the presence of TMSOTf gave the disaccharide **24**. This example has been
selected because the Tebbe reagent is not compatible with azido function; thus the
glycosyl donor **25** cannot be easily prepared by the Tebbe procedure. This
successful reverse condensation circumvents this problem, and offers an interesting
and selective way to introduce a derivative of galactosamine onto a carbohydrate
(Scheme 14).

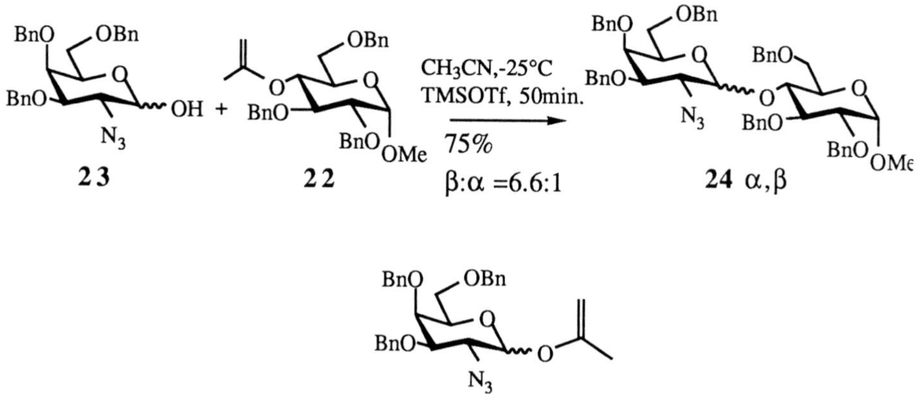

Scheme 14.

2.D. Variation Two: The Isopropenyl Glycoside Carbonate Route to Disaccharides.

The use of anomeric carbonates as glycosyl donors has been attempted with very
limited success.[39] It was shown[40] that heating ethyl 2,3,4,6-tetra-*O*-benzyl-D-
glucopyranosyl carbonate with a large excess of simple alcohols such as methanol,
isopropanol, or benzyl alcohol, gave the expected glycosides, but the reaction failed
when a carbohydrate acceptor was selected. This procedure offers no distinct
advantage when compared to historical Fischer glycosylation or transglycosylation
reactions. On the basis of our aforementioned results, and taking into consideration
the commercial availability of the isopropenyl chloroformate, we anticipated that the
glycosylating properties of the isopropenyl glycosyl carbonates (e.g. **27**) should be
of valuable synthetic utility.

Carbonates **27-30** were prepared from the corresponding hemiacetal
derivatives in nearly quantitative yields by treatment with isopropenyl chloroformate

(1.1 equiv) in dichloromethane in the presence of pyridine at 0°C for 1 h (Scheme 15).

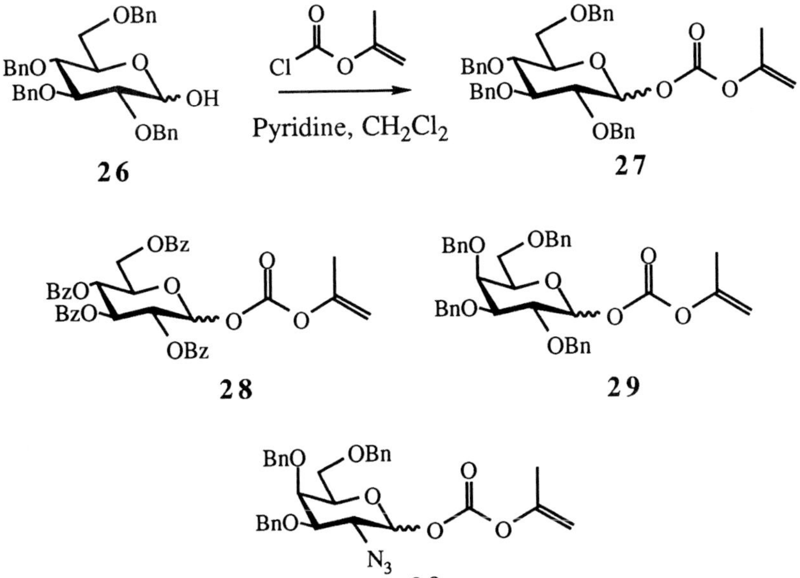

Scheme 15.

Condensation of **27** with the primary alcohol **4** (1 equiv) in acetonitrile at -25°C delivered the disaccharides **5** in 92% yield with a good β-selectivity (Scheme 16). The results achieved with various secondary carbohydrate acceptors (**6, 31**) are provided in Table 3.

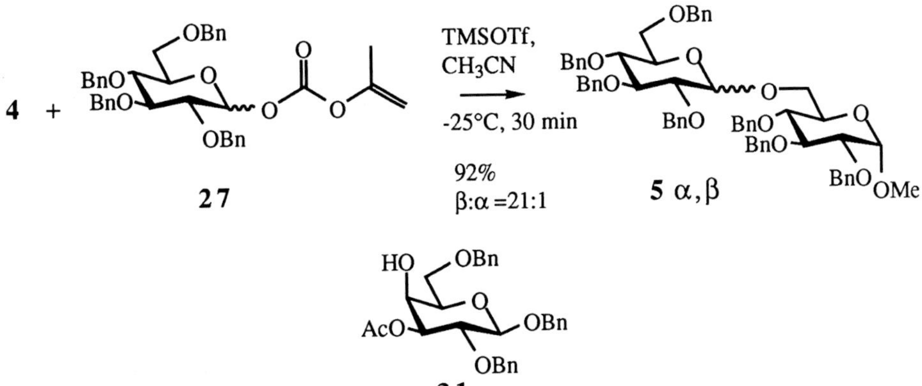

Scheme 16.

Table 3. Condensation of isopropenyl glycosyl carbonates with secondary alcohols in the presence of TMSOTf.

Donor	Acceptor	Solvent	T (°C)	Yield, %	α:β ratio
27	6	CH$_3$CN	-25	85	1:5.1
28	6	CH$_3$CN	-25	81	0:1
29	6	CH$_2$Cl$_2$	-25	79	4:1
29	6	CH$_3$CN	-25	80	1:1.6
30	6	CH$_2$Cl$_2$	-25	77	2.4:1
30	6	CH$_3$CN	-25	81	1:5
30	31	CH$_3$CN	-25	78	1:3.9
30	31	CH$_3$CH$_2$CN	-45	78	1:5.5

The carbonate 32 (α,β)[41] derived from L-fucose proved to be an efficient fucosyl donor, as recently shown in our group by the highly selective chemical synthesis of the disaccharides 34α and 36α (Scheme 17).

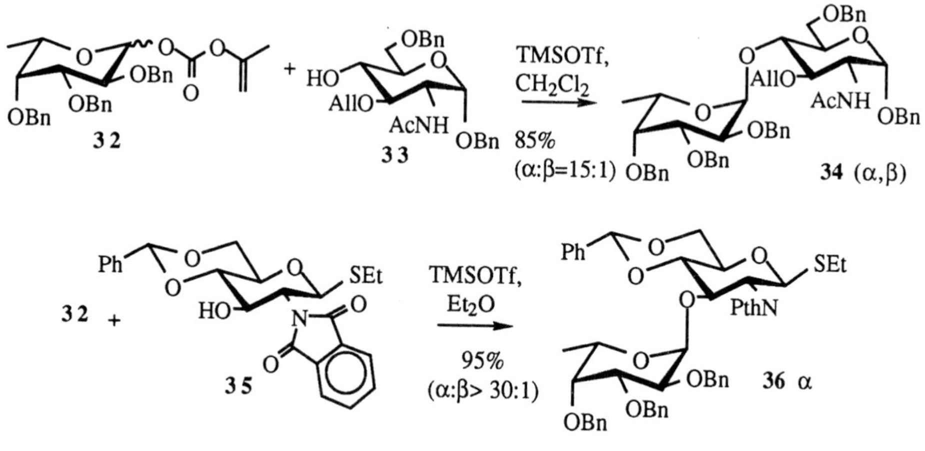

Scheme 17.

In our opinion, the isopropenyl glycosyl carbonate strategy affords an efficient route to disaccharides. In terms of reactivity and selectivity, it seems particularly adapted to carbohydrates such as L-fucose. In this respect, the carbonate 32 (α,β) derived from tri-O-benzyl L-fucose is a good building block for the synthesis of such important trisaccharides as Lewis[x] or other L-fucose containing oligosaccharides.

3.   A NOVEL CLASS OF GLYCOSYL DONORS: ANOMERIC S-XANTHATES OF 2-AZIDO-2-DEOXY-D-GALACTOPYRANOSYL DERIVATIVES

Anomeric S-xanthates are a class of S-glycosides (O-alkyl S-glycosyl dithiocarbonates) which have been used for the preparation of 1-thio sugars[42] and thio-glycosides.[43]

The azidonitration[44] of galactals, despite its rather moderate efficiency, constitutes a welcome means of preparing the anomeric nitrates of 2-azido-2-deoxy-D-galactopyranose. The problem lies in the high-yield conversion of these nitrates into potentially efficient glycosyl donors for the stereoselective synthesis of protected precursors of biologically important galactosamine-containing oligosaccharides. We demonstrated[9] that anomeric S-xanthates do indeed provide a solution to this problem.

Azidonitration of tri-O-benzyl-D-galactal (**37**) gave a crude mixture[45,46] which was directly treated with commercially available O-ethyl S-potassium dithiocarbonate in ethanol at room temperature to give O-ethyl S-(2-azido-3,4,6-tri-O-benzyl-2-deoxy-β-D-galactopyranosyl) dithiocarbonate (**39β**) (37% yield, based on **37**) together with 6% of the crystalline α-anomer **39α** (Scheme 18).

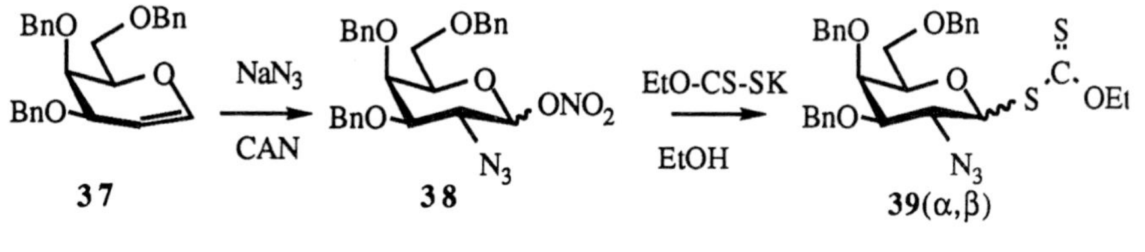

43% (β:α 6:1)

Scheme 18.

3-O-Acetyl-1,5-anhydro-4,6-O-benzylidene-2-deoxy-D-*lyxo*-hex-1-enitol (**40**) was next submitted to azidonitration. After column chromatography of the crude mixture, the α-D-galacto azidonitrate **42** was obtained in crystalline form in 44%

yield. A small amount (4%) of α-D-talo azidonitrate was also isolated. A comparison of these results with those originally obtained by Lemieux and Ratcliffe[44] on tri-*O*-acetyl-D-galactal shows that the presence of the 4,6-*O*-benzylidene fused ring, introduces rigidity into the galactal ring, and significantly increases the selectivity of the addition of the azido group at C-2. Similarly, azidonitration of 1,5-anhydro-3-*O*-benzyl-4,6-*O*-benzylidene-2-deoxy-D-*lyxo*-hex-1-enitol[47] **41** gave, after column chromatography, the α-D-galacto azidonitrate **43**, in crystalline form, in 40% yield. No trace of β-D-galacto or D-talo azidonitrates was found.

Displacement of α-nitrates **42** and **43** with *O*-ethyl *S*-potassium dithiocarbonate in acetonitrile for 5 h at room temperature gave the β-*S*-xanthates **44** and **45** in 97 and 91% yield, respectively (Scheme 19).

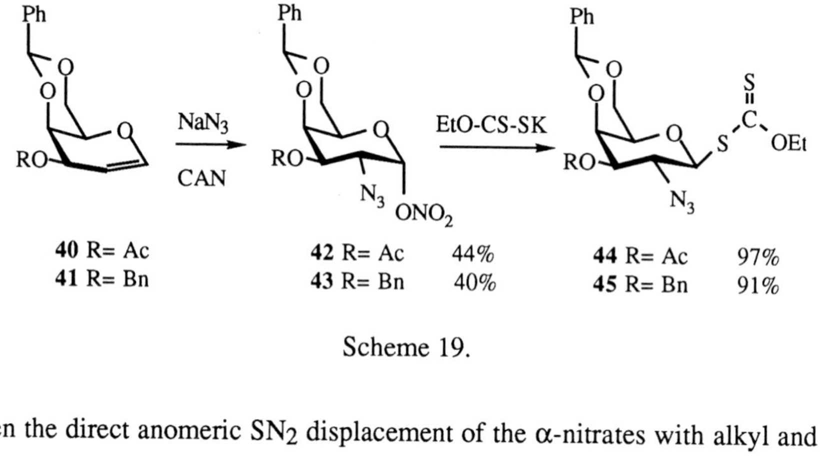

| **40** R= Ac | **42** R= Ac 44% | **44** R= Ac 97% |
| **41** R= Bn | **43** R= Bn 40% | **45** R= Bn 91% |

Scheme 19.

When the direct anomeric SN_2 displacement of the α-nitrates with alkyl and aryl thiolates was attempted, almost quantitative denitration was observed. This reaction, which is in sharp contrast with the successful anomeric SN_2 displacement of nitrates with sodium alkoxides,[48,49] was in retrospect not unexpected in the light of the mechanism of denitration of nitrate esters by sulfide or polysulfide ions which was previously studied and discussed[50]. Anomeric denitration is usually achieved[45] with sodium nitrite in aqueous dioxane at 80°C for about 6 h. We found that 3,4,6-tri-*O*-acetyl -2-azido-2-deoxy-α-D-galactopyranosyl nitrate[44] (**46**) is conveniently denitrated using the conditions shown in Scheme 20 to give known[51] **47**. The generalization of this useful methodology has been reported.[52] We observed a similar reluctance for anomeric displacement of nitrates with potassium thioacetate in dichloromethane at room temperature when high yields of 1-*O*-acetyl derivatives[53] were obtained (Scheme 20).

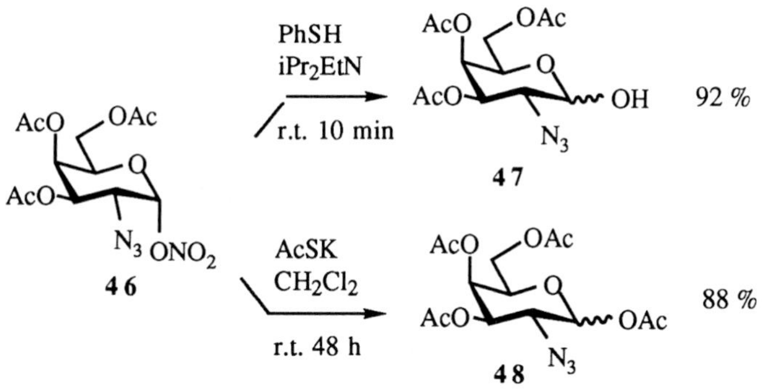

Scheme 20.

The glycosylation properties of these xanthates was next investigated. The previously reported[54] xanthate **49** was reacted at room temperature with methyl 2,3,4-tri-*O*-benzyl-α-D-glucopyranoside (**4**) in acetonitrile in the presence of copper(II) triflate and 4 Å molecular sieves to give selectively the crystalline β-disaccharide **50β** (Scheme 21). The formation of the corresponding α-disaccharide **50α** was also observed (β:α ratio of 5:1). As previously explained, the β-selectivity obtained in acetonitrile is due to the stereoselective kinetic formation of a reactive α-nitrilium intermediate.

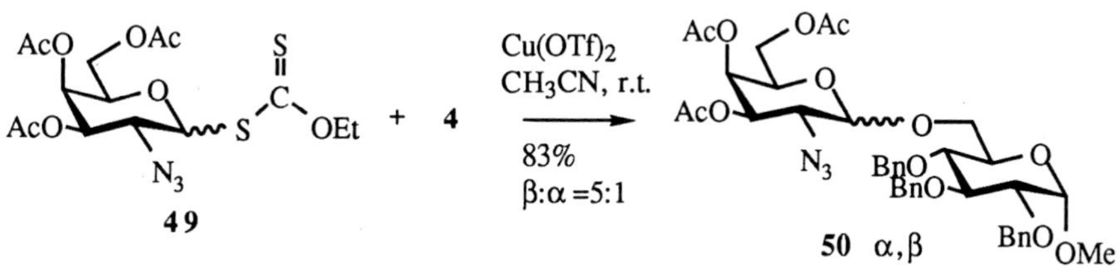

Scheme 21.

Table 4 demonstrates that anomeric *S*-xanthates **39**, **44** and **45** are interesting β-selective glycosyl species.

Table 4. Glycosylation of **4** in acetonitrile.

Xanthate[a]	Promotor	Disaccharide	Total yield (%)	α:β ratio
39	Cu(OTf)$_2$	BnO, OBn, O, BnO, N$_3$, BnO, BnO, BnO, OMe **51**	92	1:6
44	DMTST	Ph, O, O, AcO, N$_3$, BnO, BnO, BnO, OMe **52**	86	1:6
45	Cu(OTf)$_2$	Ph, O, O, BnO, N$_3$, BnO, BnO, BnO, OMe **53**	85	1:5.5
39	Ar$_3$N$^{+\circ}$	**51**	93	1:4.3

[a] donor/acceptor ratio 1.5:1

Dimethyl(methylthio)sulfonium triflate[55] (DMTST), copper(II) triflate, or commercially available tris (4-bromophenyl)ammoniumyl hexachloroantimonate[21] (Ar$_3$N$^{+\circ}$, SbCl$_6^-$) were found to be suitable promotors in CH$_3$CN. α-Disaccharides were obtained when dichloromethane was used as a solvent, for

M. V. Ovchinnikov, L. V. Bachinowsky, and N. K. Kotchekov, *Carbohydr. Res.*, **76**, 252 (1979); (c) N. K. Kotchekov, *Tetrahedron*, **43**, 2389 (1987).

4. R. U. Lemieux, K. B. Hendricks, R. V. Stick, and K. James, *J. Am. Chem. Soc.*, **97**, 4056 (1975).

5. P. Sinaÿ, *Pure Appl. Chem.*, **50**, 1437 (1978).

6. R. R. Schmidt and J. Michel, *Angew. Chem. Int. Ed. Engl.*, **19**, 731 (1980).

7. (a) K. C. Nicolaou, S. P. Seitz, and D. P. Papahatjis, *J. Am. Chem. Soc.*, **105**, 2430 (1983); (b) H. Lönn, *Carbohydr. Res.*, **139**, 105 (1985); (c) P. Fugedi and P. J. Garegg, *Carbohydr. Res.*, **149**, C9 (1986); (d) S. Sato, M. Mori, Y. Ito, and T. Ogawa, *Carbohydr. Res.*, **155**, C6 (1986); (e) Y. Ito and T. Ogawa, *Tetrahedron Lett.*, 4701 (1987); (f) H. Lönn, *Glycoconjugate J.*, **4**, 117 (1987); (g) F. Dasgupta and P. J. Garegg, *Carbohydr. Res.*, **177**, C13 (1988); (h) Y. Ito and T. Ogawa, *Tetrahedron Lett.*, 1061 (1988); (i) G. H. Veeneman and J. H. van Boom, *Tetrahedron Lett.*, 275 (1990); (j) G. H. Veeneman, S. H. van Leeuwen, and J. H. van Boom, *Tetrahedron Lett.*, 1331 (1990); (k) F. Dasgupta and P. J. Garegg, *Carbohydr. Res.*, **202**, 225 (1990); (l) Y. Ito, T. Ogawa, M. Numata, and M. Sugimoto, *Carbohydr. Res.*, **202**, 165 (1990); (m) P. Konradsson, U. E. Udodong, and B. Fraser-Reid, *Tetrahedron Lett.*, 4313 (1990); (n) R. J. Ferrier, R. W. Hay, and N. Vethaviyasar, *Carbohydr. Res.*, **27**, 55 (1973); (o) J. W. van Cleve, *Carbohydr. Res.*, **70**, 161 (1979); (p) T. Mukaiyama, T. Nakatsuka, and S. Shoda, *Chem. Lett.*, 487 (1979); (q) S. Hanessian, C. Bacquet, and N. Lehong, *Carbohydr. Res.*, **80**, C17 (1980); (r) P. J. Garegg, C. Henrichson, and T. Norberg, *Carbohydr. Res.*, **116**, 162 (1983); (s) C. Amatore, A. Jutand, J.-M. Mallet, G. Meyer, and P. Sinaÿ, *J. Chem. Soc., Chem. Commun.*, 718 (1990).

8. A. Marra, J. Esnault, A. Veyrières, and P. Sinaÿ, *J. Am. Chem. Soc.*, **114**, 6354 (1992).

9. (a) A. Marra, L.-K. Shi Shun, F. Gauffeny, and P. Sinaÿ, *Synlett*, 445 (1990); (b) A. Marra, F. Gauffeny, and P. Sinaÿ, *Tetrahedron*, **47**, 5149 (1991).

10. B. Fraser-Reid, P. Konradsson, D. R. Mootoo, and U. Udodong, *J. Chem.*

Soc., Chem. Commun., 823 (1988).

11. (a) T. D. Perrine, C. P. J. Glaudemans, R. K. Ness, J. Kyle, and H. G. Fletcher, *J. Org. Chem.,* **32**, 664 (1967); (b) L. Cottier, G. Remy, and G. Descotes, *Synthesis* 711 (1979); (c) P. Rollin, V. Verez-Bencomo, and P. Sinaÿ, *ibid.,* 134 (1984); (d) A. de Raadt and R. J. Ferrier, *J. Chem. Soc Chem. Commun.,* 1009 (1987).

12. For a review see P. Boullanger, P. Chatelard, G. Descotes, M. Kloostermann, and J. M. van Boom, *J. Carbohydr. Chem.,* **5**, 541 (1986).

13. M. A. Nashed, C. W. Slife, M. Kiso, and L. Anderson, *Carbohydr. Res.,* **58**, C13 (1977).

14. J. Gigg and R. Gigg, *J. Chem. Soc.(C) ,* 82 (1966).

15. (a) R. Eby and C. Schuerch, *Carbohydr. Res.,* **34**, 79 (1974); (b) A. Liptak, I. Jodal, and P. Nanasi, *Carbohydr. Res.,* **44**, 1 (1975)

16. J. M. Küster and I. Dyong, *Liebigs Ann. Chem.,* 2179 (1975).

17. A. J. Ratcliffe and B. Fraser-Reid, *J. Chem. Soc., Perkin Trans. 1,* 747 (1990).

18. I. Braccini, C. Derouet, J. Esnault, C. Hervé du Penhoat, J.-M. Mallet, V. Michon, and P. Sinaÿ, *Carbohydr. Res.,* **246**, 23 (1993).

19. J.-R. Pougny and P. Sinaÿ, *Tetrahedron Lett.,* 4073 (1976).

20. (a) R. U. Lemieux and R. M. Ratcliffe, *Can. J. Chem.,* **57**, 1244 (1979); (b) A. A. Pavia, S. N. Ung-Chhun, and J.-L. Durand, *J. Org. Chem.,* **46**, 3158 (1981); (c) A. Klemer and M. Kohla, *J. Carbohydr. Chem.,* **7**, 785 (1988); (d) D. M. Gordon and S. J. Danishefsky, *J. Org. Chem.,* **56**, 3713 (1991).

21. A. Marra, J.-M. Mallet, C. Amatore, and P. Sinaÿ, *Synlett ,* 572 (1990).

22. S. Hashimoto, M. Hayashi, and R. Noyori, *Tetrahedron Lett.,* 1379 (1984).

23. R. R. Schmidt, M. Behrendt, and A. Toepfer, *Synlett ,* 694 (1990).

24. S. Hashimoto, T. Honda, and S. Ikegami, *J. Chem. Soc., Chem. Commun.,* 685 (1989).

25. (a) S. Koto, S. Inada, T. Narita, N. Morishima, and S. Zen, *Bull. Chem. Soc. Jpn.,* **55**, 3665 (1982); (b) L. F. Tietze, and M. Beller, *Angew. Chem. Int., Ed. Engl.* **30**, 868 (1991).

26. A. de Raadt, and R. J. Ferrier, *Carbohydr. Res.,* **216**, 93 (1991).

27. F. N. Tebbe, G. W. Parshall, and G. S. Reddy, *J. Am. Chem. Soc.,* **100**,

3611 (1978).

28. T. S. Chou, S.-B. Huang, and W.-H. Hsu, *J. Chinese Chem. Soc., (Taipei)* **30**, 277 (1983).

29. A. G. M. Barrett, B. C. B. Bezuidenhout, A. F. Gasiecki, A. R. Howell, and M. A. Russell, *J. Am. Chem. Soc.*, **111**, 1392 (1989).

30. A. A. Pavia, J.-M. Rocheville, and S. N. Ung, *Carbohydr. Res.*, **79**, 79 (1980).

31. S. Koto, N. Morishima, and S. Zen, *Chem. Lett.*, 61 (1976).

32. P. W. Austin, F. E. Hardy, J. G. Buchanan, and J. Baddiley, *J. Chem. Soc.*, 1419 (1965).

33. B. Wegmann and R. R. Schmidt, *J. Carbohydr. Chem.*, **6**, 357 (1987).

34. G. O. Olah, A. Husain, B. G. B. Gupta, G. F. Salem, and S. C. Narang, *J. Org. Chem.*, **46**, 5212 (1981).

35 G. J. Boons and S. Isles, *Tetrahedron Lett.*, **35**, 3593 (1994)

36. (a) L. F. Tietze and M. Beller, *Liebigs Ann. Chem.*, 587 (1990); (b) L. F. Tietze, R. Fischer, M. Lögers, and M. Beller, *Carbohydr. Res.*, **194**, 155 (1989); (c) L. F. Tietze, and M. Lögers, *Liebigs Ann. Chem.*, 261 (1990).

37. M. Blanc-Muesser, J. Defaye, and J. Lehmann, *Carbohydr. Res.*, **108**, 103 (1982).

38. F. Gauffeny, A. Marra, L. K. Shi Shun, P. Sinaÿ, and C. Tabeur, *Carbohydr. Res.*, **219**, 237 (1991).

39. For glycosylation of phenols by phenyl or methyl glycosyl carbonates under pyrolytic conditions see: Y. Ishido, S. Inaba, A. Matsuno, T. Yoshino, and H. Umezawa, *J. Chem. Soc., Perkin Trans. 1,* 1382 (1977).

40. M. Boursier and G. Descotes, *C. R. Acad. Sci.* **308**, 919 (1989).

41. J. Esnault, Y-M. Zhang, J.-M. Mallet, and P. Sinaÿ, Unpublished Results.

42. D. Horton, and J. Wander, In *"The Carbohydrate Chemistry and Biochemistry,"* (W.Pigman and D. Horton, Eds.), 2nd ed., Academic Press, New York, Vol. IB, p. 799 (1980).

43. M. Sakata, M. Haga, and S. Tejima, *Carbohydr. Res.*, **13**, 379 (1970).

44. R. U. Lemieux and R. M. Ratcliffe, *Can. J. Chem.*, **57**, 1244 (1979).

45. G. Grundler and R. R. Schmidt, *Liebigs Ann. Chem.*, 1826 (1984).

46. K. Briner and A. Vasella, *Helv. Chim. Acta ,* **70**, 1341 (1987).

47. A. Fernandez-Mayoralas, A. Marra, M. Trumtel, A. Veyrières, and P. Sinaÿ, *Carbohydr. Res.*, **188**, 81 (1989).

48. (a) H. Paulsen and M. Paal, *Carbohydr. Res.*, **135**, 53 (1984); (b) J.-C. Jacquinet, and P. Sinaÿ, *Carbohydr. Res.*, **159**, 229 (1987).

49. (a) H. Paulsen, U. von Deessen, and H. Tietz, *Carbohydr. Res.*, **137**, 63 (1985); (b) G. Catelani, A. Marra, F. Paquet, and P. Sinaÿ, *Carbohydr. Res.*, **155**, 131 (1986).

50. R. T. Merrow, S. J. Cristol, and R. W. Van Dolah, *J. Am. Chem. Soc.*, **75**, 4259 (1953).

51. H. Paulsen and B. Sumfleth, *Chem. Ber.*, **112**, 3203 (1979).

52. F. Gauffeny, A. Marra, L. K. Shi Shun, P. Sinaÿ, and C. Tabeur, *Carbohydr. Res.*, **219**, 237 (1991).

53. H. Paulsen, A. Richter, V. Sinnwell, and W. Stenzel, *Carbohydr. Res.*, **64**, 339 (1978).

54. H. Paulsen, W. Rauwald, and U. Weichert, *Liebigs Ann. Chem.*, 75 (1988).

55. M. Ravenscroft, R. M. G. Roberts, and J. G. Tillett, *J. Chem. Soc., Perkin Trans. 2*, 1569 (1982).

56. F. Machetto, *Thèse de Doctorat*, University Paris VI, France, (1992).

57. L. K. Shi Shun, *Thèse de Doctorat*, University Paris VI, France, (1991).

58. (a) P. de Pouilly, A. Chénedé, J.-M. Mallet, and P. Sinaÿ, *Tetrahedron Lett.*, **33**, 8065 (1992) ; (b) P. de Pouilly, A. Chénedé, J.-M. Mallet, and P. Sinaÿ, *Bull. Soc Chim. Fr.*, **130**, 256 (1993).

Chapter 7

n-PENTENYL GLYCOSIDES IN OLIGOSACCHARIDE SYNTHESIS

ROBERT MADSEN AND BERT FRASER-REID

Department of Chemistry, Duke University, Durham, NC 27708-0346, USA

Abstract *n*-Pentenyl Glycosides (NPGs) may be prepared directly by Fischer glycosidation of aldoses, or indirectly by most of the standard procedures for synthesizing alkyl glycosides. NPGs are stable at room temperature and are unaffected by most reagents and laboratory practices employed in carbohydrate transformations. However they react readily and chemospecifically with halonium ions under oxidative conditions that leave other alkenyl residues and oxidizable functional groups unaffected. *N*-Iodo-succinimide/triethylsilyl triflate is highly recommended as the promoter, and under its agency, NPG reactions are usually complete within the time required to take a TLC sample. NPG activity can be sidetracked by bromination. The resulting vicinal dibromide is, in effect, a latent glycosyl donor since activity can be restored by reductive elimination.

1. INTRODUCTION

The discovery of *n*-pentenyl glycosides (NPGs)[1] originated in a serendipitous observation in our laboratory during an approach to the synthesis of the ansa chain of Streptovaricin A.[2] The first report appeared in 1988,[3] and since then NPGs have been explored as mechanistic probes for anomeric activation, for oligosaccharide syntheses, and for the preparation of novel protecting groups.[4]

155

Oligosaccharide synthesis presents multiple and various difficulties, two of which deserve special mention. Firstly, in order to differentiate between the multitude of hydroxyl groups, extensive protection strategies are often required which may make the synthesis lengthy and cumbersome. Secondly, prior to the coupling event, the anomeric center of the predestined glycosyl donor has to be selectively activated which, in itself, often requires several synthetic operations.

The use of NPGs can help to reduce the number of bothersome protecting group manipulations that have to be carried out. As a simple alkyl residue, the n-pentenyl group can be installed at the outset by carrying out a Fischer glycosidation upon the aldose of interest. In the ensuing transformations, the n-pentenyl moiety may serve dual functions. It can protect the anomeric center during synthetic manipulations elsewhere in the molecule, and then in the presence of an electrophile, the n-pentenyl group can be chemospecifically activated to provide a leaving group transforming it into a glycosyl donor for coupling with an acceptor[1] (Scheme 1).

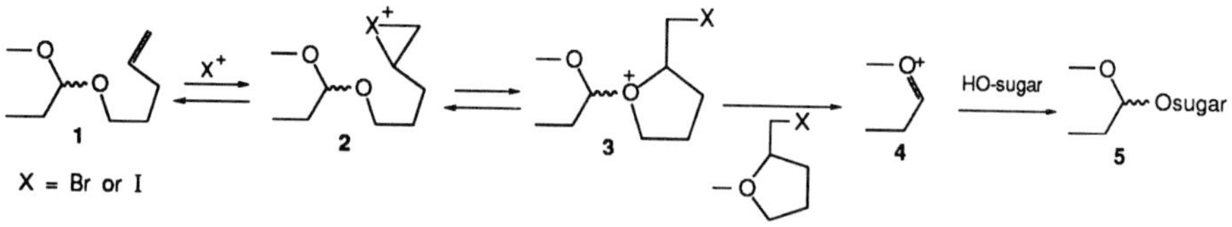

Scheme 1.

In this chapter we describe some recent developments in the use of NPGs for oligosaccharide syntheses.

2.   PREPARATION

Since NPGs are normal alkyl glycosides, methods for their preparation may be adapted from standard protocols. The most direct method of preparation is Fischer glycosidation where the aldose is treated with pent-4-enyl alcohol[5] and an acid catalyst[6] (Schemes 2a,b). At the completion of the reaction, the pentenyl alcohol can be recovered on a rotary evaporator and saved for future glycosidations. As expected the reaction gives a mixture of anomers, the ratio of which is approximately the same as would be found with any other Fischer glycosidation.

Separation of the anomers can be accomplished by flash chromatography or by crystallization of suitable derivatives.[1]

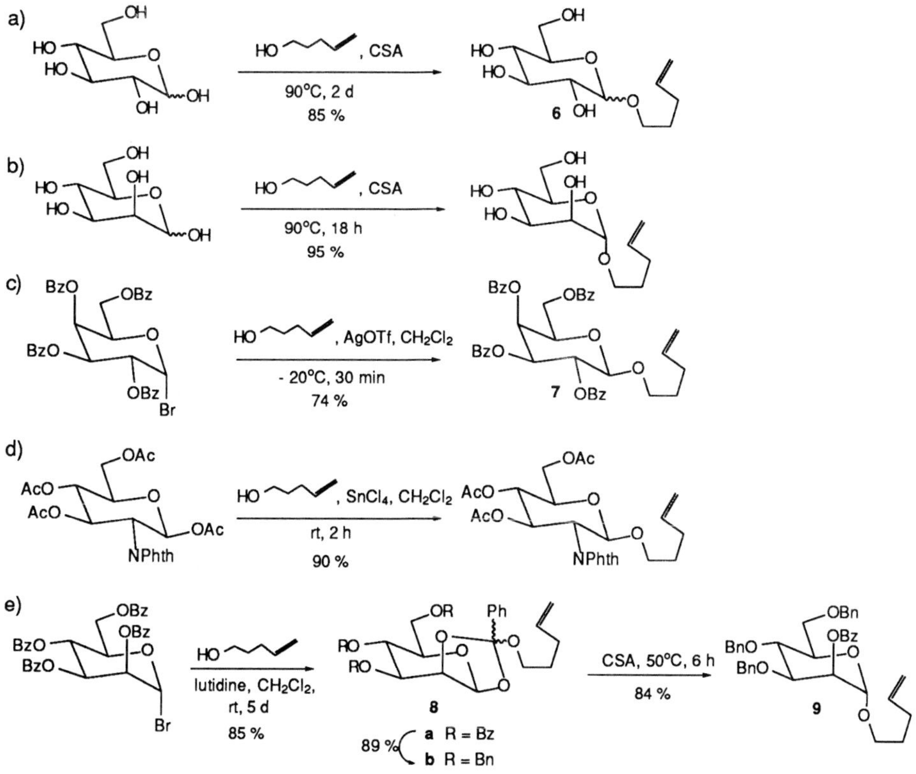

Scheme 2.

NPGs of galactose and glucosamine are most conveniently prepared by the Koenigs-Knorr procedure, or from the glycosyl acetate (Schemes 2c,d). Glycosyl bromides can serve as precursors for *n*-pentenyl 1,2-orthoesters e.g. **8** (Scheme 2e), with simultaneous protection of the anomeric center and the C2 hydroxyl group. Acid-induced rearrangement of **8** gives the NPG **9** with an ester group at C2 (*vide infra*).

Many NPGs are crystalline materials, and like other alkyl glycosides, they are stable compounds that can be stored indefinitely at room temperature. Although stable to most reagents, NPGs are incompatible with reductive hydrogenation conditions used to cleave benzyl ethers. However this particular deprotection can be readily achieved by Birch reduction.[1]

3. PROMOTERS

N-Bromo and iodosuccinimides (NBS, NIS) have been the most widely investigated electrophiles for activation of NPGs. Used by themselves the reactions
are relatively slow, often requiring hours or days for completion.[7] However
addition of a protic or Lewis acid, such as triflic acid or triethylsilyl triflate
(TESOTf), greatly enhances the rate of reaction.[8] With such catalysts, the reaction
is often complete within the time it takes to sample the mixture by TLC.

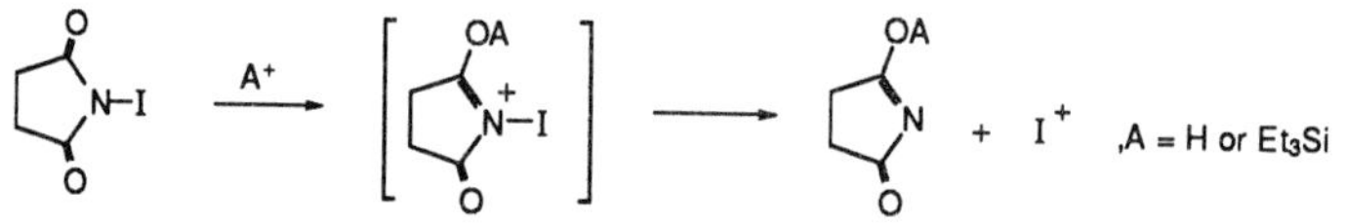

Scheme 3.

The process can be rationalized by the acid-induced heterolysis[9] shown in
Scheme 3. In most cases, only a catalytic amount of the acid is required.
Nevertheless the use of triflic acid has several disadvantages. Firstly it is critical
that the reagent must be fresh, pure and dry. Secondly it is operationally difficult to
add a precise amount of triflic acid, and because of its high potency, a slight excess
can cause severe decomposition. On the other hand, commercially available
TESOTf can be added directly by syringe, and this has been shown to work
exceedingly well.[10]

Currently NIS/TESOTf is our preferred promoter for NPG couplings. Prior to
its development, iodonium dicollidine perchlorate (IDCP) had been used in our
laboratory for couplings of NPGs.[11] IDCP is of intermediate potency, being more
reactive than NIS used by itself, but less reactive than NIS/TESOTf. IDCP is not
commercially available, but can be prepared as a stable crystalline salt by the
procedure of Lemieux and Morgan.[12,13] It has found success for the couplings of
reactive (armed) NPGs,[11] but is not potent enough to be used for less reactive
(disarmed) NPGs.

A puzzling observation is that some IDCP promoted reactions have a tendency
to stop abruptly, leaving substantial amounts of unreacted starting materials.[1] This

is probably due to an inhibiting effect of the collidine liberated during the reaction. However, the slower promoters, IDCP and *N*-halosuccinimides, made it possible for the armed/disarmed methodology of oligosaccharide synthesis to be conceived.[11] Thus it was found that oxidative hydrolysis (NBS/H$_2$O) of NPGs required minutes when the C2 protecting group was an ether, e.g. **10** (armed), but hours when it was an ester e.g. **11** (disarmed).[3]

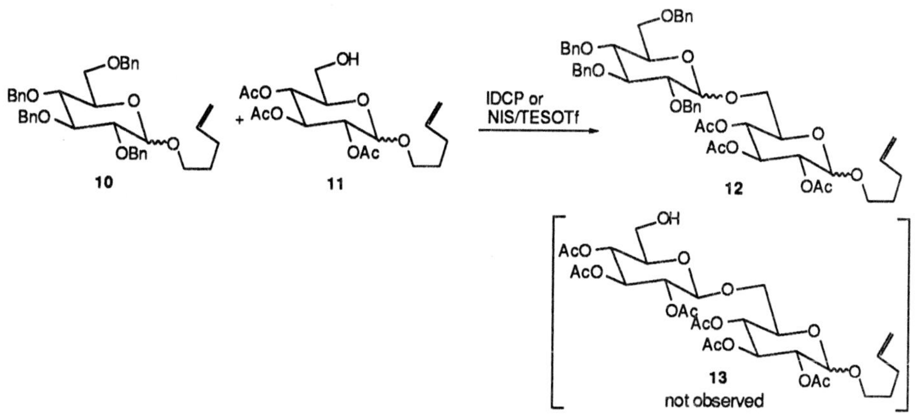

Scheme 4.

These observations are in accordance with trends that had been reported earlier for other glycosyl donors,[14] namely that acyl protecting groups depress anomeric reactivity relative to their ether counterparts. However, that these reactivity differences could be exploited synthetically for chemospecific coupling of two partners bearing the *same* anomeric activating group, was first demonstrated with NPGs.[11,15] Thus coupling of the armed donor **10** and the disarmed analog **11** afforded only the cross-coupled product **12**, with none of the self-coupled disaccharide **13** observed (Scheme 4).

It should be noted that the result in Scheme 4 is independent of the promoter used. NIS/Et$_3$SiOTf reacts with each of **10** and **11** independently within two minutes. Nevertheless coupling of **10** with **11** under the agency of NIS/TESOTf still gives **12** as the only product of coupling.[1]

The selectivity in Scheme 4 results from electronic effects arising partly, but not exclusively, from the C2 substituent. However it has also been shown that the reactivity of NPGs can be dramatically affected by torsional effects. Thus the torsional strain inflicted by cyclic acetal protecting groups has been shown to disarm NPGs in much the same way as ester protecting groups do.[16]

Since the disclosure of the armed/disarmed protocol for NPGs, this strategy has been extended to other glycosyl donors.[17]

4.   OLIGOSACCHARIDE ASSEMBLY

Several procedures are available for coupling of NPGs[1] (Scheme 5), direct coupling (option a) being the most widely used. However in special cases conversion into the glycosyl bromide followed by traditional Koenigs-Knorr coupling (option b), or use of the *n*-pentenyl 1,2-orthoester (option c) can be advantageous. In this section application of the direct coupling (option a) will be demonstrated, while options b and c will be discussed below.

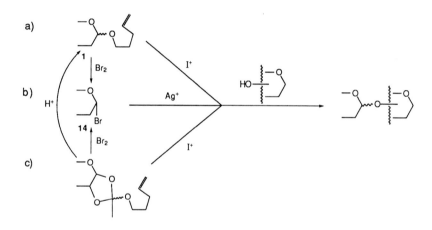

Scheme 5.

The use of the armed/disarmed coupling methodology for rapid assembly of oligosaccharides is illustrated by preparation of the trisaccharide glycan segment of the nephritogenic glycopeptide isolated from rat glomerular basement membrane[18] (Scheme 6).

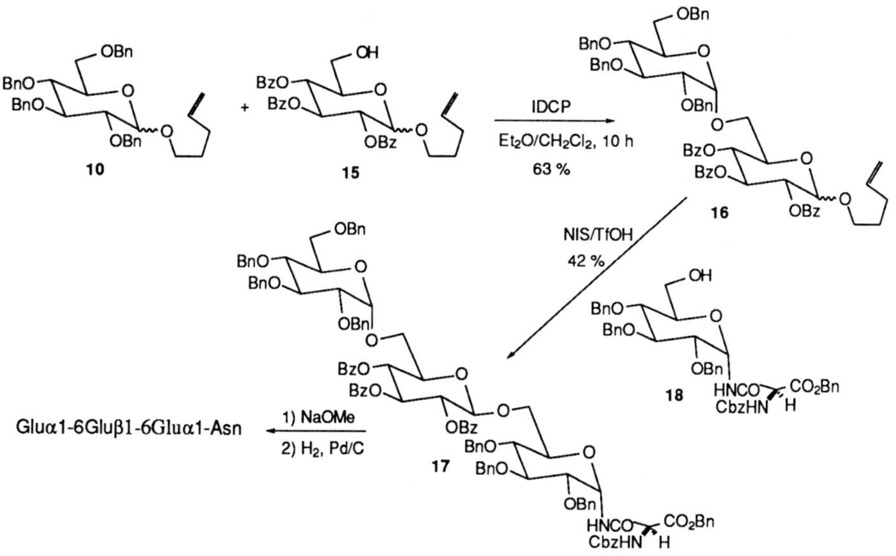

Scheme 6.

Coupling of the armed and disarmed NPGs, **10** and **15** respectively, under the agency of IDCP afforded the cross coupled compound **16** as the only product. Since **16** is a disarmed NPG, IDCP could not be used for the next coupling. However in the presence of NIS/TfOR (R=H or Et₃Si), coupling with acceptor **18** occurs within minutes to give trisaccharide **17**.[18]

This application of the armed/disarmed protocol made it possible to carry out two consecutive couplings without changing any protecting groups. In addition, all three substrates **10, 15** and **18**, were prepared from the same precursor, *n*-pentenyl glucoside **6** (Scheme 2a).

In Scheme 6, the growing oligosaccharide served as the glycosyl donor. A complementary approach for linear oligosaccharide assembly is shown in Scheme 7 where the growing oligosaccharide now functions as the glycosyl acceptor. This is illustrated by synthesis of the blood group substance B tetrasaccharide **25**.[19] Coupling of NPG **19** with acceptor **20** followed by deprotection, yields disaccharide **21** which was then coupled with NPG **22**. The resulting trisaccharide **23** was deacetylated and then coupled with NPG **24** to obtain tetrasaccharide **25**.[19] It is seen that NIS/TESOTf was used as promoter for all couplings whether the

NPG was armed or disarmed. This stepwise approach involves virtually iterative practices, couple... deprotect...couple...deprotect...couple... .

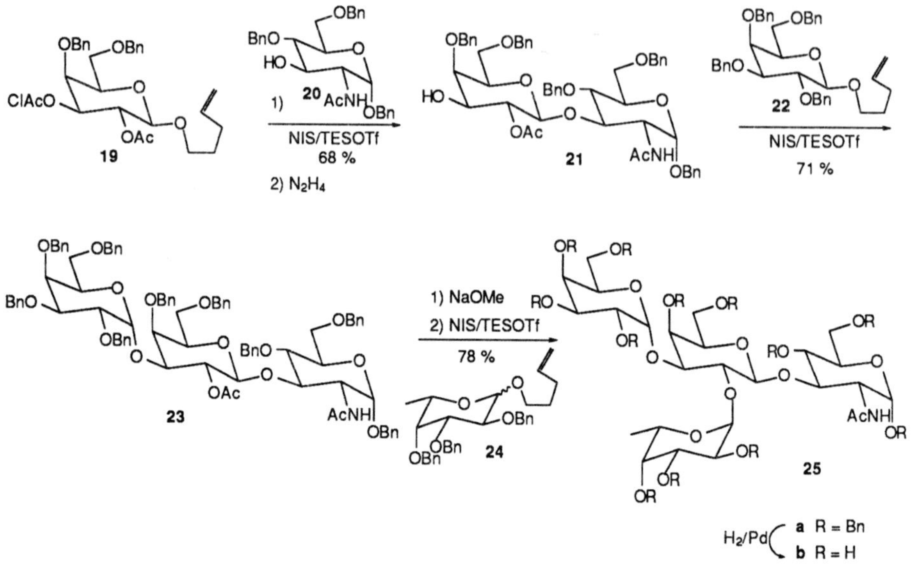

Scheme 7.

However for preparation of larger oligosaccharides, the stepwise (linear) approach is often unattractive due to the number of manipulations that have to be carried out on relatively large molecules. Instead, a block (convergent) approach is preferred, where smaller units are fabricated and then assembled to form the target oligosaccharide.

An example of this approach is the convergent synthesis of the heptasaccharide of the Thy-1 glycophosphatidyl inositol anchor[20] (Scheme 8). Thus preparation of heptasaccharide 29 from the readily available subunits 26, 27, and 28 (see Scheme 12), required only three synthetic operations, NIS/TESOTf being used for the two NPG couplings.[20]

The block synthesis approach has also been used for assembling the nonasaccharide portion of high mannose glycoproteins[21] (vide infra).

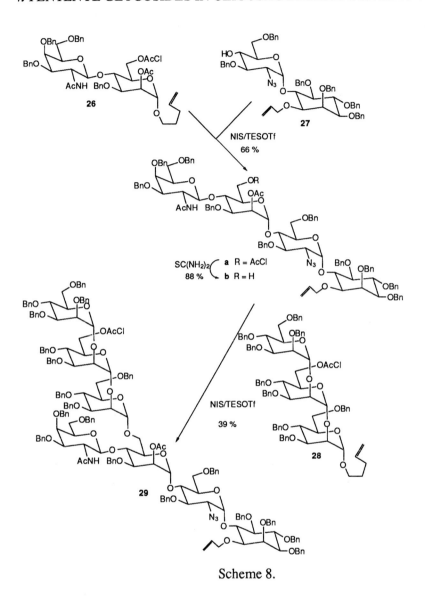

Scheme 8.

5.   BROMINATION REACTIONS OF NPGs

An attractive alternative to the armed/disarmed protocol for chemoselective coupling of NPGs involves dibromination of the double bond to give **30** whereby glycosyl donor activity is "switched off". However depending on how the bromination reaction is carried out, one can obtain either a glycosyl bromide **14**, or a vicinal dibromide **30**.[15] (Scheme 9).

    These reactions can be rationalized by focusing on the fate of the intermediate
cyclic bromonium ion **2**. In the absence of other nucleophiles, ring closure to the
furanylium ion **3** occurs. Ejection of 2-bromomethyltetrahydrofuran then gives the
oxocarbenium ion **4** which is trapped by Br[-] (the only nucleophile in the reaction
mixture) to give glycosyl bromide **14**.[22]

    However, if an excess of Br[-] is provided by addition of Et_4NBr, the
bimolecular reaction is able to dominate over the intramolecular process leading to
the dibromide **30**.[15,23] This dibromide can be considered as a latent[24] NPG from
which the active form, **1**, can be regenerated by reductive debromination.

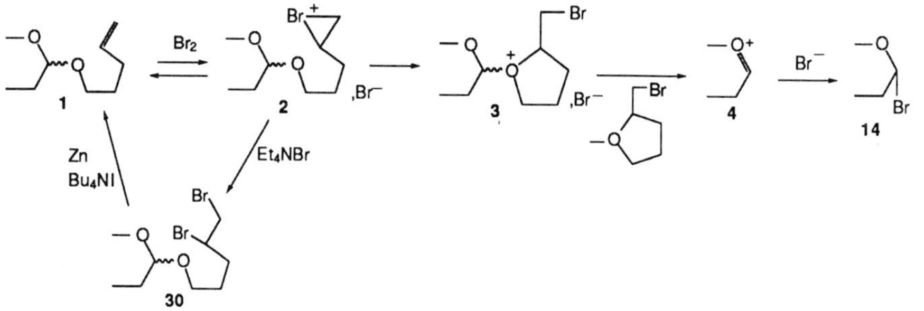

Scheme 9.

    The ability to convert NPGs into glycosyl bromides or the related dibromides
is a valuable asset for coupling of two NPGs whose reactivities are too similar for
the armed/disarmed strategy to be employed. For example the coupling of **31** and
33[25] (Scheme 10a) cannot be carried out directly because the latter does not have an
electron withdrawing substituent at C2. To circumvent this problem, the donor **31**
is titrated with bromine to obtain the galactosyl bromide **32** which is then coupled
in situ under standard Koenigs-Knorr conditions to afford disaccharide **34**.[25]

    Alternatively the acceptor NPG can be dibrominated, which masks its ability to
act as a glycosyl donor until the pentenyl moiety has been restored. This strategy is
demonstrated in Scheme 10b for preparation of the trimannan **38**, a protected form
of the lowest antenna in the nonamannose moiety of high mannose glycoprotein.[21]
Coupling of the disarmed NPG **35** to dibromo acceptor **36** under the agency of
NIS/TESOTf gives the disaccharide **37** which, after deprotection, is again coupled

with **35** to give **38a**. Reductive debromination yields **38b** as a glycosyl donor.[21]

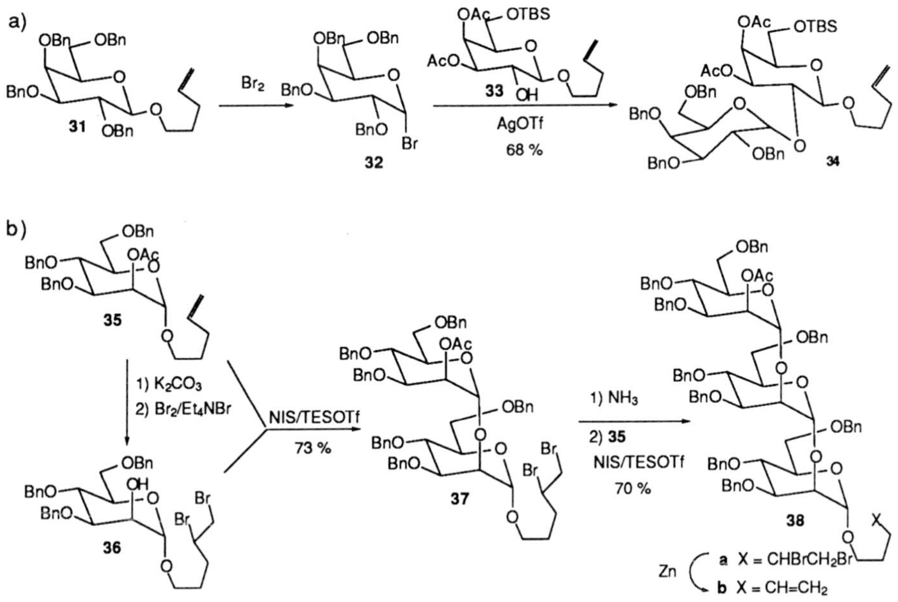

Scheme 10.

Another advantage of the dibromination option is that the number of starting materials requiring preparation are effectively reduced. Thus **31** and **33** are both prepared from the same galactoside **7** (Scheme 2c), while only the mannoside **35** is necessary for the preparation of **38**.

6. *n*-PENTENYL ORTHOESTERS

Orthoesters are readily prepared derivatives[26] (Scheme 2e) which are stable to base. In the presence of mild acids, they undergo a stereoelectronically directed rearrangement whereby the alkoxy moiety is transferred to the anomeric center e.g. **39**--->**40**[27] (Scheme 11, option a). In the case of an *n*-pentenyl orthoester, an additional mode of reaction is possible. In the presence of a halonium ion the orthoester **39** could give the furanylium ion **43** and hence the same dioxolenium ion **41**. If the electrophile is Br$_2$, the Br$^-$ released upon formation of **43** (X=Br) would react with **41** to give the glycosyl bromide **42**[22] (option b). Alternatively, if a partially protected sugar (HO-sugar) is present, coupling would lead to the oligosaccharide **44** (option c).

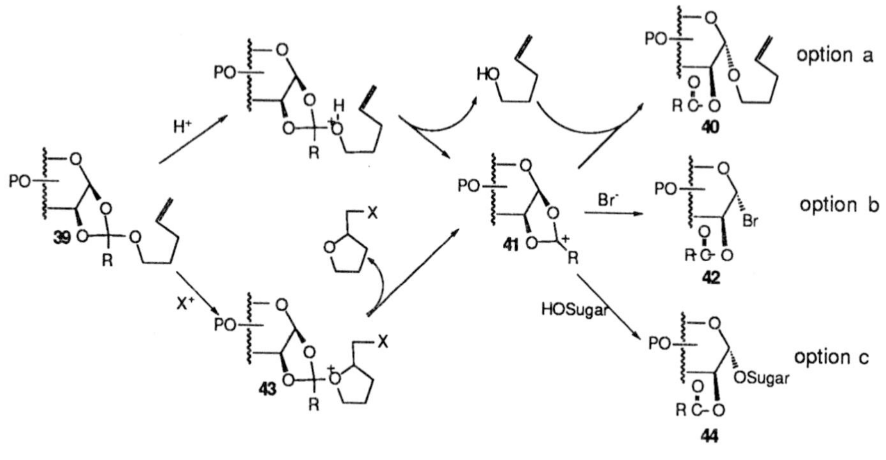

Scheme 11.

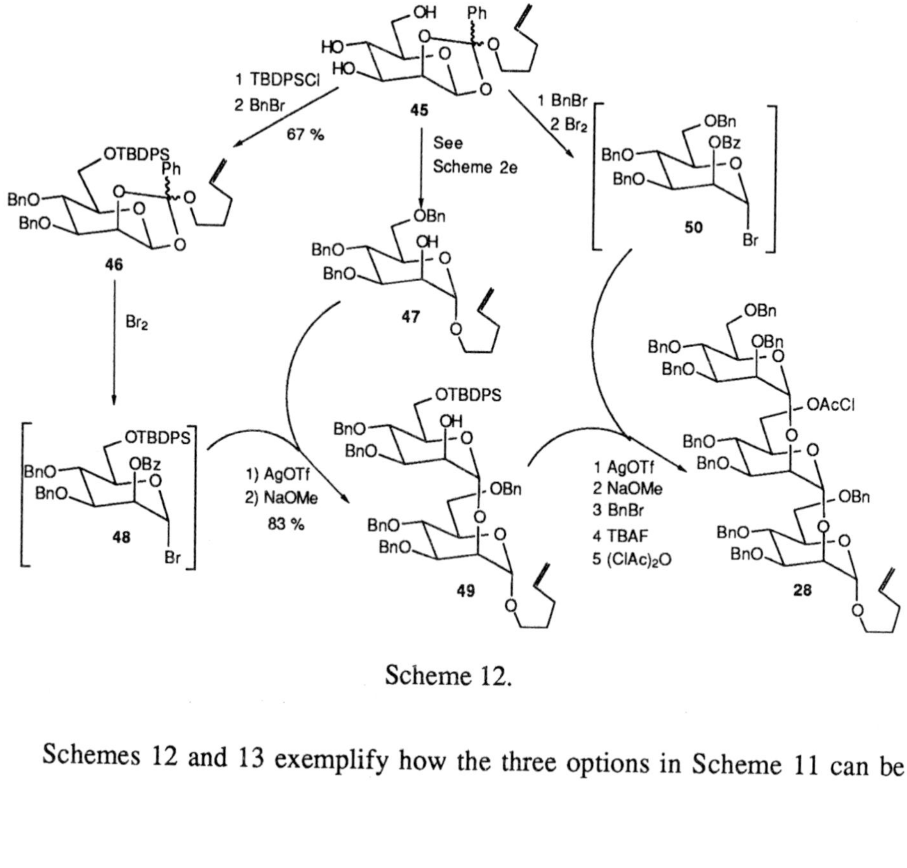

Scheme 12.

Schemes 12 and 13 exemplify how the three options in Scheme 11 can be utilized for different approaches to the trimannan **28**.[20] The orthester **45** was benzylated and then subjected to acid-catalysed rearrangement (option a) followed by debenzoylation which afforded glycosyl acceptor **47** (Scheme 12). Application of option b to the differentiated orthoester **46** gave the glycosyl bromide **48**, which was used for *in situ* coupling to acceptor **47** under standard Koenigs-Knorr conditions, and then debenzoylated to give disaccharide acceptor **49**.

Another portion of orthoester **45** now served as precursor for the bromide **50** which was coupled to acceptor **49** to furnish a trisaccharide whose protecting groups were adjusted to give **28**.

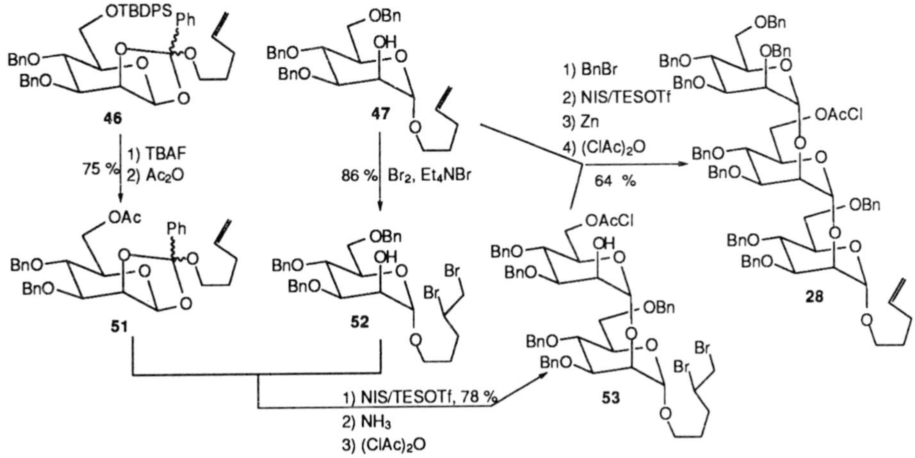

Scheme 13.

These "protecting group adjustments" had to be postponed until the end of the synthetic sequence, because glycosyl bromides **48** and **50** would not be able to tolerate the required chemical manipulations. An alternative plan designed entirely around *n*-pentenyl glycoside chemistry, which is depicted in Scheme 13, avoids this problem. The previously prepared orthoester **46** leads to the 6-O-acetyl derivative **51**. One portion of previously prepared **47** was dibrominated to give acceptor **52**, which was coupled with orthoester **51** in keeping with option c, to give a disaccharide whose protecting groups could be safely adjusted as shown in the glycosyl acceptor **53**.[28] The required donor was obtained by benzylation of **47**, and coupling to **53** gave a trisaccharide smoothly, which was subjected to reductive debromination and rechloroacetylation to afford previously prepared **28**.[28]

7.   GLYCOPEPTIDES

Mechanistic investigations on NPGs had indicated that in the presence of acetonitrile, a Ritter reaction was occurring leading to nitrilium ions[29,30] such as **54** (Scheme 14a). We have subsequently shown that this intermediate could be trapped by inclusion of carboxylic acids in the reaction medium.[30] Thus using a protected aspartic acid we could obtain *N,N*-diacyl derivatives such as **55**. Protracted experimentation revealed that piperidine in DMF allowed chemoselective cleavage of the *N*-acetyl residue, leaving the *N*-aspartoyl group intact in the product **56**.[18]

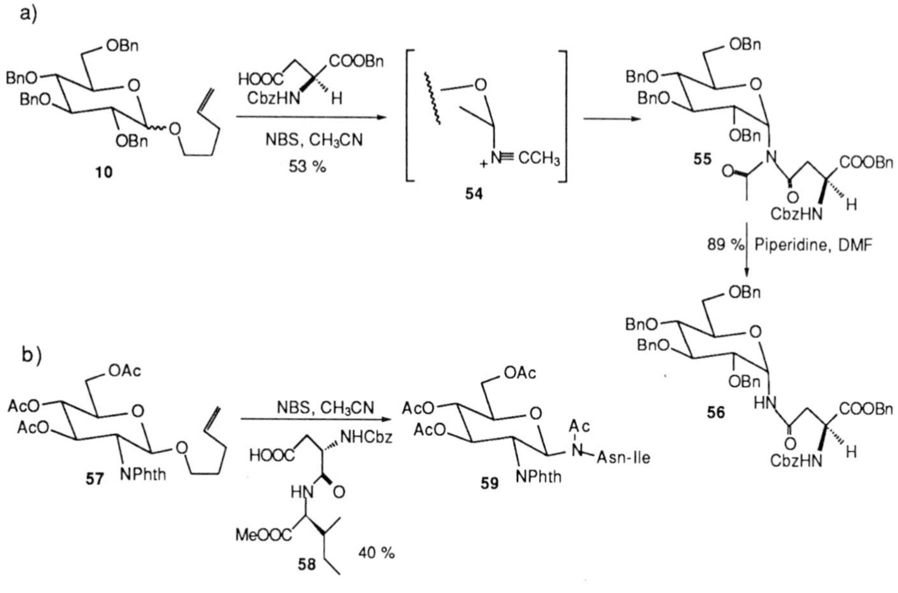

Scheme 14.

The methodology can be extended to protected glucosaminides and to higher peptides. Thus reaction of the NPG **57** with dipeptide **58** afforded the glycopeptide **59** (Scheme 14b).[31]

REFERENCES

1. B. Fraser-Reid, U.E. Udodong, Z. Wu, H. Ottosson, J.R. Merritt, C.S. Rao, C. Roberts and R. Madsen, *Synlett*, 927 (1992).

2. (a) B. Fraser-Reid, B.F. Molino, L. Magdzinski and D.R. Mootoo, *J. Org. Chem.*, **52**, 4505, (1987); (b) D.R. Mootoo, V. Date and B. Fraser-Reid, *J. Chem. Soc. Chem. Comm.*, 1462 (1987).

3. D.R. Mootoo, V. Date and B. Fraser-Reid, *J. Am. Chem. Soc.*, **110**, 2662 (1988).

4. (a) Z. Wu, D.R. Mootoo and B. Fraser-Reid, *Tetrahedron Lett.*, **29**, 6549 (1988); (b) R. Madsen and B. Fraser-Reid, *J. Chem. Soc., Chem. Comm.*, 749 (1994).

5. For preparation of *n*-pentenyl alcohol see: L.A. Brooks and H.R. Snyder, *Org. Synth. Coll.* Vol. 3, J. Wiley & Sons, New York, p. 698 (1955).

6. P. Konradsson, C. Roberts and B. Fraser-Reid, *Recl. Trav. Chim. Pays-Bas*, **110**, 23 (1991).

7. (a) B. Fraser-Reid, P. Konradsson, D.R. Mootoo and U. Udodong *J. Chem. Soc. Chem. Comm.*, 823 (1988); (b) A.J. Ratcliffe, D.R. Mootoo, C.W. Andrews and B. Fraser-Reid, *J. Am. Chem. Soc.*, **111**, 7661 (1989).

8. P. Konradsson, D.R. Mootoo, R.E. McDevitt and B. Fraser-Reid, *J. Chem. Soc. Chem. Comm.*, 270 (1990).

9. F.L. Lambert, W.D. Ellis and R.J. Parry, *J. Org. Chem.*, **30**, 304 (1965).

10. P. Konradsson, U.E. Udodong and B. Fraser-Reid, *Tetrahedron Lett.*, **31**, 4313 (1990).

11. D.R. Mootoo, P. Konradsson, U. Udodong and B. Fraser-Reid, *J. Am. Chem. Soc.*, **110**, 5583 (1988).

12. R.U. Lemieux and A.R. Morgan, *Can. J. Chem.*, **43**, 2190 (1965).

13. H. Carlsohn, *Ber. Dtsch. Chem. Ges.*, **68**, 2209 (1935).

14. H. Paulsen , *Angew. Chem. Int. Ed. Engl.*, **21**, 155 (1982).

15. B. Fraser-Reid, Z. Wu, U.E. Udodong and H. Ottosson *J. Org. Chem.*, **55**, 6068 (1990).

16. B. Fraser-Reid, Z. Wu, C.W. Andrews, E. Skowronski and J.P. Bowen *J. Am. Chem. Soc.*, **113**, 1434 (1991).

17. Glycals: (a) R.W. Friesen and S.J. Danishefsky, *J. Am. Chem. Soc.*, **111**,

6656 (1989); thioglycosides: (b) G.H. Veeneman and J.H. van Boom, *Tetrahedron Lett.*, **31**, 275 (1990); selenoglycosides: (c) H.M. Zuurmond, P.H. van der Meer, P.A.M. van der Klein, G.A. van der Marel and J.H. van Boom, *J. Carbohydr. Chem.* **12**, 1091 (1993).

18.   (a) A.J. Ratcliffe, P. Konradsson and B. Fraser-Reid, *J. Am. Chem. Soc.*, **112**, 5665 (1990); (b) A.J. Ratcliffe, P. Konradsson and B. Fraser-Reid, *Carbohydr. Res.*, **216**, 323 (1991).

19.   U.E. Udodong, C.S. Rao and B. Fraser-Reid, *Tetrahedron*, **48**, 4713 (1992)

20.   U.E. Udodong, R. Madsen, C. Roberts and B. Fraser-Reid, *J. Am. Chem. Soc.*, **115**, 7886 (1993).

21.   J.R. Merritt and B. Fraser-Reid, *J. Am. Chem. Soc.*, **114**, 8334 (1992).

22.   P. Konradsson and B. Fraser-Reid, *J. Chem. Soc., Chem. Comm.*, 1124 (1989).

23.   With some strongly disarmed NPGs, e.g. *n*-pentenyl mannosides, the dibromide is also formed by Br_2 titration in absence of Et_4NBr.

24.   The active/latent protocol has also been developed for thioglycosides: (a) R. Roy, F.O. Andersson and M. Letellier, *Tetrahedron Lett.*, **33**, 6053 (1992); anomeric sulfoxides: (b) S. Raghavan and D. Kahne *J. Am. Chem. Soc.*, **115**, 1580 (1993); and vinyl glycosides: (c) G.-J. Boons and S. Isles, *Tetrahedron Lett.*, **35**, 3593 (1994).

25.   U.E. Udodong and B. Fraser-Reid, Unpublished Results.

26.   (a) K. Kochetkov, E.M. Klimov, N.N. Malysheva and A.V. Demchenko, *Carbohydr. Res.*, **212**, 77 (1991); (b) K. Kochetkov, V.M. Zhuin, E.M. Klimov, N.N. Malysheva, Z.G. Makarova and A.Y. Ott, *Carbohydr. Res.*, **164**, 241 (1987).

27.   For a historical perspective see: "Chemistry of the O-Glycosidic Bond" A.F. Bochkov and G.E. Zaikov, Chapter 2, Pergamon Press, Oxford (1979).

28.   C. Roberts, C.L. May and B. Fraser-Reid, *Carbohydr. Lett.*, **1**, 89 (1994).

29.   A.J. Ratcliffe and B. Fraser-Reid, *J. Chem. Soc. Perkin Trans. I*, 1805 (1989).

30.   (a) J.R. Pougny and P. Sinay, *Tetrahedron Lett.*, 4073 (1976); (b) R.R. Schmidt and J. Michel, *J. Carbohydr. Chem.*, **4**, 141 (1985).

31.   A.L. Handlon and B. Fraser-Reid, *J. Am. Chem. Soc.*, **115**, 3796 (1993).

Chapter 8

COUPLING OF GLYCALS: A NEW STRATEGY FOR THE RAPID ASSEMBLY OF OLIGOSACCHARIDES

MARK T. BILODEAU* AND SAMUEL J. DANISHEFSKY*,†

*Laboratory of Bio-organic Chemistry, Sloan-Kettering Institute for Cancer Research, 1275 York Avenue, New York, NY 10021, USA and †Department of Chemistry, Columbia University, New York, NY 10027, USA

Abstract The use of glycals in the synthesis of oligosaccharides is reviewed. Efficient methods have been developed which provide α- and β-deoxyglycosides, α- and β-glycosides and 2-amino-deoxyglycosides. The use of these procedures in the construction of naturally ocurring carbohydrates is discussed. Recent results in the solid-supported synthesis of carbohydrates using glycals are also presented.

1. INTRODUCTION

It is becoming increasingly apparent that oligosaccharides are intimately involved in important biological processes. Carbohydrates have been implicated as specific recognition elements in the immune response, cell-cell adhesion and in the attachment of pathogenic species.[1] Given the significant roles these structures play in biology, the ability to assemble them in homogeneous form continues to be of substantial importance. The development of methods that allow for efficient construction of the complex architectures exhibited in polysaccharides remains a fruitful area of research.[2]

The utilization of glycals has emerged as a powerful method for the assemblage of oligosaccharides. This approach can provide several advantages over more

171

classical pyranose coupling strategies. The relative ease of differential hydroxyl group protection is a significant benefit. Furthermore, when glycals are used as acceptors, the inherent functionality enables an efficient reiterative approach to carbohydrate construction.

In coupling strategies employing fully oxygenated pyranose derivatives a particular hydroxyl (one of five) must be identified for glycosylation. Further extensions require the selective elaboration of hydroxyl groups to reveal further sites for attachment. The protecting group manipulations required for this approach can be a significant impediment. In contrast, the reiterative coupling of glycals avoids several of the most difficult problems of differential protection. Thus the activated olefinic linkage of uniformly protected D functions as the donor (Scheme 1). One of three hydroxyls of acceptor A is to be presented for glycosylation with activated D. After coupling, the olefinic linkage of DA is activated (either *in situ* or in a discrete process) to produce a DA donor for a new acceptor glycal, A'. In this way, trisaccharide DAA' is obtained. It is ready for further priming of the A' sector en route to tetrasaccharide (DAA'A") or higher oligomers. For the reiterative method described in Scheme 1 to be viable and widely applicable, glycals must function as glycosyl acceptors. Furthermore, glycal linkages in larger oligosaccharides must function as viable donors or donor precursors. For maximum applicability, it is necessary to fashion a menu of coupling methods with glycals functioning both as donors and as acceptors.

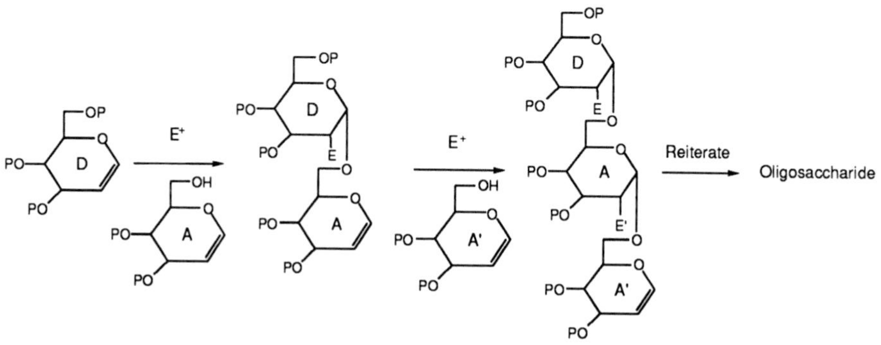

Scheme 1.

2.   SYNTHESIS OF 2-DEOXYGLYCOSIDES

The possibility of utilizing glycals as glycosyl donors in disaccharide synthesis was originally demonstrated in the pioneering research of Lemieux[3] and Thiem[4] by way of halonium mediated coupling to suitably disposed acceptors. Lemieux found that the reaction of a glycal in the presence of I_2, a silver salt and base affords 2-iodoglycosides in good yield and high α/β ratios. In subsequent studies other promoters have been employed, including: IDCP,[5,6] NBS,[7] and NIS.[4] The products so obtained have been readily converted to the corresponding 2-deoxy-α-glycosides by reductive dehalogenation. This methodology has been widely employed in the synthesis of glycoside-containing natural products.[2]

Originally, this reaction was carried out with a glycal (such as a glucal) serving as a donor (Scheme 2). Reversible attack upon the double bond of **1** by an 'I$^+$ equivalent' reagent (cf *N*-iodosuccinimide or sym-collidine iodonium perchlorate) generates the intemediate iodonium species. In the ordinary case the intermediate is attacked by an acceptor (**2**) whose reducing end is capped. The stereochemistry of glycosylation is governed by *trans*-diaxial addition and the α-linked disaccharide **3** is produced. The utility of this method for the synthesis of oligosaccharides was demonstrated in Thiem's synthesis of kijanimicin.[8] Differentially protected pyranoses were used as acceptors and a particular hydroxyl would then be exposed to allow for reaction with another glycal donor.

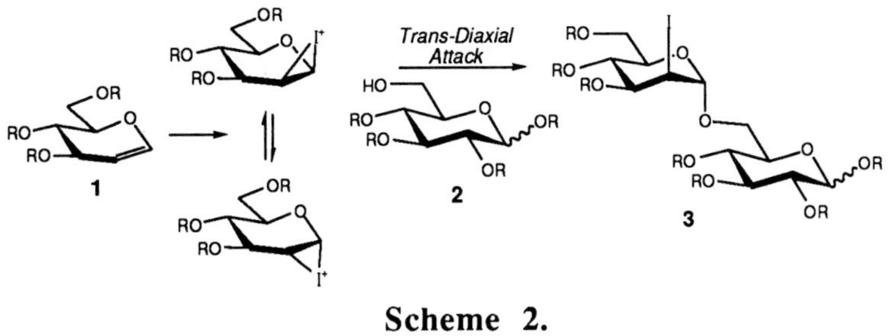

Scheme 2.

The iodoglycosylation reaction seemed to represent a particularly difficult challenge to the concurrent use of glycals as acceptors and donors. The presence of a glycal linkage in the intended acceptor itself provides the capacity to serve as the glycosyl donor. Therefore, even if one of the two glycals were ineligible as a glycosyl acceptor, significant problems (symmetrical coupling and polymerization) can be encountered in regulating the glycosyl acceptor. Addition of I$^+$ to one glycal followed by addition of the 'acceptor' glycal failed to provide selective coupling due to the reversibility of formation of the iodonium intermediate. However it was found that exploitation of electronic differences in the respective glycal double bonds, engendered by differential protection, could lead to selective formation of disaccharides.

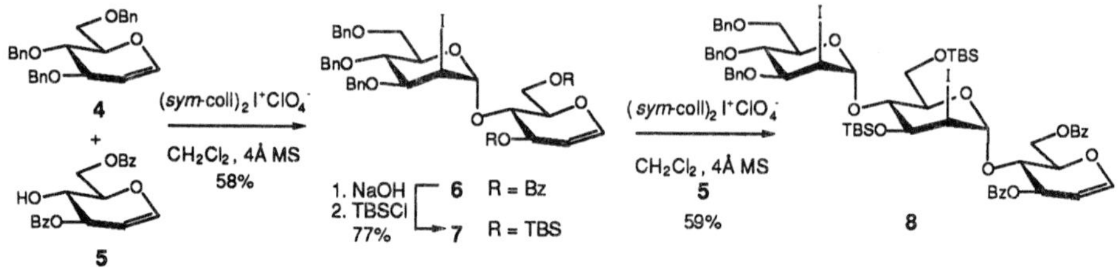

Scheme 3.

In the event, it was found that 3,4,6-tribenzyl glucal (**4**) in the presence of dibenzoate **5** and an iodonium source would give selectively the disaccharide **6** derived from activation of the more electron-rich glucal with the electron-poor glycal acting as the acceptor (Scheme 3).[6] The benzoate protecting groups could then be interchanged for silyl protecting groups thus preparing the disaccharide glycal as a suitable donor for a reiterative coupling with 3,6-dibenzoate glucal (**5**) to provide the trisaccharide **8**. This methodology has culminated in a synthesis of ciclamycin 0 trisaccharide and the ultimate attachment of the anthraquinone portion of the natural product.[6b]

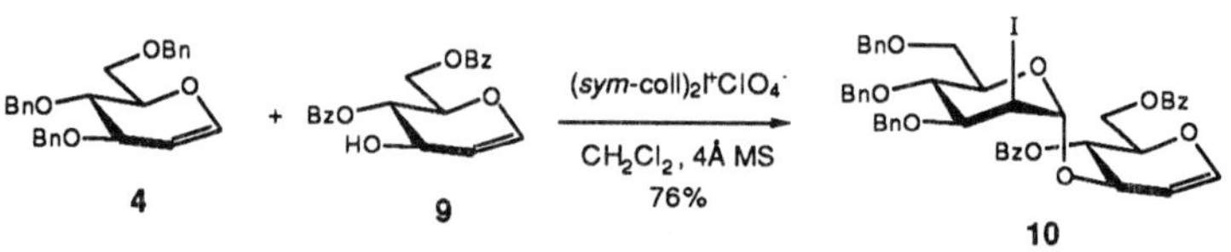

Scheme 4.

The issues involved in regulating the iodonium mediated coupling of two glycals *via* selection of resident substituents have not yet been elucidated. The most obvious position where the electronic difference between an acyloxy and an alkoxy group are apt to be decisive, would be at the C-3 hydroxyl allylic to the double bond. However, the clean and successful coupling of **4** and **9** (Scheme 4) indicates that even without an acyloxy group at C-3, subtle effects can be exploited to bring about an orderly progression leading to disaccharide **10**.

Other activation methods have been employed in the synthesis of 2-deoxy-α-glycosides. Sinaÿ employed PhSeCl as an activating agent to afford 2-seleno-2-deoxy-α-glycosides and showed that the selenium could be reductively removed.[9] Various acid promoters have been utilized to afford 2-deoxy-α-glycosides in moderate stereoselectivity directly from glycals. Reagents employed in this connection include: CSA,[10] TsOH,[11] triphenylphosphine hydrobromide[12] and AG50 WX2-resin.[13]

There has also been some success in the formation of 2-deoxy-β-glycosides from glycals. Ogawa reacted glycals with phenyl sulfenate ester in the presence of TMSOTf to produce 2-deoxy-2-(phenylthio)-β-glycosides in low stereo-selectivity.[14] Franck generated 2-deoxy-2-(phenylthio)-β-glycosides in moderate steroeselectivity by the electrophilic activation of glycal by a phenylbis(phenylthio)sulfonium salt and trapping with stannyl ethers (Scheme 5).[15] Thus treatment of glucal **4** with the sulfonium salt in the presence of acceptor **11** provided the disaccharide **12**. Phenylthioglycoside **12** has been converted into the 2-deoxy-β-glycoside **13** by treatment with Raney-Ni.

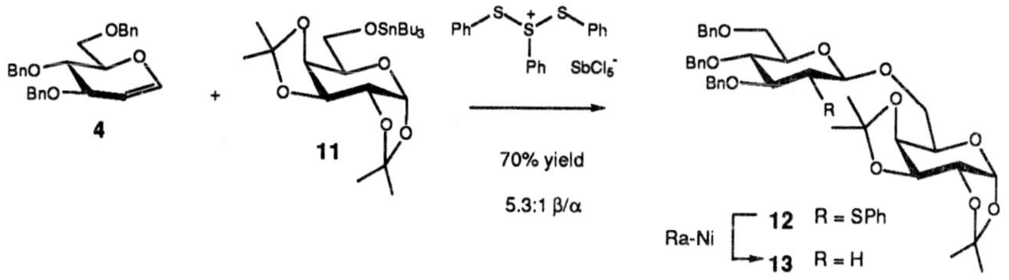

Scheme 5.

As discussed later, we have developed deoxygenations of β-glycosides (obtained *via* 1,2-anhydrosugars) to provide 2-deoxy-β-glycosides.

3.   SYNTHESIS OF 2-AMINO-2-DEOXYGLYCOSIDES

For the glycal method of assembly to be broadly useful, a number of activation methods are needed which impart varied functionality to the eventual products. Several approaches to amination and subsequent glycosylation of glycals have been explored.[2b] Of course, given the iodoglycosidation chemistry discussed above, the most obvious approach to introduce an equatorially disposed nitrogen at C-2 would be by displacement of the axial iodine. However, serious difficulties are encountered in attempted S_N2 displacements of axially disposed leaving groups in a 1,2-anti relationship to an axial glycoside bond. Such displacements tend to be low yielding and are plagued by competing elimination reactions.

Important advances in the use of glycals for the introduction of 2-amino functionality have been divulged by Lemieux. Thus, nitrosochlorination[16] or, more powerfully, azidonitration[17] served to convert glycals into eventual 2-aminopyranosyl donor equivalents. While the nitrosochlorination reaction was a major advance, the methods which were developed to convert the oximino products to desired goal structures were not fully satisfactory with regards to yield and stereoselectivity. Greater progress was achieved via azidonitration (Scheme 6). Treatment of a galactal 14 with ceric ammonium nitrate (CAN) and sodium azide in acetonitrile leads to 2-azido-2-deoxyglycosyl nitrate 15 which can then be manipulated further. For example, Schmidt has hydrolyzed the nitrates and converted them to trichloroacetimidates (16)[18] and Sinaÿ has performed direct displacements of a xanthate salt on the nitrates to afford the corresponding anomeric xanthate 17.[19] These procedures allow the formation of 2-azido-2-deoxy glycosyl donors in a couple of steps from glycals, and their subsequent use in α- or β-glycosylation reactions. The major limitation to these methods is the fact that the azidonitration of glucals is unselective at the azide-bearing stereocenter, although Sinaÿ has reported selective azidonitration of a benzylidene rigidified glucal.[20]

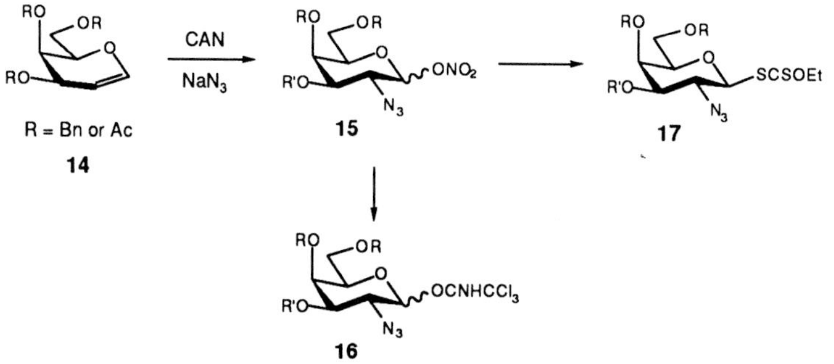

Scheme 6.

the disaccharide **41** in 42% yield.[38]   In some applications where this standard protocol is not effective for the glycosylation it has been found that the process can be achieved by reaction of the epoxide with the tributylstannyl ether derivatives of the acceptor promoted by Zn(OTf)$_2$ (*vide infra*).

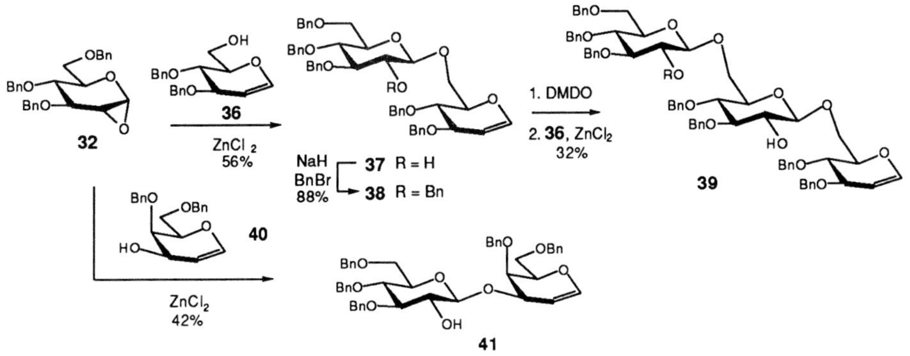

Scheme 12.

Subsequent to the experiments summarized here, it was found that the particular oxirane **32** is among the poorest of the 1,2-anhydrosugar donors.   With some acceptors,[39] modest amounts of α-glycosides have been observed.   We have recently found that in some of these cases the use of 4,6-benzylidene glucals as donors provides excellent stereoselectivity for β-glycoside formation.   Nonetheless, the method has already charted an efficient pathway from glycal derivatives to β-glycosides.

An illustrative example of the utility of this chemistry in the construction of

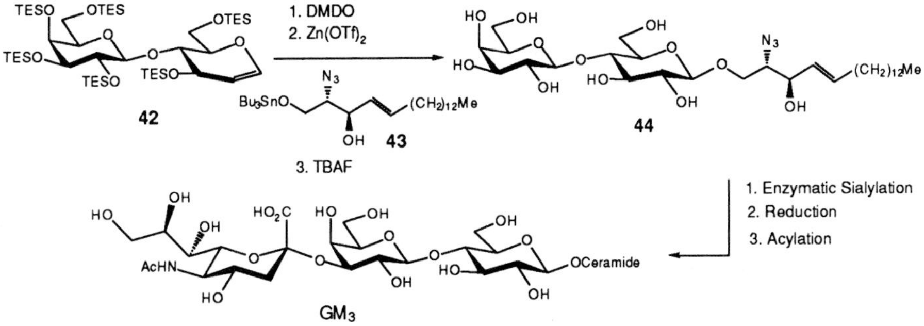

Scheme 13.

oligosaccharides is shown in a recent synthesis of GM₃ (Scheme 13).[40] A combination of chemical and biological means was used to achieve a very straightforward synthesis. Epoxidation of per-triethylsilylated lactal (**42**) followed by the Zn(OTf)₂ promoted reaction of the the stannylated diol **43** and then desilylation provided stereoselectively the adduct **44**. Enzymatic sialylation followed by azide reduction and acylation afforded the target ganglioside.

Thus far we have focused on the direct use of glycal epoxides in the synthesis of β-glycosides. It is also possible to convert such epoxides to other glycosylating agents (Scheme 14).[41] For instance, compound **32** could be converted to the thiophenyl glycoside **45**, the pentenyl glycoside **46** and the fluoroglycoside **47**. Benzylation of the C-2 hydroxyl of **47** gives rise to the fluoro-sugar **48** which can serve as a glycosyl donor in a conventional Mukaiyama reaction to produce α-glycosides (**49**). This route from glycals to α-glycosides found application in our recently completed synthesis of a cyanobacterial sulfolipid.[42]

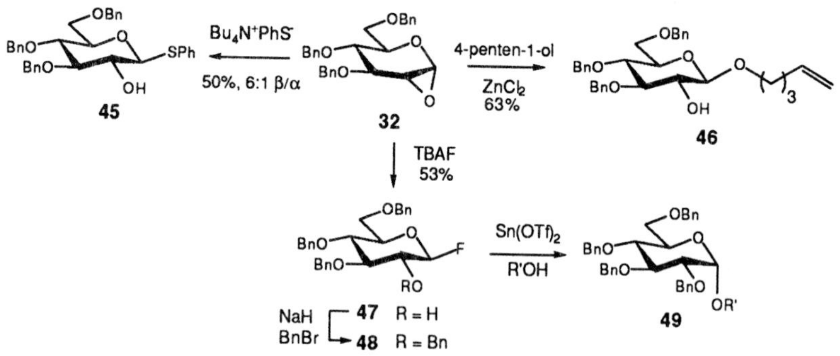

Scheme 14.

In a recent report we have shown that, in certain cases, an α-glucoside can be obtained directly from a glucal epoxide (Scheme 15).[43] Reaction of the epoxide **32** with stannylether **50**, promoted by AgBF₄, followed by acetylation affords the α-glycoside **51**. The process can be readily reiterated to provide the trisaccharide **52**. This method is presently limited to primary hydroxyl acceptors.

A potentially significant advantage of the epoxide coupling reaction is the fact that the reaction affords a unique, unprotected 2'-hydroxyl, which can be exploited in a variety of ways. Deoxygenation sequences have been developed to provide 2-deoxy-β-glycosides (Scheme 16).[38] Conversion of the hydroxyl group of **41** to

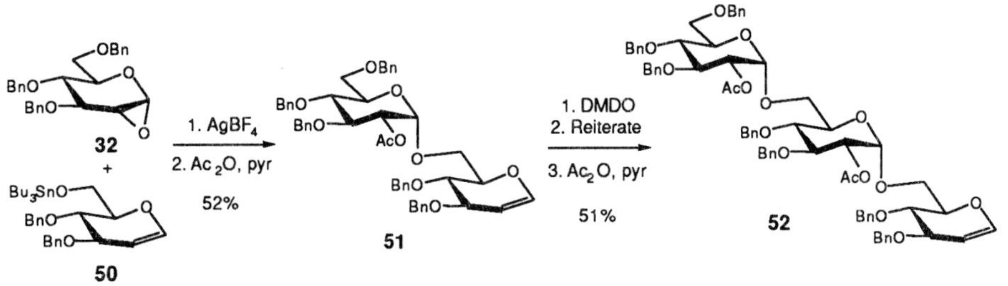

Scheme 15.

the corresponding thiocarbonate and treatment with Ph$_3$SnH-AIBN provided the 2-deoxy-β-glycoside **53**. The unique 2'-hydroxyl has also been subjected to oxidation and stereoselective reduction to provide the corresponding β-mannoside (Scheme 17).[44] Thus, oxidation of **37** with acetic anhydride and DMSO followed by reduction of the unpurified product with NaBH$_4$ and acetylation provided selectively the β-mannoside **54** in 89% yield.

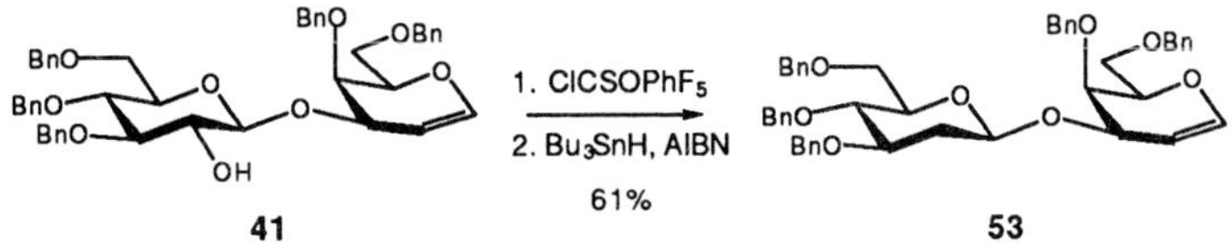

Scheme 16.

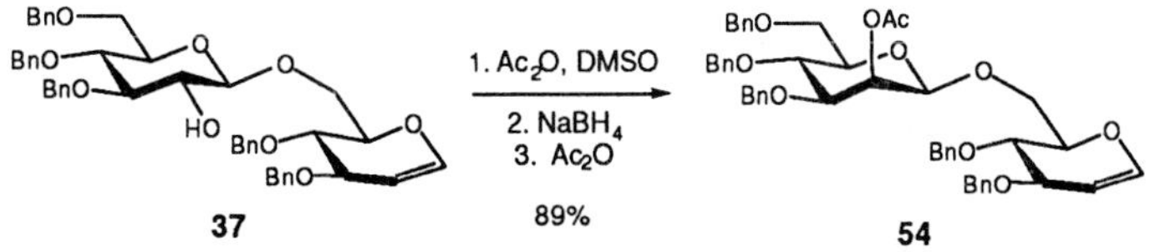

Scheme 17.

Moreover, the uniquely distinguished hydroxyl group can be utilized in the synthesis of branched sugars, as demonstrated in a recently completed synthesis of a complex branched saponin (Scheme 18).[45] Reaction of the galactal epoxide **55** with the steroid tigogenin provided the adduct **56**. Protecting group manipulation afforded the product **57** with the axial hydroxyl exposed. Epoxide **59**, generated from the corresponding glycal in 4:1 selectivity, was reacted with 4,6-benzylidene glucal **60** to afford the dissacharide **61**. Benzylation and epoxidation provided **62**, and reaction with the stannyl ether (**58**) derived from **57** afforded the steroid trisaccharide **63** selectively. Conveniently, the newly formed hydroxyl group at C2 of the erstwhile donor unit is now disposed for glycosylation. This hindered hydroxyl was unreactive towards 1,2-anhydrosugars, however the fluoro sugar **64**, generated *via* an epoxide opening with fluoride (see Scheme 14) followed by benzoylation, reacted cleanly with the acceptor **63** with Sn(OTf)$_2$ promotion to afford the tetrasaccharide **65**, which was then deprotected to afford the final product **66**.

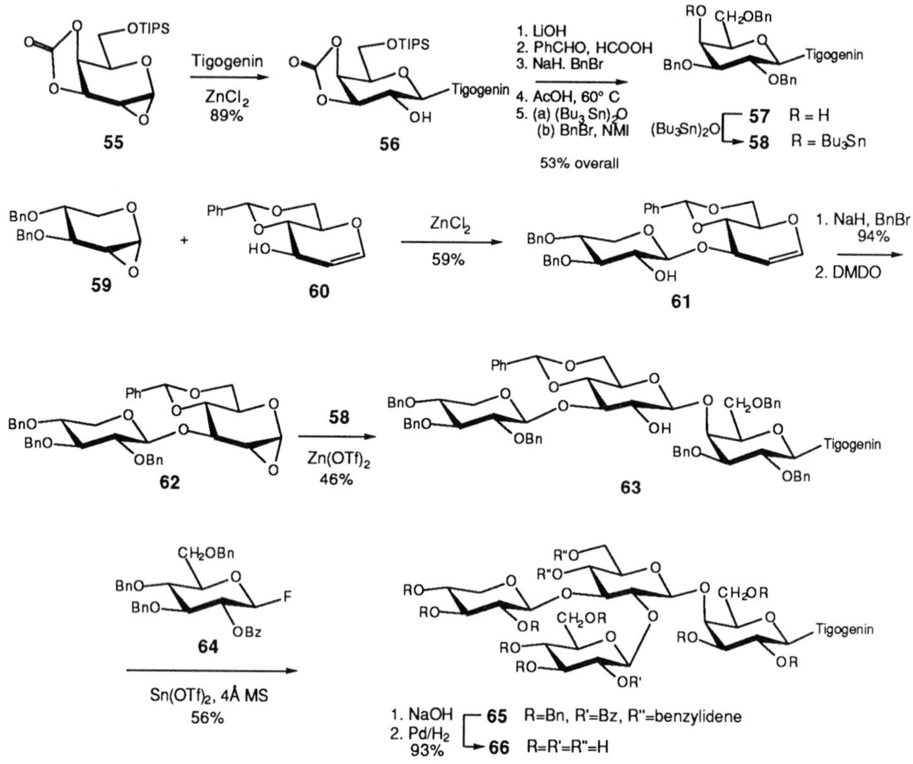

Scheme 18.

Glycal chemistry also played a significant role in a recently completed synthesis of the Lewis^y blood-group determinant (Scheme 19).[46] The primary positions of lactal (**67**) were selectively silylated and the 3',4'-positions of the galactose unit were engaged as a carbonate (see compound **68**). The remaining free hydroxyl groups were α-fucosylated with fluoro-sugar **69** using established conditions to afford the tetrasaccharide **70**. Treatment with (*sym*-coll)$_2$I$^+$ClO$_4^-$ and PhSO$_2$NH$_2$

afforded the sulfonamide **71**. Reaction of iodosulfonamide **71** with the stannyl ether **72** with the agency of AgBF$_4$ provided the pentasaccharide **73**. Deprotection and peracetylation followed by epoxidation and reaction with allyl alcohol provided the allyl glycoside, which was deacetylated to give the Lewis^y determinant **74**. The allyl unit is now poised for the formation of protein conjugates. The biological properties of such systems is a current area of research.

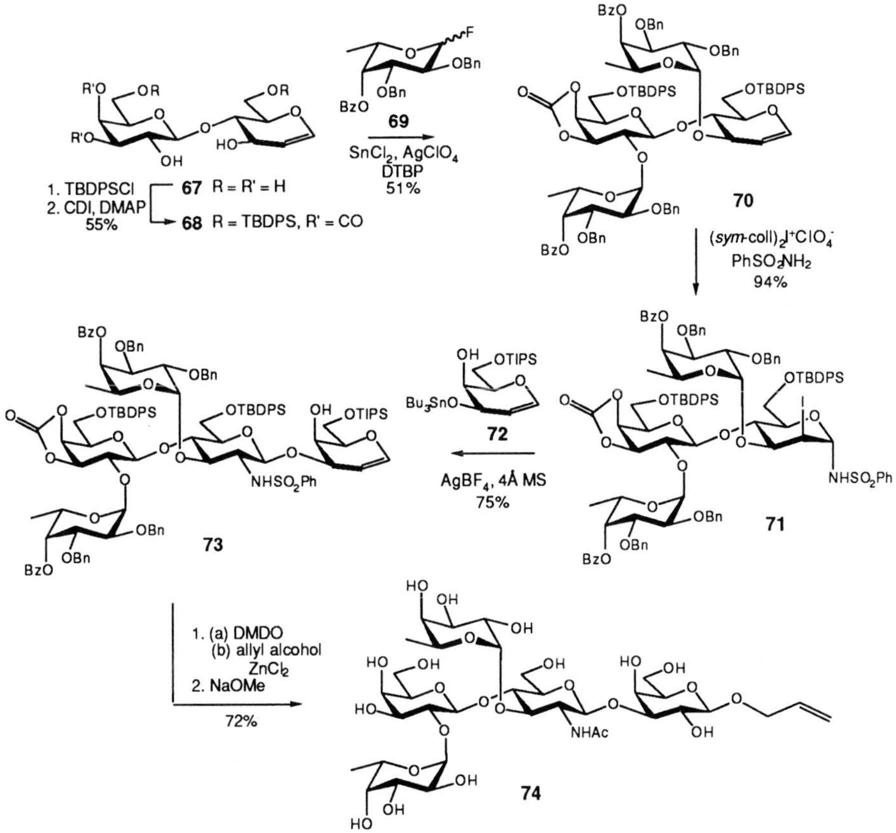

Scheme 19.

5.   SYNTHESIS ON A SOLID SUPPORT

The synthesis of oligosaccharide sequences on a solid support is a burgeoning area of research.[47] However, these efforts have yet to provide a generally applicable approach which may be developed into an automated procedure, such as those which have revolutionized the chemical synthesis of oligopeptides and oligonucleotides. We have recently begun to study an approach to solid-phase synthesis of oligosaccharides and glycoconjugates using the glycal assembly method.[48]

Our method involves attachment of a glycal to a solid support through a silyl ether appendage to a glycal. Employing the menu of glycal activation methods discussed above could lead to efficient syntheses of oligosaccharides on a solid support. The polymer support which we have employed in our studies is polystyrene crosslinked with 1% divinylbenzene. This resin was functionalized as shown in Scheme 20, using the procedure of Chan and Huang.[49] Polystyrene was lithiated using a solution of butyllithium in cyclohexane, and then quenched with a dialkyldichlorosilane. Two different silylated polymers have been prepared. Polymer **75** (R = phenyl) and **76** (R = isopropyl) were each reacted with **77**, derived from D-galactal, in the presence of Hunig's base to give support-bound glycals **78** (0.5–0.6 mmol **77**/gram **78**), and **79** (0.6–0.9 mmol **77**/gram **79**), respectively.

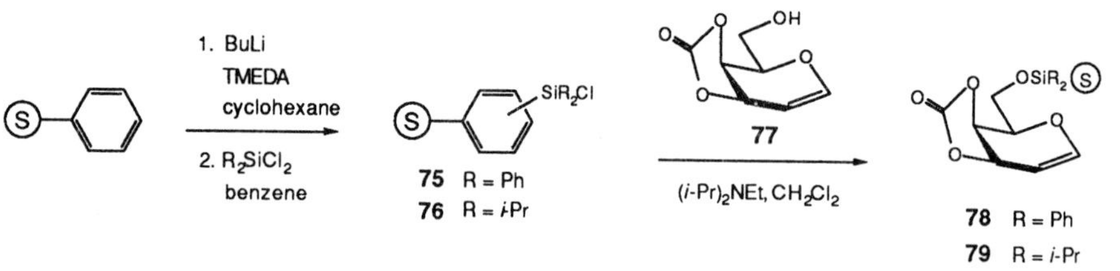

Scheme 20.

An application of this method to the synthesis of a tetrasaccharide is shown in Scheme 21. Polymer-bound galactal **78** was converted to the 1,2-anhydrosugar **80** by stirring in a solution of 3,3-dimethyldioxirane. Compound **80** acted as a glycosyl donor when treated with a solution of **77** in the presence of zinc chloride, thereby providing **81**. Treatment of **81** with tetrabutylammonium fluoride (TBAF) in the presence of acetic acid (to maintain the cyclic carbonate protecting group) afforded **82** in 77% yield from **78**. Compound **81** was resubjected to the two-step glycosidation procedure to give **83**, which was then epoxidized and reacted with glucal derivative **36**, to give **84**. Tetrasaccharide glycal **85** was retrieved from the support by the action of TBAF, and was found to have been produced in a 32% overall yield from **78**.

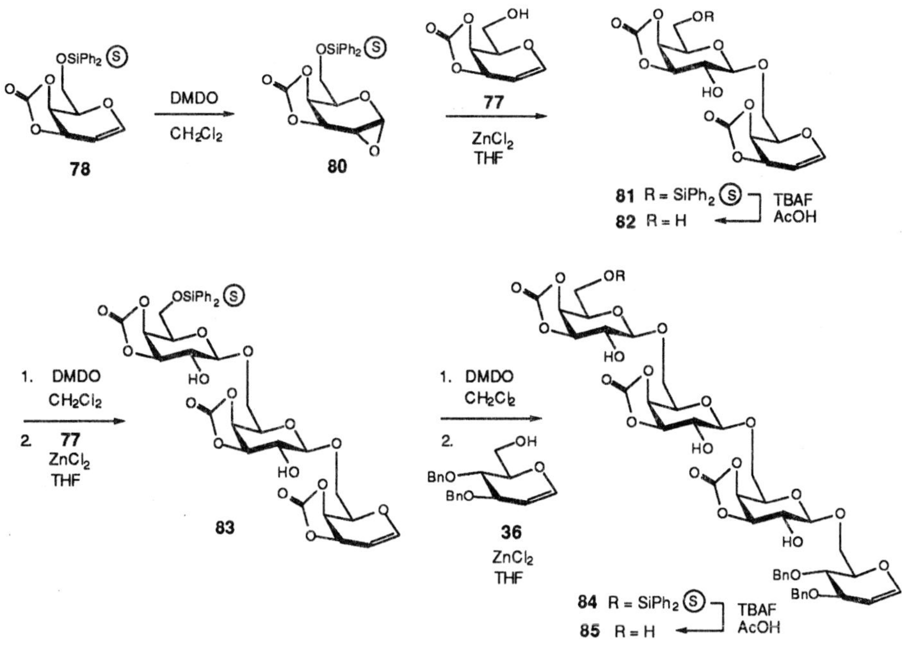

Scheme 21.

Further experimentation has demonstrated versatility in the method. Thus, secondary alcohol glycosyl acceptors, as well as disaccharide acceptors, are readily accommodated. In addition, polymer-bound glucal derivatives can also function as glycosyl donors following epoxidation with dimethyldioxirane. These findings have led to the assembly of a variety of oligosaccharides, including a hexasaccharide (Scheme 22).

We have also been investigating an extension of our solid-support method to include synthetic approaches to branched oligosaccharides (Scheme 23). Thus, polymer-bound glycosyl donor **80** was converted, upon reaction with solution-based acceptor **86**, into polymer-bound glycosyl acceptor **87**. Glycosylation of **87** with solution-based glycosyl donor **64** provided trisaccharide **88**, which was retrieved from the support by the action of TBAF to give **89** in a 52% overall yield from **78**. This method provides a straightforward solution to the problem of assembling oligosaccharide patterns which contain branching at C-2.

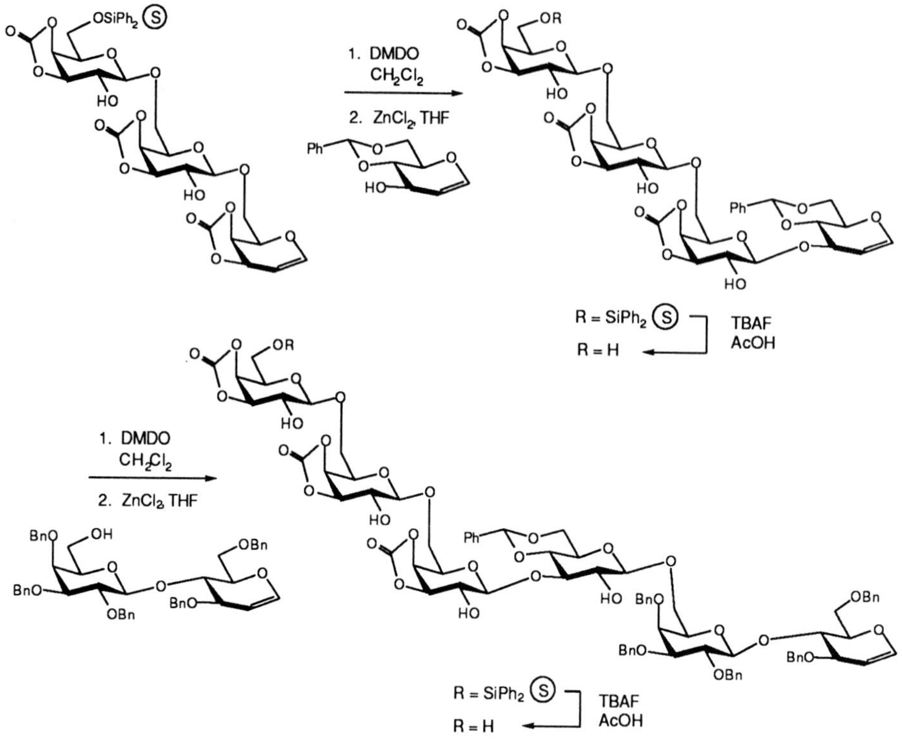

Scheme 22.

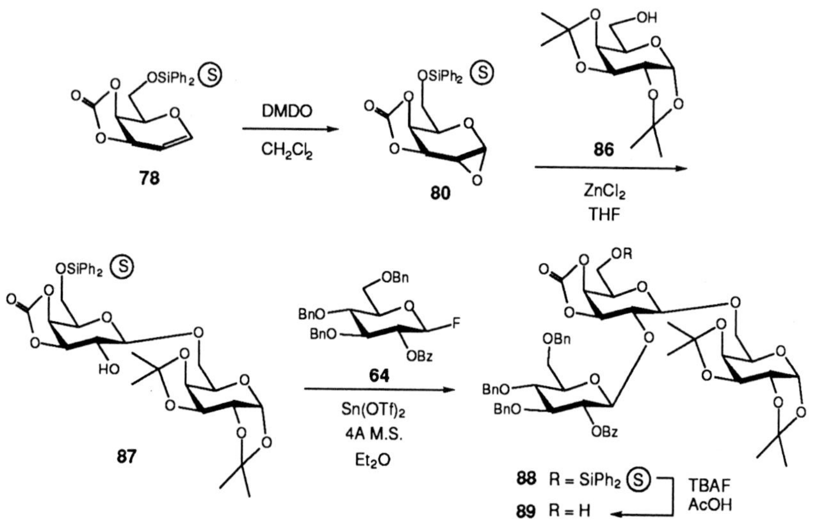

Scheme 23.

Our most recent efforts have involved an application of the method to the synthesis of carbohydrate domains having blood group determining specificities.[50] A representative example is the synthesis of neoglycoproteins having Lewis[b] specificity using a combination of solid- and solution-phase techniques. Our approach to Le[b] first addressed the problem of the core tetrasaccharide, which was assembled on the polymer support as shown in Scheme 24. Polymer-bound galactal **79** was epoxidized with dimethyldioxirane and the resultant epoxide was reacted with a solution of glucal derivative **90** to give polymer-bound disaccharide diol **91**. This reaction proceeded in high regioselectivity for glycosylation at C-3 of the acceptor. Bisfucosylation of **91** using donor **69** provided polymer-bound tetrasaccharide glycal **92**. Treatment of **92** with TBAF gave **93**, which was obtained in a 40% overall yield from **79**.

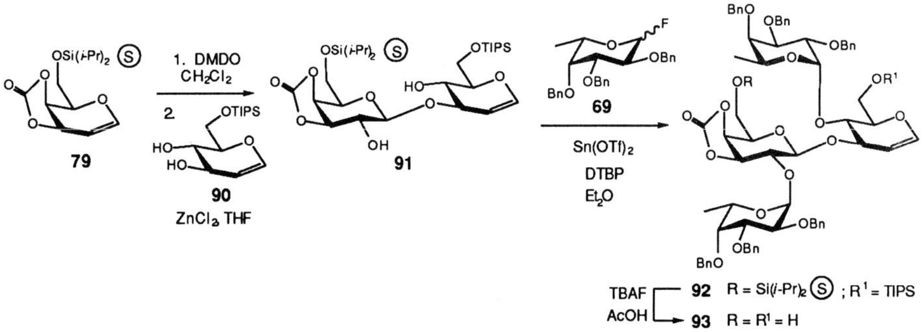

Scheme 24.

At this point the azaglycosylation methodology is not compatible with the solid-support so the remaining transformations were performed in solution. Glycal **93** was further converted into a hexasaccharide of the Le[b] system as shown in Scheme 25. Glycal **93** was silylated with triisopropylsilyl chloride to give **94**, which was converted to benzenesulfonamide **95**. Compound **95** was coupled with lactal derivative **96** in the presence of silver tetrafluoroborate to provide hexasaccharide glycal **97**. Deprotection of **97**, followed by peracetylation, gave **98**, which was epoxidized using dimethyldioxirane. Stirring of this epoxide in allyl alcohol, followed by treatment with a solution of sodium methoxide, provided allyl glycoside **99**, which was converted to aldehyde **100** by ozonolysis. Compound **100** was later conjugated with human serum albumin by the action of sodium cyanoborohydride to provide the desired neoglycoprotein.[51] The biological properties of this glycoconjugate are currently being investigated.

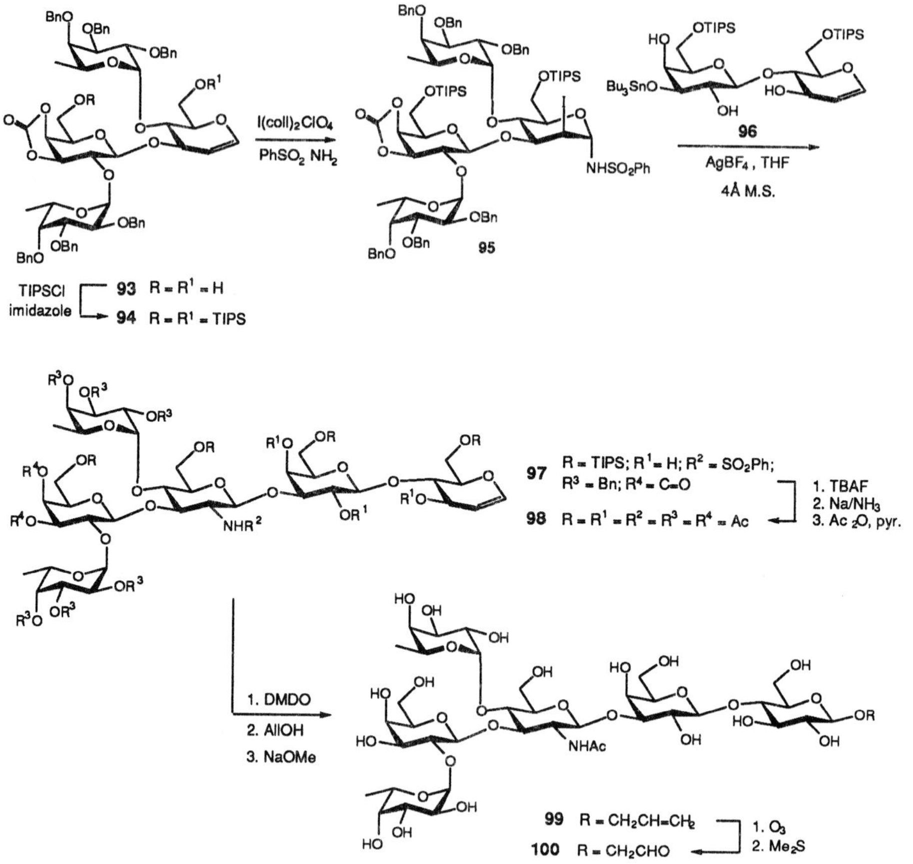

Scheme 25.

6. CONCLUSION

Glycals are proving to be of great utility in the construction of oligosaccharides. A variety of activation methods have been developed to provide different polysaccharide structural types. For example 2-deoxyglycosides, 2-amino-2-deoxyglycosides and simple glycosides can be efficiently assembled, although obtaining certain linkages in a straightforward manner still remains a challenge. The efficiency of a reiterative approach to carbohydrate construction employing glycals as acceptors has been demonstrated and this process is showing promise for the solid-supported synthesis of oligosaccharides.

REFERENCES

1. (a) M. J. Polley, M. L. Phillips, E. Wagner, E. Nudelman, A. K. Singhal, S. Hakomori and J. C. Paulson, *Proc. Natl. Acad. Sci. USA*, **88**, 6224 (1991); (b) M. L. Phillips, E. Nudelman, F. C. A. Gaeta, M. Perez, A. K. Singhal, S. Hakomori and J. C. Paulson, *Science*, **250**, 1130 (1990); (c) Y. Hirabayashi, A. Hyogo, T. Nakao, K. Tsuchiya, Y. Suzuki, M. Matsumoto, K. Kon and S. Ando, *J. Biol. Chem.*, **265**, 8144 (1990); (d) U. Spohr and R. U. Lemieux, *Carboydr. Res.*, **174**, 211 (1988).

2. For recent reviews of available methods see: (a) K. Toshima and K. Tatsuta, *Chem. Rev.*, **93**, 1503 (1993); (b) J. Banoub, P. Boullanger and D. Lafont, *Chem. Rev.*, **92**, 1167 (1992).

3. R. U. Lemieux and S. Levine, *Can. J. Chem.*, **42**, 1473 (1964).

4. (a) J. Thiem, H. Karl and J. Schwentner, *Synthesis*, 696 (1978); J. Thiem and H. Karl, *Tetrahedron Lett.*, **19**, 4999 (1978); (b) J. Thiem and W. Klaffke, *J. Org. Chem.* **54**, 2006 (1989).

5. R. U. Lemieux and A. R. Morgan, *Can. J. Chem.*, **43**, 2190 (1965).

6. (a) R. W. Friesen and S. J. Danishefsky, *J. Am. Chem. Soc.*, **111**, 6656 (1989); (b) K. Suzuki, G. A. Sulikowski, R. W. Friesen and S. J. Danishefsky, *J. Am. Chem. Soc.*, **112**, 8895 (1990); (c) R. W. Friesen and S. J. Danishefsky, *Tetrahedron*, **46**, 103 (1990).

7. K. Tatsuta, K. Fujimoto, M. Kinoshita and S. Umezawa, *Carbohydr. Res.*, **54**, 85 (1977).

8. J. Thiem and S. Köpper, *Tetrahedron*, **46**, 113 (1990).

9. G. Jaurand, J.-M. Beau and P. Sinaÿ, *J. Chem. Soc., Chem. Commun.*, 572 (1981).

10. T. Wakamatsu, H. Nakamura, E. Naka and Y. Ban, *Tetrahedron Lett.*, **27**, 3895 (1986).

11. C. J. Tu and D. Lednicer, *J. Org. Chem.*, **52**, 5624 (1987).

12. V. Bolitt, C. Mioskowski, S.-G. Lee and J. R. Falck, *J. Org. Chem.*, **55**, 5812 (1990).

13. S. Sabesan and S. Neira, *J. Org. Chem.*, **56**, 5468 (1991).

14. Y. Ito and T. Ogawa, *Tetrahedron Lett.*, **28**, 2723 (1987).

15. G. Grewal, N. Kaila and R. W. Franck, *J. Org. Chem.*, **57**, 2084 (1992).

16. R. U. Lemieux and T. L. Nagabhushan, *Can. J. Chem.*, **46**, 401 (1968).

17. R. U. Lemieux and R. M. Ratcliffe, *Can. J. Chem.*, **57**, 1244 (1979).

18. G. Grundler and R. R. Schmidt, *Liebigs Ann. Chem.*, 1826 (1984).

19. A. Marra, F. Gauffeny and P. Sinaÿ, *Tetrahedron*, **47**, 5149 (1991).

20. P. Sinaÿ, *Pure Appl. Chem.*, **63**, 519 (1991).

21. S. Czernecki and D. Randriamandimby, *Tetrahedron Lett.*, **34**, 7915 (1993).

22. (a) B. J. Fitzsimmons, Y. Leblanc, N. Chan and J. Rokach, *J. Am. Chem Soc.*, **110**, 5229 (1988); (b) Y. Leblanc and B. J. Fitzsimmons, *Tetrahedron Lett.*, **30**, 2889 (1989).

23. E. Kozlowska-Gramsz and G. Descotes, *Can. J. Chem.*, **60**, 558 (1982).

24. H. Driguez, J.-P. Vermes and J. Lessard, *Can. J. Chem.*, **56**, 120 (1978).

25. F. W. Lichtenthaler, E. Kaji and S. Weprek, *J. Org. Chem.*, **50**, 3505 (1985).

26. D. Lafont and G. Descotes, *Carbohydr. Res.*, **175**, 35 (1988).

27. D. A. Griffith and S. J. Danishefsky, *J. Am. Chem. Soc.*, **112**, 5811 (1990).

28. D. A. Griffith and S. J. Danishefsky, *J. Am. Chem. Soc.*, **113**, 5863 (1991).

29. S. J. Danishefsky, J. Gervay, J. M. Peterson, F. E. McDonald, K. Koseki, T. Oriyama and D. A. Griffith, *J. Am. Chem. Soc.*, **114**, 8329 (1992).

30. S. J. Danishefsky, K. Koseki, D. A. Griffith, J. Gervay, J. M. Peterson, F. E. McDonald and T. Oriyama, *J. Am. Chem. Soc.*, **114**, 8331 (1992).

31. F. R. McDonald and S. J. Danishefsky, *J. Org. Chem.*, **57**, 7001 (1992).

32. P. Z. Brigl, *Physiol. Chem.*, **122**, 245 (1922).

33. (a) R. U. Lemieux and G. Huber, *J. Am. Chem. Soc.*, **78**, 4117 (1956); (b) G. Maghin, *Bull. Soc. Chim. Belg.*, **77**, 575 (1968); (c) P. C. Wyss, J. Kiss and W. Arnold, *Helv. Chim. Acta* , **58**, 1847 (1975).

34. For other syntheses of 1,2-anhydrosugars see: (a) S. J. Sondheimer, H. Yamaguchi and C. Schuerch, *Carbohydr. Res.*, **74**, 327 (1979); (b) P. F. Sharkey, R. Eby and C. Schuerch, *Carbohydr. Res.*, **96**, 223 (1981).
35. D. L. Trumbo and C. Schuerch, *Carbohydr. Res.*, **135**, 195 (1985).
36. R. W. Murray and R. Jeyaraman, *J. Org. Chem.*, **50**, 2847 (1985).
37. R. L. Halcomb and S. J. Danishefsky, *J. Am. Chem. Soc.*, **111**, 6661 (1989).
38. J. Gervay, S. Danishefsky, *J. Org. Chem.*, **56**, 5448 (1991).
39. C. M. Timmers, G. A. van der Marel and J. H. van Boom, *Recl. Trav. Chim. Pays-Bas*, **112**, 609 (1993).
40. K. K.-C. Liu and S. J. Danishefsky, *J. Am. Chem. Soc.*, **115**, 4933 (1993).
41. D. M. Gordon and S. J. Danishefsky, *Carbohydr. Res.*, **206**, 361 (1990).
42. D. M. Gordon and S. J. Danishefsky, *J. Am. Chem. Soc.*, **114**, 659 (1992).
43. K. K.-C. Liu and S. J. Danishefsky, *J. Org. Chem.*, **59**, 1895 (1994).
44. K. K.-C. Liu and S. J. Danishefsky, *J. Org. Chem.*, **59**, 1892 (1994).
45. J. T. Randolph and S. J. Danishefsky, *J. Am. Chem. Soc.*, **115**, 8473 (1993).
46. V. Behar and S. J. Danishefsky, *Angew. Chem. Int. Ed. Eng.*, **33**, 1468 (1994).
47. J. J. Krepinsky, *Advances in Polymer-Supported Solution Synthesis of Oligosaccharides*, In this volume.
48. S. J. Danishefsky, K. F. McClure, J. T. Randolph and R. B. Ruggeri, *Science*, **260**, 1307 (1993).
49. T.-H. Chan and W.-Q. Huang, *J. Chem. Soc., Chem. Commun.*, 909 (1985).
50. J. T. Randolph and S. J. Danishefsky, *Angew. Chem. Int. Ed. Eng.*, **33**, 1470 (1994).
51. M. A. Berstein and L. D. Hall, *Carbohydr. Res.*, **78**, C1 (1980).

Chapter 9

ADVANCES IN POLYMER-SUPPORTED SOLUTION SYNTHESIS OF OLIGOSACCHARIDES

JIRI J. KREPINSKY

Department of Molecular and Medical Genetics, Carbohydrate Research Centre, and Protein Engineering Network of Centres of Excellence (PENCE), University of Toronto, Toronto, Ontario, Canada M5S 1A8

Abstract   Oligosaccharides can be efficiently synthesized using poly(ethylene glycol) monomethylether (MPEG)-supported methodology employing both the succinoyl diester linker and the α,α'-**DiOxyXy**lyl diether (DOX) linker. The latter allows more flexibility in reaction conditions than the former. After the oligosaccharide has been completed, either the entire MPEG-DOX support can be removed, or MPEG alone can be removed. This results in the formation of a *p*-tolylmethyl group (TM) protecting the carbohydrate hydroxyl to which MPEG-DOX was originally bound. MPEG of MW 5,000 and 12,000 is suitable and the method is exemplified by syntheses of oligosaccharides up to decasaccharides.

1. INTRODUCTION

The chemistry of oligosaccharides, which occurr naturally in glycolipids, glycoproteins and proteoglycans, is undergoing a transformation from a scientific curiosity into a matter of immediate practical importance. This is primarily due to the envisioned use of complex oligosaccharides in human therapeutics. Since these oligosaccharides cannot be obtained from natural sources in sufficient quantities, the efficient preparation of

194

oligosaccharides and their further elaboration into final formulations, e.g. glycopeptides and glycolipids, must be carried out by synthesis.

Unfortunately, oligosaccharides are not easy to synthesize. This is primarily due to three factors: a) monosaccharides, from which oligosaccharides are generally constructed, contain several hydroxyl groups of similar reactivity; b) glycosidic linkages must be formed stereospecifically at anomeric carbons, and; c) glycosidic linkages are acid-labile hemi-acetals. Several approaches have been developed to deal with these problems.

Historically, the first strategy for oligosaccharide synthesis was realized in solution and it is exemplified by the Koenigs-Knorr reaction and its numerous variations. This strategy is referred to in this chapter as the *classical* solution synthesis even if some important recent variants are described in this volume. In essence, it can be described as a nucleophilic substitution at the anomeric carbon with a free (or activated) hydroxyl group of a future aglycon (Scheme 1). A critical feature of this approach is that the other hydroxyls present must be protected to prevent them from taking part in such substitutions, so the synthetic strategy usually requires elaborate protection-deprotection schemes.[1] Many reactions involved in oligosaccharide synthesis, glycosylations in particular, are often both incomplete and accompanied by side-reactions resulting in numerous by-products. This results in the need for an extensive purification, usually by chromatography, after each synthetic step. The whole process is quite laborious and time consuming. Therefore it is understandable that albeit this strategy has been immensely successful[2] in providing a variety of oligosaccharides, alternative approaches have been sought. It should be noted that classical chemical synthesis of oligosaccharides is a costly process, mainly because of the need for purification. If the chromatographic purification is defined as the crucial bottleneck operation in the chemical syntheses, routes should be devised that do not require chromatography, making chemical synthesis more competitive.

As a possible solution to this problem, methods which employ enzymes have been investigated. Glycosyltransferases[3] and glycosidases working in reverse,[4] or combined enzyme and chemical solution strategies[5] have been examined. Enzymic syntheses also have a number of their own problems. For instance, glycosyltransferases require relatively unstable intermediates, the yields are limited, and unnatural

196

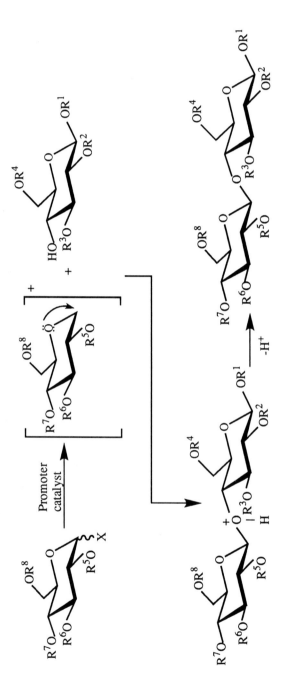

Scheme 1.

oligosaccharides are unlikely to be formed.   Glycosidase-catalyzed glycosylation reactions exhibit low regiospecificity, in addition to low yields.[4]

An example of enzymic syntheses of oligosaccharides using glycosyltransferases in combination with chemical synthesis of a peptide on a solid support was recently published.[6]

That alternative designs are worthy of exploring was manifested by the solution of analogous problems in the syntheses of oligonucleotides and oligopeptides through polymer-supported strategies. Solid-phase syntheses of oligopeptides and oligonucleotides on polystyrene resins and controlled-pore glass have been spectacularly successful. The main benefit of solid phase procedures is that the purification of desired intermediates from by-products, excess reagents and products of their decomposition is achieved by simply washing them from the solid support, obviating the need for time-consuming chromatography.   Furthermore, the reaction itself is quenched almost instantaneously and the work-up is fast, thus eliminating overexposure of reaction products to excess reagents and by-products.

Therefore it is natural that this approach has been examined for the synthesis of oligosaccharides. The first attempted solid phase synthesis based on Koenigs-Knorr chemistry was studied in Schuerch's laboratory using an *immobilized glycosyl acceptor* on an organic polymer support.[7]   It was followed by investigations of an inorganic support, namely porous glass, which has been zirconized and silanized.[8] Unfortunately, many difficulties were encountered. Among the problems experienced were decreased reaction rate and correspondingly lower yields compared to solution strategies, incomplete coupling, and lack of complete anomeric specificity.   Although it was speculated that the use of a soluble polymer could remedy the situation,[9] the only further attempt to synthesize a saccharide derivative using a polymer support was done on controlled-pore glass in van Boom's laboratory.[10]

Different approaches to solid phase synthesis of oligosaccharides were reported recently from Kahne's and Danishefsky's laboratories.[11]   In the latter approach, an organic polymer was used to *immobilize glycosyl donors*, and glycal chemistry was employed[12] instead of Koenigs-Knorr reaction. The merits of this polymer-supported solid phase vs. solution synthesis were commented upon.[13] Solid-phase synthesis using enzymes has been recently reported by Wong's group.[14]

In order to become successful, solid-phase oligopeptide and oligonucleotide syntheses required a complete reformulation of chemistry to satisfy the characteristics unique to the solid-phase processes. An analogous requirement could be expected in the case of oligosaccharides. Unfortunately, as stated above, oligosaccharide synthesis is an even more daunting task than the synthesis of oligonucleotides or oligopeptides. Oligosaccharides often contain branched structures requiring at least three kinds of hydroxyl protecting groups, with clearly differentiated stability, two of them being removed at different times during the course of the synthesis. Thus, in oligosaccharide synthesis, the delicate glycosidic bonds together with a host of steric and electronic factors conspire to keep yields in the key glycosylation steps in the 80% range, and often much less. And this is even after recent dramatic advances in classical synthesis.[2] So it is not surprising that the first attempts at solid-phase synthesis of oligosaccharides were found to have serious limitations.

2. POLYMER SUPPORTED SOLUTION SYNTHESES

Solid phase syntheses of oligopeptides and oligonucleotides in automated machines furnish the products only up to micromolar quantities. This is due to the high molecular weight of the support compared to the synthesized substance. To synthesize relatively short sequences in larger quantities, a polymer-supported solution synthesis has been developed.[15] It utilizes as the supporting polymer polyethyleneglycol ω-mono-methylether (MPEG) which has substantially higher loading capacity than the solid polymers.[16] The method is based on the solubility of MPEG in most solvents including water, but not ethers. In principle, the polymer-bound synthon which is soluble under reaction conditions is rendered insoluble during the work-up. Since all other components of the reaction mixture remain in the solution, they can be separated from the desired MPEG-bound product by simple filtration (Scheme 2). Under these conditions the reactions are faster and more complete, and are more easily controlled than with the reactants immobilized on insoluble matrices. We have applied this strategy to the synthesis of oligosaccharides with the anticipation that the solubility of the reactants should allow for reaction kinetics and anomeric control similar to those observed in solution chemistry.[17]

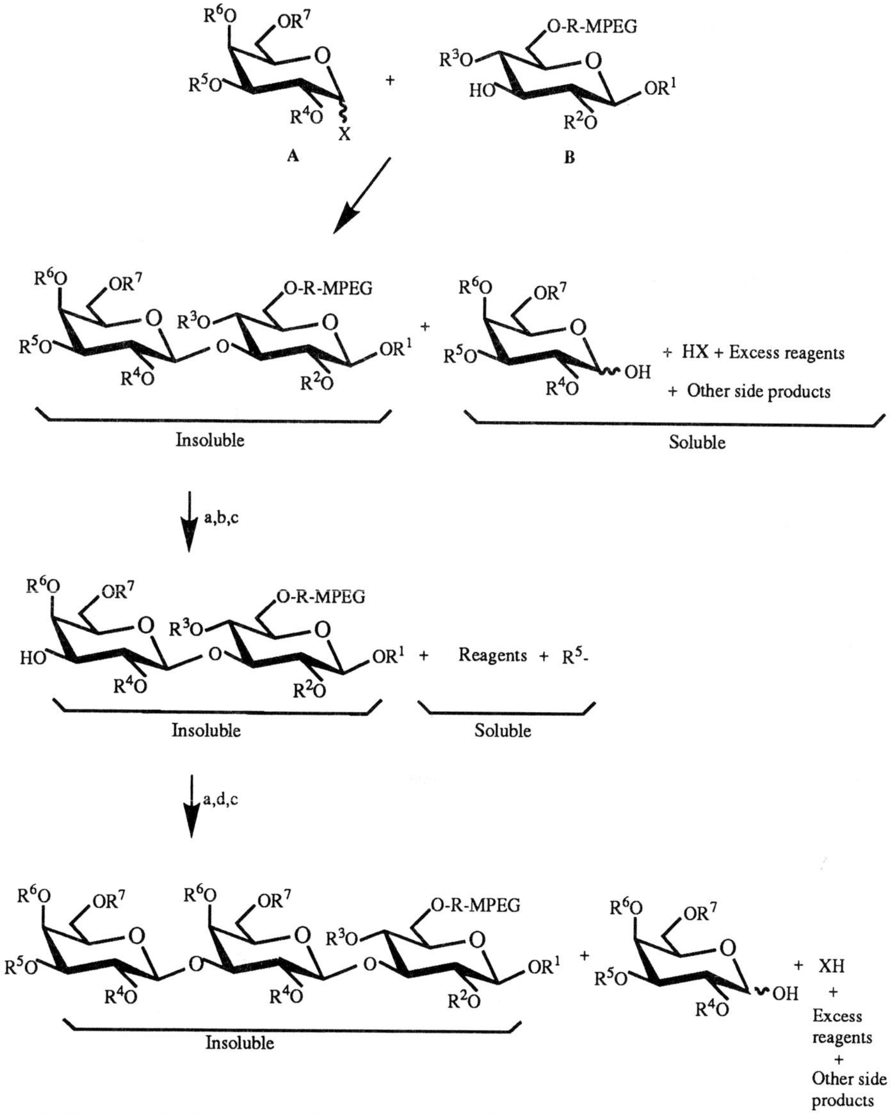

(a) Dissolve; (b) Deprotect; (c) Add ether; (d) Glycosylate with **A**
X = Leaving group; R ≠ H; MPEG = -$(CH_2CH_2O)_nCH_3$

Scheme 2.

In the synthesis of oligosaccharides, MPEG is bound to the first member of the oligosaccharide chain by a linker permitting easy removal of MPEG from the completed oligosaccharide, after the completion of the synthesis. Most physical properties of such a synthon, i.e. carbohydrate-linker-MPEG, are dictated by MPEG, and are not much different from those of MPEG. The properties relevant to its role in the oligosaccharide synthesis are summarized in the following paragraph.

2.A. Properties of Polyethylene ω-Monomethylether (MPEG)

Commercially[*] available MPEG has the average molecular weight 5,000 and it forms rigid rod-like linear structures stabilized by intramolecular interactions.[18] It is very soluble in solvents such as benzene, toluene, dichloromethane, chloroform, acetonitrile, and acetone, as well as in water. It is insoluble in hexane, diethylether and t-butylmethyl ether, and soluble in hot terahydrofuran and ethanol. MPEG retains water easily because of the formation of hydrate complexes, but the majority of the retained water can be removed by azeotropic distillation, e.g. with toluene or benzene. It can be purified by crystallization from ethanol, since it is insoluble in cold ethanol, or by precipitation with an ether from its solutions in the organic solvents cited above. t-Butylmethyl ether is preferable for the precipitation because its higher boiling and flash points make its use safer than the use of diethyl ether.[†] Recovery of MPEG derivatives is usually no less than 95%. However, MPEG avidly forms strong complexes with metal cations and albeit in most cases it is still possible to precipitate MPEG to which the cations are bound, or recrystallize such MPEG complexes from ethanol, recoveries are usually low. This property of MPEG imposes limitations on its use with many reagents, including salts of silver, mercury, chromium and cerium.

The recovered yields of MPEG diminish seriously upon prolonged exposure to humid air, e.g. during the filtration. MPEG derivatives carrying large synthons, the molecular weight of which approaches the molecular weight of MPEG itself, also

---

[*]Suitable MPEG 5,000 was obtained from Fluka AG, Buchs, Switzerland. MPEG always contains a certain proportion of molecules of higher and lower molecular weight than 5,000. Union Carbide's name for PEG is Carbowax.

[†]t-Butylmethyl ether of sufficient quality and degree of dryness for this purpose is available inexpensively in bulk quantities from ARCO (Atlantic Richfield) Chemicals.

precipitate less easily and in smaller particles, more difficult to filter. This can be remedied by using MPEG mw=12,000.[‡]

MPEG derivatives can be purified from small reactants and reaction by-products by ultrafiltration. However, the filters commonly available for ultrafiltration (commercially available from Amicon, Inc.) do not support the use of apolar solvents, and it is more difficult and time-consuming to dry aqueous solutions.

Although MPEG is sufficiently stable under many reaction conditions that it can be considered just another protecting group, it should be kept in mind that it is a polyether and its stability is limited. It may undergo a partial degradation under certain reaction conditions (cf. the review by Harris[16]). Pyridinium salts are particularly effective in peeling off one or more ethylene glycol units, usually from the non-methylated end of MPEG.[19] Furthermore, MPEG cannot be linked to the anomeric carbon. When an MPEG glycoside, for instance 2,3,4,6-tetra-O-acetyl-ß-D-galactoside (1) is treated with a glycosylating agent such as acetobromomannose (2), a glycosyl exchange occurs and MPEG 2,3,4,6-tetra-O-acetyl-α-D-mannoside (4) is precipitated. Glycolyl or (bis)glycolyl 2,3,4,6-tetra-O-acetyl-ß-D-galactoside (3) can be isolated from mother liquors[20] (Scheme 3).

Since MPEG contains a single OCH_3 group (δ=3.380 ppm), the course of reactions can be monitored by ^{1}HNMR spectrometry using this signal as an internal standard. The glycol peak of MPEG at δ=3.640 ppm is very strong and must be suppressed by irradiation for 3.0 sec in order to increase the dynamic range and obtain a good quality spectrum. Sufficient structural information to identify the product can be usually obtained from ^{1}HNMR spectra of MPEG-bound substances despite the glycol peak suppression in the spectral area of interest.

For the purity control of MPEG-bound substances, we have recently investigated supercritical fluid chromatography on silica gel with methanol-containing CO_2 as the mobile phase. Residual impurities elute as narrow single peaks with 5% MeOH in CO_2 while MPEG-bound compounds elute with higher methanol concentrations as broad, sometimes comb-shaped peaks.[21] For a possible use of other analytical methods the reader is referred to the review of Harris.[16]

[‡]MPEG 12,000 is manufactured by Union Carbide, Inc. The loading capacity of MPEG 12,000 is naturally less than 50% of the loading capacity of MPEG 5,000.

202

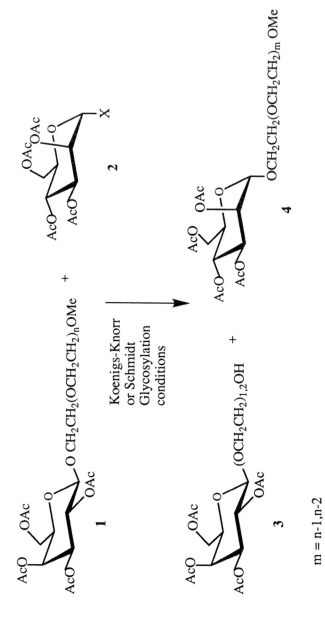

Scheme 3.

2.B.   The Linker

2.B.i.   Succinoyl diester

MPEG cannot be bound directly to the carbohydrate with which the synthesis would start for several reasons: 1) it is too labile in the anomeric position as it was mentioned above, and 2) too stable in all other positions and MPEG cannot be removed after the synthesis is completed since under acidic conditions required for MPEG's removal, the hemiacetal glycosidic bonds are more labile than ether bonds between MPEG and carbohydrate hydroxyls. Furthermore, the closeness of MPEG to the carbohydrate may create interactions detrimental to the course of the synthesis. For these reasons, a tether is needed, which does not change the basic physico-chemical and chemical properties of MPEG-saccharide, but allows an easy release of the carbohydrate from MPEG-linker while maintaining a safe distance between MPEG and the carbohydrate. Succinoyl diester has been extensively used in polymer-supported solution syntheses of oligopeptides and oligonucleotides and there seems to be no compelling reason which would disqualify its application in polymer-supported synthesis of oligosaccharides.

Thus in the oligosaccharide synthesis MPEG[17] was first bound to a hydroxyl of the first monosaccharide through the succinoyl diester linker (Scheme 4). Another hydroxyl of the monosaccharide is then activated by deprotection, and glycosylated. Examples of syntheses of di- and tri-saccharides are shown in Scheme 5. A hydroxyl group is then deblocked, and the above process is repeated until the required oligosaccharide is synthesized.   As a general rule, unfinished glycosylations are minimized by using glycosylating agents in excess, or better still by repeated additions of the glycosylating agent.

As it was mentioned earlier, one of the advantages of the polymer-supported solution synthesis is that familiar procedures from solution chemistry can be applied directly to the MPEG-supported strategy. We have often observed an additional bonus: the increased anomeric specificity in MPEG-supported glycosylations. For instance, glycosylations leading to the disaccharide 21 without MPEG support always gave the desired ß-anomer accompanied with small amounts of the α-anomer. The identical glycosylation performed on an MPEG-supported acceptor yielded the correct disaccharide 22 without any identifiable trace of the α-anomer (Scheme 6).[22]   In another example, the MPEG-supported synthetic sequence gave the final product 23 in

204

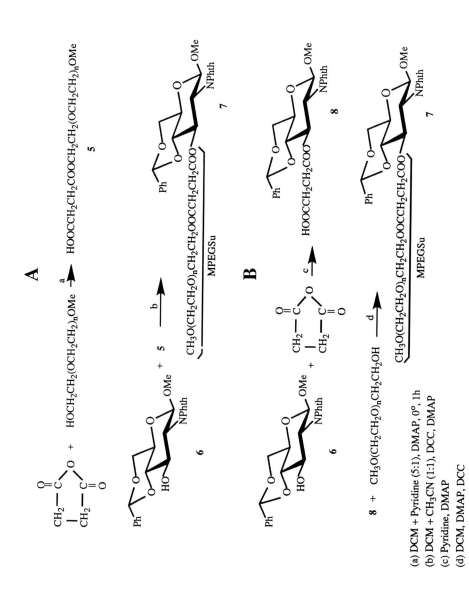

(a) DCM + Pyridine (5:1), DMAP, 0°, 1h
(b) DCM + CH₃CN (1:1), DCC, DMAP
(c) Pyridine, DMAP
(d) DCM, DMAP, DCC

Scheme 4.

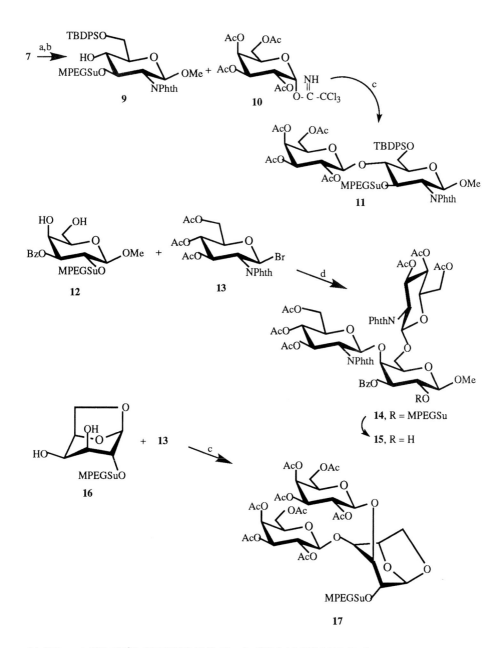

(a) 60% aq. AcOH, 100°C; (b) TBDPS-Cl, Imidazole, DCM; (c) DCM, BF₃.Et₂O;
(d) AgOTf, DMAP, DCM, 4A MS.

Scheme 5.

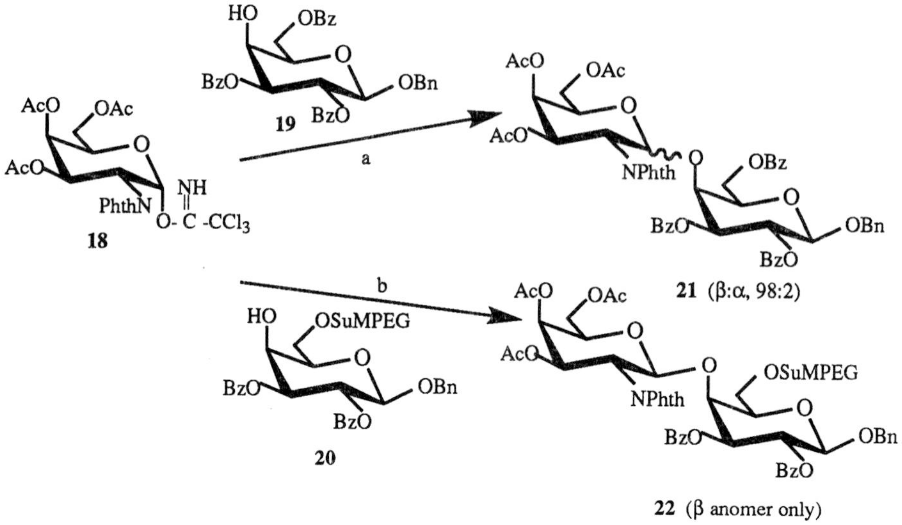

(a) AgOTf, DCM, 20°C, 72H, 86%.
(b) Et₃SiOTf, DCM, 0-5 C, 3h, 75%.

Scheme 6.

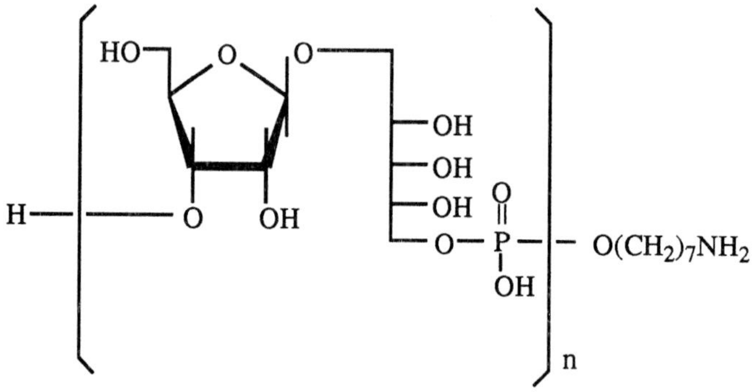

23   (n=2,3,5,6)

four times better yield than the procedure not employing MPEG.[23] In the synthesis of a methyl heptaglucoside **24** (Scheme 7), the regioselective glycosylation of the primary hydroxyl in 4,6-diol worked well in both MPEG-unsupported acceptors[24] and the MPEG-supported solution synthesis,[25] but the MPEG-supported sequences were always much faster. The MPEG-bound first unit, **A**, is glycosylated with the second unit **B** of the disaccharide **B-C** using thioglycoside activated by *N*-iodosuccinimide. The 4,6-benzylidene in the unit **B** in the resulting trisaccharide **A-B-C** is removed, and the primary hydroxyl in the unit **B** is glycosylated by the unit **D** as in the preceding glycosylation. The *t*-butyldimethylsilyl protective group in the unit **D** in the branched tetrasaccharide **A-B(C)-D** thus formed is then removed and the free primary hydroxyl in the unit **D** is then glycosylated with the same disaccharide synthon **B-C** as in the first glycosylation. This last disaccharide moiety is marked **E-F** in **24**. The 4,6-benzylidene in the unit **E** in the resulting hexasaccharide **A-B(C)-D-E-F** is removed, and the primary hydroxyl in the unit **E** is glycosylated by the unit **G** to give the heptasaccharide **B(C)-D-E(F)-G**. Since OH-4 remained free after the regiospecific glycosylation of 4,6 diols in the units **B** and **E**, it was acetylated (capped) to protect it from being glycosylated during the following steps of the synthesis. The synthesis shows that the mild character of iodonium ion-promoted glycosylations with thioglycosides is compatible with MPEGSu-supported design. In another example of an iterative synthesis of a decasaccharide model of heparan sulfate-like structure, levulinoyl protective group was used as a temporary protection, removable in the presence of other ester groups (Scheme 8).[26] It is worth noting that the trimethylsilyl triflate promoted glycosylations with the glycosyl trichloroacetimidate (unit **B**) used in minimal excess showed excellent α-anomeric specificity. Also, elevated temperature (+10°) was required.

It is absolutely essential that both the protective groups and the linker must be stable in all reactions of the synthetic sequence. The base-lability of the ester linkages of the succinoyl linker excludes any use of temporary ester protecting groups (except levulinoyl) that need to be removed before the completion of the synthesis. Furthermore, suitably located succinoyl-MPEG groups are prone to migrations in acidic environment, for instance 2-O-SuMPEG → 3-O-SuMPEG in ß-mannosides.[27]

Even despite these restrictions, MPEG-succinoyl-methodology has been successfully used in several laboratories.[22,23,25,26] Some examples, summarized in

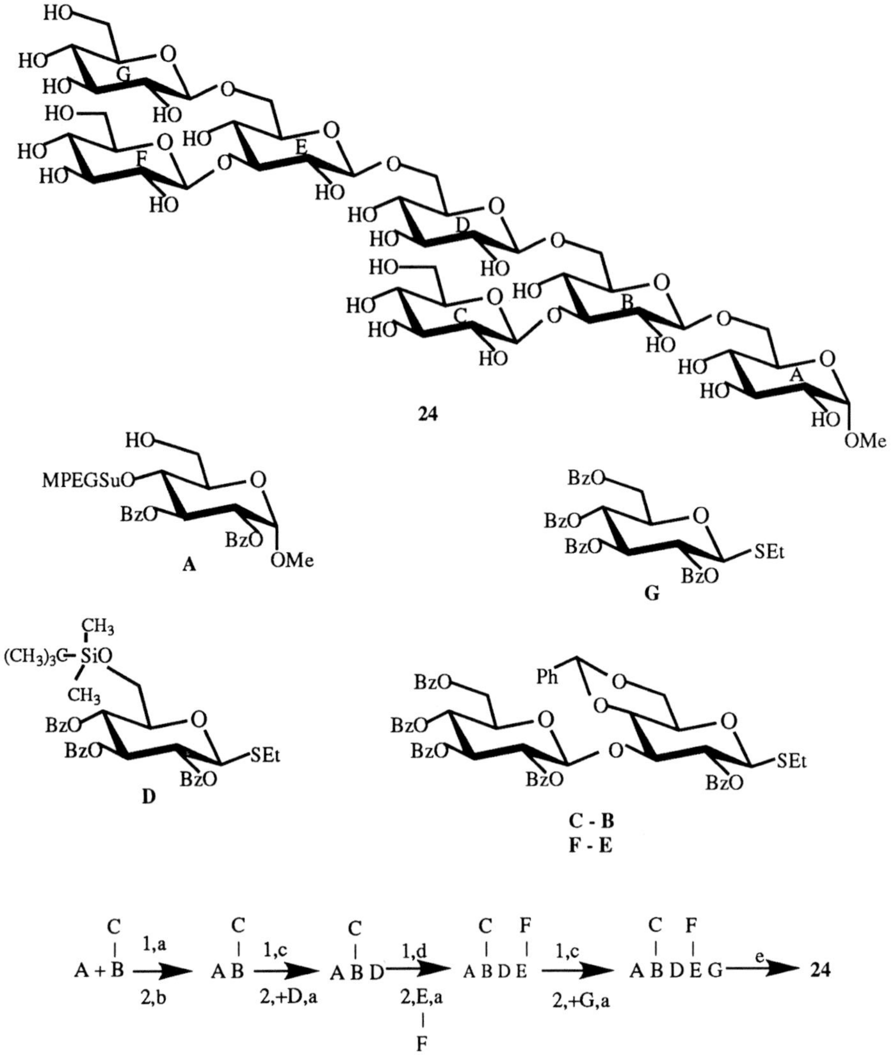

24

A

G

D

C - B
F - E

$$
\begin{array}{c}
\overset{\displaystyle C}{\underset{}{|}} \\
A + B \xrightarrow[2,b]{1,a} AB \xrightarrow[2,+D,a]{1,c} ABD \xrightarrow[2,E,a]{1,d} \overset{\displaystyle C\ \ F}{\underset{}{|\ \ |}}{ABDE} \xrightarrow[2,+G,a]{1,c} \overset{\displaystyle C\ \ F}{\underset{}{|\ \ |}}{ABDEG} \xrightarrow{e} 24 \\
\end{array}
$$

(a)   N-iodosuccinimide, Trifluoroacetic acid
(b)   Removal of benzylidine
(c)   Acetylation
(d)   Removal of t-butyldimethylsilyl
(e)   Sodium methoxide/Methanol

Scheme 7.

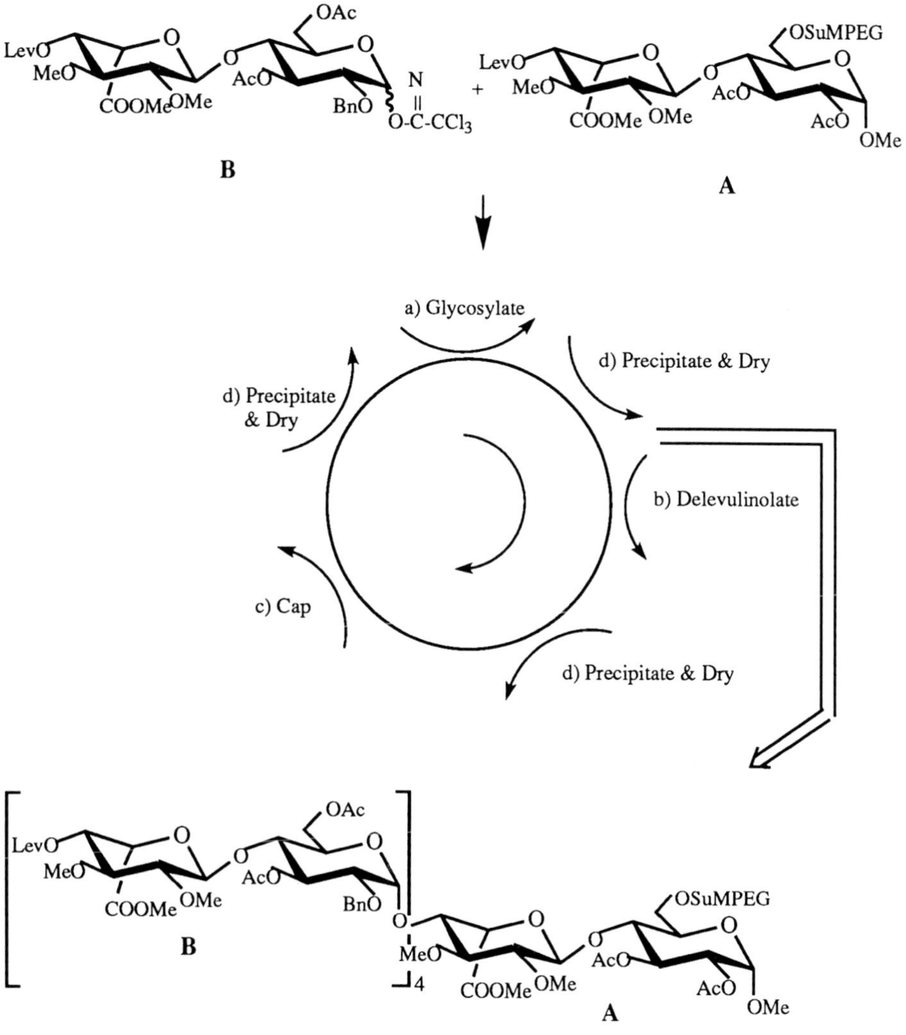

25

(a) B, TMSOTf, DCM, 10°C, 4A MS,4h; (b) H_2NNH_2, AcOH, Pyridine; (c) Ac_2O, DMAP, Pyridine

Scheme 8.

Schemes 5-8, were briefly discussed above. As it is true for any organic chemical synthetic scheme, reagents and protecting groups must act in harmony. Glycosylations with trichloroacetimidates promoted by boron trifluoride, triflic anhydride, and trimethylsilyl or triethylsilyltriflates are compatible with ester, allylic, benzylic (including benzylidene-type acetals), and some silicon-based protecting groups. Iodonium ion promoted glycosylations with thioglycosides show a similar compatibility and appear to be even milder. Metal-based promoters usually cannot be used, although silver triflate has been used on several occassions.

2.B.ii. Dioxyxylyl diether (DOX)

Despite the accomplishments of the succinoyl diester linker-based synthetic design, its sensitivities described above restrict the use of the methodology. The availability of many glycosylation conditions and protecting groups to modern carbohydrate chemists led us to search for a more stable linker, compatible with the wider variety of reaction conditions than the presence of the diester linker would allow. The search was successfully concluded with Douglas' discovery[28] of α,α'-DiOxyXylyl diether,- $OCH_2C_6H_4CH_2O$-, (DOX) which satisfies the following critical criteria: (a) bound via a hydroxyl or as an O-glycosidic linkage, it is stable under many reaction conditions including glycosylation; (b) it is easily removable from a finished oligosaccharide by hydrogenolysis under mildly acidic conditions; (c) it is easily accessible from α,α'-dichloro-p-xylene. Furthermore, the proper choice of hydrogenolytic conditions allows for removal of MPEG *only*, leaving behind a p-methylbenzyl (tolylmethyl) protected oligosaccharide.

The MPEG-DOX support-linker system is readily synthesized from the monomethylether of PEG and α,α'-dichloro-p-xylene by Williamson ether synthesis[29] giving the monochloro compound MPEG-DOX-Cl (**26**) or, after hydrolysis, the alcohol MPEG-DOX-OH (**27**); cf. Scheme 9.

The alcohol MPEG-DOX-OH (**27**) can be glycosylated to link the support through the anomeric oxygen; such a connection would not be stable under glycosylation reactions using succinoyl linker. In the example worked out by Whitfield[30] and shown in Scheme 9, the MPEG-DOX-OH is glycosylated using the trichloroacetimidate procedure[2c] by 2-O-acetyl-3,4,6-tri-O-benzyl-α-D-mannopyranosyl trichloroacetimidate (**28**). The mannosyl donor has an acyl group at O-2 that not

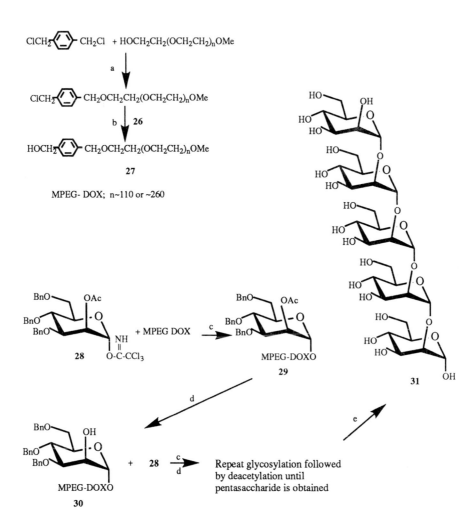

(a) ClCH$_2$C$_6$H$_4$CH$_2$Cl, THF, NaH, 96h, RT, 96%; (b) 10% aq. Na$_2$CO$_3$, 70°C, 16h, 83%;
(c) DCM, TESOTf, 4A MS, 0-4°C, 4h, 92%; (d) MeOH, DBU, RT, 16h, 97%; (e) Raney
Nickel W-2, EtOH, Reflux, 16h, or Pd/C, 50% aq. AcOH, 50°C, 72h.

Scheme 9.

only allows for stereocontrol via neighbouring group participation but also is readily cleaved by a base to yield the acceptor for the next glycosylation. Repetition of the glycosylation and hydrolytic steps gave the protected pentasaccharide which, after hydrogenolysis, acetylation, purification and deacetylation gave pentamannopyranoside [Manp(α1-2)]$_4$Manp (**31**). As expected, no ß-anomer was observed.

The attachment of MPEG-DOX to a carbohydrate hydroxyl other than the anomeric one is achieved by reaction of such hydroxyl with monochloro compound MPEG-DOX-Cl (**26**) under conditions of Williamson reaction (Scheme 10).

The MPEG-DOX in **37** can be transformed in p-methylbenzyl group in **38** by hydrogenation under neutral conditions (Scheme 11).

MPEG-DOX is completely removed by more vigorous hydrogenation, benzyl protecting groups being removed at the same time, and peracetatylated oligosaccharides are formed after acetylation. Peracetylated oligosaccharides are usually more suitable for final purification by chromatography than deprotected compounds; after deacetylation pure oligosaccharides are readily obtained.[31]

2.C. Capping

Capping, or protection of the unreacted hydroxyl to prevent it from reacting in the next condensation step, is a very important operation in oligonucleotide automatic synthetic protocols. It appears that in the succinoyl diester protocol, acetylation of the unreacted hydroxyl is feasible.[25,26] In DOX protocol, methylation with methyl iodide appeared initially to be the method of choice. It was found later, unfortunately, that under these methylation conditions, the bond between MPEG and DOX is easily broken. The capping in the DOX protocol thus remains unresolved.

3. EXPERIMENTAL CONDITIONS

In order to illustrate the simplicity of the polymer supported solution synthesis,[§] preparations of several compounds are briefly described in this section. More details can be found in cited publications.

[§]The reactions can be performed in most solvents except ethers; products are isolated from the reaction mixtures by precipitation after addition of an ether, generally *t*-butylmethyl ether.

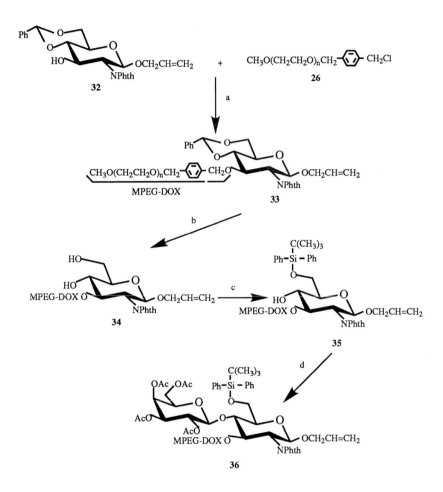

(a) NaH, THF, 82%; (b) 60% aq. AcOH, 100°C, 40 min, 96%; (c) TBDPSCl, Imidazole, DCM, RT, 16h, 93%; (d) 2,3,4,6-Tetra-O-acetyl-D-galactopyranosyl trichloroacetimidate, Tf$_2$O, DCM, 0-5°C, 2h, 88%.

Scheme 10.

214

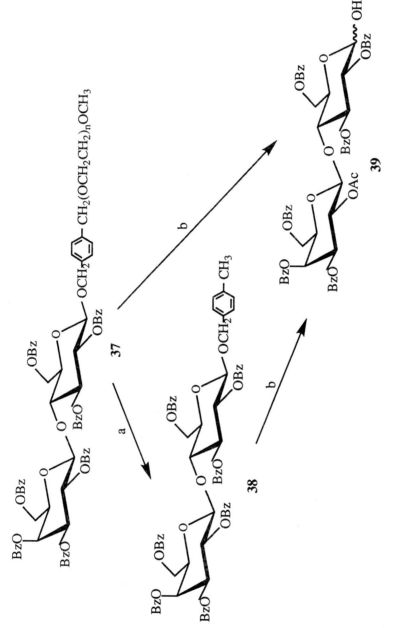

(a) Pd-H$_2$, EtOH, 48h, RT, 95%; (b) Pd-H$_2$, 50% aq. AcOH, 72h, 90%.

Scheme 11.

3.A. Succinoyl linker[17]

MPEGSu-Sugar. Method A. MPEG[¶] (m.w. 5,000; 20 g) was dried overnight at high vacuum with succinic anhydride (2 g, 5 eq.) and DMAP (200 mg). To this mixture was added DCM (140 mL) and pyridine (30 mL). After stiring overnight, the mixture was concentrated to 75 mL, cooled to 0° in ice, and it was diluted with stirring to 1.0 L with cold ether. It was allowed to stand 1 h on ice, the solid was filtered off by suction, washed with ether, and air-dried for 1 h. It was further purified by recrystallization from hot EtOH (700mL).

To MPEGSu (**5**; 5g) was added **6**, (1.5 eq.) and DMAP (100 mg), and the mixture was dried at high vacuum overnight. DCM (25 mL), CH_3CN (25 mL), and DCC (1.5 mL of a 1 M solution in DCM, 1.5 eq.) were added under argon, and the reaction mixture was stirred overnight at room temperature. The precipitated urea was filtered off, washed with DCM, and the volume of the combined filtrates was reduced to its original size. It was cooled to 0°, anhydrous ether was added with vigorous stirring, and **7** precipitated. The solid was filtered off, dissolved in hot absolute EtOH, and after filtering off the insoluble impurities, it was cooled to 4°. Compound **7** crystallized out and after filtering, it was washed with ether, and dried. The unreacted sugar was recovered from the combined filtrates.

MPEGSu-Sugar. Method B.   Methyl 4,6-benzylidene-2-deoxy-2-*N*-phthalimido-ß-D-glucopyranoside (**6**; 0. 44g, 1.07 mM), succinic anhydride (0.54 g, 5.3 mM), and DMAP (50 mg) were stirred in dry pyridine (50 mL) at room temperature. After completion of the reaction (monitored by TLC), pyridine was removed by evaporation in vacuo, and the residue subjected to flash chromatography in EtOAc to give 3-O-hemisuccinate **8** (0.4 g, 70%).

MPEG (3.2 g; 0.8 eq.), mixed with **8** was dried overnight at high vacuum over P_2O_5. This mixture was dissolved in anhydrous DCM (25 mL), and a catalytic amount of DMAP, followed by DCC (0.16 g, 0.77 mM), was added. The solution became

---

[¶]Abbreviations: Ac, acetyl; Bz, benzoyl; DBMP, 2,6-di-*t*-butyl-4-methylpyridine; DBU, 1,8-diazabicyclo[5.4.0]undec-7-ene; DCC, 1,3-dicyclohexylcarbodiimide; DCM, dichloromethane; DMAP, 4-(dimethylamino)pyridine; Lev, levulinoyl; MPEG, polyethyleneglycol monomethylether; Phth, phthalimido; RM, room temperature; TBDPS, *t*-butyl-diphenylsilyl; TESOTf, triethylsilyl triflate; Tf₂O, triflic anhydride; TMSOTf, trimethylsilyl triflate. Temperature is given in degrees Celsius.

cloudy in 15 min and was stirred overnight at room temperature. The precipitated urea was removed by filtration, washed with DCM, and the volume of the combined filtrates was reduced to its original size. It was cooled to 0°, ether (250 mL) was added with vigorous stirring, and **7** precipitated out. After filtration, the solid was dissolved in hot absolute EtOH (50 mL), the solution was filtered, cooled to 4°, and the recrystallized **7** was filtered, washed with ether and dried.

Glycosylation: *Preparation of trisaccharide 14*. Diol **12** with attached SuMPEG (312 mg, 0.057 mM) was mixed with bromide **13** (56 mg, 2 eq.), AgOTf (28 mg, 2 eq.), DBMP (11 mg, 1 eq.), and a small amount of powdered 4A ms, and the mixture was dried at high vacuum overnight. Then the flask was cooled in ice water under argon, DCM (4 mL) was added and the reaction mixture was stirred for two h. At this point another portion of dried bromide **13** (2 eq.), AgOTf (2 eq.), and DBMP (1 eq.), were added, and an identical addition was made after another 2 h. The stirring was continued overnight, the reaction mixture was diluted with DCM (10 mL) and the molecular sieves and precipitated silver salts were filtered off. The filtrate was evaporated to dryness, the residue was redissolved in DCM (4 mL), the solution was cooled to 0° in an ice bath, and the product was precipitated by the addition of Et_2O (40 mL) with vigorous stirring. After standing for 1 h, the precipitate was collected by filtration, washed with Et_2O, and dried on air for at least 1 h. The dry solid was dissolved in warm EtOH (15 mL), filtered from undissolved solids, and the solution let crystallize at 4°. The solid was washed with cold EtOH and ether, and dried in vacuo to give **14**.

Preparation of disaccharide 11. To a cold solution (-10°) of **9** (77 mg, 0.013 mM) and imidate **10** (0.02 g, 0.04 mMol) in DCM (1 mL), $BF_3.Et_2O$ (0.08 M in DCM, 6 μL, 0.048 mM) was added. The reaction mixture was allowed to warm up slowly to room temperature and stirring was continued overnight. The reaction mixture was worked up as described for **14**. This procedure was repeated two more times, and complete galactosylation gave **11**.

Saccharide cleavage from the polymer: *Cleavage of trisaccharide 14*. The trisaccharide moiety was removed from the polymer in **14** (330 mg) by treatment with $N_2H_4.H_2O$ (1 mL) and EtOH (2 mL) at 70° for 2 h. The liquids were removed by co-evaporation with toluene (2x10 mL) and the resulting solid was dried at high vacuum for 2 h, cooled on ice, and pyridine (2 mL) and Ac_2O (1 mL) were added under argon and the reaction mixture was stirred overnight at room temperature. The liquids were removed under high vacuum, the residue was dissolved in hot EtOH (15mL), filtered, and allowed to precipitate at 4°. The precipitated MPEG was filtered off, rinsed with cold EtOH, the combined filtrate and washings were evaporated to dryness and purified by chromatography on silica gel to yield peracetylated trisaccharide **15**.

3.B. Diether (DOX) linker

4-(Chloromethyl)benzyl ω-methylpoly(ethyleneglycol)yl ether (MPEG-DOX-Cl) (**26**). Dry MPEG (m.w. 5,000; 5 g, 1 mmol) was dissolved with heating in dry THF (50 mL) with exclusion of humidity and cooled to room temperature. NaH (0.12 g; 3 mmol) was added with stirring, followed after 10 min by NaI (0.17 g, 1.15 mmol) and α,α'-dichloro-p-xylene (5.25 g., 30 mmol). The reaction mixture was stirred for 96 h. Then the reaction mixture was filtered through a celite bed which was then washed with a small amount of DCM (5 mL). The combined filtrate and washings were cooled on an ice-bath, the MPEG derivative was precipitated with ether (300 mL). The solid was filtered off, resuspended in ether (100 mL), and filtered again. Then it was redissolved in hot absolute EtOH (80 mL), and cooled to 4° until precipitation was completed. After filtration and drying **26** (5 g; 96%) was obtained.

4-(Hydroxymethyl)benzyl ω-methylpoly(ethyleneglycol)yl ether (MPEG-DOX-OH) (**27**). MPEG-DOXCl (**26**, 21.95 g; 4.3 mmol) was dissolved in 10% aqueous Na_2CO_3 under argon. This solution was heated at 70° using an air condenser stoppered with a rubber septum for 16 h. The solution was concentrated *in vacuo* and the residual water removed by co-evaporation with toluene (2 x 250 mL), and the residue was dried *in vacuo*. Then the residue was taken up in DCM, filtered, and rinsed three times with DCM (total volume 350 mL). Toluene (100 mL) was added to the filtrate, the solvents were evaporated, and the residue was dried *in vacuo*. Then it was redissolved in dry DCM (75 mL), filtered, and the filtrate was cooled in an ice bath and precipitated with *t*-butylmethyl ether (900 mL). The precipitate was filtered off, rinsed with *t*-butylmethyl ether (100 mL) followed by diethyl ether (100 mL) and dried *in vacuo*. The residue was recrystallized from absolute EtOH (500 mL), filtered off, rinsed successively with EtOH and *t*-butylmethyl ether and diethyl ether (100 mL of each) and dried *in vacuo* as a white powder **27** (18.2 g; 83%).

Glycosylation: MPEG-DOXyl 2-O-acetyl-3,4,6-tri-O-benzyl-α-D-mannopyranoside (**29**). DCM (15 mL) was added to a mixture of MPEG-DOXOH (**27**, 4.0 g, 0.78

mmol) and 4A ms (1 g) followed by a solution of 2-O-acetyl-3,4,6-tri-O-benzyl-α-D-mannopyranosyl trichloroacetimidate[32] **28** (875 mg, 1.36 mmol) in DCM (5 mL) under argon with stirring. This mixture was cooled in an ice bath and after 30 min triethylsilyl trifluoromethanesulfonate (TESOTf) (177 μL, 0.78 mmol) was added by syringe. After 4 h at this temperature, diisopropylethylamine (10 drops, about 90 mg) was added to the solution and after 5 min excess *t*-butylmethyl ether (230 mL) was added to precipitate the polymer. The white solid was separated by filtration and after rinsing with *t*-butylmethyl ether (50 mL) it was recrystallized from absolute ethanol (200 mL). The white precipitate was collected by filtration and after rinsing with ethanol and *t*-butyl methylether (50 mL each) it was dried *in vacuo* to yield **29** (4.11 g, 92%).

The disaccharide *MPEG-DOXyl 2-O-(2-O-acetyl-3,4,6-tri-O-benzyl-α-D-mannopyranosyl)-3,4,6-tri-O-benzyl-α-D-mannopyranoside* and the following tri, tetra, and pentasaccharides, were prepared by glycosylations with **28** in a manner analogous to the preparation of **29**. The deprotection to free accepting hydroxyl were done as shown for the preparation of **30**.

MPEG-DOXyl 3,4,6-tri-O-benzyl-α-D-mannopyranoside (**30**). Polymer bound monosaccharide **29** (4.11 g, 0.72 mmol) was dissolved in dry MeOH (20 mL) while it was gently heated with a heat gun. After cooling to room temperature, DBU (12 drops from a Pasteur pipette, about 200 mg) was added and the solution stirred tightly stoppered for 16 h. The solution was cooled in an ice bath and the polymer was precipitated by the addition of *t*-butyl methylether (350 mL). The solid was recovered by filtration and after rinsing with *t*-butylmethyl ether it was recrystallized from absolute EtOH (250 mL). The white solid was collected by filtration and after rinsing with EtOH and *t*-butylmethyl ether (50 mL each) was dried *in vacuo* to yield **30** (3.83 g, 97%).

Peracetylated **31**. To polymer bound perbenzylated pentasaccharide was added Raney Ni W-2 (Aldrich 50% aqueous slurry, 5 g). Before use, the catalyst was washed with water until neutral with pH paper (3x), then with EtOH (3x) and without allowing it to dry, added as an ethanol slurry to the reaction flask (approximately 50 mL of ethanol total) and the reaction mixture was refluxed under argon for 16 h. The hot solution was filtered through a celite bed and rinsed well with warm EtOH (100 mL). Toluene (35 mL) was added to the combined filtrate and washings, and the solution was evaporated to dryness. The residue was acetylated, evaporated to dryness, and the residue was dissolved in hot EtOH. After cooling on ice, the precipitated MPEG was removed by filtration, rinsed with ethanol, the combined filtrate and washings were evaporated to dryness and subjected to chromatography on silica gel to give peracetylated **31**.

Allyl 4,6-O-benzylidene-2-deoxy-3-O-MPEG-DOXyl-2-phthalimido-β-D-glucopyrano- side (**33**). Compound **32** (765 mg, 1.75 mmol) was dissolved in THF (50 mL) at 50° under argon. To this solution was added NaH (60% dispersion in oil, 152 mg, 4 mmol) and after stirring for 30 min, MPEG-DOXCl (**26**; 5.1 g, 1 mmol) and NaI (225 mg, 1.5 mmol) were added as solids. The flask was stoppered tightly and the stirring at 50° was continued for 48 h. Then solid NH₄Cl (214 mg, 4 mmol) was added, and after 5 min the mixture was filtered and the filtrate concentrated to dryness. The residue was taken up in DCM (3 x 50 mL), filtered and evaporated to dryness. This residue was again taken up in dry DCM (50 mL), filtered quickly, cooled in an ice-bath under argon, and precipitated with *t*-butylmethyl ether (500 mL). The solids were collected by filtration and recrystallized from EtOH (250 mL) to yield **33** (4.1 g, 82%).

Allyl 2-deoxy-3-MPEG-DOXyl-2-phthalimido-β-D-glucopyranoside (**34**). The benzylidene group of 3-O-MPEG-DOXyl ether **33** (3.94 g, 0.71 mmol) was cleaved by heating at 100° for 40 min in 60% aqueous AcOH (40 mL). The liquids were removed by evaporation and subsequent co-evaporation with toluene (50 mL). The product was then obtained by recrystallization from EtOH to yield diol **34** (3.74 g, 96%).

Allyl 2-deoxy-6-O-(t-butyldiphenylsilyl)-3-O-MPEG-DOXyl-2-phthalimido-β-D-gluco- pyranoside (**35**). Diol **34** (3.74 g, 0.64 mmol) was dissolved in dichloromethane (10 mL) and imidazole (204 mg, 3.0 mmol) was added as a solid followed by *t*- butyldiphenylsilyl chloride (390 μL, 1.5 mmol) by syringe under argon. The mixture was tightly stoppered and stirred for 16 h. Then the mixture was cooled in an ice bath, precipitated with *t*-butylmethylether (150 mL) and recrystallized twice from EtOH (150 mL) to yield **35** (3.83 g, 93%).

Allyl 4-O-(2,3,4,6-tetra-O-acetyl-β-D-galactopyranosyl)-2-deoxy-6-O-(t-butyldi- phenylsilyl)-3-MPEG-DOXyl-2-phthalimido-β-D-glucopyranoside (**36**, R₁R₂=Phth). Alcohol **35** (0.55 g, 0.10 mmol) was glycosylated with 2,3,4,6-tetra-O-acetyl- galactopyranosyl trichloracetimidate (98 mg, 0.20 mmol) in DCM (5 mL) in the presence of 3A ms (500 mg) at 0-5° for 2 h using trifluoromethanesulfonic anhydride (25 μL, 0.15 mmol) as promoter. After precipitation with *t*-butylmethyl ether and recrystallization from EtOH the polymer-bound disaccharide **36** (512 mg, 88%) was obtained.

Removal of MPEG from DOX: *p-Tolyl 4-O-(2,3,4,6-tetra-O-benzoyl-ß-D-galacto-pyranosyl)-2,3,6-tri-O-benzoyl-ß-D-glucopyranoside* (**38**). A solution of MPEG-DOXyl 4-O-(2,3,4,6-tetra-O-benzoyl-ß-D-galactopyranosyl)-2,3,6-tri-O-benzoyl-ß-D-glucopyranoside (**37**; 0.3 g) in ethanol (50 mL) was hydrogenated at atmospheric pressure over Pd black (30 mg) for 48 h while monitoring hydrogen consumption. The catalyst was filtered off, ethanol evaporated to dryness, and the residue was chromatographed on silica gel using 40% ethyl acetate in hexane, yielding **38** (28 mg).[**]

4. CONCLUDING REMARKS

In summary, the polymer-supported solution synthesis of protected oligosaccharides appears to be a suitable methodology for the synthesis of oligosaccharides in quantities required for biomedical applications. It is clear, that the methodology is in its infancy, and that more efficient reaction conditions will be discovered as it is used for a prolonged period of time. This, of course, applies to any novel methodology. Finally, the MPEG-DOX-saccharides as such, because of their stability and solubility in both polar and apolar media, offer interesting biomedical applications.

ACKNOWLEDGEMENTS

The work from the author's laboratory reviewed in this chapter would not be possible without dedicated excellent organic chemists: Dr. S.P. Douglas, Dr. O.T. Leung, Dr. N. Lupescu, Dr. M.Y. Meah, Dr. D.M. Whitfield, T.J. Kunath and M.D. To. MPEG 12,000 has been made available to the author's laboratory through the courtesy of Dr. S. Verma. The financial support was provided by the Protein Engineering Network of Centres of Excellence, Natural Sciences and Engineering Research Council of Canada, and Medical Research Council of Canada. The author thanks to Dr. N. Lupescu, Dr.

[**]MPEG-DOX is removed completely by hydrogenating **37** (0.25 g) over Pd black (30 mg) in 50% aqueous acetic acid at 50° for 72 h to give **39**.

S.P. Douglas, and Ms. Linda Houston for their help in the preparation of this manuscript.

REFERENCES

1. For recent reviews, see (a) K. Toshima and K. Tatsuta, *Chem. Revs.*, **93**, 1503 (1993); (b) S.H. Khan and O. Hindsgaul in *Molecular Glycobiology* (M. Fukuda and O. Hindsgaul, Eds.) IRL Press, Oxford, p. 206 (1994).

2. (a) H. Paulsen, *Angew. Chem. Int. Ed. Engl.*, **21**, 155 (1982); (b) *ibid.*, **29**, 823 (1990); (c) R.R. Schmidt, *Angew. Chem. Int. Ed. Engl.*, **25**, 212 (1986); (d) P. Fügedi, P.J. Garegg, H. Lönn, and T. Norberg, *Glycoconjugate J.*, **4**, 97 (1987); (e) D.R. Mootoo, V. Date, and B. Fraser-Reid, *J. Am. Chem. Soc.*, **110**, 2662 (1988); (f) P. Konradsson and B. Fraser-Reid, *J. Chem. Soc. Chem., Commun.*, 1124 (1989); (g) B. Fraser-Reid, Z. Wu, U.K. Udodong, and H. Ottoson, *J. Org. Chem.*, **55**, 6068 (1990); (h) G.H. Veeneman, S.H. van Leeuwen, H. Zuurmond, and J.H. van Boom, *J. Carbohydr. Chem.*, **9**, 783 (1990); (i) O. Kanie, M. Kiso, and A. Hasegawa, *J. Carbohydr. Chem.*, **7**, 501 (1988); (j) G.V. Reddy, V.R. Kulkarni, and H.B. Mereyala, *Tetrahedron Lett.*, **30**, 4283 (1989); (k) R.W. Friesen and S.J. Danishefsky, *J. Am. Chem. Soc.*, **111**, 6656 (1989); (l) R.W. Friesen and S.J. Danishefsky, *Tetrahedron,* **46**, 103 (1990); (m) Y.D. Vankar, P.S.Vankar, M. Behrendt, and R.R. Schmidt, *Tetrahedron,* **47**, 9985 (1991); (n) U.E. Udodong, R. Madsen, C. Roberts, and B. Fraser-Reid, *J. Am. Chem. Soc.*, **115**, 7886 (1993); (o) S. Raghavan and D. Kahne, *J. Am. Chem. Soc.*, **115**, 1580 (1993); (p) S.H. Kim, D. Augeri, D. Yang, and D. Kahne, *J. Am. Chem. Soc.*, **116**, 1766 (1994); (q) H. Kondo, S. Aoki, Y. Ichikawa, R.L. Halcomb, H. Ritzen, and C.-H. Wong, *J. Org. Chem.*, **59**, 864 (1994); (r) K.C. Nicolaou, N.J. Bockovich, and D.R. Carcanague, *J. Am. Chem. Soc.*, **115**, 8843 (1993); (s) P. Sinaÿ, *Pure Appl. Chem.*, **63**, 519 (1991); (t) T. Slaghek, Y. Nakahara, and T. Ogawa, *Tetrahedron Lett.*, **33**, 4971 (1992).

3. For instance: (a) K.J. Kaur, G. Alton, and O. Hindsgaul, *Carbohydr. Res.*, **210**, 145 (1991); (b) C.H. Wong, Y. Ichikawa, T. Krach, C. Gautheron-Le Narvor, D.P. Dumas, and G.C. Look, *J. Am. Chem. Soc.*, **113**, 8137 (1991); (c) S.

Roth, *US Pat.*, **5,180,674** (1993); (d) Y. Ichikawa, Y.C. Lin, D.P. Dumas, G.J. Shen, E. Garcia-Junceda, M.A. Williams, R. Bayer, C. Ketcham, L.E. Walker, J.C. Paulson, and C.H. Wong, *J. Am. Chem. Soc.*, **114**, 9283 (1992); (e) for a review, see Y. Ichikawa, G.C. Look, and C.H. Wong, *Anal. Biochem.*, **202**, 215 (1992).

4. (a) J. Thiem and B. Sauerbrei, *Angew. Chem. Int. Ed. Engl.*, **30**, 1503 (1991); (b) B. Sauerbrei and J. Thiem, *Tetrahedron Lett.*, **33**, 201 (1992).

5. For instance, G.C. Look, Y. Ichikawa, G.J. Shen, P.W. Cheng, and C.H. Wong, *J. Org. Chem.*, **58**, 4326 (1993).

6. M. Schuster, P. Wang, J.C. Paulson, and C.-H. Wong, *J. Am. Chem. Soc.*, **116**, 1135 (1994).

7. (a) J.M.J. Fréchet in *Polymer-Supported Reactions in Organic Synthesis* (P. Hodge and D.C. Sherrington, Eds.); Wiley, Chichester, p. 293 (1980); (b) *ibid.*, p. 407; (c) J.M. Fréchet and C. Schuerch, *J. Am. Chem. Soc.,* **93**, 492 (1971); (d) J.M. Fréchet and C. Schuerch, *Carbohydr. Res.*, **22**, 399 (1972); (e) N.K. Mathur, C.K. Narang, and R.E. Williams, *Polymers as Aids in Organic Chemistry*; Academic Press, New York, p. 105 (1980); (f) U. Zehavi, *Adv. Carbohydr. Chem. Biochem.*, **46**, 179 (1988); (g) J.M.J. Fréchet, *Tetrahedron*, **37**, 663 (1981).

8. R. Eby and C. Schuerch, *Carbohydr. Res.*, **69**, 151 (1975).

9. R.D. Guthrie, A.D. Jenkins, and J. Stehlicek, *J. Chem. Soc. (C)*, 2690 (1971).

10. (a) P. Westerduin, G.H. Veeneman, Y. Pennings, G.A. van der Marel, and J.H. van Boom, *Tetrahedron Lett.*, **28**, 1557 (1987); (b) G.H. Veeneman, H.F. Brugghe, H. van den Elst, and J.H. van Boom, *Carbohydr. Res.*, **195**, C1 (1990).

11. (a) L. Yan, L.C.M. Taylor, R. Goodnow, Jr., D. Kahne, *J. Am. Chem. Soc.*, **116**, 6953 (1994); (b) S.J. Danishefsky, K.F. McClure, J.T. Randolph, and R.B. Ruggeri, *Science*, **260**, 1307 (1993).

12. For example, K. Suzuki, G.A. Sulikowski, R.W. Friesen, and S.J. Danishefsky, *J. Am. Chem. Soc.*, **112**, 8895 (1990).

13. C.M. Timmers, G.A. van der Marel, and J.H. van Boom, *Recl. Trav. Chim Pays-Bas,* **112**, 609 (1993).

14.   M. Schuster, P. Wang, J.C. Paulson, and C.-H. Wong, *J. Am. Chem. Soc.*, **116**, 1135 (1994).

15.   (a) E. Bayer and M. Mutter, *The Peptides* (E. Gross and J. Meienhofer, Eds.), Academic Press, New York, volume **2**, p. 286 (1980); (b) G.M. Bonora, C.L. Scremin, F.P. Colonna, and A. Garbesi, *Nucl. Acids Res.*, **18**, 3155 (1990); (c) G.M. Bonora, G. Biancotto, M. Maffini, and C.L. Scremin, *Nucl. Acids Res.*, **21**, 1213 (1993).

16.   J.M. Harris, *J. Macromol. Sci.-Rev. Macromol. Chem. Phys.*, **C25**, 326 (1985).

17.   (a) S.P. Douglas, D.M. Whitfield, and J.J. Krepinsky, *J. Am. Chem. Soc.*, **113**, 5095 (1991); (b) J. J. Krepinsky, S.P. Douglas, and D.M. Whitfield, *US Pat.*, **5,278,303** (1994).

18.   (a) Technical Bulletin: *Polyglycols Hoechst. Polyethylene Glycols: Properties and Applications*, Hoechst, Frankfurt (1983); (b) J. Dale, *Isr. J. Chem.*, **20**, 3 (1980).

19.   S.P. Douglas, D.M. Whitfield, and H.Y.S. Pang, Unpublished Results.

20.   (a) D.M. Whitfield, S.P. Douglas, and J.J. Krepinsky, *Tetrahedron Lett.*, **33**, 6795 (1992); (b) S.P.Douglas and H.Y.S. Pang, Unpublished Results.

21.   J.A. Baptista, J. Chociej, and E. Manawadu, Unpublished Results.

22.   O.T. Leung, D.M. Whitfield, S.P. Douglas, H.Y.S. Pang, and J.J. Krepinsky, *New J. Chem.*, **18**, 349 (1994).

23.   A.A. Kandil, N. Chan, P. Chong, and M. Klein, *Synlett*, 555 (1992).

24.   R. Verduyn, M. Douwes, P.A.M. van der Klein, E. M. Mösinger, G.A. van der Marel, and J.H. van Boom, *Tetrahedron*, **49**, 7301 (1993).

25.   R. Verduyn, P.A.M. van der Klein, M. Douwes, G.A. van der Marel, and J.H. van Boom, *Recl. Trav. Chim. Pays-Bas*, **112**, 464 (1993).

26.   C.M. Dreef-Tromp, H.A.M. Willems, J.E.M. Basten, P. Westerduin, and C.A.A. van Boeckel, *Abstracts, 17th Int. Carbohydr. Symp.*, Ottawa, p. 511 (1994).

27.   M.Y. Meah, D.M.Whitfield, S.P. Douglas, and J.J. Krepinsky, Unpublished Results.

28.   S.P. Douglas, D.M. Whitfield, and J.J. Krepinsky, *J. Am. Chem. Soc.*, **117**, 2116 (1995).

29.   (a) H. Feuer and I. Hooz, In *The Chemistry of the Ether Linkage* (S. Patai, Ed.), Interscience, New York, p. 446 (1967); (b) *ibid.*, p. 460.

30.   D.M. Whitfield, S.P. Douglas, and J.J. Krepinsky, *Abstracts of the Annual Meeting of the Chemical Institute of Canada*, Winnipeg (1994).

31.   D.M. Whitfield, H.Y.S. Pang, J.P. Carver, and J.J. Krepinsky, *Can. J. Chem.*, **68**, 942 (1990).

32.   (a) F. Yamazaki, S. Sato, T. Nukada, T. Ito, and T. Ogawa, *Carbohydr. Res.*, **201**, 31 (1990); (b) H. Paulsen and B. Helpap, *Carbohydr. Res.*, **216**, 289 (1991).

Chapter 10

PROTECTING GROUPS IN OLIGOSACCHARIDE SYNTHESIS

T. BRUCE GRINDLEY

Department of Chemistry, Dalhousie University, Halifax, Nova Scotia,
Canada B3H 4J3

Abstract   A summary of the chemistry of those *O*- and *N*-protecting groups
that are useful for oligosaccharide synthesis is provided. Significant new
groups and methods are highlighted. Particular emphasis is placed on the
regioselectivity of preparative methods. Regioselective methods for removal
are also emphasized where they exist. The compatibility of each protecting
group with common reagents is outlined.

1.   INTRODUCTION

A variety of protecting groups are required to perform oligosaccharide synthesis
in the most concise manner. There are demanding geometric constraints; it is
necessary to place groups regioselectively at specific locations: on primary
alcohols, on *cis*-diols, on *trans*-diols, on 1,2-diols, on 1,3-diols, or on particular
secondary alcohols. The presence of certain functional groups, such as alkenes
or esters or features such as furanose rings in the target oligosaccharide may
require that the protecting groups used for a synthesis not be sensitive to acid, or
base, or reduction, or other commonly used cleavage techniques. Thus, different
categories of target molecules will need different sets of protecting groups.

The properties of the group next to the anomeric center are also important.[1]
Whether this group is participating or non-participating plays a significant role in

control of glycoside stereochemistry. The electronic nature of this group is also critical. Because most reactions at the glycosidic center proceed via electron-deficient intermediates, electron-releasing substituents on O-2, normally ethers, tend to accelerate reactions at the glycosidic center while electron-withdrawing substituents, normally esters or amides, slow reactions. These ideas have been formalized with the introduction of the armed/disarmed concept.[2] Thus, protecting groups on sites near the anomeric center must be chosen carefully both for how they affect reactivity and stereochemistry at the anomeric center and for how readily they can be manipulated.

Excellent extensive general reviews of the chemistry of protecting groups are available.[3,4] In addition, many reviews have appeared of particular aspects of protecting group chemistry as applied to carbohydrates.[5,6,7,8,9,10] This article will emphasize protecting groups that are particularly useful for oligosaccharide synthesis and groups and methods that have been introduced recently.

Particular classes of protecting groups are normally formed under the same conditions but individual members of the class will have particular advantages with respect to cleavage conditions or other properties. Therefore, protecting groups are classified here under the parent member of each class; individual members are highlighted when they are commonly used and where they have different properties than the parent. The general reviews contain many variations in structure that cannot be outlined here because of space limitations.[3,4]

2. ACETALS AND KETALS

Cyclic acetals and ketals have been the key protecting groups in many syntheses of carbohydrate derivatives because they block two hydroxyl groups with excellent regioselectivity.[8,9,11,12] These compounds are normally prepared under acid-catalyzed conditions, using aldehydes or ketones either under thermodynamic or kinetic control. Under thermodynamic control, aldehydes prefer to form six-membered rings, whereas ketones prefer to form five-membered rings.[8,12] *Cis*-fusion to pyranose rings or furanose rings is preferred to *trans*-fusion for both aldehydes and ketones. Under kinetic conditions, the initial bond is formed from the least hindered oxygen atom, a primary oxygen atom if it is available. Ring closure takes place from the next most accessible oxygen atom, to form a five-membered ring preferentially. Since for hexopyranoses the initial bond is to O-6, the only available oxygen atom is O-4 and 4,6-*O*-acetals result for both aldehydes and ketones.[13]

2.A. Isopropylidene Acetals

Isopropylidene acetals are traditionally formed under thermodynamic conditions with acetone and acid.[9,14] Five-membered rings are preferred over six-membered rings because formation of the latter ring size requires that a methyl group adopt an axial orientation at C-2 in the chair conformation of a 1,3-dioxane ring. This orientation is very much destabilized.[15,16] Larger groups on the acetal carbon atom result in even greater preference for 1,3-dioxolane rings.[17,18] Isopropylidene acetals are also commonly formed under kinetic conditions in DMF with 2,2-dimethoxypropane[19] or 2-ethoxypropene[13,20] and a catalytic amount of an acid, often *p*-toluenesulfonic acid. Isopropylidene rings are hydrolyzed with aqueous acid.   Cyclopentylidene acetals are hydrolyzed about twice as fast as isopropylidene acetals while cyclohexylidene acetals are hydrolyzed about six times slower.[21]

2.B. Benzylidene Acetals

Benzylidene acetals are normally prepared under thermodynamic conditions with benzaldehyde and zinc chloride.[22] Higher yields are obtained if the benzaldehyde-zinc chloride complex is formed first.[23] They can also be prepared under kinetic conditions with α,α-dimethoxytoluene in DMF in the presence of an acid catalyst, usually *p*-toluenesulfonic acid.[24] Recently, it was found that *endo*-2,3- or 3,4-*O*-benzylidene acetals are formed very stereoselectively when *cis*-diols are reacted in neat α,α-dimethoxytoluene in the presence of catalytic *p*-toluenesulfonic acid for 5 to 10 min at room temperature.[25]

Benzylidene acetals can be removed completely by acid hydrolysis or by hydrogenolysis.[6,7,8] In addition, a number of reagents have been developed that result in regioselective cleavage of these acetals. 4,6-*O*-Benzylidene acetals are opened under reductive conditions with LiAlCl$_4$/AlCl$_3$. Good yields of the 4-*O*-benzyl ethers are obtained if O-3 is protected, particularly if the protecting group is more bulky than a methyl ether.[7,26,27] 4,6-O-Benzylidene acetals are also reductively cleaved with very good to excellent regioselectivity to give 6-O-benzyl ethers by reaction at room temperature with NaCNBH$_3$ in THF to which HCl in ether is added dropwise.[28] 4,6-O-(4-Methoxybenzylidene) acetals have been shown to react in a similar fashion with NaCNBH$_3$/trifluoroacetic acid.[29] 4,6-O-Benzylidene acetals having benzoate or benzyl protecting groups at O-2 and O-3 are opened with borane-trimethylamine-aluminum trichloride in toluene to give 4-O-benzyl ethers in moderate yields (22 to 54%), but in THF give 6-O-benzyl ethers in 62 to 86% yields.[30] Reductive cleavage of endo-2,3- or 3,4-O-benzylidene acetals with LiAlH$_4$/AlCl$_3$ gives the axial benzyl ether with excellent regioselectivity for the 3,4-O-acetals[31,32] and good regioselectivity (4/1) for the 2,3-O-acetals.[33] Exo-O-benzylidene acetals give equatorial benzyl ethers preferentially. However, these latter products can usually be made more efficiently directly from the diol via stannylene acetal chemistry (see later).[34,35] The NaCNBH$_3$-HCl conditions give the same regiochemistry for cleavage of the exo- and endo- isomers of 2-phenyl-1,3-dioxolane rings as does LiAlCl$_4$/AlCl$_3$.[28]

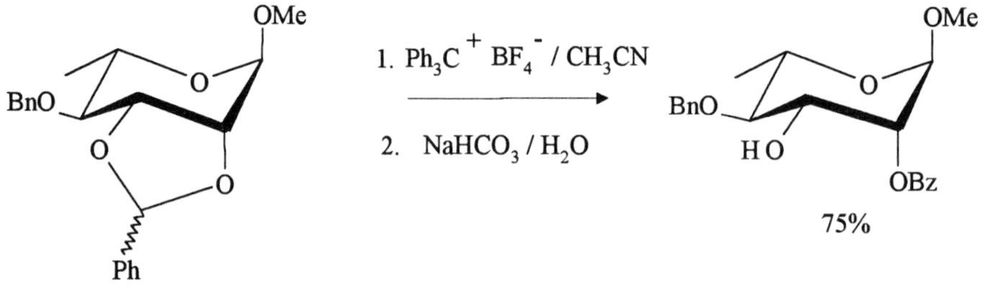

Scheme 1.

Benzylidene acetals derived from cis-diols on pyranose rings are oxidatively cleaved with triphenylcarbenium tetrafluoroborate in acetonitrile followed by aqueous workup to give axial benzoates (see Scheme 1).[36,37] Photochemical cleavage of exo or endo 2-nitrobenzylidene acetals followed by mild oxidation also results in axial 2-nitrobenzoate esters in high yield.[10,38] 4,6-O-Benzylidene acetals open oxidatively with N-bromosuccinimide in carbon tetrachloride at reflux to give 4-O-benzoyl-6-bromo-6-deoxy derivatives in yields of about 70%.[7,39,40]

forming the dibutylstannylene acetal.[125] They are hydrolyzed much faster than acetates by amines.[126]   They are selectively removed by thiourea,[124] 2-mercaptoethylamine,[124] pyridine and water,[127] hydrazine, and hydrazine-dithiocarbonate[128] in lutidine/acetic acid at room temperature. The latter reagent cleaves chloroacetates in the presence of acetates without causing migration. Bromoacetates have been used to avoid migration during removal of chloroacetates with thiourea.[129]

4.C. Benzoates

Benzoates are normally formed with benzoyl chloride in pyridine at room temperature or at 0°C.[130] Mono-O-benzoates can be formed regioselectively by reaction of dibutylstannylene acetals with benzoyl chloride either in the presence or absence of an added nucleophile.[34,35] The regioselectivity obtained is normally the same as that obtained in benzylation reactions (see above) but in the absence of an added nucleophile unusual selectivity is occasionally obtained, i.e. methyl 4,6-O-benzylidene-α-D-mannopyranoside gave 85% 2-O-benzoate and 15% 3-O-benzoate.[131] The 2-O-benzoyl derivative of methyl 4,6-O-benzylidene-α-D-galactopyranoside was obtained in 69% yield after recrystallization by phase transfer catalysis.[132]   1-(Benzyloxy)benzotriazole is easily prepared from 1-hydroxybenzotriazole and benzoyl chloride. It gives good selectivity for reaction at primary alcohols over secondary. It also reacted selectively at O-2 of methyl 4,6-O-benzylidene-α-D-glucopyranoside and -altropyranoside, but the β-D-glucopyranoside anomer gave a 50:43 mixture.[133] Benzoyl cyanide with triethylamine in acetonitrile at -20°C gives good selectivity for reaction at primary alcohols.[125,134,135]

Benzoates are less easily hydrolyzed than acetates. Some perbenzoates, particularly per-p-nitrobenzoates, can be regioselectively mono-de-O-benzoylated at O-2.[136]

4.D. Levulinates

Levulinate esters are introduced with levulinic anhydride in pyridine[137] or with levulinic acid, p-dimethylaminopyridine, and dicyclohexylcarbodiimide.[138] This group can be put on primary positions over secondary with very good selectivity with levulinic acid in the presence of 2-chloro-1-methyl-pyridinium iodide and

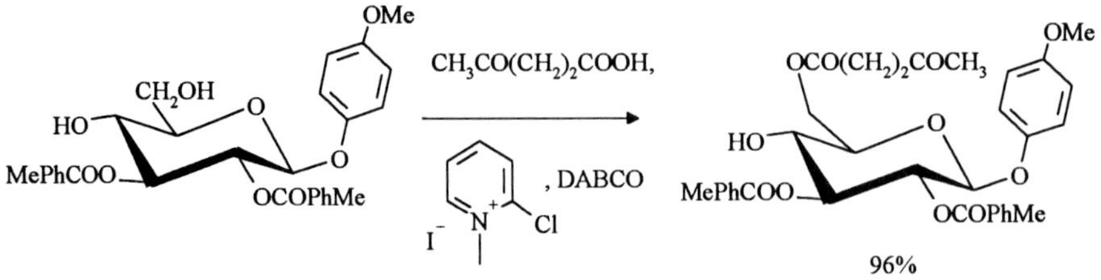

Scheme 11.

DABCO (see Scheme 11).[139,140] It survives hydrolysis of cyclohexylidene acetals with trifluoroacetic acid in dichloromethane at 0°C.[141]

Levulinate esters are cleaved with hydrazine in pyridine-acetic acid in 2 min,[138] conditions that do not cause cleavage of normal esters.[139,142]

4.E. Pivaloates

Pivaloyl or 2,2-dimethylpropanoyl esters can be formed very selectively with pivaloyl chloride in pyridine at -20°C.[143] Methyl α-D-glucopyranoside in ether reacted with pivaloyl chloride and pyridine at 4°C to give the 2,6-di-O-pivaloate in 89% yield.[144] The pivaloate ester has a much decreased tendency to give orthoesters during Koenig's-Knorr glycosidation.[145]

Pivaloates are cleaved by aqueous methylamine at room temperature in about 12 h and by aqueous methanolic tetramethylammonium hydroxide in 2 to 3 h.[146] Pivaloates are not cleaved by hydrazine in ethanol at reflux, conditions that remove phthalimides.[119] Aqueous ammonia removes acetates but not pivaloates.[146]

4.F. Allyloxycarbonates

These esters are formed with allyloxycarbonyl chloride or allyl 1-benzotriazoyl carbonate[147] and pyridine or triethylamine. If sodium hydroxide is used as the base, diols give carbonates.[98] 4,6-O-TIPS acetals of methyl hexopyranosides give 2-O-allyloxycarbonates with excellent regioselectivity.[57] Allyloxycarbonates can be selectively removed in the presence of other esters with tetrakis-

(triphenylphosphine)palladium in THF and morpholine[140,147,148] or with an aqueous or two phase system using Pd(0) and diethylamine.[149]

4.G. Cyclic Carbonates

Cyclic carbonates are very stable to acidic hydrolysis and are more stable to basic hydrolysis than acyclic esters.[150] They were prepared with phosgene in pyridine but trichloromethyl formyl chloride in pyridine[151] or in THF with collidine[56] have now been found to be more convenient. Formation of five-membered rings is strongly favored over six-membered rings and *cis*-fused rings are favored over *trans*-fused rings. For another example of selective formation, see Scheme 12.[56] This group is conveniently cleaved with aqueous barium hydroxide at 70°C or aqueous pyridine at reflux in 15 min.[152]

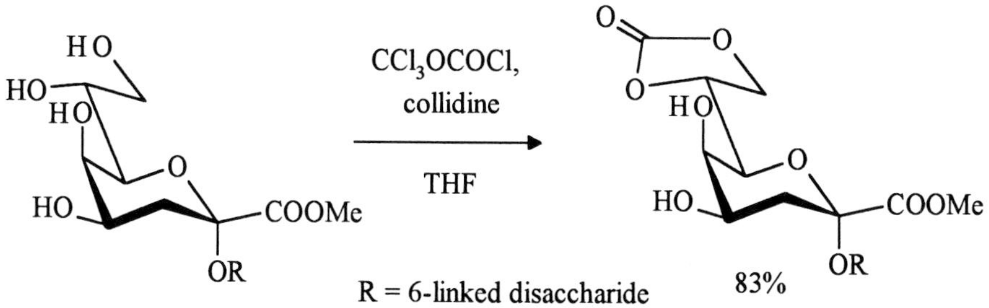

R = 6-linked disaccharide   83%

Scheme 12.

5. AMINO GROUP PROTECTION

Many oligosaccharides contain amino groups, or acetamido or other amide-containing groups. Most of the alcohol protecting groups discussed above have also been employed for amino group protection. Those most commonly used in oligosaccharide synthesis are considered below.

5.A. Acetamides

Acetamides can be made selectively from the amines by reaction with acetic anhydride in water without added base. Pentafluorophenyl acetate in DMF without added base *N*-acetylates selectively, but in the presence of added triethylamine at 80°C, *N*- and *O*-acetylates.[153] This group can be removed with vigorous acidic or basic conditions, but other sensitive groups will also be removed.[118,154] Acetyl groups can be removed selectively in the presence of *O*-acetates from 2-acetamido sugars with HBr in acetic acid or from any acetamide with triethyloxonium fluoroborate in dichloromethane.[154,155]

2-Chloroacetamides are more reactive glycosyl donors than acetamides or oxazolines and can be converted to acetamides with zinc and acetic acid in THF without affecting other protecting groups.[56,156]

5.B. Allyloxycarbonyl Amides

Allyoxycarbonyl amides are formed with allyloxycarbonyl chloride or allyl 1-benzotriazoyl carbonate and triethylamine in THF.[147,157] These derivatives are about as activated as glycosyl donors under Lewis acid catalyzed conditions as phthalimides.[157] They can be removed efficiently under mild conditions with tetrakis(triphenylphosphine) palladium in THF and mild base[147] (see Scheme 13[157]) or with an aqueous or two phase system using Pd(0) and diethylamine.[149]

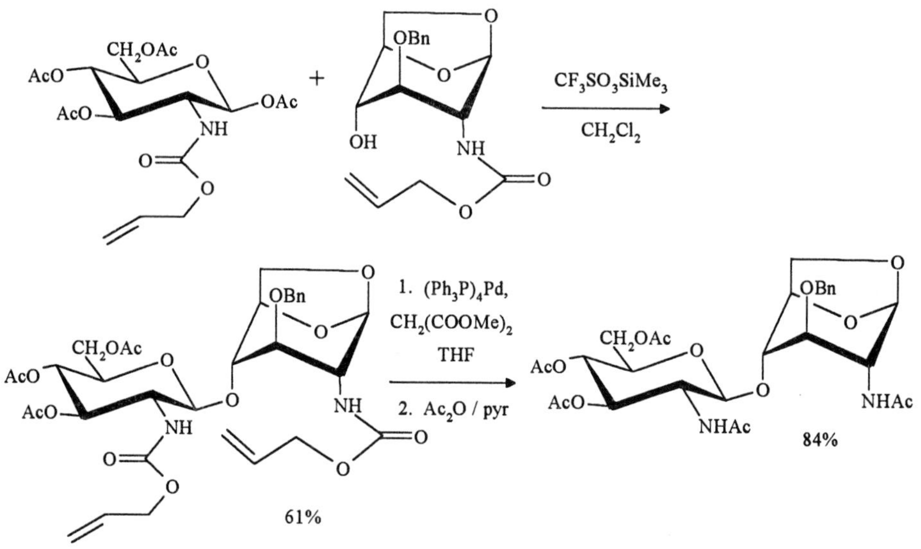

Scheme 13.

5.C. 2,2,2-Trichloroethoxycarbonyl Amides

This group is formed with 2,2,2-trichloroethoxycarbonyl chloride and a very weak base, such as aqueous sodium hydrogen carbonate.[158,159] It is a strongly participating group that does not deactivate adjacent hydroxyls as glycosyl acceptors as the phthalimide group does.[160] It can be removed under mild selective conditions with zinc dust in acetic acid. These conditions do not affect other protecting groups. 2,2,2-Trichloroethoxycarbonyl amides do not give oxazoline by-products during glycoside formation.[158] However, benzyl groups cannot be introduced without loss of this group.[160]

5.D. Phthalimides

Phthalimides are made with phthalic anhydride and a base, often triethylamine.[161,162] 1-Halosugars bearing 2-phthalimdo groups are very reactive glycosyl donors. Phthalimide groups participate strongly during glycoside formation and give good stereochemical control of *trans*-glycoside formation.[162] They are often crystalline. Although they participate at the glycosidic center, they do not form stable orthoamides[162] and cannot form oxazolines as acetamides[1] do. The disadvantage of using phthalimides is that they require vigorous conditions for removal: heating with methanolic or ethanolic hydrazine.[162]

5.E. Azides

Azides are not protecting groups but they have been included here because they serve as masked non-participating amino groups and have become very important for the preparation of α-linked 2-amino-2-deoxy-glycosides.[1] 2-Azido groups are introduced by azidonitration of glycals[163,164] or by addition to epoxides.[165,166] Hydrogen sulfide in pyridine reduces azides to amines in the presence of other reducible groups, such as double bonds or benzyl groups.[167]

REFERENCES

1. H. Paulsen, *Angew. Chem. Int. Ed. Engl.*, **21**, 155 (1982).
2. (a) D.B. Mootoo, P. Konradson, U. Udodong, and B. Fraser-Reid, *J. Am. Chem. Soc.*, **110**, 5583 (1988); (b) C.S. Burgey, R. Vollerthun, and B. Fraser-Reid, *Tetrahedron Lett.*, **35**, 2637 (1994).

3. T.W. Greene and P.G.M. Wuts, *Protecting Groups in Organic Synthesis*, 2nd Ed., John Wiley & Sons, New York (1991).

4. J.F.W. McOmie (Ed.), *Protecting Groups in Organic Chemistry*, Plenum, New York (1973).

5. A.H. Haines, *Adv. Carbohydr. Chem. Biochem.*, **33**, 11 (1976).

6. A.H. Haines, *Adv. Carbohydr. Chem. Biochem.*, **39**, 13 (1981).

7. J. Gelas, *Adv. Carbohydr. Chem. Biochem.*, **39**, 71 (1981).

8. D.M. Clode, *Chem. Rev.*, **79**, 491 (1979).

9. (a) A.N. De Belder, *Adv. Carbohydr. Chem.*, **20**, 219 (1965); (b) A.N. De Belder, *Adv. Carbohydr. Chem. Biochem.*, **34**, 199 (1977).

10. V. Zehavi, *Adv. Carbohydr. Chem. Biochem.*, **46**, 179 (1988).

11. R.F. Brady, Jr., *Adv. Carbohydr. Chem. Biochem.*, **26**, 197 (1971).

12. J.F. Stoddart, *Stereochemistry of Carbohydrates*, Wiley-Interscience, New York, p. 186 (1971).

13. J. Gelas and D. Horton, *Heterocycles*, **16**, 1587 (1981).

14. O.T. Schmidt, In *Methods Carbohydr. Chem.* (R.L. Whistler and M.L. Wolfrom, Eds.), Vol. 2, Academic Press, New York, p. 318 (1963).

15. K. Pihlaja and S. Luoma, *Acta Chem. Scand.*, **22**, 2401 (1968).

16. E.L. Eliel and M.C. Knoeber, *J. Am. Chem. Soc.*, **90**, 3444 (1968).

17. S. Masamune, P. Ma, H. Okumoto, J.W. Ellingboe, and Y. Ito, *J. Org. Chem*, **49**, 2834 (1984).

18. A.I. Myers and J.P. Lawson, *Tetrahedron Lett.*, **23**, 4883 (1982).

19. M.E. Evans, F.W. Parrish, and L. Long, Jr., *Carbohydr. Res.*, **3**, 453 (1967).

20. S. Chládek and J. Smrt, *Collect. Czech. Chem. Commun.*, **28**, 1301 (1963).

21. W.A.R. van Heeswijk, J.B. Goedhart, and J.F.G. Vliegenthart, *Carbohydr. Res.*, **58**, 337 (1977).

22. H.G. Fletcher, Jr., In *Methods in Carbohydr. Chem.* (R.L. Whistler and M.L. Wolfrom, Eds.), Vol. 2, Academic Press, New York, p. 307 (1963).

23. D.H. Hall, *Carbohydr. Res.*, **86**, 158 (1980).

24. M.E. Evans, *Carbohydr. Res.*, **21**, 473 (1972).

25. J. Kerékgýartó and A. Lipták, *Carbohydr. Res.*, **248**, 365 (1993).

26. S.S. Bhattacharjee and P.A.J. Gorin, *Can. J. Chem.*, **47**, 1195 (1969).

27. A. Lipták, I. Jodál, and P. Nánási, *Carbohydr. Res.*, **44**, 1 (1975).

28. P.J. Garegg, H. Hultberg, and S. Wallin, *Carbohydr. Res.*, **108**, 97 (1982).

29. K. von dem Bruch and H. Kunz, *Angew. Chem. Int. Ed. Engl.*, **33**, 101 (1994).

30. M. Ek, P.J. Garegg, H. Hultberg, and S. Oscarson, *J. Carbohydr. Chem.*, **2**, 305 (1983).

31. (a) A. Lipták, *Tetrahedron Lett.*, 3551 (1976); (b) A. Lipták, *Carbohydr. Res.*, **63**, 69 (1978).
32. P. Rollin and P. Sinaÿ, *C. R. Acad. Sci. Ser. C*, **284**, 65 (1977).
33. (a) A. Lipták, P. Fügedi, and P. Nánási, *Carbohydr. Res.*, **65**, 209 (1978); (b) A. Lipták, J. Imre, J. Harangi, and P. Nánási, *Tetrahedron*, **38**, 3721 (1982).
34. (a) S. David and S. Hanessian, *Tetrahedron*, **41**, 643 (1985); (b) M. Pereyre, J. P. Quintard, and A. Rahm, In *Tin in Organic Synthesis*, Butterworths, London, Ch. 11 (1987).
35. T.B. Grindley, *ACS Symp. Ser.*, Vol. 560, Ch. 4 (1994).
36. S. Jacobsen and C. Pederson, *Acta Chem. Scand., Ser. B*, **28**, 866 (1974).
37. H.P. Wessel and D.R. Bundle, *J. Chem. Soc., Perkin Trans. I*, 2251 (1985).
38. P.M. Collins and V.R.N. Munsinghe, *J. Chem. Soc., Perkin Trans. I.*, 921 (1983).
39. (a) S. Hanessian and N.R. Plessas, *J. Org. Chem*, **34**, 1035 (1969); (b) *ibid*, p. 1045; (c) *ibid*, p. 1053.
40. T.L. Hullar and S.B. Siskin, *J. Org. Chem.*, **35**, 225 (1970).
41. (a) S.V. Ley, R. Leslie, P.D. Tiffin, and M. Woods, *Tetrahedron Lett.*, **33**, 4767 (1992); (b) G.-J. Boons, P. Grice, S.V. Ley, and L.L.Yeung, *Tetrahedron Lett.*, **34**, 8523 (1993).
42. D.A. Entwistle, A.B. Hughes, S.V. Ley, and G. Visentin, *Tetrahedron Lett.*, **35**, 777 (1994).
43. R. Schwesinger, C. Hasenfratz, H. Schlemper, L. Walz, E.-M. Peters, K. Peters, and H.G. von Schnering, *Angew. Chem. Int. Ed. Engl.*, **32**, 1361 (1993).
44. R.U. Lemieux and H. Driguez, *J. Am. Chem. Soc.*, **97**, 4069 (1975).
45. S. Hanessian and R. Roy, *Can. J. Chem.*, **63**, 163 (1985).
46. S. Oscarson and A.-K. Tidén, *Carbohydr. Res.*, **247**, 323 (1993).
47. S. Oscarson and M. Szönyi, *J. Carbohydr. Chem.*, **8**, 663 (1989).
48. F.-I. Auzanneau and D.R. Bundle, *Carbohydr. Res.*, **212**, 13 (1991).
49. (a) C.A.A. van Boeckel and J.H. van Boom, *Tetrahedron Lett.*, **21**, 3705 (1980); (b) C.A.A. van Boeckel and J.H. van Boom, *Tetrahedron*, **41**, 4545 (1985).
50. W.T. Markiewicz, *J. Chem. Research (S)*, 24 (1979).
51. H. Paulsen and E.C. Höffgen, *Tetrahedron Lett.*, **32**, 2747 (1991).
52. T. Ziegler, K. Neumann, E. Eckhardt, G. Herold, and G. Pantkowski, *Synlett*, 699 (1991).
53. M. Heuer, K. Hohgardt, F. Heinemann, H. Kuhne, W. Dietrich, D. Grzelak, D. Muller, P. Welzel, A. Markus, Y. van Heijenoort, and J. van Heijenoort, *Tetrahedron*, **50**, 2029 (1994).

54. T. Ziegler, E. Eckhardt, and G. Pantkowski, *J. Carbohydr. Chem.*, **13**, 81 (1994).

55. C.H.M. Verdegaal, P.L. Janese, J.F.M. de Rooij, and J.H. van Boom, *Tetrahedron Lett.*, **21**, 1571 (1980).

56. P. Kosma, M. Strobl, G. Allmaier, E. Schmid, and H. Brade, *Carbohydr. Res.*, **254**, 105 (1994).

57. J.J. Oltvoort, M. Kloosterman, and J.H. van Boom, *Recl. Trav. Chim. Pays-Bas*, **102**, 501 (1983).

58. C.M. McCloskey, *Adv. Carbohydr. Chem.*, **12**, 137 (1957).

59. (a) H.G. Fletcher, Jr., In *Methods in Carbohydr. Chem.* (R.L. Whistler and M.L. Wolfrom, Eds.), Vol. 2, Academic Press, New York, p. 166 (1963); (b) R.E. Wing and J.N. BeMiller, In *Methods in Carbohydr. Chem.* (R.L. Whistler and J.N. BeMiller, Eds.), Vol. 6, Academic Press, New York, p. 368 (1972); (c) J.S. Brimacombe, *ibid*, p. 376.

60. M.E. Tate and C.T. Bishop, *Can. J. Chem.*, **41**, 1801 (1963).

61. R. Kuhn, I. Löw, and H. Trishmann, *Chem. Ber.*, **90**, 203 (1957).

62. I. Croon and B. Lindberg, *Acta Chem. Scand.*, **13**, 593 (1959).

63. (a) T. Iversen and D.R. Bundle, *J. Chem. Soc., Chem. Commun.*, 1240 (1981); (b) H.P. Wessel, T. Iversen, and D.R. Bundle, *J. Chem. Soc., Perkin Trans. I*, 2247 (1985).

64. P.J. Garegg, T. Iversen, and S. Oscarson, *Carbohydr. Res.*, **50**, C12 (1976).

65. V. Pozsgay, *Carbohydr. Res.*, **69**, 284 (1979).

66. R. Eby, K.T. Webster, and C. Schuerch, *Carbohydr. Res.*, **129**, 111 (1984).

67. S. David, A. Thiéffry, and A. Veyrières, *J. Chem. Soc., Perkin Trans. I*, 1796 (1981).

68. S.J. Danishefsky and R. Hungate, *J. Am. Chem. Soc.*, **108**, 2486 (1986).

69. (a) N. Nagashima and M. Ohno, *Chem. Lett.*, 141 (1987); (b) N. Nagashima and M. Ohno, *Chem. Pharm. Bull.*, **39**, 1972 (1991).

70. F.A.W. Koeman, J.W.G. Meissner, H.R.P. van Ritter, J.P. Kamerling, and J.F.G. Vliegenthart, *J. Carbohydr. Chem.*, **13**, 1 (1994).

71. M.E. Haque, T. Kikuchi, K. Yoshimoto, and Y. Tsuda, *Chem. Pharm. Bull.*, **33**, 2243 (1985).

72. Y. Ichikawa, R. Monden, and H. Kuzuhara, *Carbohydr. Res.*, **172**, 37 (1988).

73. C. Monneret, R. Gagnet, and J.-C. Florent, *J. Carbohydr. Chem.*, **6**, 221(1987).

74. S.-H. Chen, R.F. Horvath, J. Joglar, M. Fisher, and S.J. Danishefsky, *J. Org. Chem.*, **56**, 5834 (1991).

75. H. Qin and T.B. Grindley, *J. Carbohydr. Chem.*, **13**, 475 (1994).

76. T. Ogawa, T. Kitajima, and T. Nukada, *Carbohydr. Res.*, **123**, C5 (1983).
77. A. Veyrières, *J. Chem. Soc., Perkin Trans. I*, 1626 (1981).
78. A. Toepfer and R.R. Schmidt, *J. Carbohydr. Chem.*, **12**, 809 (1993).
79. C. Cruzado and M. Martín-Lomas, *Carbohydr. Res.*, **175**, 193 (1988).
80. T. Ogawa and M. Matsui, *Carbohydr. Res.*, **62**, C1 (1978).
81. H.G. Fletcher, Jr., In *Methods in Carbohydr. Chem.* (R.L. Whistler and M.L. Wolfrom, Eds.), Vol. 2, Academic Press, New York, p. 386 (1963).
82. S. Hanessian, T.J. Liak, and B. Vanasse, *Synthesis*, 396 (1981).
83. T. Bieg and W. Szeja, *Synthesis*, 76 (1985).
84. J.R. Hwu, v. Chua, J.E. Schroeder, R.E. Barrans Jr., K.P. Khoudary, N. Wang, and J.M. Wetzel, *J. Org. Chem.*, **51**, 4733 (1986).
85. D. Beaupre, I. Boutbaiba, G. Demailly, and R. Uzan, *Carbohydr. Res.*, **180**, 152 (1988).
86. K. Horita, T. Yoshioka, T. Tanaka, Y. Oikawa, and O. Yonemitsu, *Tetrahedron*, **42**, 3021 (1986).
87. K.P.R. Kartha, F. Dasgupta, P.P. Singh, and H.C. Srivastava, *J. Carbohydr. Chem.*, **5**, 437 (1986).
88. A.B. Smith III, K.J. Hale, H.A. Vaccaro, and R.A. Rivero, *J. Am. Chem. Soc.*, **113**, 2112 (1991).
89. R. Eby, S.J. Sondheimer, and C. Schuerch, *Carbohydr. Res.*, **73**, 273 (1979); M.M. Ponpipom, *Carbohydr. Res.*, **59**, 311 (1977).
90. H. Hori, Y. Nishida, H. Ohrui, and H. Meguro, *J. Org. Chem.*, **54**, 1346 (1989).
91. R. Johansson and B. Samuelsson, *J. Chem. Soc., Perkin Trans. I*, 2371 (1984).
92. Y. Oikawa, T. Tanaka, K. Horito, T. Yoshioka, and O. Yonemitsu, *Tetrahedron Lett.*, **25**, 5393 (1984).
93. P.J. Garegg, L. Olsson, and S. Oscarson, *J. Carbohydr. Chem.*, **12**, 955 (1993).
94. N. Nakajima, R. Abe, and O. Yonemitsu, *Chem. Pharm. Bull.*, **36**, 4244 (1988).
95. (a) R. Gigg, *ACS Symp. Ser.*, **39**, 253 (1977); (b) *ibid*, **77**, 44 (1978).
96. P.A. Manthorpe and R. Gigg, In *Methods in Carbohydr. Chem.* (R.L. Whistler and J.N. BeMiller, Eds.), Vol. 8, Academic Press, New York, p. 305 (1980).
97. J.-C. Jacquinet and P. Sinaÿ, *J. Org. Chem.*, **42**, 720 (1977).
98. F. Guibe and Y. Saint M'Leux, *Tetrahedron Lett.*, **37**, 3591 (1981).
99. R. Lakhmiri, P. Lhoste, and D. Sinou, *Tetrahedron Lett.*, **30**, 4669 (1989).
100. P.J. Garegg, H. Hultberg, and S. Oscarson, *J. Chem. Soc., Perkin Trans. I*, 2395 (1982).
101. J. Gigg and R. Gigg, *J. Chem. Soc. C*, 82 (1966).

102. E.J. Corey and W.J. Suggs, *J. Org. Chem.*, **38**, 3224 (1973).
103. P.A. Gent and R. Gigg, *J. Chem. Soc., Chem. Commun.*, 277 (1974).
104. J. Cunningham, R. Gigg, and C.D. Warren, *Tetrahedron Lett.*, 1191 (1964).
105. R. Gigg and C.D. Warren, *J. Chem. Soc. C*, 1903 (1968).
106. C. Lamberth and M.D. Bednarski, *Tetrahedron Lett.*, **32**, 7369 (1991).
107. R. Boss and R. Schefford, *Angew. Chem. Int. Ed. Engl.*, **15**, 558 (1976).
108. R. Gigg, *J. Chem. Soc., Perkin Trans. I*, 738 (1980).
109. P.A. Gent, R. Gigg, and R. Conant, *J. Chem. Soc., Perkin Trans. I*, 1535 (1972).
110. P.A. Gent, R. Gigg, and R. Conant, *J. Chem. Soc., Perkin Trans. I*, 1858 (1973).
111. E.J. Corey and A. Venkateswarlu, *J. Am. Chem. Soc.*, **94**, 6190 (1972).
112. A. Glen, D.A. Leigh, R.P. Martin, J.S. Smart, and A.M. Truscello, *Carbohydr. Res.*, **248**, 365 (1993).
113. W. Kinzy and R.R. Schmidt, *Tetrahedron Lett.*, **28**, 1981 (1987).
114. J.F. Cormier, *Tetrahedron Lett.*, **32**, 187 (1991).
115. S. Hanessian and P. Lavallee, *Can. J. Chem.*, **53**, 2975 (1975).
116. K. Jansson, S. Ahlfors, T. Frejd, J. Kihlberg, G. Magnusson, J. Dahmén, G. Noori, and K. Stenvall, *J. Org. Chem.*, **53**, 5629 (1988).
117. M.L. Wolfrom and A. Thompson, In *Methods in Carbohydr. Chem.* (R.L. Whistler and M.L. Wolfrom, Eds.), Vol. 2, Academic Press, New York, p. 211 (1963).
118. A. Thompson and M.L. Wolfrom , In *Methods in Carbohydr. Chem.* (R.L. Whistler and M.L. Wolfrom, Eds.), Vol. 2, Academic Press, New York, p. 215 (1963).
119. T. Nakano, Y. Ito, and T. Ogawa, *Carbohydr. Res.*, **243**, 43 (1993).
120. M.K. Gurjar and U.K. Saha, *Tetrahedron*, **48**, 4039 (1992).
121. A. Nudelman, J. Herzig, H.E. Gottlieb, E. Kerinan, and J. Sterling, *Carbohydr. Res.*, **162**, 145 (1987).
122. J. Herzig, A. Nudelman, and H.G. Gottlieb, *Carbohydr. Res.*, **177**, 21 (1988).
123. C.P.J. Glaudemans and M.J.Bertolini, In *Methods in Carbohydr. Chem.* (R.L. Whistler and J.N. BeMiller, Eds.), Vol. 8, Academic Press, New York, p. 271 (1980).
124. A.F. Cook and D.T. Maichuk, *J. Org. Chem.*, **35**, 1940 (1970).
125. S. Rio, J.-M. Beau, and J.-C. Jacquinet, *Carbohydr. Res.*, **255**, 103 (1994).
126. C.B. Reese, J.C.M. Stewart, J.H. van Boom, H.P.M. de Leeuw, J. Nagel, and F.M. de Rooy, *J. Chem. Soc., Perkin Trans. I*, 934 (1975).
127. P.M. Åberg, L. Blomberg, H. Lönn, and T. Norberg, *J. Carbohydr. Chem.*, **13**, 141 (1994).

128. C.A.A. van Boeckel and T. Beetz, *Tetrahedron Lett.*, **24**, 3775 (1983).
129. T. Ziegler, P. Kováč, and C.P.J. Glaudemans, In *Carbohydrates - Synthetic Methods and Applications in Medicinal Chemistry* (H. Ogura, A. Hasagawa, and T. Suami, Eds.) Kodansha, Tokyo, p. 357 (1992).
130. (a) H.G. Fletcher, Jr., In *Methods in Carbohydr. Chem.* (R.L. Whistler and M.L. Wolfrom, Eds.), Vol. 2, Academic Press, New York, p. 231 (1963); (b) J.W. van Cleve, *ibid*, p. 237.
131. C.W. Holzapfel, J.M. Koekemoer, and C.F. Marais, *S. Afr. J. Chem.*, **37**, 19 (1984).
132. W. Szeja, *Synthesis*, 821 (1979).
133. S. Kim, H. Chang, and W.J. Kim, *J. Org. Chem*, **50**, 1751 (1985).
134. M. Havel, J. Velek, J. Pospísek, and M. Souček, *Coll. Czech. Chem. Commun.*, **44**, 2443 (1979).
135. R.M. Soll and S.P. Seitz, *Tetrahedron Lett.*, **28**, 5457 (1987).
136. Y. Ishido, N. Sakairi, M. Sekiya, and N. Nakazaki, *Carbohydr. Res.*, **97**, 51 (1981).
137. A. Hassner, G. Strand, M. Rubinstein, and A. Patchornik, *J. Am. Chem. Soc.*, **97**, 1614 (1975).
138. J.H. van Boom and P.M.J. Burgers, *Tetrahedron Lett.*, 4875 (1976).
139. H.J. Koeners, J. Verhoeven, and J.H. van Boom, *Recl. Trav. Chim. Pays-Bas*, **100**, 65 (1981).
140. T.M. Slaghek, Y. Nakahara, T. Ogawa, J.P.Kamerling, and J.F.G. Vliegenthart, *Carbohydr. Res.*, **255**, 61 (1994).
141. Y. Watanabe, H. Hirofuji, and S. Ozaki, *Tetrahedron Lett.*, **35**, 123 (1994).
142. N. Jeker and C. Tamm, *Helv. Chim. Acta*, **71**, 1895 (1988).
143. S.J. Angyal and M.E. Evans, *Austral. J. Chem.*, **25**, 1495 (1972).
144. S. Tomić-Kulenović and D. Keglević, *Carbohydr. Res.*, **85**, 302 (1980).
145. H. Kunz and A. Harreus, *Liebigs Ann. Chem.*, 41 (1982).
146. B. E. Griffin, M. Jarman, and C. B. Reese, *Tetrahedron*, **24**, 639 (1968).
147. Y. Hayakawa, H. Kato, M. Uchiyama, H. Kajino, and R. Noyori, *J. Org. Chem.*, **51**, 2400 (1986).
148. H. Kunz and H. Waldman, *Angew. Chem. Int. Ed. Engl.*, **23**, 71 (1984).
149. J.P. Genêt, E. Blart, M. Savignac, S. Lemeune, S. Lemaire-Audoire, J.-M. Paris, and J.-M. Bernard, *Tetrahedron*, **50**, 497 (1994).
150. L. Hough, J.E. Priddle, and R.S. Theobald, *Adv. Carbohydr. Chem.*, **15**, 91 (1960).
151. K. Tatsuta, K. Akimoto, M. Annaka, Y. Ohno, and M. Kinoshita, *Bull. Chem. Soc. Jpn.*, **58**, 1699 (1985).
152. R.L. Letsinger and K.K. Ogilvie, *J. Org. Chem.*, **32**, 296 (1967).
153. L. Kisfaludy, T. Mohacsi, M. Low, and F. Drexler, *J. Org. Chem.*, **44**, 654 (1979).

154. S. Hanessian, In *Methods in Carbohydr. Chem.* (R.L. Whistler and J.N. BeMiller, Eds.), Vol. 6, Academic Press, New York, p. 208 (1972).

155. S. Hanessian, *Tetrahedron Lett.*, 1549 (1967).

156. F. Dasgupta and L. Anderson, *Carbohydr. Res.*, **202**, 239 (1990).

157. (a) P. Boullanger, J. Banoub, and G. Descotes, *Can. J. Chem.*, **65**, 1343 (1987); (b) D. Lafont, P. Boullanger, and B. Fenet, *J. Carbohydr. Chem.*, **13**, 565 (1994).

158. M. Imoto, H. Yoshimura, T. Shimamoto, N. Sakaguchi, S. Kusumoto, and T. Shiba, *Bull. Chem. Soc. Jpn.*, **60**, 2205 (1987).

159. H. Paulsen and C. Krogmann, *Liebigs Ann. Chem.*, 1203 (1989).

160. V. Pozsgay, In *Carbohydrates - Synthetic Methods and Applications in Medicinal Chemistry* (H. Ogura, A. Hasagawa, and T. Suami, Eds.) Kodansha, Tokyo, p. 188 (1992).

161. T. Sasaki, K. Minamoto, and H. Itoh, *J. Org. Chem.*, **43**, 2320 (1978).

162. (a) R.U. Lemieux, T. Takeda, and B.Y. Chang, *ACS Symp. Ser.*, **39**, 90 (1976); (b) R.U. Lemieux, S.Z. Abbas, and B.Y. Chung, *Can. J. Chem.*, **60**, 58 (1982).

163. (a) R.U. Lemieux and R.M. Ratcliffe, *Can. J. Chem.*, **57**, 1244 (1979); (b) S. Sabesan and R.U. Lemieux, *Can. J. Chem.*, **62**, 644 (1984).

164. J. Broddefalk, U. Nilsson, and J. Kihlberg, *J. Carbohydr. Chem.*, **13**, 129 (1994).

165. H. Gnichtel, D. Rebentisch, T.C. Tompkins, and P.H. Gross, *J. Org. Chem.*, **47**, 2691 (1982).

166. C.A.A. van Boeckel and G. Petitou, *Angew. Chem. Int. Ed. Engl.*, **32**, 1671 (1993).

167. (a) T. Adachi, Y. Yamada, I. Inoue, and M. Saneyoshi, *Synthesis*, 45 (1977); (b) R.U. Lemieux, S.Z. Abbas, M.H. Burzynska, and R.M. Ratcliffe, *Can. J. Chem.*, **60**, 63 (1982).

Chapter 11

SYNTHESIS OF β-D-MANNOSE CONTAINING OLIGOSACCHARIDES[*]

FRANK BARRESI AND OLE HINDSGAUL

Department of Chemistry, University of Alberta,
Edmonton, Alberta, Canada T6G 2G2

Abstract The chemical synthesis of β-D-mannopyranosides is reviewed with emphasis placed on naturally occurring structures. The insoluble promoter strategy and oxidation-reduction method are discussed as the two most often used procedures to synthesize β-D-mannosides. In addition, the use of 2-oxo glycosyl halides and inter and intramolecular nucleophiles are examined. Finally, intramolecular aglycon delivery is discussed as an alternative method to make β-D-mannopyranosides with complete stereocontrol.

1. INTRODUCTION

The study of the biological role of oligosaccharides has spawned a new area of research termed "glycobiology." There is overwhelming evidence that this class of biopolymer is implicated in numerous biological functions that range from intercellular communication to viral adhesion and cancer.[1-13] As a result, numerous academic and industrial research teams have been exploring what has been called "one of the last great frontiers of biochemistry."[2]

One area of glycobiology that has shown dramatic growth in recent years is

---

[*] Taken in part from F. Barresi, Ph.D. dissertation, University of Alberta, Edmonton, (1994).

the field of synthetic carbohydrate chemistry. In order to fully explore the functions of carbohydrates, methods of synthesizing sufficient amounts of these oligosaccharides are necessary. Consequently, great ingenuity has been applied to the development of new synthetic methods for the synthesis of complex carbohydrates.[14-29]

One of the major challenges in carbohydrate chemistry is the synthesis of the β-D-mannopyranosidic linkage. This linkage has been considered one of the most difficult anomeric linkages to construct,[22] and only the synthesis of the α-D-sialic acid linkage can compare in difficulty.[14,16] In this chapter, the various strategies for the formation of β-D-mannopyranosides will be discussed. The synthesis of structures in which the aglycon is another carbohydrate unit, especially naturally occurring molecules, will be emphasized. For other reviews on the subject, refer to Paulsen[21,22] and Kaji and Lichtenthaler.[30]

2. β-MANNOSIDES THAT OCCUR IN NATURE

Several oligosaccharide structures that contain β-mannopyranosidic linkages have been identified to date. The most important of these structures are the N-linked glycoproteins. All N-linked glycoproteins consist of a non-variable pentasaccharide covalently attached via an asparagine (Asn) residue to the protein portion of the molecule. The core pentasaccharide contains a β-D-Man-(1→4)-β-D-GlcNAc,* as is emphasized in Figure 1.

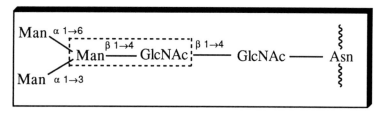

Figure 1.   The pentasaccharide core of N-linked glycoproteins.

Because of the importance of N-linked glycoproteins in biological systems, most of the research conducted in the area of β-mannoside synthesis has been directed

* Monosaccharide Abbreviations: Man = D-mannose; GlcNAc = N-acetyl-D-glucosamine; Glc = D-glucose; Xyl = D-xylose; Rha = L-rhamnose; Gal = D-galactose; Fuc = L-fucose; Neu5Ac = Neuramimic acid, sialic acid.

towards this class of structure.

The β-mannopyranosidic linkage is also found in the spermatozoa and ova of the fresh water bivalve *Hyriopsis schlegellii* [31,32] (Figure 2).

α-D-Man-(1→3)-β-**D-Man-(1→4)-β-D-Glc**-Ceramide

α-D-Man-(1→3)
　　　　　　　＼
　　　　　　　　　β-**D-Man-(1→4)-β-D-Glc**-Ceramide
　　　　　　　／
β-D-Xyl-(1→2)

Figure 2.　Oligosaccharides from *Hyriopsis schlegelii*.

Another source for the β-mannosidic linkage are the *O*-polysaccharides of a variety of *Salmonella* serogroups. For example, the serogroups E_1, E_2, E_3, E_4 and D_2 all contain polymeric units which contain a β-linkage between mannose and the 4-position of L-rhamnose[33] (Figure 3A). In addition, *Salmonella thompson* (serogroup C_1) contains two β-mannosidic linkages, as shown in Figure 3B.[34,35]

A) [6)–β-**D-Man-(1→4)-α-L-Rha**-(1→3)-α-D-Gal(1]

B) [2)–β-**D-Man-(1→2)-α-D-Man**-(1→2)-α-D-Man-(1→2)-
　　β-**D-Man-(1→3)-β-D-GlcNAc**-(1]

Figure 3.　Polymeric units from *Salmonella* serotypes that contain β-mannosidic linkages.

Candida albicans is another organism that contains β-mannoside linkages. In this organism, there exists a polymeric β-Man-(1→2)-β-Man linkage that is part of a phosphomannan-protein complex in the cell wall of these organisms.[36]

3.　THE CHEMICAL SYNTHESIS OF β-MANNOPYRANOSIDES

A variety of approaches have been taken in order to synthesize β-mannosides.

Despite this, no single method can reliably construct this linkage. Every time an oligosaccharide is constructed in which a β-mannoside is present, the focus of the synthetic strategy must center on creating the β-linkage first, and building around this problematic step. There are two main reasons why the β-mannosidic linkage is so difficult to synthesize:

    1.   The anomeric effect[37] stereoelectronically favors the α-mannosidic linkage both thermodynamically and kinetically. Thus, any generation of an oxocarbenium ion (**Z**) (Figure 4) gives predominantly the α-product, in which the anomeric group is axial and;

    2.   Neighboring group participation via an ester (i.e. an acetyl or benzoyl group) gives α-mannopyranosides. This is the method of choice to stereo-selectively form 1,2-*trans* glycosidic linkages, as is the case with β-*gluco* or *galacto* linkages. However, in mannose the 2-OH group is axial, thus participation from the 2-position leads to only the α-product. For this reason a non-participating group such as a benzyl group must be used at C-2.

Most of the strategies developed so far for the synthesis of the β-D-mannosidic linkage can be categorized into several groups:

    A.   The insoluble promoter strategy;[38]

    B.   The oxidation followed by reduction of the 2-position of a β-glucoside to convert it to a β-mannoside;[39-41]

    C.   The use of 2-oxo glycosyl halides that combines both the insoluble promoter and oxidation-reduction strategies;[30,42]

    D.   The use of intermolecular[43-47] and intramolecular nucleophiles[48-51] to epimerize the 2-position of β-glucosides into β-mannosides and;

    E.   Intramolecular aglycon delivery[52-56] that can synthesize β-manno-pyranosides with complete stereocontrol.

These strategies are discussed below and cover most of the methods that have been used to make at least disaccharide linkages. Other methods such as the use of mannosidase enzymes in transmannosylations[57,58] anomeric radicals,[59] dibutylstannylene complexes,[60] anomeric,[61] and 2-*O*-sulfonates,[62] and α-mannosyl thiocyanates[63] have also been reported. However, these methods have not yet been as successful as the five groupings listed above in reliably producing high yields of β-mannosides with acceptable stereoselectivity.

3.A. The Use of Insoluble Promoters

The choice of an insoluble promoter to activate a glycosyl halide (bromide or chloride) is the most often used procedure for the synthesis of β-mannosides (Figure 4).

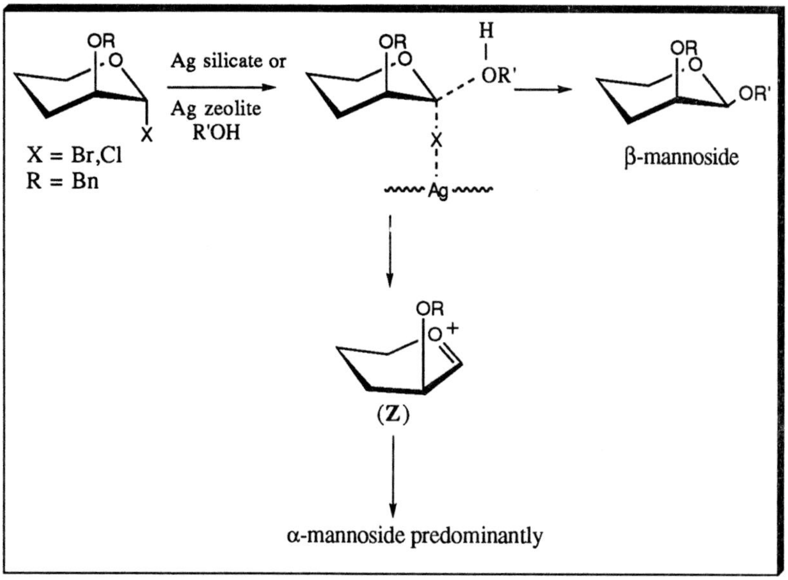

Figure 4. Insoluble promoter strategy for making β-mannopyranosides.

Two crucial factors must be controlled in order for the reaction to give primarily the 1,2-*cis* configuration. The first factor is that the displacement of the glycosyl halide must occur in an S_N2-like fashion. This is because any oxocarbenium ion formation (the S_N1 reaction) will lead to a preponderance of α-glycoside due to a kinetic anomeric effect.[37] The second factor is that the more stable α-halide must be activated in preference to the more reactive β-halide. This allows for the S_N2 displacement to give the β-glycoside. The use of an insoluble promoter is ideal for this scenario since the surface of the promoter can preferentially complex to the α-anomer thereby protecting the α-side of the molecule from attack by the alcohol. The incoming nucleophile can therefore attack the activated α-halide in an S_N2 fashion to give a β-mannoside.

The first reported case using insoluble silver salts for the formation of β-mannosides was by Gorin and Perlin in 1961 (Table 1, entry A).[64] They used traditional Koenigs-Knorr[38] chemistry with silver oxide as the promoter and the linkage formed was with the highly reactive 6-position of glucose. Other uses of silver oxide are shown in entries B,[65] C,[66] and D[67] to form the β-D-Man-(1→4)-α-L-Rha linkage that is present in the *Salmonella* serotypes D$_2$ and E.

The use of other insoluble promoters such as silver salicylate[68] (Table 1, entry E), silver tosylate,[61] silver imidazolate,[69] and thallium zeolite[70] have also been used to construct β-mannosides. In these cases, only the reactive primary position of carbohydrate alcohols or highly reactive non-carbohydrate alcohols were used. Silver carbonate (entry F)[71] is another common, first generation insoluble silver salt and has been used frequently with reactive alcohols.[72-74]

A significant breakthrough that enabled this procedure to be applied to less reactive alcohols came in 1981 by Paulsen and Lockoff (Table 1, entry G).[74] They utilized the much more reactive silver silicate as an insoluble promoter and obtained satisfactory yields with secondary carbohydrate alcohols. Prior to this discovery, attempted couplings with the notoriously unreactive 4-hydroxyl position of GlcNAc derivatives could not be achieved with adequate results. The introduction of this catalyst set the stage for the elegant syntheses of several N-linked oligosaccharides (Figure 5).[21,26,75-85] Further examples of Paulsen's silver silicate method are shown in Table 1 (entries H-P). In many cases, the acceptor used is an anhydro sugar that locks the hydroxyl group into the axial position thus making it more reactive. Ogawa and coworkers utilized the more direct acceptor in entry M of Table 1 to give the β-linked product in 40% yield.[76] The realization that an electron withdrawing group at the 4-position of the donor increases the amount of β-mannoside by reducing the amount of oxocarbenium ion that forms[86] is shown in entries J and K. Ogawa and coworkers took advantage of this fact to obtain an impressive yield of 48% β-mannoside with only 19% of the corresponding α-anomer for the difficult β-D-Man-(1→4)-β-D-GlcNAc linkage (entry N).[83]

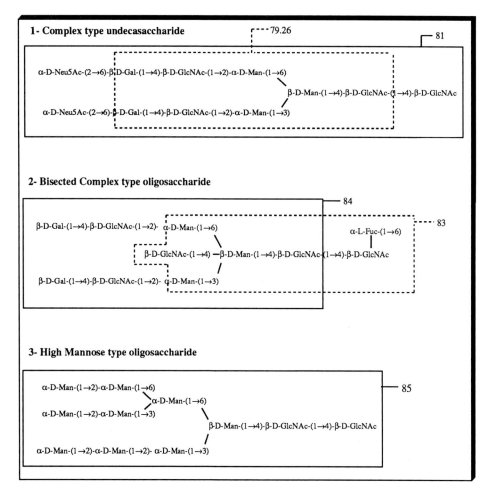

Figure 5. *N*-Linked oligosaccharide structures. (Boxed structures with reference numbers pertain to chemical syntheses).

Garegg and coworkers have used the promoter silver zeolite[88] to achieve similar success in the formation of β-mannopyranosides with secondary carbohydrate alcohols (Table 1, entry T). Silver zeolite is the second most common promoter used to make β-mannosides (after Paulsen's silver silicate) and is considered less reactive.[86] Takeda and coworkers[32,89] have successfully used this promoter in the synthesis of structures from *Hyriopsis schlegellii* (entries R, S). Silver zeolite has also been used to form β-mannopyranosides with simple alcohols.[70,90]

Table 1. Insoluble silver salt promoters used in β-mannoside formation.[a]

Donor	Acceptor	Procedure	Yield (%) α	Yield (%) β	Reference
(A)		Ag_2O/I_2	trace	45	64
(B)		1. Ag_2O 2. TFA 3. Ac_2O, Pyr.	<5	73	65
(C)		Ag_2O	0	91	66
(D)		Ag_2O / $AgClO_4$	11	77	67
(E)		Ag salicylate	4	48	68
(F)		$AgCO_3$	14	81	71
(G)		Silver Silicate	0	81	74
(H)		Silver Silicate	10	67	75
(I)		Silver Silicate	11	65	75
(J)		Silver Silicate	5	40	86
(K)		Silver Silicate	40	40	86

Table 1 con't. Insoluble silver salt promoters used in β-mannoside formation.[a]

Donor	Acceptor	Procedure	Yield (%) α	β	Reference
(L)		Silver Silicate	11	65	82
(M)		Silver Silicate	36	40	76
(N)		Silver Silicate	19	48	83
(O)		Silver Silicate	37	40	87
(P)		Silver Silicate	18	77	34
(Q)		Silver Silicate	17	46	32
(R)		Silver Zeolite	10	70	32
(S)		Silver Zeolite	--	72	89
(T)		Silver Zeolite	21	50	88

[a] Abbreviations: Ac = acetate; Me = methyl; Bn = benzyl; Ph = phenyl; NPhth = N-phthalimido

3.B. The Oxidation-Reduction Method

The oxidation-reduction method is the second most commonly used strategy for the synthesis of β-mannosides. This reliable but laborious[39] procedure creates a β-glucoside and then epimerizes the 2-position by deprotection, oxidation and reduction (Figure 6). One obvious advantage is the replacement of the difficult β-mannosylation step with the simpler β-glucosylation step. After deprotection and oxidation, the reduction step gives a predominance of *manno* configuration, because equatorial hydride attack on the 2-oxo intermediate is favored over axial hydride attack.

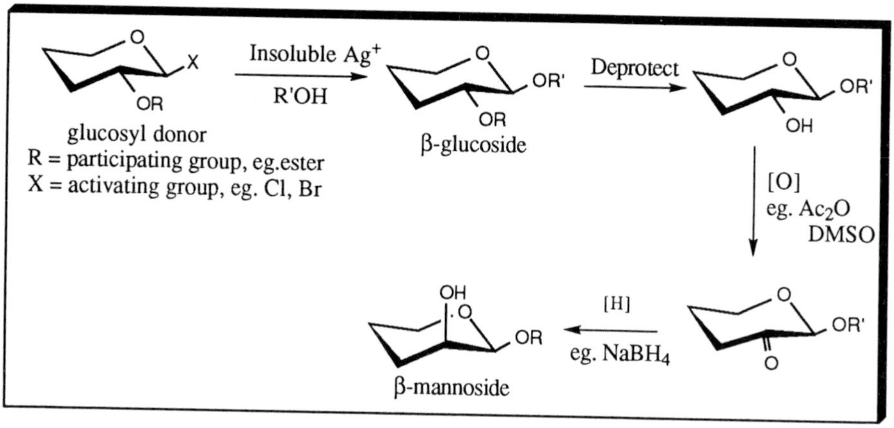

Figure 6. The oxidation-reduction method.

Theander[41] first demonstrated in 1958 that the stereoselective reduction of methyl β-D-arabino-hexopyranosidulose gave preferentially the *manno* configuration over the *gluco* configuration. However, it was not until 1972 that Lindberg and coworkers utilized this fact to synthesize β-mannosides[39] (Table 2, entry A). Jeanloz and coworkers applied this strategy to the synthesis of the β-D-Man-(1→4)-β-D-GlcNAc linkage present in the core pentasaccharide of N-linked glycoproteins[91-93] (entries B, C and D in Table 2). This was the first synthesis of this linkage, although an earlier report synthesized the linkage via the degradation of a naturally occurring disaccharide.[94] Kotchetkov and coworkers also used this oxidation-reduction procedure to synthesize a β-D-Man-(1→4)-α-L-Rha linkage present in the O-specific polysaccharide of *Salmonella* strains (Table 2, entry E).[95,96] Garegg and Hallgren[34] utilized this procedure in the synthesis of the β-

Table 2. Oxidation reduction method for the synthesis of β-mannosides.[a]

Donor	Acceptor	Procedure	Product	Overall Yield (% β)	Ref.
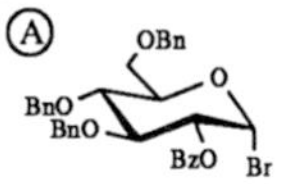 Ⓐ	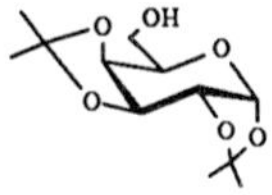	1.HgBr$_2$ (77%) 2.NaOMe (89%) 3.Ac$_2$O/DMSO (79%) 4.H$_2$/Pt,H$_2$/Pd/C (90%) (19:1 *manno:gluco*)	A	49	39
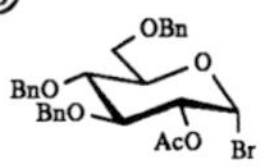 Ⓑ	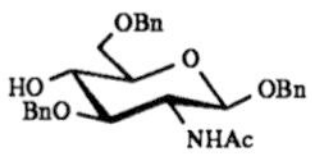	1.Hg(CN)$_2$ (67%) 2.NaOMe (89%) 3.Ac$_2$O/DMSO (56%) 4.NaBH$_4$ (69%)	B	23	91
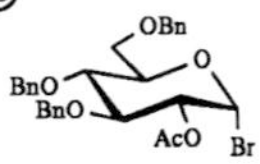 Ⓒ	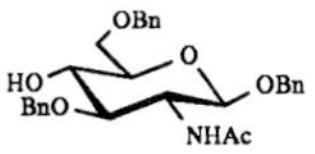	1.AgOTf (62%) 2.NaOMe (74%) 3.Ac$_2$O/DMSO (82%) 4.NaBH$_4$ (84%)	C	31	92
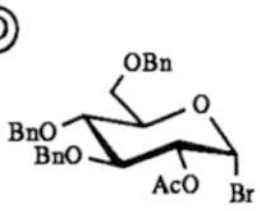 Ⓓ	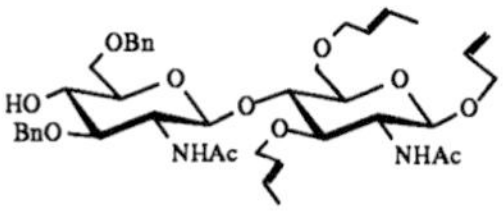	1.AgOTf (44%) 2.NaOMe (98%) 3.Ac$_2$O/DMSO 4.NaBH$_4$ (72%)	D	31	93
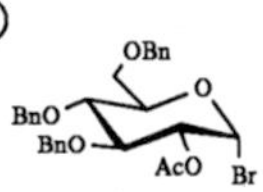 Ⓔ	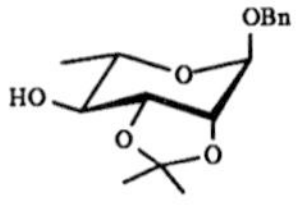	1.Hg(CN)$_2$ (80%) 2.NaOMe (80%) 3.Ac$_2$O/DMSO (75%) 4.NaBH$_4$ (88%) (15:1 *manno:gluco*)	E	42	95
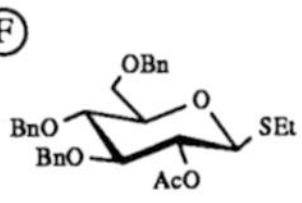 Ⓕ	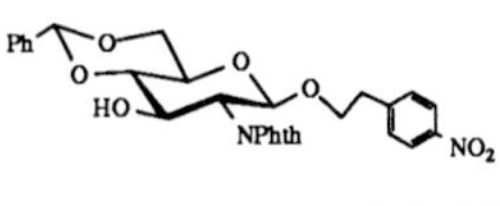	1.NIS/AgOTf (64%) 2.NH$_2$NH$_2$, Ac$_2$O, PtO$_2$, TFAA, NaOMe (61%) 3.Ac$_2$O/DMSO 4.NaBH$_4$ (80%) (10:1 *manno:gluco*)	F	31	34

Products:

A =

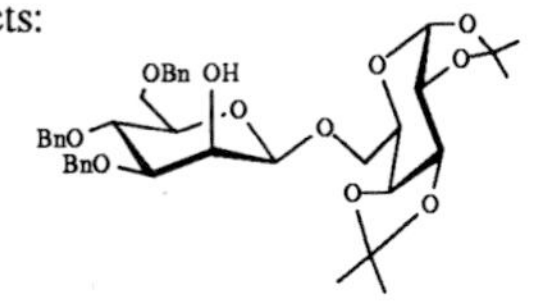

E =

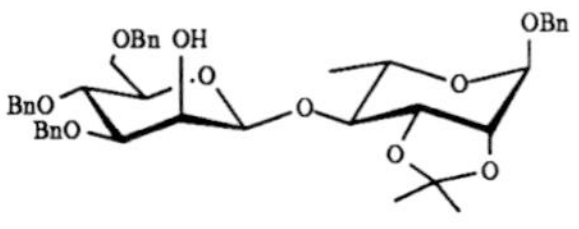

B = C =

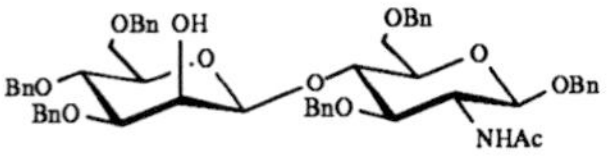

F =

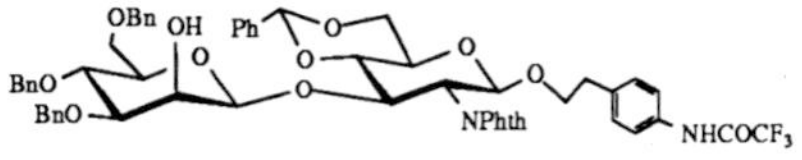

D =

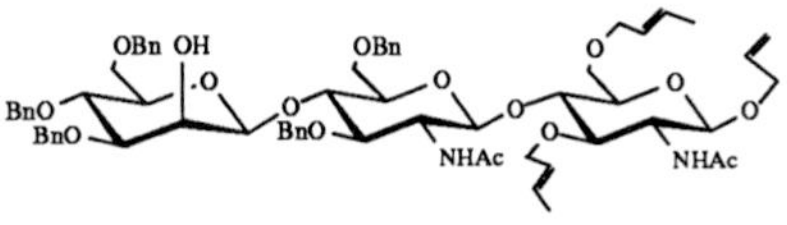

[a]Abbreviations: Ac$_2$O = acetic anhydride; DMSO = dimethyl sulfoxide; Tf = trifluoromethane sulphonyl; NIS = *N*-iodosuccinimide; TFAA = trifluoroacetic anhydride; Et = ethyl

D-Man-(1→3)-β-D-GlcNAc linkage present in *Salmonella thompson* (entry F). Danishefsky and coworkers have recently combined their glycal glycosylation strategy[27,28] with the oxidation-reduction method to give high yields of primary linked di- and trisaccharides. Other examples of the use of this strategy for more reactive alcohols have also been reported.[98-101]

3.C. The Use of 2-Oxo Glycosyl Halides

An extension of the insoluble promoter and oxidation-reduction strategies has been recently disclosed by Lichtenthaler and coworkers.[30,42] They utilized 2-oxo glycosyl bromide derivatives as donors to selectively make β linkages, followed by reduction of the keto functionality to give β-mannosides (Figure 7).

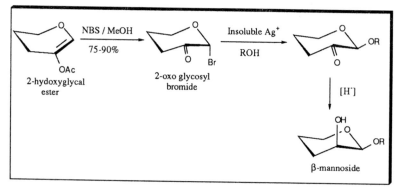

Figure 7. The use of 2-oxo glycosyl donors for the synthesis of β-mannosides.

These glycosyl donors can be obtained in high yield (75-90%) from the treatment of 2-hydroxy glycal esters with *N*-bromosuccinimide (NBS) in methanol. The unique aspect of this procedure is the electron withdrawing 2-keto group which limits oxocarbenium ion formation.[86] This enables an S_N2 displacement to occur at the anomeric center to give preferentially the β-isomer. Reduction of the keto group then gives a preponderance of β-mannoside.

Reactions employing 2-oxo glycosyl halides have given promising results. For example, the best results to date are in the synthesis of the disaccharide component of *Hyriopsis schlegelii*,[31] which was achieved in a combined yield of 61% for the glycosylation and reduction steps (Figure 8). The use of the promoter silver aluminosilicate (Van Boeckel's Catalyst[86]) gave superior results over

Paulsen's silver silicate catalyst. Despite the success of this procedure, the inherent low reactivity of the 2-oxo glycosyl bromides may prove to be a problem when applied to more complex syntheses such as *N*-linked structures.

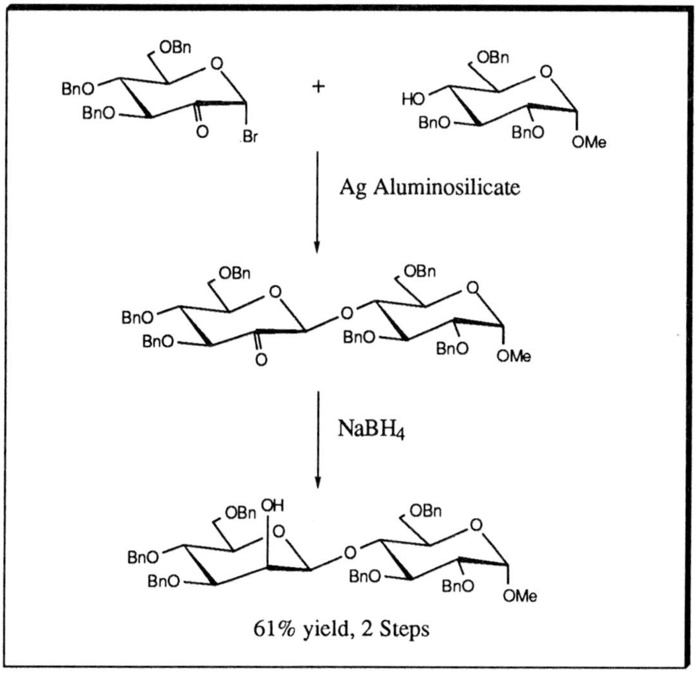

Figure 8. The 2-oxo glycosyl donor approach in the synthesis of a disaccharide.

3.D. The Use of Intermolecular and Intramolecular Nucleophiles

3.D.i. Intermolecular nucleophiles

Another approach for the synthesis of β-mannosides utilizes intermolecular nucleophiles. The best results have been obtained by epimerizing the 2-position of a β-glucose residue[43,45] or both the 2- and 4-position of a β-galactose residue[44,46] to give a β-mannoside. The strategy relies upon creating a highly reactive leaving group at the position that is to be inverted and treating it with a strong nucleophile. The nucleophile causes an S_N2 displacement that epimerizes the position on the carbohydrate ring. The displacement at the 2-position with intermolecular nucleophiles is especially difficult since the nucleophile must approach the 2-position carbon from an axial orientation.

The progress of the intermolecular nucleophile method is summarized in Table 3.

Table 3. The use of intermolecular nucleophiles for the synthesis of β-mannosides.[a]

Starting Material	Product	Nucleophile	Yield (%)	Reference
(A)		KOBz	62	43
(B)		Bu₄NOBz	64	44
(C)		Bu₄NOBz	90	45
(D)		Bu₄NOBz	47	46

R =   R' =   Ims =

[a]Abbreviations: Bz = benzoyl; Bu = butyl

In entry A, Miljkovic and coworkers[43] displaced a mesylate leaving group with a benzoate to give the methyl β-mannoside in 62% yield. In entry C, David and coworkers[45] extend this concept to disaccharides, making the β-D-Man-(1→4)-β-D-Glc derivative. In this example, an imidazylate is used as a leaving group. This group is made from *N,N'*-sulfuryldi-imidazole and is easier to handle than the more commonly used triflic anhydride. David *et al*[44,46] also carried out the double displacement of a galactose residue to a mannose residue in entries B and D by using the highly reactive *O*-triflate leaving groups. Note that entry D represents a protected form of the β-D-Man-(1→4)-β-D-GlcNAc linkage present in *N*-linked carbohydrates. The elegance of the double inversion is exemplified by starting the reaction at room temperature (25°C) to first allow displacement of the more reactive 4-position to give the *gluco* configuration, then heating to 100°C to displace the less accessible 2-position to give the desired product.

3.D.ii. Intramolecular nucleophiles

Kunz and Gunther have recently published a novel method to create β–mannoside linkages.[48-50] They utilize an intramolecular epimerization of the 2-position of a glucose residue from the neighboring 3-position (Figure 9). This approach again replaces the difficult β-mannosylation with the easier β-glucosylation.

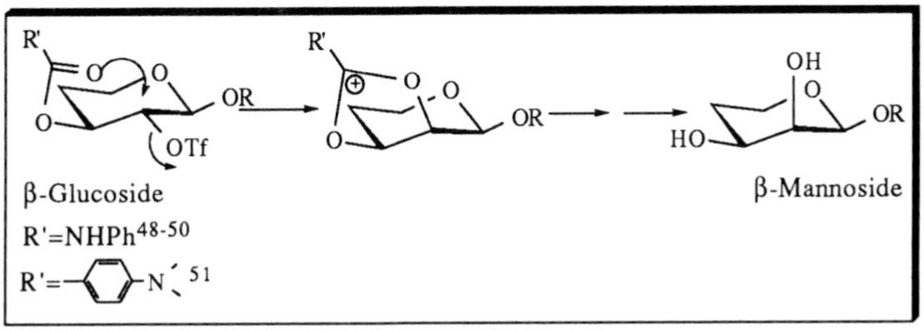

Figure 9.   The synthesis of β-mannosides by intramolecular inversion.

The procedure is outlined in Figure 10 in the synthesis of a tetrasaccharide block of the core pentasaccharide of *N*-linked glycoproteins. The methodology involves making a β-glucosidic linkage with a phenylurethane on the 3-position of the glycosyl donor. The phenylurethane serves as an intramolecular nucleophile to invert the neighboring 2-triflate. After optimization, the overall yield for the indirect formation of the β-mannoside linkage is quite high despite the numerous steps required. One disadvantage is the necessity of the benzylidene ring in order for the displacement to occur. A similar strategy was attempted by Griffith[51] in which a *p-N,N'*-dimethylaminobenzoyl group was used at the 3-position as the intramolecular nucleophile. Ring contraction produced many side products when a 4,6-*O*-benzylidene group was not present on the glucose unit.

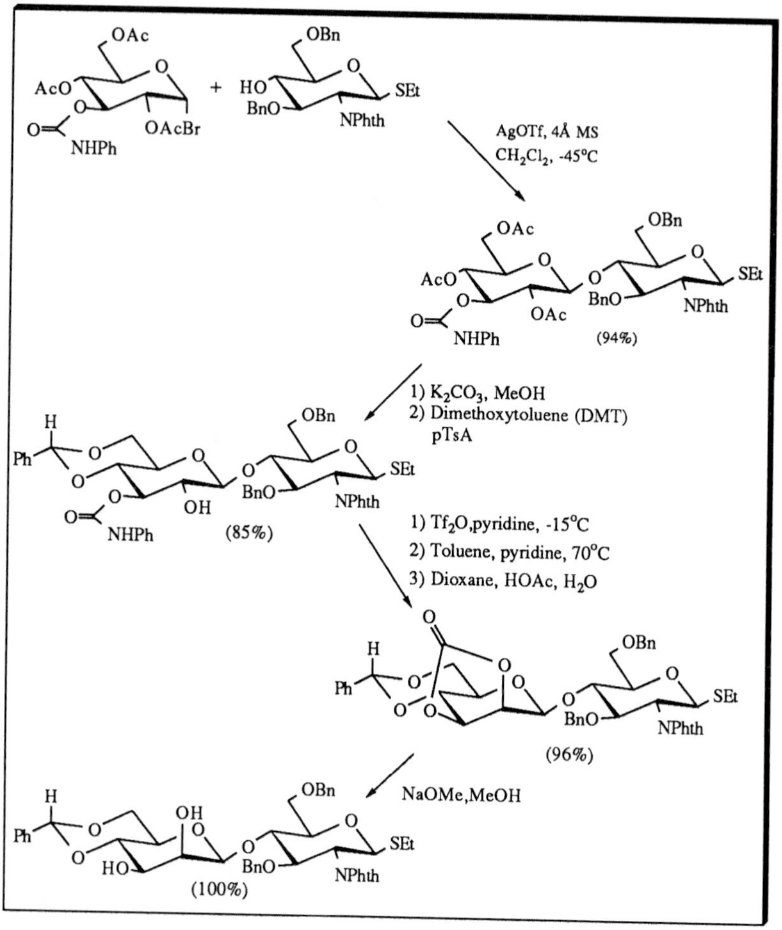

Figure 10. Intramolecular inversion via participation from the 3-position.

3.E. Intramolecular Aglycon Delivery

The primary disadvantage of all of the strategies described so far is that none of the reactions proceed with complete stereocontrol, thus leading to often difficult anomer or epimer separations. Recently, an approach termed intramolecular aglycon delivery (IAD)[52-56] has been used that solves this problem by creating β-mannosidic linkages with complete stereoselectivity.

The general strategy for IAD capitalizes on the principle of intramolecular transfer within a pyranose ring (Figure 11).

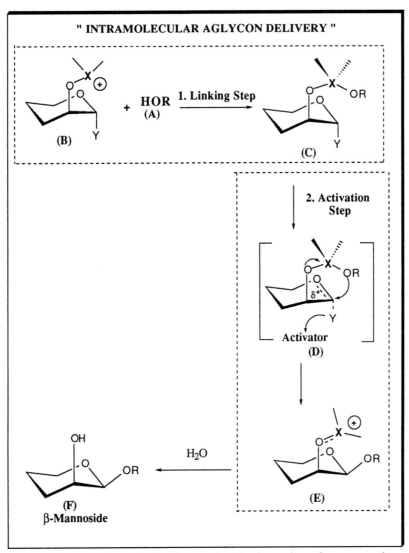

Figure 11. General strategy for the intramolecular inversion of an anomeric center via participation from the 2-position.

For D-mannose, covalent attachment of an aglycon (A) to the 2-position of an appropriately derivatized sugar (B) yields the adduct C. Activation of the anomeric position of C leads to stereocontrolled intramolecular delivery proceeding through a five membered transition state (D) to give the cis-linked intermediate E. Quenching of E with water gives β-mannoside F. The intramolecular nature of the reaction controls the stereochemistry, ensuring only the *cis*-linked product is formed. As can be seen in Figure 11, there are two critical steps involved in IAD:

1. The Linking Step: This step involves covalently attaching the aglycon to *O*-2 of the pyranose ring. In order for IAD to be successful, this first step must be a high yielding, convenient procedure that does not interfere with any other groups in the carbohydrate molecule. Ideally, no new stereogenic atoms should be created in the process or **C** will represent two diastereomers.

2. The Activation Step: This step involves activation of the anomeric carbon. It is essential that the anomeric substituent be stable enough to withstand the earlier modification of the 2-position in step 1. Furthermore, there must also be a wide range of activation conditions available that can be used to enhance the viability of this procedure. Most importantly, activation procedures must be mild enough not to cause any destruction of the covalent linker formed in step 1 prior to stereocontrolled delivery of the aglycon to the anomeric carbon.

Two similar strategies have been reported so far for the synthesis of β-mannopyranosides by IAD. One approach utilizes a carbon acetal (Figure 11, X = C)[52-55] and the other approach uses a silicon acetal[56] (Figure 11, X = Si). To date, only the carbon acetal strategy has been successful in making β-mannosides with highly unreactive secondary carbohydrate alcohols. Although the silicon strategy has been unsuccessful so far in coupling secondary carbohydrate alcohols,[102] it has been extended to the stereocontrolled synthesis of α-gluco- and galactopyranosides.[103-106] The progress of IAD in the synthesis of *cis*-linked sugars has recently been reviewed.[107]

In the carbon acetal strategy, the linking step involves the acid catalyzed coupling of the vinyl ether (Figure 12, structure **A**) with the aglycon alcohol to produce the isopropylidene ketal (**B**). The vinyl ether can be prepared from the corresponding 2-*O*-acetate in good yield (70-90%) by the use of a methylene transfer complex such as Tebbe's reagent.[108-110] Electrophilic activation of the resulting thioglycoside results in delivery of the aglycon to yield the β-mannoside (**C**).

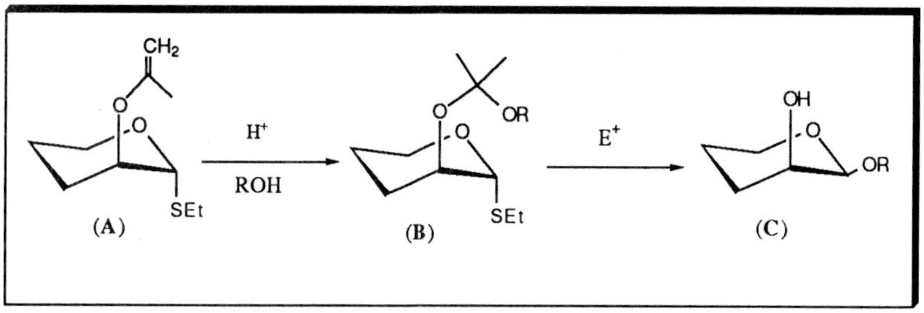

Figure 12.   Synthesis of disaccharide β-mannosides by IAD.

The systems studied thus far are shown in Table 4. In all cases the reactions proceed with complete stereoselectivity, with no α-mannopyranosides being obtained. Entries A and B show that the procedure gives good yields for simple systems. Entries C and D show the procedure is also viable for highly unreactive secondary carbohydrate alcohols, including a precursor to the β-D-Man-(1→4)-β-D-GlcNAc linkage of the core pentasaccharide (entry D). Entries E and F show limitations to the methodology when applied to larger sections of the core pentasaccharide.

A modification of the carbon acetal strategy has been reported by Ito and Ogawa.[55] In this procedure, a *p*-methoxy benzyl group on the 2-*O*-position of mannose is oxidized and the intermediate carbocation trapped by an alcohol. Activation of the anomeric group (a glycosyl fluoride) then leads to stereocontrolled β-mannopyranosides in yields that rival the isopropylidene strategy. More importantly, the procedure is a one-pot strategy and can be applied to precursors of the naturally occuring β-D-Man-(1→4)-β-D-GlcNAc linkage.

The approach used by Stork and Kim[56] to make β-mannosides utilized a silicon tether (Figure 11, X=Si) to covalently attach the aglycon to the 2-position of D-mannose (Figure 13). Their scheme involves the reaction of a primary alcohol with n-butyl lithium (n-BuLi) in tetrahydrofuran (THF) to generate an alkoxide ion. This is followed by the addition of dichlorodimethylsilane to form the chlorodimethylsiloxane compound. Reaction of this structure with the thiophenylglycoside gave the silicon tethered structure in quantitative yield. Activation of the anomeric group using Kahne's sulfoxide strategy[111] gave the β-mannoside in 68% yield. Unfortunately, this method was not successful when applied to secondary carbohydrate alcohols due to the inability in coupling the

Table 4. β-Mannosides synthesized by IAD.[52-54]

Coupling Step		Activation Step	
Mixed Acetal Product	% Yield, Conditions[a]	β-Mannoside Product	% Yield, Conditions[a]
(A)	88, (i)		60, (iii)
(B)	74, (i)		60, (iv)
(C)	57, (i)		77, (v)
(D)	55, (i)		51, (v)
(E)	38, (ii)		27, (v)
(F)	25, (ii)		28, (v)

[a]Conditions: (i) pTSA, CH_2Cl_2, -40°C
(ii) CSA (camphorsulphonic acid), CH_2Cl_2, -40°C
(iii) NBS (5.0 eq.), CH_2Cl_2
(iv) NIS (5.0 eq.), CH_2Cl_2
(v) NIS (5.0 eq.), 4-Methyl-Di-t-Butyl Pyridine (4-Me-DTBP, 5.0 eq.), CH_2Cl_2

two unreactive sterically hindered alcohols with a silicon tether.[102]

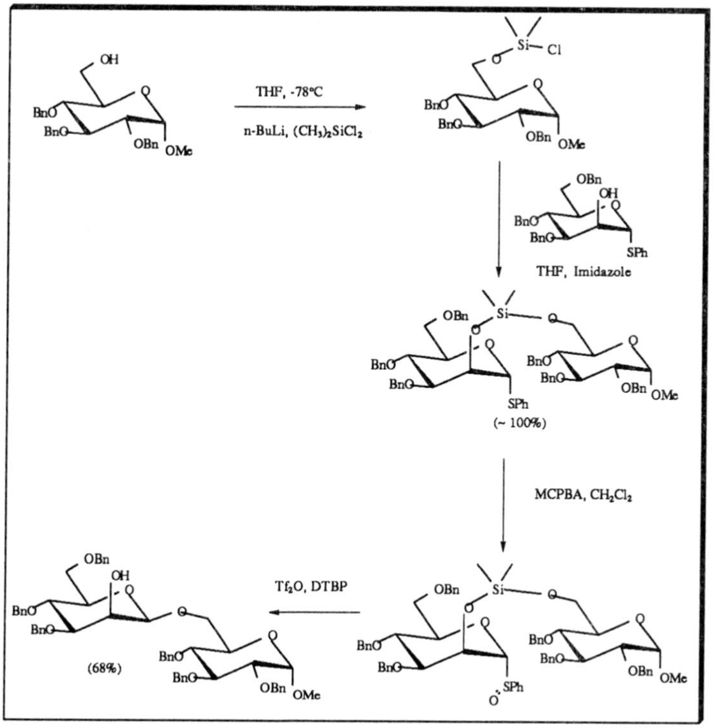

Figure 13.   The stereocontrolled synthesis of β-mannosides using a silicon tether.[56]

4. CONCLUSION

The synthesis of β-D-mannopyranosides is extremely challenging as can be seen from the great volume of research reported in this area. Although there remains no absolute method of choice to synthesize this linkage, it seems that the collective research efforts of the numerous workers in this field will eventually solve this long-standing problem in carbohydrate chemistry.

REFERENCES

1.   A. Varki, *Glycobiology*, 3, 97 (1993).
2.   G. W. Hart, *Curr. Opin. Cell Biol.*, 4, 1017 (1992).
3.   N. Sharon and H. Lis, *Sci. Am.*, 268, 82 (1993).
4.   B. K. Brandley, *Cell Biology*, 2, 281 (1991).

5.   H. Schachter, *Trends Glycosci. Glycotech.*, **4**, 241 (1992).

6.   A. Kobata, *Acc. Chem. Res.*, **26**, 319 (1993).

7.   S. Borman, *Chem. Eng. News*, **71**, 27 (1993).

8.   S. Borman, *Chem. Eng. News*, **70**, 25 (1992).

9.   Y. C. Lee, *Trends Glycosci. Glycotech.*, **4**, 251 (1992).

10. J. Hodgson, *Biotechnology*, **9**, 609 (1991).

11. J. C. Paulson, *Trends Biochem. Sci.*, **14**, 272 (1989).

12. P. Knight, *Biotechnology*, **7**, 35 (1989).

13. T. Feizi, *Nature*, **314**, 35 (1985).

14. O. Kanie and O. Hindsgaul, *Curr. Opin. Struct. Biol.*, **2**, 674 (1992).

15. O. Hindsgaul, *Semin. Cell Biol.*, **2**, 319 (1991).

16. S. H. Khan and O. Hindsgaul, In *Frontiers in Molecular Biology*,
     (M. Fukuda and O. Hindsgaul, Eds.) IRL Press, Oxford, p. 206 (1994).

17. B. Fraser-Reid, U. E. Udodong, Z. Wu, H. Ottosson, J. R. Merritt, C. S.
     Rao, C. Roberts and R. Madsen, *Synlett*, 927 (1992).

18. P. J. Garegg, *Acc. Chem. Res.*, **25**, 575 (1992)

19. K. C. Nicolaou, T. J. Caulfield and R. D. Groneberg, *Pure Appl. Chem.*,
     63, 555 (1991).

20. P. Sinay, *Pure Appl. Chem.*, **63**, 519 (1991).

21. H. Paulsen, *Angew. Chem. Int. Ed. Eng.*, **29**, 823 (1990).

22. H. Paulsen, *Angew. Chem. Int. Ed. Eng.*, **21**, 155 (1982).

23. R. R. Schmidt, *Angew. Chem. Int. Ed. Eng.*, **25**, 212 (1986).

24. R.U. Lemieux, *Chem. Soc. Rev.*, **7**, 423 (1978).

25. F. Goto and T. Ogawa, *Pure Appl. Chem.*, **65**, 793 (1993).

26. T. Ogawa, H. Yamamoto, T. Nukada, T. Kitajima and M. Sugimoto, *Pure
     Appl. Chem.*, **56**, 779 (1984).

27. R. W. Friesen and S. J. Danishefsky, *J. Am. Chem. Soc.*, **111**, 6656 (1989).

28. R. L. Halcomb and S. J. Danishefsky, *J. Am. Chem. Soc.*, **111**, 6661
     (1989).

29. K. Toshima and K. Tatsuta, *Chem. Rev.*, **93**, 1503 (1993).

30. E. Kaji and F. W. Lichtenthaler, *Trends Glycosci. Glycotech.*, **5**, 121
     (1993)

31. T. Hori, M. Sugita, S. Ando, M. Kuwahara, K. Kumauchi, E. Sugie and O.

Itasaka, *J. Biol. Chem.*, **256**, 10979 (1981).

32. O. Kanie, T. Takeda, N. Hada and Y. Ogihara, *J. Carbohydr. Chem.*, **10**, 561 (1991).

33. D. R. Bundle, *Topics in Current Chemistry*, **154**, 1 (1990).

34. P. J. Garegg and C. Hallgren, *J. Carbohydr. Chem.*, **11**, 425 (1992).

35. B. Lindberg, K. Leontein, U. Lindquist, S. B. Svensson, G. Wrangsell, A. Dell and M. Rogers, *Carbohydr. Res.*, **174**, 313 (1988).

36. N. Shibata, S. Fukasawa, H. Kobayashi, M. Tojo, T. Yonezu, A. Ambo, Y. Ohkubo and S. Suzuki, *Carbohydr. Res.*, **187**, 239 (1989).

37. R. U. Lemieux and S. Koto, *Tetrahedron*, **30**, 1933 (1974).

38. W. Koenigs and E. Knorr, *Chem. Ber.*, **34**, 957 (1901).

39. G. Ekborg, B. Lindberg and J. Lonngren, *Acta. Chem. Scan.*, **26**, 3287 (1972).

40. H. Boren, G. Ekborg, K. Eklind, P. J. Garegg, A. Pilotti and C. Swahn, *Acta. Chem. Scan.*, **27**, 2639 (1973).

41. O. Theander, *Acta. Chem. Scan.*, **12**, 1883 (1958).

42. F. W. Lichtenthaler, U. Klares, M. Lergenmuller and S. Schwidetzky, *Synthesis*, 179 (1992).

43. M. Miljkovic, M. Gligorijevic and D. Glisin, *J. Org. Chem.*, **39**, 3223 (1974).

44. S. David and A. Fernandez-Mayoralas, *Carbohydr. Res.*, **165**, C11 (1987).

45. S. David, A. Malleron and C. Dini, *Carbohydr. Res.*, **188**, 193 (1989).

46. J. Alais and S. David, *Carbohydr. Res.*, **201**, 69 (1990).

47. H. P. Kleine and R. Sidhu, *Carbohydr. Res.*, **182**, 307 (1988).

48. W. Gunther and H. Kunz, *Carbohydr. Res.*, **228**, 217 (1992).

49. W. Gunther and H. Kunz, *Angew. Chem. Int. Ed. Engl.*, **29**, 1050 (1990).

50. W. Gunther and H. Kunz, *Angew. Chem. Int. Ed. Engl.*, **27**, 1086 (1988).

51. M. H. E. Griffith, *Ph.D. Dissertation*, University of Alberta, Edmonton, Canada (1991).

52. F. Barresi and O. Hindsgaul, *J. Am. Chem. Soc.*, **113**, 9376 (1991).

53. F. Barresi and O. Hindsgaul, *Synlett,* 759 (1992).

54. (a) F. Barresi and O. Hindsgaul, *Can. J. Chem.*, **72**, 1447 (1994); (b) F. Barresi, *Ph.D. Dissertation*, University of Alberta, Edmonton, Canada

(1994).

55.   Y. Ito and T. Ogawa, *Angew. Chem. Int. Ed. Engl.,* **33**, 1765 (1994).

56.   G. Stork and G. Kim, *J. Am. Chem. Soc.,* **114**, 1087 (1992).

57.   N. Taubken, B. Sauerbrei and J. Thiem, *J. Carbohydr. Chem.,* **12**, 651 (1993).

58.   N. Taubken and J. Thiem, *Synthesis,* 517 (1992).

59.   D. Kahne, D. Yang, J. Lim, R. Miller and E. Papuaga, *J. Am. Chem. Soc.,* **110**, 8716 (1988).

60.   V. K. Srivastava and C. Schuerch, *Tetrahedron Lett.,* **20**, 3269 (1979).

61.   V. K. Srivastava and C. Schuerch, *J. Org. Chem.,* **46**, 1121 (1981).

62.   L. F. Awad, El Sayed H. El Ashry and C. Schuerch, *Bull. Chem. Soc. Jpn.,* **59**, 1587 (1986).

63.   E. M. Klimov, A. V. Demchenko, N. N. Malysheva and N. K. Kochetkov, *Biorg. Khim.,* **17**, 1660 (1991).

64.   P. A. J. Gorin and A. S. Perlin, *Can. J. Chem.,* **39**, 2474 (1961).

65.   G. M. Bebault and G. Dutton, *Carbohydr. Res.,* **37**, 309 (1974).

66.   N. K. Kochetkov, V. I. Torgov, N. N. Malysheva and A. S. Shaskov, *Tetrahedron,* **36**, 1099 (1980).

67.   V. I. Betaneli, M. V. Ovchinnikov, L. V. Backinowsky and N. K. Kochetkov, *Carbohydr. Res.,* **84**, 21 (1980).

68.   G. Wulff and J. Wichelhaus, *Chem. Ber.,* **112**, 2847 (1979).

69.   P. J. Garegg, R. Johansson and B. Samuelsson, *Acta. Chem. Scan.,* **B35**, 635 (1981).

70.   D.M. Whitfield, R. N. Shah, J. P. Carver and J. J. Krepinsky, *Synth. Comm.,* **15**, 737 (1985).

71.   P. J. Garegg, T. Iversen and R. Johansson, *Acta. Chem. Scan.,* **B34**, 505 (1980).

72.   T. Sugawara, K. Irie, H. Iwasawa, T. Yoshikawa, S. Okuno, H. Watanabe, T. Kato, M. Shibukawa and Y. Ito, *Carbohydr. Res.,* **230**, 117 (1992).

73.   P. J. Garegg and T. Iversen, *Carbohydr. Res.,* **70**, C13 (1979).

74.   H. Paulsen and O. Lockhoff, *Chem. Ber.,* **114**, 3102 (1981).

75.   H. Paulsen and R. Lebuhn, *Liebigs Ann. Chem.,* 1047 (1983).

76.   T. Ogawa, T. Kitajima and T. Nukada, *Carbohydr. Res.,* **123**, C5 (1983).

77.   H. Paulsen and R. Lebuhn, *Carbohydr. Res.*, **130**, 85 (1984).

78.   H. Paulsen and R. Lebuhn, *Angew. Chem. Int. Ed. Engl.*, **21**, 926 (1982).

79.   T. Ogawa, K. Katano, K. Sasajima and M. Matsui, *Tetrahedron*, **37**, 2779 (1981).

80.   T. Ogawa, T. Kitajima and T. Nukada, *Carbohydr. Res.*, **123**, C8 (1983).

81.   T. Ogawa, M. Sugimoto, T. Kitajima, K. K. Sadozai and T. Nukada, *Tetrahedron Lett.*, **27**, 5739 (1986).

82.   H. Paulsen, M. Heume, Z. Gyorgydeak and R. Lebuhn, *Carbohydr. Res.*, **144**, 57 (1985).

83.   F. Yamazaki, T. Nukada, Y. Ito, S. Sato and T. Ogawa, *Tetrahedron Lett.*, **30**, 4417 (1989).

84.   H. Paulsen, M. Heume and H. Nurnberger, *Carbohydr. Res.*, **200**, 127 (1990).

85.   T. Nukada, T. Kitajima, Y. Nakahara and T. Ogawa, *Carbohydr. Res.*, **228**, 157 (1992).

86.   C. A. A. van Boeckel, T. Beetz and S. F. van Aelst, *Tetrahedron*, **40**, 4097 (1984).

87.   M. Mori, Y. Ito and T. Ogawa, *Carbohydr. Res.*, **195**, 199 (1990).

88.   P. J. Garegg and P. Ossowski, *Acta. Chem. Scand.* **B37**, 249 (1983).

89.   T. Takeda, N. Hada and Y. Ogihara, *Chem. Pharm. Bull.*, **40**, 1930 (1992).

90.   R. N. Shah, D. A. Cumming, A. A. Grey, J. P. Carver and J. Krepinsky, *Carbohydr. Res.*, **153**, 155 (1986).

91.   M. Shaban and R. W. Jeanloz, *Carbohydr. Res.*, **52**, 115 (1976).

92.   C. D. Warren, C. Auge, M. L. Laver, S. Suzuki, D. Power and R. Jeanloz, *Carbohydr. Res.*, **82**, 71 (1980).

93.   C. Auge, C. D. Warren, R. W. Jeanloz, M. Kiso and L. Anderson, *Carbohydr. Res.*, **82**, 85 (1980).

94.   G. Johnson, R. T. Lee and Y. C. Lee, *Carbohydr. Res.*, **39**, 271 (1975).

95.   N. K. Kochetkov, B. A. Dmitriev, N. N. Malysheva, A. Chernyak, E. M. Klimov, N. E. Bayramova and V. I. Torgov, *Carbohydr. Res.*, **45**, 283 (1975).

96.   N. K. Kochetkov, B. A. Dmitriev, O. S. Chizhov, E. M. Klimov, N. N. Malysheva, A. Chernyak, N. E. Bayramova and V. I. Torgov, *Carbohydr.*

Res., **33**, C5 (1974).

97.   K. K.-C. Liu and S. J. Danishefsky, *J. Org. Chem.*, **59**, 1892 (1994).

98.   G. Ekborg, J. Lonngren and S. Svensson, *Acta. Chem. Scand.*, **B29**, 1031 (1975).

99.   H. H. Lee, L. N. Congson, D. M. Whitfield, L. R. Radics and J. J. Krepinsky, *Can. J. Chem.*, **70**, 2607 (1992).

100. J. Kerekgyarto, J. P. Kamerling, J. B. Bouwstra, J. F. G. Vliegenthart and A. Liptak, *Carbohydr. Res.*, **186,** 51 (1989).

101. E. E. Lee, G. Keaveney and P. S. O'Colla, *Carbohydr. Res.*, **59**, 268 (1977).

102. Gilbert Stork, *Personal Communication* (1993).

103. M. Bols, *J. Chem. Soc., Chem. Commun.*, 913 (1992).

104. M. Bols, *J. Chem. Soc., Chem. Commun.*, 791 (1993).

105. M. Bols, *Acta. Chem. Scan.*, **47**, 829 (1993).

106. M. Bols, *Tetrahedron*, **49**, 10049 (1993).

107. P. J. Garegg, *Chemtracts-Organic Chemistry*, **5**, 389 (1992).

108. F. N. Tebbe, G. W. Parshall and G. S. Reddy, *J. Am. Chem. Soc.*, **100**, 3611 (1978).

109. L. F. Cannizzo and R. H. Grubbs, *J. Org. Chem.*, **50**, 2386 (1985).

110. M. H. Ali, P. M. Collins and W. G. Overend, *Carbohydr. Res.*, **205**, 428 (1990).

111. D. Kahne, S. Walker, Y. Cheng and D. Van Engen, *J. Am. Chem. Soc.*, **111**, 6881 (1989).

Chapter 12

SYNTHESIS OF SIALOGLYCOCONJUGATES

AKIRA HASEGAWA

Department of Applied Bioorganic Chemistry

Gifu University, Gifu 501-11, Japan

Abstract A facile, α-stereoselective glycoside synthesis of sialic acids, α-sialyl-(2→8)-sialic acid, and α-sialyl-(2→8)-α-sialyl-(2→8)-sialic acid by use of their protected methyl or phenyl 2-thioglycosides as the glycosyl donor and the suitably protected sugar acceptors in acetonitrile solution is described. This procedure is effectively applied to the systematic synthesis of sialoglycoconjugates such as a variety of sialo-oligosaccharides and gangliosides, in order to elucidate their functions at the molecular level.

1. INTRODUCTION

Sialooligosaccharides are important constituents of gangliosides and glycoproteins of cell membranes. Biologically, these membrane components are considered to be responsible for many physiological activities.[1-8] An approach toward the systematic understanding of structure-function relationships of sialoglycoconjugates necessitates efficient regio- and stereoselective synthetic routes, giving various sialooligosaccharides, their derivatives and analogs. The focal point in the synthesis of sialo-oligosaccharides has been α-stereoselective glycoside synthesis of sialic acids with various sugar residues. Recently, we have developed[9-11] a facile, α-stereoselective glycosylation of sialic acids using the 2-thioglycosides of sialic acids as the glycosyl donors and the suitably protected sugar acceptors with dimethyl(methylthio)sulfonium triflate (DMTST)[12] or *N*-iodosuccinimide

(NIS)[13, 14] as the glycosyl promoter in acetonitrile solution under kinetically controlled conditions. The α-sialyloligosaccharides thus obtained have been effectively employed as the building blocks for the systematic synthesis of ganglio,[15] lacto,[16] neolacto,[16] globo and isoglobo,[17] polysialo,[18,19] and KDN[20] series of gangliosides and their analogs.[21, 22]

In the first part of this article we describe the efficient method for the α-glycoside synthesis of sialic acids, α-sialyl-(2→8)-sialic acid, and α-sialyl-(2→8)-α-sialyl-(2→8)-sialic acid. A systematic synthesis of sialyl Le^x and sialyl Le^a oligosaccharides, various types of analogs, and their ceramide derivatives is then described. These compounds have been used to determine the structural requirements necessary for recognition of selectins, a family of cell adhesion molecules involved in leukocyte trafficing[8,23] and tumor metastasis.[24]

2. REGIO- AND α-STEREOSELECTIVE GLYCOSIDE SYNTHESIS OF SIALIC ACIDS

Kuhn et al.[25] was the first to obtain an α-glycoside of N-acetylneuraminic acid (Neu5Ac) with sugar derivatives, using the 2-chloro derivatives of Neu5Ac as the glycosyl donor . With the 2-halo derivatives of Neu5Ac, the yield and stereoselectivity of the glycosides are generally poor, especially of those with the secondary hydroxyl groups of sugar derivatives. The competitive elimination reaction due to the deoxy center at position 3 of Neu5Ac gives rise to 2,3-dehydro derivatives. The bottleneck in these reactions is how to control the formation of this elimination product by selection of suitably designed glycosyl donors and acceptors, and glycosyl promoters. In addition, achieving α-glycosides of sialic acids with hydroxyl groups of sugar derivatives, especially with the secondary hydroxyl, in high yield is very difficult because the β-glycosides are thermodynamically favored. Recently, several new methods have been developed to obtain mainly α-glycosides, using the 2-halo-3-substituted Neu5Ac derivatives,[26,27] Neu5Ac phosphites,[28, 29] S-glycosyl xanthates [30, 31] of Neu5Ac, and the 2-thioglycosides of sialic acids.[9, 32]

A facile sialylation of the suitably protected sugar derivatives with thioglycoside derivatives of sialic acids by the use of thiophilic promoters will now be described. With the wide use of thioglycosides in oligosaccharide synthesis, preliminary attempts [9] using the methyl α-2-thioglycoside[33] of Neu5Ac derivative as a suitable glycosyl donor, DMTST [dimethyl(methylthio)sulfonium triflate] as a promoter and various alcohols as the acceptors, indicated that the reactions conducted in acetonitrile solution under kinetically controlled conditions gave predominantly α-glycosides in good yields. The use[10] of a 1:1 anomeric mixture of the methyl 2-thioglycoside of Neu5Ac, which is easily obtained on a large scale, in this reaction was effective as a glycosyl donor affording the α-glycosides of Neu5Ac regio- and stereoselectively in high yields. This method was successfully extended to the synthesis of sialyl α-(2→3)- and sialyl α-(2→6)-sugar derivatives in high yields (50 - 70%) which may be the highest among those reported so far. Iodonium-ion-promoted glycosylations[13,14] are also attractive for oligosaccharide syntheses. We have examined its application to the sialylation involving the 2-thioglycoside of Neu5Ac and reported[11] the comparative reactivities of DMTST and N-iodosuccinimide (NIS)-p-toluenesulfonic acid (TfOH) in acetonitrile or dichloromethane, with the objective of obtaining predominantly the α-glycoside (Table 1; Figure 1). Notably, with DMTST in acetonitrile, secondary hydroxyls were glycosylated to give exclusively the α-configuration (40 - 50%), while an anomeric mixture (α:β = 4:1; 65 - 90%) was formed with primary hydroxyls. On the other hand, with NIS / TfOH in acetonitrile, even hindered primary and less reactive secondary hydroxyls were glycosylated in high yields (~77%), but an increased amount of the β-glycoside was formed in some cases. Further, the reactions in dichloromethane with either DMTST or NIS / TfOH showed poor stereoselectivity. Based on the above results, a possible reaction mechanism for affording the thermodynamically unfavored α-glycosides of Neu5Ac stereoselectively can be postulated as follows (Figure 2): A less reactive acceptor nucleophile and stable donor anomeric-intermediate (d) are the two probable factors leading to the formation of α-glycosides of Neu5Ac . On the other hand, the reactive alcohol can attack other intermediates in addition to (d), consequently increasing amounts of the β-glycosides are formed. When dichloromethane is used, the nucleophile also reacts with the intermediates (a),

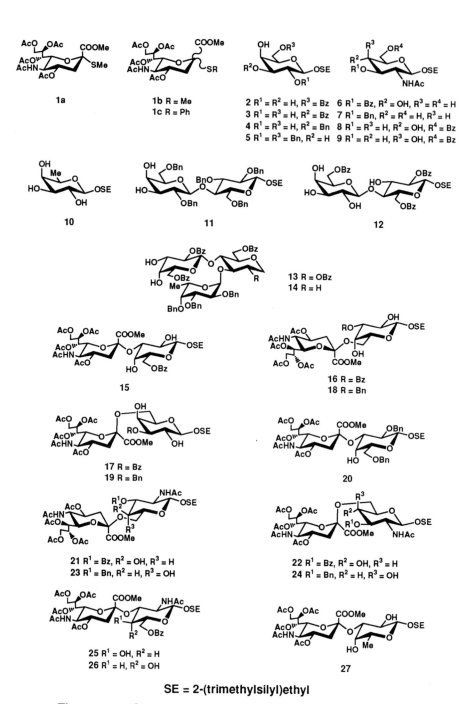

SE = 2-(trimethylsilyl)ethyl

Figure 1. α-Stereoselective glycoside synthesis of Neu5Ac.

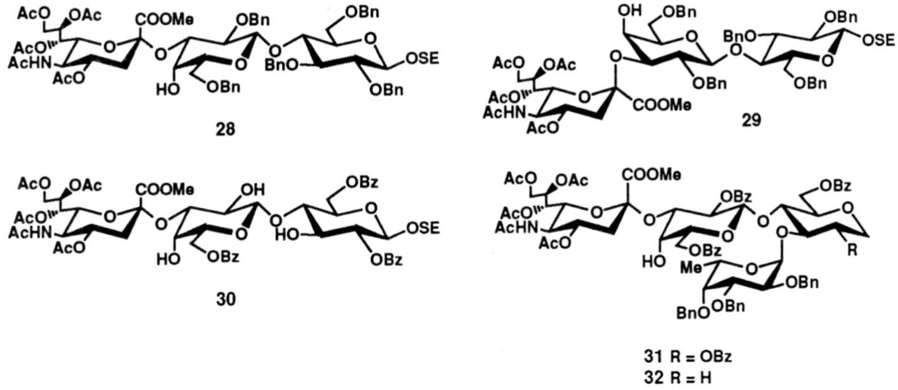

Figure 1. (continued) α-Stereoselective glycoside synthesis of Neu5Ac.

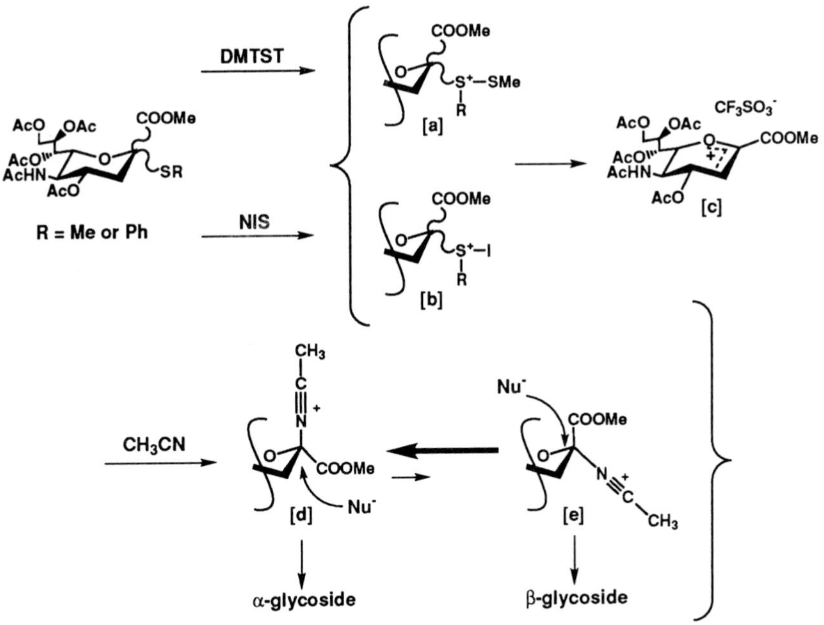

Figure 2. The reaction mechanism suggested for α-predominant glycoside formation of Neu5Ac.

Table 1. DMTST[a]- and NIS-TfOH[b]-promoted glycosylation by use of the 2-thioglycosides of Neu5Ac.

Acceptor	Donor	Promoter	Solvent	Product	Yield[c](%)	
					α	β
2	1a	DMTST	CH_3CN	15	47	0
2	1b	DMTST	CH_3CN	15	52	0
2	1b	NIS	CH_3CN	15	61	0
2	1c	NIS	CH_3CN	15	77	0
3	1a	DMTST	CH_3CN	16	70	0
3	1b	NIS	CH_3CN	16	59	0
3	1b	NIS	CH_2Cl_2	16, 17	49	25
4	1b	DMTST	CH_3CN	18, 19	50	15
4	1b	NIS	CH_3CN	18, 19	51	26
4	1b	NIS	CH_2Cl_2	18, 19	43	45
5	1c	NIS	CH_3CN	20	70	0
6	1a	DMTST	CH_3CN	21, 22	71	20
7	1a	DMTST	CH_3CN	23, 24	63	24
8	1a	DMTST	CH_3CN	25	49	0
9	1a	DMTST	CH_3CN	26	46	0
10	1c	NIS	CH_3CN	27	70	0
11	1b	DMTST	CH_3CN	28, 29	30	0
11	1b	NIS	CH_3CN	28, 29	59	10
12	1b	DMTST	CH_3CN	30	47	0
12	1b	NIS	CH_3CN	30	55	0
13	1c	NIS	CH_3CN	31	45	0
14	1c	NIS	CH_3CN	32	40	0

[a]Reactions were performed at -15°C
[b]Reactions were performed at -30 ~ -40°C
[c]Isolated yield

(b) and (c) to give an anomeric mixture non-stereoselectively. This method was extended[20,22,34-36] to the α–glycoside synthesis (40-70%) of a variety of modified Neu5Ac analogs (Figure 3), demonstrating its usefulness. There are many polysialoglycoconjugates containing α-sialyl-(2→8)-sialic acid or α-sialyl-(2→8)-α-sialyl-(2→8)-sialic acid units, or both in their structures, and these have many important biological roles.[5,37-39] We have employed this procedure for the synthesis of disialyl and trisialyl α-(2→3)- and α-(2→6)-sugar derivatives, in a systematic approach to the synthesis of polysialogangliosides. Treatment[40,41] of sialyl-α-(2→8)-sialic acid or sialyl-α-(2→8)-sialyl-α-(2→8)-sialic acid, which was easily prepared by hydrolysis of colominic acid, with Amberlite IR-120 (H⁺) resin in methanol, followed by acetylation, gave the corresponding anomeric mixture of the ester-lactones **51** and **52** in 84 and 50% yields, respectively. The conversion of the O-acetyl into the phenyl thioglycosides **53** (89%) and **54** (80%) was achieved by treatment[31] with thiophenol and boron trifluoride etherate in dichloromethane. Glycosylation[40,41] of the galactose or lactose acceptor used before, with **53** and **54** in acetonitrile with NIS-TfOH present, gave the expected α–glycosides (**55-59**) as the predominant products in good yields (30 - 50%) (Figure 4). The α-glycosides of α-sialyl-(2→8)-sialic acid have been effectively employed as the building blocks for the systematic synthesis of polysialogangliosides such as GD2,[42] GD3,[40] and GQ1b.[43]

3. SYNTHESIS OF sLe^x AND sLe^a GANGLIOSIDES RECOGNIZED BY THE SELECTIN FAMILY

Sialyl Lewis X (sLe^x) ganglioside was first isolated[44] from the human kidney and found[45] to be widespread as a tumor-associated ganglioside antigen. Recently, it has been demonstrated[8,23,46-49] that the selectin family that includes E-selectin (endothelial leukocyte adhesion molecule-1, ELAM-1), P-selectin (granule membrane protein, GMP-140), and L-selectin (leukocyte adhesion molecule-1), recognize the sLe^x determinant, α-Neu5Ac-(2→3)-β-D-Gal-(1→4)-[α-L-Fuc-(1→3)]-β-D-GlcNAc. This sequence is found as the terminal carbohydrate structure in both cell membrane glycolipids and glycoproteins. In view of these new findings, the synthesis of sLe^x and various types of analogs

284

A. HASEGAWA

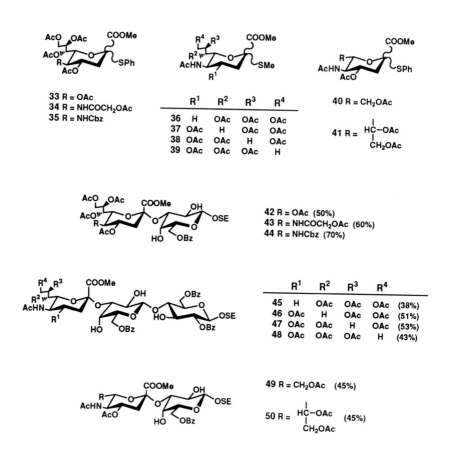

Figure 3. α-Glycoside synthesis of Neu5Ac analogs.

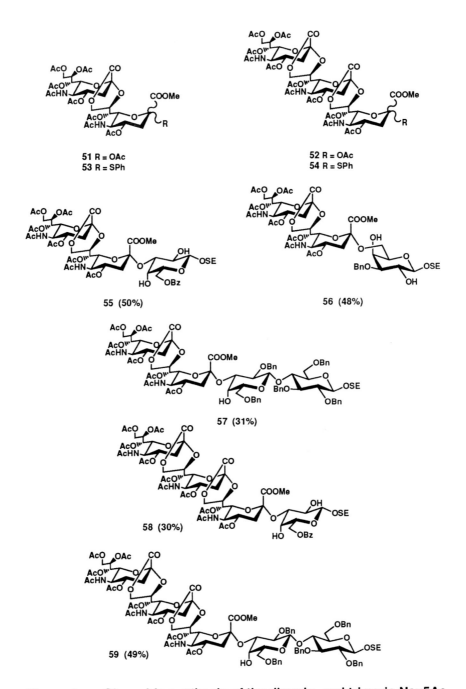

Figure 4. α-Glycoside synthesis of the dimeric- and trimeric-Neu5Ac.

is critical for progress toward the goal of elucidating the structural features of this carbohydrate ligand required for the selectin recognition which is related to cell-cell adhesion, tumor-metastasis, inflammation, and thrombosis.

3.A.   Synthesis of sLe[x], sLe[x] Gangliosides, and its Position Isomer

Sialyl Le[x] ganglioside[50] and its sialyl α-(2→6) positional isomer[51] with regard to the substitution of the Gal residue by Neu5Ac were synthesized as shown in Figure 5. The trisaccharide acceptor[16] 61 was first coupled with methyl α-L-thiofucopyranoside derivative 60 using DMTST in benzene to afford the desired α-tetrasaccharide 62 (95%). Reductive ring-opening of the benzylidene acetal in 62 with sodium cyanoborohydride-hydrogen chloride[52] gave 63, which was glycosylated with 64[16] (easily derived from 15) afforded the hexasaccharide 66 (70%). Hydrogenolytic removal of the benzyl groups and subsequent acetylation afforded the fully acylated derivative 67 (81%). Selective removal[53] of the 2-(trimethylsilyl)ethyl group in 67 using trifluoroacetic acid, and subsequent treatment[54] with trichloroacetonitrile in the presence of 1,8-diazabicyclo[5.4.0]undec-7-ene (DBU) gave the α-imidate 69 (91%). The final glycosylation of (2S, 3R, 4E)-2-azido-3-O-benzoyl-4-octadecene-1,3-diol[54,55] (70) with 69 using boron trifluoride etherate, afforded only the expected β-glycoside 71 (56%). This compound was transformed into the title sLe[x] ganglioside 74 in high yield by the following sequence: selective reduction[55,57] of the azido group with H_2S in aq. 83% pyridine solution, coupling of the amine 72 with octadecanoic acid in dichloromethane by use of 1-(3-dimethylaminopropyl)-3-ethylcarbodiimide hydrochloride (WSC), O-deacylation, and saponification of the methyl ester group. Similarly, by the coupling of 61 with sialyl α-(2→6)-galactose donor[58] 65 obtained from 18 by O-benzoylation, replacement of the 2-(trimethylsilyl)ethyl group by acetyl, and introduction of the methylthio group by reaction with methylthio(trimethyl)silane, and subsequent reactions as described for the synthesis of 74, a positional isomer 75 of sLe[x] ganglioside was synthesized. Sialyl Le[x] oligosacchaside[59] 76 (69%) was also synthesized from 68 by O-

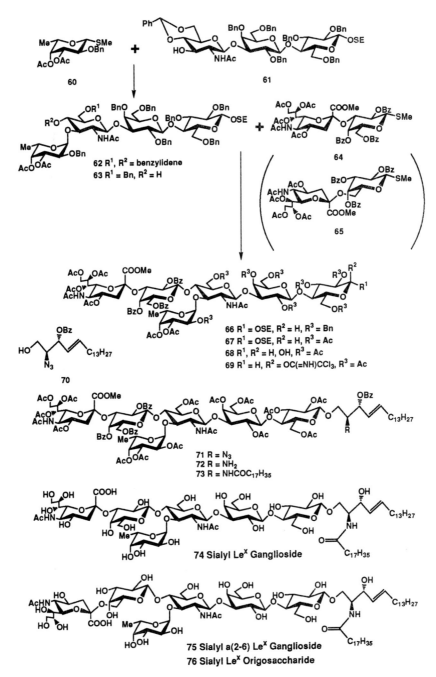

Figure 5. Synthesis of sLe^x, sLe^x ganglioside and its position isomer.

(tetrahydropyran-2-y1)ation, O-deacylation, saponification of the methyl ester group, and hydrolysis of the tetrahydropyranyl group.

3.B.   Synthesis of sLe^a   Ganglioside

Sialyl Le[a] ganglioside, which is a position isomer of sLe[x] ganglioside with regard to the substitution of the GlcNAc residue by fucose and sialyl α-(2→3)-galactose, was isolated[60] as the tumor-associated glycolipid antigen of digestive organs.   Recent reports[23,24,61] revealed that this ganglioside was also recognized by the selectin family. Figure **6** shows the synthesis[62] of the sLe[a] ganglioside. Glycosylation of the trisaccharide acceptor **77**, derived by reductive ring-opening of the benzylidene acetal in **61**, with **64** using DMTST, afforded the expected pentasaccharide **78** (45%) and its position isomer **79** (36%). Glycosylation of **78** with methyl 2,3,4-tri-O-benzyl-1-thio-β-L-fucopyranoside (**80**) by use of NIS-TfOH gave the hexasaccharide **81** (60%), which, after the reactions described for the synthesis of sLe[x] ganglioside, yielded sLe[a] ganglioside **88** in good yield.

4.   SYSTEMATIC SYNTHESIS OF sLe[x] GANGLIOSIDE ANALOGS CONTAINING THE MODIFIED FUCOSE, SIALIC ACID, OR GALACTOSE RESIDUE

An energy minimized conformation[49] demonstrated using sLe[x], sLe[a], and sialyl α-(2→6)-Le[x], thus synthesized, has shown that the configuration of fucose, sialic acid, and galactose residues is critical for the selectin recognition. Based on this three-dimentional structure, we have synthesized various types of sLe[x] ganglioside analogs according to our procedure, in order to clarify the structural features required for the selectin recognition.

4.A.   Synthesis of the Deoxy-Fucose-Containing sLe^x Ganglioside Analogs

The synthesis[63] of sLe[x] ganglioside analogs containing the three possible pyranosyl deoxy-L-fucoses is described. For the synthesis of the target sLe[x] analogs, we employed the methyl 1-thioglycosides[63] **89-91** of the deoxy-fucose derivatives as the glycosyl donors and 2-(trimethylsilyl)ethyl O-(2-

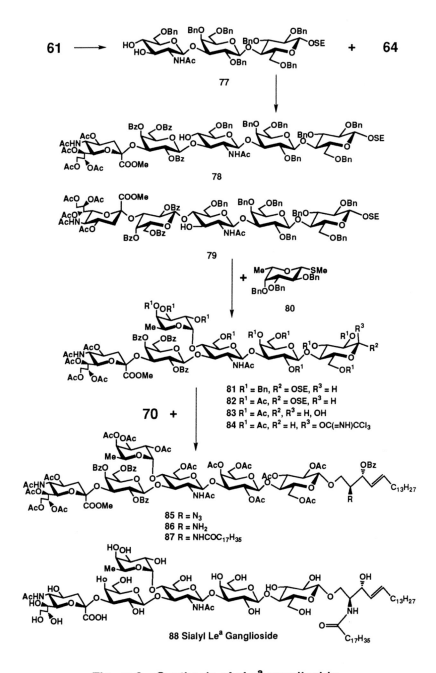

Figure 6.   Synthesis of sLe^a ganglioside.

acetamido-4,6-O-benzylidene-2-deoxy-β-D-glucopyranosyl)-(1→4)-2,4,6-tri-
O-benzyl-β-D-galactopyranoside[64] as a suitably protected glycosyl acceptor
(Figure 7). Glycosylation of the disaccharide acceptor with **89-91** in benzene
for 10h at 5-10°C using DMTST as a promoter, gave exclusively the α-
glycosides **92** (86%), **93** (82%), and **94** (57%), respectively. These were
converted by reductive ring-opening of the benzylidene acetal into the
corresponding 6-O-benzyl derivatives of the GlcNAc residue, which was
glycosylated with **64** to afford the pentasaccharides **95, 97,** and **99**.
Compounds **95, 97,** and **99** were transformed into the corresponding
trichloroacetimidates **96, 98, 100** by reductive removal of the benzyl groups,
O-acetylation, selective removal of the 2-(trimethylsilyl)ethyl group, and
subsequent imidate formation, all in good yields. By glycosylation of the 2-
azido-sphingosine derivative **70** with the pentasaccharide donors described
above, and subsequent manipulation as described for the synthesis of sLe[x]
ganglioside **74**, the three target deoxy-fucose-containing sLe[x] ganglioside
analogs **101-103** were synthesized in good yields.

4.B. Synthesis of sLe[x] Ganglioside Analogs Containing the Stereoisomers and Derivatives of L-Fucopyranose

Sialyl Le[x] analogs containing the 2-epi-, 4-epi-, 2,4-di-epi-, and 2-methoxy-L-
fucopyranoses were synthesized by our synthetic procedure, in order to clarify
the role of the hydroxyls in the L-fucose residue required for selectin
recognition. For the synthesis of the target sLe[x] analogs (Figure 8), we
employed the methyl 1-thioglycosides[65] **104-107** of the 6-deoxy-
hexopyranoses which were prepared from L-fucose or L-rhamnose by
conventional procedures. These are appropriately derivatized for use in the α-
glycoside synthesis. Glycosylation of 2-(trimethylsilyl)ethyl O-(2-acetamido-
4,6-O-benzylidene-2-deoxy-β-D-glucopyranosyl)-(1→4)-2,4,6-tri-O-benzyl-β-
D-galactopyranaoside[64] with **104-107** in benzene for 3-10h at 5°C, using
DMTST as a promoter, gave exclusively the α-glycosides **108** (86%), **109**
(92%), **110** (70%) and **111** (79%), respectively. After reductive ring-opening
of the benzylidene acetal in **108-111**, the glycosyl acceptors obtained were
coupled with **64** in the presence of DMTST to afford the desired

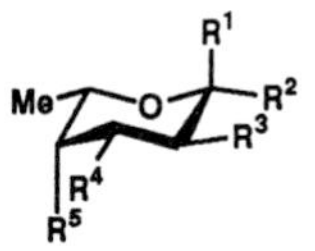

	R¹	R²	R³	R⁴	R⁵
89	H	SMe	H	OBz	OBz
90	H	SMe	OBn	H	OBn
91	H	SMe	OBn	OBn	H

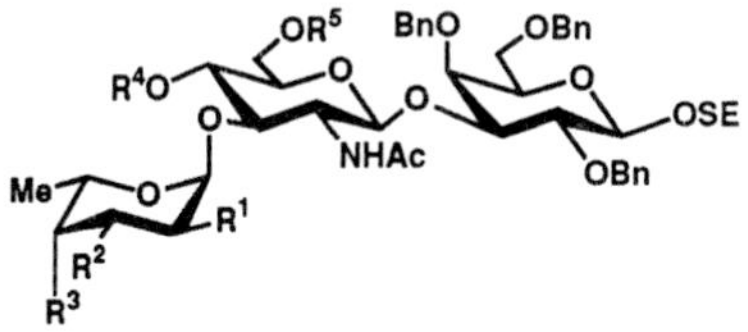

	R¹	R²	R³	R⁴	R⁵
92	H	OBz	OBz	benzylidene	
93	OBn	H	OBn	benzylidene	
94	OBn	OBn	H	benzylidene	

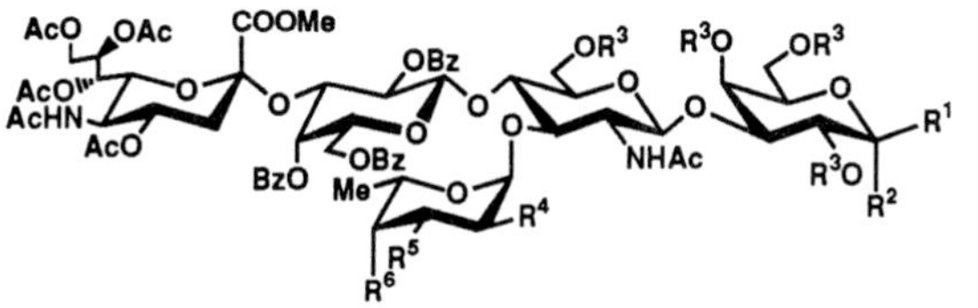

	R¹	R²	R³	R⁴	R⁵	R⁶
95	OSE	H	Bn	H	OBz	OBz
96	H	OC(=NH)CCl₃	Ac	H	OBz	OBz
97	OSE	H	Bn	OBn	H	OBn
98	H	OC(=NH)CCl₃	Ac	OAc	H	OAc
99	OSE	H	Bn	OBn	OBn	H
100	H	OC(=NH)CCl₃	Ac	OAc	OAc	H

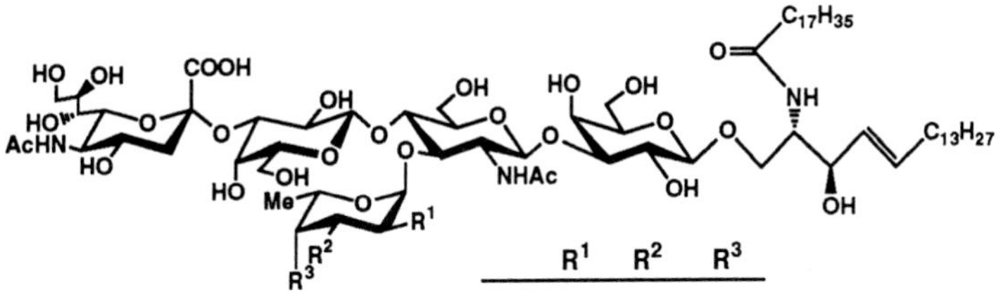

	R¹	R²	R³
101	H	OH	OH
102	OH	H	OH
103	OH	OH	H

Figure 7. Synthesis of the deoxy-fucose-containing sLeˣ ganglioside analogs.

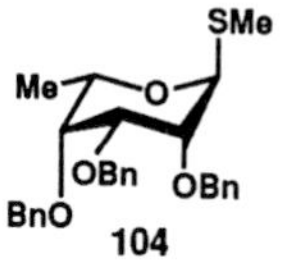

104

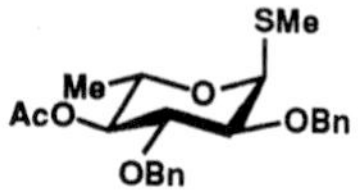

105

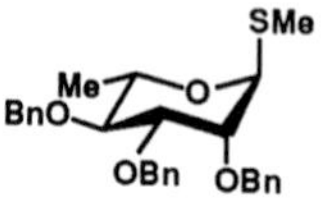

106

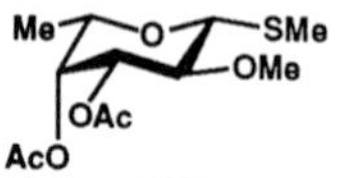

107

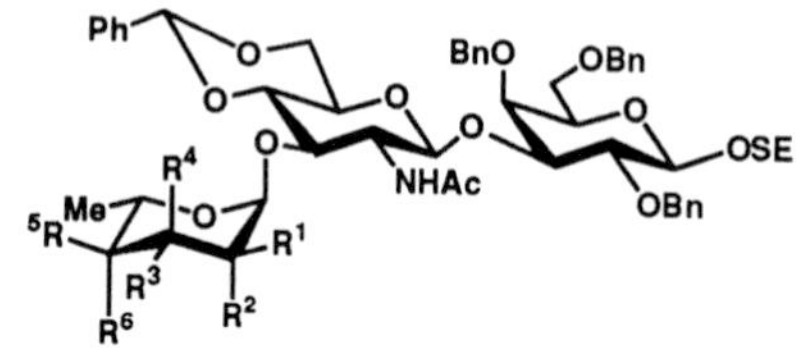

	R¹	R²	R³	R⁴	R⁵	R⁶
108	H	OBn	OBn	H	H	OBn
109	OBn	H	OBn	H	OAc	H
110	H	OBn	OBn	H	OBn	H
111	OMe	H	OAc	H	H	OAc

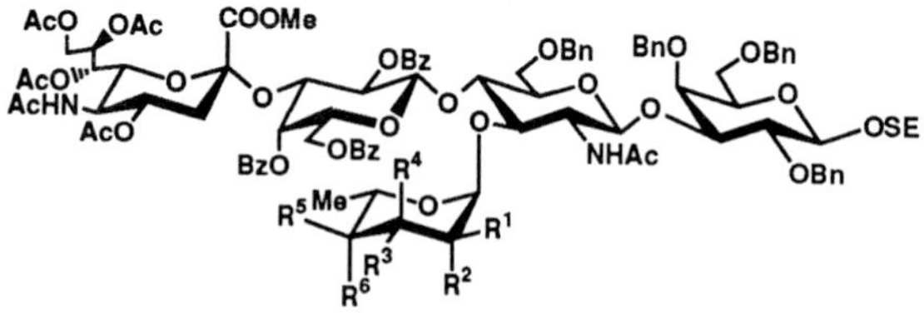

	R¹	R²	R³	R⁴	R⁵	R⁶
112	H	OBn	OBn	H	H	OBn
113	OBn	H	OBn	H	OAc	H
114	H	OBn	OBn	H	OBn	H
115	OMe	H	OAc	H	H	OAc

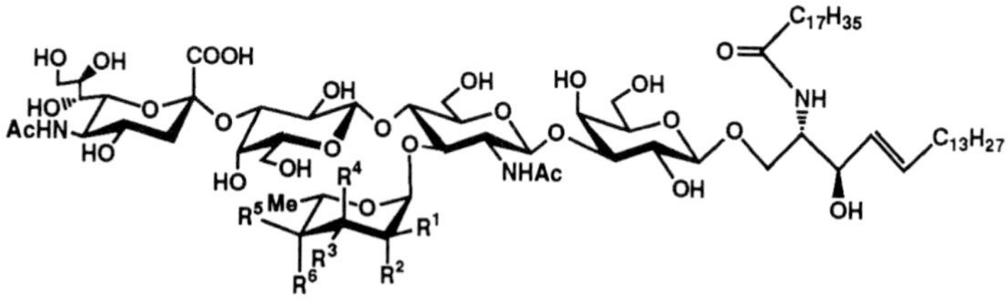

	R¹	R²	R³	R⁴	R⁵	R⁶
116	H	OH	OH	H	H	OH
117	OH	H	OH	H	OH	H
118	H	OH	OH	H	OH	H
119	OMe	H	OH	H	H	OH

Figure 8. Synthesis of sLe^x ganglioside analogs containing the chemically modified fucose residues.

pentasaccharides **112-115**. These were converted by a series of reactions as described for the synthesis of sLe^x ganglioside **74**, into the target compounds in good yields.

4.C. Synthesis of sLe^x Ganglioside Analogs Containing the Chemically Modified Sialic Acids

The synthesis of sLe^x ganglioside analogs containing the C7-Neu5Ac, C8-Neu5Ac, Neu5Gc, and KDN (2-keto-3-deoxy-D-*glycero*-D-*galacto*-nonulosonic acid), to explore the structural requirements of sialic acid moiety for the selectin recognition, was completed as follows (Figure **9**). Glycosylation of the trisaccharide acceptor[64] **124** with the methyl 1-thioglycosides[20,35,36,66] **120-123** of the sialyl α-(2→3)-galactose derivatives, which were prepared from compounds **42, 43, 49, 50** by *O*-benzoylation, selective transformation of the 2-(trimethylsilyl)ethyl group to acetyl with acetic anhydride-boron trifluoride etherate, and introduction of the methylthio group with methylthio(trimethyl)silane, at 5°C in the presence of DMTST, yielded the corresponding β-glycosides **125**[35] (53%), **126**[35] (53%), **127**[36] (45%), **128**[66] (46%), respectively. Similarly, by a series of reactions as described for the synthesis of sLe^x ganglioside **74**, the target sLe^x ganglioside analogs **129-132** were synthesized in good yields.

4.D. Synthesis of the Deoxy-Galactose-Containing sLe^x Ganglioside Analogs

The 4- and 6-deoxy-galactose-containing sLe^x gangliosides(pentasaccharide) were synthesized[67] in order to clarify the role of the hydroxyls of galactose residue for the selectin recognition. For the synthesis (Figure **10**) of the target gangliosides, we employed the methyl 1-thioglycosides derivatives[67] **133, 134**, of sialyl α-(2→3)-4- or -6-deoxy-galactose as the glycosyl donors and **124** as the acceptor. Compound **133** was prepared from **20** by reduction (82%) of a (phenoxy)thiocarbonyl group, introduced at the C-4 position of **20**, selective removal (97%) of the 2-(trimethylsilyl)ethyl group, subsequent imidate formation (97%), and replacement (98%) of the trichloroacetimidate by a methylthio group by treatment with methylthio(trimethyl)silane. Similarly, compound **134** was also prepared from **27** in good yield. Coupling of **124**

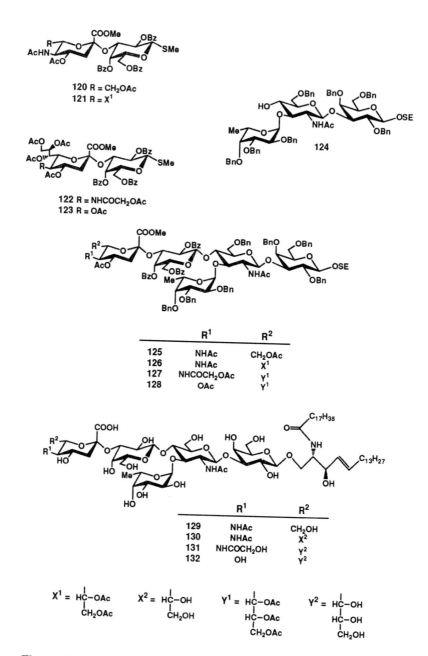

Figure 9. Synthesis of the modified Neu5Ac containing sLe^x gangliosIde analogs.

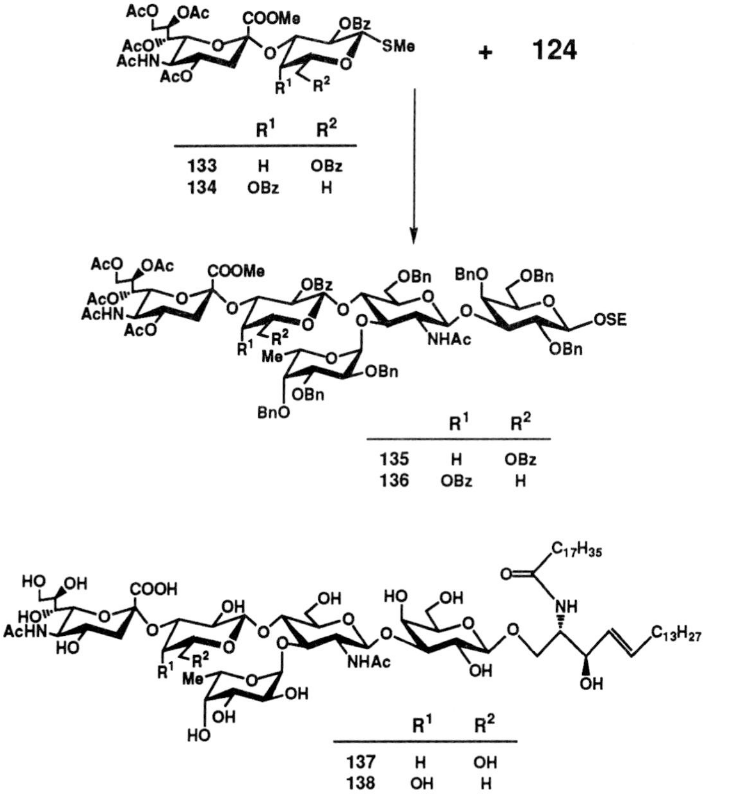

Figure 10.   Synthesis of the deoxy-galactose-containing sLe^x
ganglioside analogs.

with the glycosyl donors **133** and **134** in the presence of DMTST at 7°C gave the corresponding pentasaccharides **135** (45%) and **136** (43%), respectively. These pentasaccharides were easily transformed by the reactions described for the synthesis of sLe^x ganglioside, into the target ganglioside analogs in good yields.

As described in this article, through the facile α-stereo-controlled glycosylations of sialic acids, dimeric and trimeric sialic acids with the suitably protected sugar residues are now feasible by use of either DMTST or NIS-TfOH in acetonitrile solution under kinetically controlled conditions. Using this procedure, various types of sialyl oligosaccharides and their lipophilic derivatives including gangliosides, and their analogs have been systematically synthesized. These molecules will be effectively used to define structure-activity relationships not only in the sialyl Le^x epitope, but also in sialo-glycoconjugates at the molecular level.

ACKNOWLEDGMENTS

This work was supported in part by a Grant-in-Aid (No. 05274102) for Scientific Research on Priority Areas and a Grant-in-Aid (No. 05152053) for Cancer Research from the Ministry of Education, Science and Culture of Japan. The author would like to express his gratitude to various colleagues cited in the references.

REFERENCES

1. H. Wiegandt, In *Glycolipids* (H. Wiegandt, Ed.), New Comprehensive Biochemistry 10, Elesevier, Amsterdam, New York, Oxford, p. 199 (1985).

2. W. Reutter, E. köttgen, C. Bauer, and W. Gerok, In *Biological Significance of Sialic Acids* (R. Schauer, Ed.), Cell Biology Monographs 10, Springer-Verlag, Wien, New York, p. 263 (1983).

3. K. Furukawa and A. Kobata, In *Cell Surface Carbohydrates-Their Involvement in Cell Adhesion* (H. Ogura, A. Hasegawa and T. Suami, Eds.), Carbohydrates-Synthetic Methods and Applications in Medicinal

Chemistry, Kodansha-VCH, Tokyo, Weinheim, New York, Cambridge, Basel, p. 369 (1992).

4. S. Hakomori, *Sci. Am.*, **254**, 32 (1986).

5. S. Tsuji, T. Yamakawa, M. Tanaka, and Y. Nagai, *J. Neurochem.*, **50**, 414 (1988).

6. A. Takada, K. Ohmori, T. Yoneda, K. Tsuyuoka, A. Hasegawa, M. Kiso, and R. Kannagi, *Cancer Res.*, **53**, 354 (1993).

7. M. S. Mullingan, J. C. Paulson, S. DeFrees, Z-L. Zheng, J. B. Lowe, and P. A. Ward, *Nature*, **364**, 149 (1993).

8. C. Foxall, S. R. Watson, D. Dowbenko, C. Fennie, L. A. Lasky, M. Kiso, A. Hasegawa, D. Asa, and B. K. Brandley, *J. Cell. Biol.*, **117**, 895 (1992).

9. T. Murase, H. Ishida, M. Kiso, and A. Hasegawa, *Carbohydr. Res.*, **184**, c1 (1988).

10. A. Hasegawa, H. Ohki, T. Nagahama, H. Ishida, and M. Kiso, *Carbohydr. Res.*, **212**, 277 (1991).

11. A. Hasegawa, T. Nagahama, H. Ohki, K. Hotta, H. Ishida, and M. Kiso, *J. Carbohydr. Chem.*, **10**, 493 (1991).

12. P. Fügedi and P. J. Garegg, *Cabohydr. Res.*, **149**, c9 (1986).

13. P. Konradsson, U. E. Udodong, and B. Fraser-Reid, *Tetrahedron Lett.*, **31**, 4313, (1990).

14. G. H. Veeneman, S. H. van Leevwen, and J. H. van Boom, *Tetrahedron Lett.*, **31**, 1331 (1990).

15. A. Hasegawa, H-K. Ishida, T. Nagahama, and M. Kiso, *J. Carbohydr. Chem.*, **12**, 703 (1993), and references cited therein.

16. A. Kameyama, H. Ishida, M. Kiso, and A. Hasegawa, *Carbohydr. Res.*, **200**, 269 (1990), and references cited therein.

17. T. Miyawaki, H. Ishida, M. Kiso, and A. Hasegawa, *Abstracts of the Japanese Agricultural Chemical Society Annual Meeting*, Tokyo, Japan, p. 39 (1994).

18. H. Prabhanjan, K. Aoyama, M. Kiso, and A. Hasegawa, *Carbohydr. Res.*, **233**, 87 (1992).

19. H-K. Ishida, Y. Ohta, Y. Tsukada, M. Kiso, and A. Hasegawa, *Carbohydr. Res.*, **246**, 75 (1993).

20. T. Terada, M. Kiso, and A. Hasegawa, *J. Carbohydr. Chem.*, **12**, 425 (1993).

21. A. Hasegawa, H. Ogawa, H. Ishida, and M. Kiso, *Carbohydr. Res.*, **224**, 175 (1992).

22. A. Hasegawa, K. Adachi, M. Yoshida, and M. Kiso, *J. Carbohydr. Chem.*, **11**, 95 (1992).

23. M. Larkin, T. J. Aherm, M. S. Stoll, M. Shaffer, D. Sako, J. O'Brien, C.-T. Yuen, A. M. Lawson, R. A. Childs, K. M. Barone, P. R. Langer-Safer, A. Hasegawa, M. Kiso, G. R. Larsen, and T. Feizi, *J. Biol. Chem.*, **267**, 13661 (1992).

24. A. Tanaka, K. Ohmori, N. Takahashi, K. Tsuyuoka, A. Yago, K. Zenita, A. Hasegawa, and R. Kannagi, *Biochem. Biophys. Res. Commun.*, **179**, 713 (1991).

25. R. Kuhn, P. Lutz, and D. L. MacDonald, *Chem. Ber.*, **99**, 611 (1966).

26. T. Ogawa and Y. Ito, *Tetrahedron Lett.*, **28**, 6221 (1987).

27. T. Kondo, H. Abe, and T. Goto, *Chem. Lett.*, 1657 (1988).

28. T. J. Martin and R. R. Schmidt, *Tetrahedron Lett.*, **33**, 6123 (1992).

29. M. M. Sim, H. Kondo, and C. H. Wong, *J. Am. Chem. Soc.*, **115**, 2260 (1993).

30. H. Lönn and K. Stenvall, *Tetrahedron Lett.*, **33**, 115 (1992).

31. A. Marra and P. Sinaÿ, *Carbohydr. Res.*, **195**, 303 (1990).

32. O. Kanie, M. Kiso, and A. Hasegawa, *J. Carbohydr. Chem.*, **7**, 501 (1988).

33. A. Hasegawa, J. Nakamura, and M. Kiso, *J. Carbohydr. Chem.*, **5**, 11 (1986).

34. A. Hasegawa, K. Adachi, M. Yoshida, M. Kiso, *Carbohydr. Res.*, **230**, 273 (1992).

35. M. Yoshida, A. Uchimura, M. Kiso, and A. Hasegawa, *Glycoconjugate J.*, **10**, 3 (1993).

36. A. Uchimura, *M. Sc. Dissertation,* Gifu University, Japan (1994).

37. S. Tsuji, M. Arita, and Y. Nagai, *J. Biochem.*, **94**, 303 (1983).

38. (a) S. Tsuji, T. Yamashita, and Y. Nagai, *J. Biochem.*, **104**, 498 (1988);
     (b) T. Nakaoka, S. Tsuji, and Y. Nagai, *J. Neurosci. Res.*, **31**, 724
     (1992) ; (c) S. Tsuji, T. Yamashita, Y. Matsuda, and Y. Nagai,
     Neurochem. Int., **4**, 549 (1992).

39. Y. Yada, Y. Okano, and Y. Nozawa, *Biochem. J.*, **279**, 665 (1991).

40. (a) A. Hasegawa, H-K. Ishida, and M. Kiso, *J. Carbohydr. Chem.*, **12**,
     371 (1993) ; (b) H-K. Ishida, Y. Ohta, Y. Tsukada, M. Kiso, and A.
     Hasegawa, *Carbohydr. Res.*, **246**, 75 (1993).

41. H-K. Ishida, H. Ishida, M. Kiso, and A. Hasegawa, *J. Carbohydr.
     Chem.*, **13**, 655 (1994).

42. (a) H-K. Ishida, Y. Ohta, Y. Tsukada, Y. Isogai, H. Ishida, M. Kiso,
     and A. Hasegawa, *Carbohydr. Res.*, **252**, 283 (1994) ; (b) A.
     Hasegawa, H-K. Ishida, Y. Isogai, H. Ishida, and M. Kiso, *J.
     Carbohydr. Chem.*, **12**, 1217 (1993).

43. H-K. Ishida, H. Ishida, M. Kiso, and A. Hasegawa, *Carbohydr. Res.*,
     260, c1 (1994).

44. H. Pauvala, *J. Biol. Chem.*, **251**, 7517 (1976).

45. K. Fukushima, M. Hirota, P. I. Terasaki, A. Wakisaka, H. Togashi, D.
     Chia, N. Sayama, Y. Fukushi, S. Nudelman, and S. Hakomori, *Cancer
     Res.*, **44**, 5279 (1984).

46. G. Walz, A. Aruffo, W. Kolanus, M. Bevilacqua, and B. Seed, *Science*,
     250, 1132 (1990).

47. M. L. Phillips, E. Nudelman, F. C. A. Graeta, M. Perez, A. K. Singhal,
     S. Hakomori, and J. C. Paulson, *Science*, **250**, 1130 (1990).

48. J. B. Lowe, L. M. Stoolman, R. P. Nair, R. D. Larsen, T. L. Berhend,
     and R. M. Marks, *Cell*, **63**, 475 (1990).

49. D. Tyrrell, P. James, N. Rao, C. Foxall, S. Abbas, F. Dasgupta, M.
     Mashed, A. Hasegawa, M. Kiso, D. Asa, J. Kidd, and B. K. Brandly,
     Proc. Natl. Acad. Sci. U. S. A., **88**, 10372 (1991).

50. (a) A. Kameyama, H. Ishida, M. Kiso, and A. Hasegawa, *Carbohydr.
     Res.*, **209**, c1 (1991) ; (b) *ibid.*, *J. Carbohydr. Chem.*, **10**, 549 (1991).

51. A. Kameyama, H. Ishida, M. Kiso, and A. Hasegawa, *J. Carbohydr.
     Chem.*, **10**, 729 (1991).

52. P. J. Garegg, H. Hultberg, and S. Wallin, *Carbohydr. Res.*, **108**, 97 (1982).

53. K. Jansson, S. Ahlfors, T. Frejd, J. Kihlberg, G. Magnusson, J. Dahmen, G. Noori, and K. Stenvall, *J. Org. Chem.*, **53**, 5629 (1988).

54. R. R. Schmidt and J. Michel, *Angew. Chem. Int. Ed. Engl.*, **19**, 731 (1980).

55. Y. Ito, M. Kiso, and A. Hasegawa, *J. Carbohydr. Chem.*, **8**, 285 (1989).

56. R. R. Schmidt, P. Zimmermman, *Angew. Chem. Int. Ed. Engl.*, **25**, 725 (1986).

57. T. Adachi, I. Yamada, I. Inoue, and M. Saneyoshi, *Synthesis,* 45 (1977).

58. A. Hasegawa, K. Hotta, A. Kameyama, H. Ishida, and M. Kiso, *J. Carbohydr. Chem.*, **10**, 439 (1991).

59. A. Hasegawa, *ACS Symp. Ser.*, **560**, 184 (1994).

60. S. Hakomori and R. Kannagi, *J. Natl. Cancer Inst.*, **71**, 231 (1983).

61. E. L. Berg, M. K. Robinson, O. Mansson, E. C. Bucher, and J. L. Magnani, *J. Biol. Chem.*, **266**, 14869 (1991).

62. (a) A. Kameyama, H. Ishida, M. Kiso, and A. Hasegawa, *Abstracts of the XVth Int. Carbohydr. Symp.*, Yokohama, Japan, p. 132 (1990); (b) *ibid.*, *J. Carbohydr. Chem.*, **13**, 641 (1994).

63. A. Hasegawa, T. Ando, M. Kato, H. Ishida, and M. Kiso, *Carbohydr. Res.*, **257**, 67 (1994).

64. A. Hasegawa, T. Ando, A. Kameyama, and M. Kiso, *J. Carbohydr. Chem.*, **11**, 645 (1992).

65. M. Kato, T. Ando, H. Ishida, M. Kiso, and A. Hasegawa, *Abstracts of the Japanese Agricultural Chemical Society Annual Meeting*, Tokyo, Japan, p. 36 (1994).

66. T. Terada, M. Kiso, and A. Hasegawa, *Carbohydr. Res.*, **259**, 201 (1994).

67. S. Konba, H. Ishida, M. Kiso, and A. Hasegawa, *Abstracts of the Japanese Agricultural Chemical Society Annual Meeting*, Tokyo, Japan, p. 36 (1994).

Chapter 13

SYNTHETIC STUDIES ON CELL-SURFACE GLYCANS: AN APPROACH TO O-LINKED SIALOGLYCOPROTEIN

YOSHIAKI NAKAHARA,* HIROYUKI IIJIMA,* AND TOMOYA OGAWA*,†

*The Institute of Physical and Chemical Research (RIKEN), Wako-shi, Saitama, 351-01, Japan and †Graduate School for Animal Resource Sciences and Veterinary Medical Sciences, University of Tokyo, Yayoi, Bunkyo-ku, Tokyo, 113, Japan

Abstract The synthesis of an O-linked sialoglycoprotein fragment was studied. In view of the advantages of a benzyl-based protection strategy, synthons **4, 5, 11**, and **12** were designed, and coupled via a highly stereoselective glycosylation which led to **13** and **14** after manipulation of substituents. By the use of **13** and **14** as key building blocks, syntheses of glycopentapeptides having N-terminal amino acid sequences of glycophorin AM and AN were accomplished in solution phase.

1.   INTRODUCTION

Sialoglycoproteins are widely distributed as serum, membrane, epithelial, and other types of glycoproteins in mammalian systems.[1,2] It is currently accepted that the protein-bound carbohydrates contribute to the physicochemical properties of glycoproteins, protect glycoproteins against proteolytic attacks, and function in signal recognition in cell-cell, cell-microbe, and hormone-receptor interactions.[2,3] Because of the low availability of pure samples from natural sources, synthetic studies of a glycopeptide as the designed fragment of glycoprotein seems to be indispensable in the elucidation of the function of those carbohydrates. Synthetic approaches to *O*-linked glycoproteins have been made over a number of years since the first synthesis of *N*-(benzyloxycarbonyl)-3-*O*-(3,4,6-tri-*O*-acetyl-2-benzamido-2-deoxy-β-D-glucopyranosyl)-DL-serine methyl ester by Micheel and Köchling in 1958.[4] The first stereoselective synthesis of an α-glycoside derived from D-GalNAc and L-Ser residues, a common linkage of mucin-type *O*-glycoproteins, was performed in 1980 by Ferrari and Pavia[5] using the azido sugar strategy developed by Paulsen *et al.*[6] They used 3,4,6-tri-*O*-acetyl-2-azido-2-deoxy-β-D-galactopyranosyl chloride as glycosyl donor, which had become readily available after Lemieux and Ratcliffe's discovery of azidonitration of 3,4,6-tri-*O*-acetylgalactal.[7] This basic methodology was further developed and generalized for the synthesis of the α-D-GalNAc-(1→3)-L-Ser/Thr core structure. This improvement has coincided with the development of new glycosylation technologies such as the trichloroacetimidate-, thioglycoside-, and glycosyl fluoride-methods. Some derivatives of α-D-GalNAc-(1→3)-L-Ser/Thr and β-D-Gal-(1→3)-α-D-GalNAc-(1→3)-L-Ser/Thr thus prepared were further transformed into glycopeptide oligomers in solution or solid phase.[8]

Glycophorin A is a major sialoglycoprotein of human erythrocyte membranes and its structure was elucidated in the 1970's.[9,10] The *N*-terminal region, located outside the cell membrane, is highly *O*-glycosylated at the serine and threonine sites with the tetrasaccharide **1** containing two sialic acid residues. The polymorphic *N*-terminal pentapeptide sequences are recognized as the antigenic determinants of human MN blood type by anti-MN antibodies.[11]

R R R

Glycophorin AM **2** : Ser • Ser • Thr • Thr • Gly • • • • • •

AN **3** : Leu Glu

1 R =

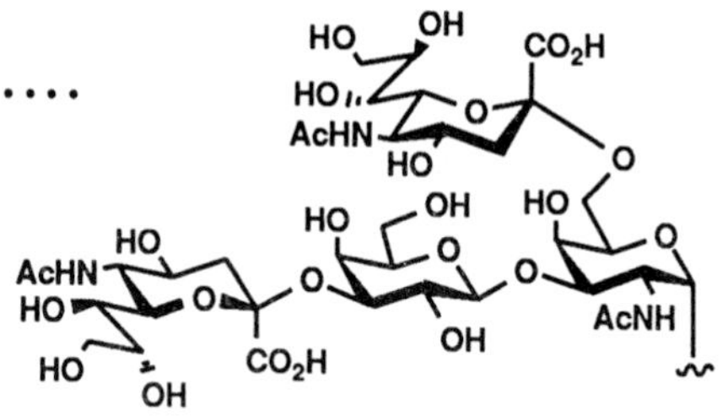

Scheme 1.

The attempt, however, was unsuccessful because of the lability of the phenacyl group to the reaction conditions [Ph₃SnH, AIBN(α,α'-azobisisobutyronitrile)]. When **27** was submitted to the same tin hydride reduction, a complex mixture was produced from which the desired compound **29** was isolated only in 16% yield. Other undesired products (~ 80%) were found to have lost their allyl ester functionality, presumably by the radical attack preferentially occuring at the olefinic bond rather than the sulfur atom. All attempts at using a modified procedure for selective desulfurization failed in our hands. Therefore, we decided to eliminate the phenylthio group after cleavage of the allyl ester, although it was anticipated that monitoring of the reaction progress in desulfurization and purification of product might be somewhat problematic owing to the highly polar carboxylic acid moiety present. Deallylation of **27** and **28** was effected with Pd(0) catalyst and N-methylaniline[30] in THF to give **7** (98%) and **30** (99%), respectively. After considerable experimentation, it was found that the phenylthio group was efficiently removed by reaction with a large excess (> 30 eq) of Ph₃SnH and repeated addition of AIBN to the mixture until the sulfide substrate had been consumed. The desulfurised products **13** and **14** were isolated in 93 and 80% yield, respectively, after chromatographic purification on silica gel and then on ODS (octadecylsilanized, C₁₈)-silica gel.

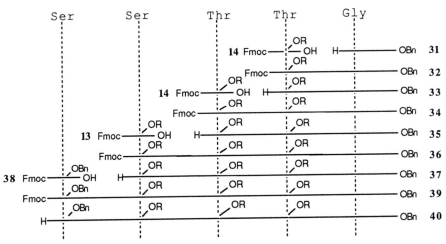

R = protected oligosaccharide, Fmoc = 9-fluorenylmethoxycarbonyl, Bn = benzyl

Scheme 5.

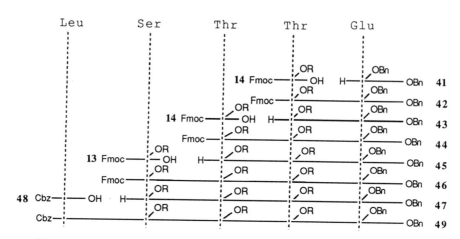

R = protected oligosaccharide, Fmoc = 9-fluorenylmethoxycarbonyl, Bn = benzyl, Cbz = benzyloxycarbonyl

Scheme 6.

With the key building blocks **13** and **14** in hand, the synthesis of the MN epitope analogs of glycophorin A was undertaken in solution according to the Fmoc strategy. The peptides were synthesized with EEDQ (2-ethoxy-1-ethoxycarbonyl-1,2-dihydroquinoline)[32] as the coupling agent, the procedure being illustrated in Schemes 5 and 6. All the coupling reactions and the subsequent cleavage of the *N*-terminal Fmoc group with morpholine proceeded smoothly. The protected intermediates were chromatographed on silica gel and characterized. To complete the synthesis, the pentapeptide **39** was now deprotected by treatment with morpholine and then by hydrogenation with 20% Pd(OH)$_2$/C in aq. MeOH. Compound **49** was hydrogenated directly. The deblocked synthetic samples were purified by gel permeation (Sephadex LH 20) and ion-exchange chromatography on Mono Q to give the pure M-epitope analog **15** (91%) and N analog **16** (88%). The structures of **15** and **16** were proved conclusively by FAB•MS and ^{1}H-NMR data, where the signals assigned to the Neu5Ac residue were in good agreement with those of the related natural glycoprotein fragment containing the α-D-Neu5Ac-(2→6)-α-D-GalNAc linkage.[33-35]

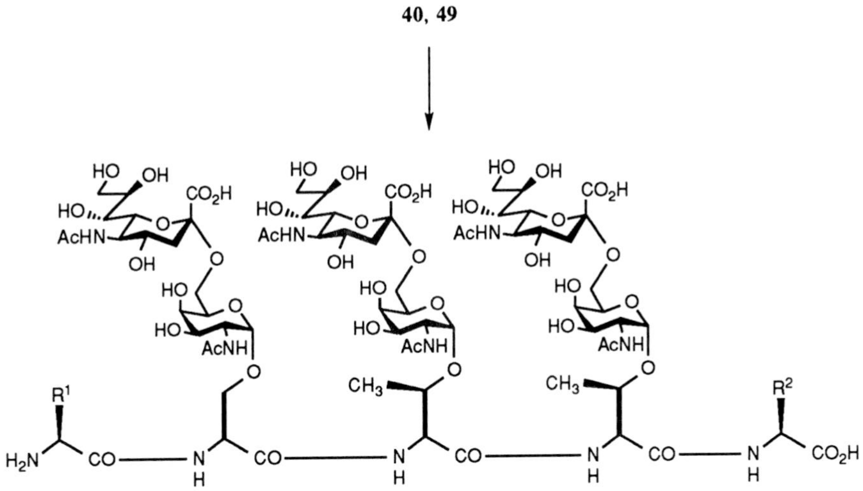

40, 49

15 R^1 = CH$_2$OH, R^2 = H

16 R^1 = CH$_2$CH(CH$_3$)$_2$, R^2 = CH$_2$CH$_2$CO$_2$H

Scheme 7.

3. EXPERIMENTAL SECTION

For the preparation of glycosylated serine and threonine derivatives, selected experimental procedures of either serine or threonine series are described. Optical rotations at 23 ± 2° C with CHCl$_3$ unless noted otherwise.[36]

Glycosylation of amino acid derivative (**11, 12**) *with glycosyl fluoride* **5** (→ **17** ~ **20**) : A mixture of Cp$_2$ZrCl$_2$ (261mg, 0.89mmol), AgClO$_4$ (370mg, 1.78mmol), and dried molecular sieves 4A powder (1g) in dry CH$_2$Cl$_2$ (6.5ml) was stirred at room temperature under argon for 30 min, then cooled on an ice-MeOH bath (- 20° C). To the mixture was added a solution of **5** (229mg, 0.59mmol) and **11** (262mg, 0.71mmol) in dry CH$_2$Cl$_2$ (10ml). After stirring at -20° ~ -10° C for 3 h, the reaction was quenched with pyridine (1ml). The reaction mixture was diluted with CHCl$_3$, filtered through Celite, washed with aq. NaHCO$_3$, water, and brine, dried on Na$_2$SO$_4$, and concentrated *in vacuo*. The residue was purified by flash-chromatography on silica gel with toluene-EtOAc (1 : 1) to give **17** (298mg, 68%) and **18** (29mg, 7%). **17** : Rf 0.56 (toluene-EtOAc, 7 : 3), mp 101° C, [α]$_D$ + 132° (c 1.1), **18** : Rf 0.33 (toluene-EtOAc, 7 : 3), [α]$_D$ + 26° C (c 1.2), **19** (71%) : Rf

0.61 (toluene-EtOAc, 3 : 1), $[\alpha]_D + 117°$ (c 2.1), **20** (7%) : R_f 0.40 (toluene-EtOAc, 3 : 1), $[\alpha]_D + 21°$ (c 2.0).

Hydrolysis of benzylidene acetal ($\rightarrow$ **21, 22**) : A solution of **17** (200mg, 0.27mmol) in 80% aq.AcOH (7ml) was heated at 60° C for 5 h and concentrated *in vacuo*. The residue was chromatographed on silica gel with EtOAc-toluene (7 : 3) to give **21** (150mg, 85%). **21** : R_f 0.15 (toluene-EtOAc, 1 : 1), $[\alpha]_D + 85°$ (c 1.1), **22** (69%) : R_f 0.10 (toluene-EtOAc, 2 : 1), $[\alpha]_D + 66°$ (c 3.9).

Glycosylation of **21, 22** *with* **4** ($\rightarrow$ **23, 24**) : To a cold (- 20° C) mixture of **22** (89mg, 0.14mmol), HgBr$_2$ (50mg, 0.14mmol), Hg(CN)$_2$ (118mg, 0.47mmol), 2-methyl-2-butene (214µl, 2.0mmol), and dried molecular sieves 4A powder (1.5g) in dry CCl$_4$ (3.5ml), was added a solution of **4** (189mg, 0.20mmol) in dry CCl$_4$ (3.5ml) under argon with stirring on an ice-MeOH bath. After stirring at -20° C ~ room temperature for 24 h, the reaction mixture was diluted with EtOAc, filtered through Celite, washed with water and brine, dried on Na$_2$SO$_4$, and concentrated *in vacuo*. The residue was chromatographed on Bio-beads S x 4 (800ml) with toluene and then on silica gel with toluene-EtOAc (4 : 1) to give **24** (171mg, 84%). **24** : R_f 0.50 (toluene-EtOAc, 7 : 3), $[\alpha]_D + 50°$ (c 1.7), **23** : R_f 0.43 (toluene-EtOAc, 7 : 3), $[\alpha]_D + 55°$ (c 1.0).

Conversion of azide **23, 24** *into acetamide* **27, 28** : Freshly distilled AcSH (0.8ml) was added to a solution of **24** (171mg, 0.11mmol) in dry pyridine (0.4ml). The mixture was stirred at room temperature for 24 h, then evaporated *in vacuo*. The residue was chromatographed on silica gel with toluene-EtOAc (1 : 1) to give **28** (156mg, 90%). **28** : R_f 0.42 (toluene-EtOAc, 1 : 1), $[\alpha]_D + 54°$ (c 0.5), **27** : R_f 0.52 (toluene-EtOH, 7 : 1), $[\alpha]_D + 45°$ (c 0.3).

Deallylation of **27, 28** ($\rightarrow$ **7, 30**) : According to the Kunz's procedure,[30] a mixture of **27** (115mg, 76µmol), Pd(Ph$_3$P)$_4$ (25mg, 22µmol) and *N*-methylaniline (224ml, 2.07mmol) in dry THF (1.6ml) was stirred under argon at room temperature for 2 days. The mixture was diluted with EtOAc, washed with dil. HCl and brine, dried on Na$_2$SO$_4$, and concentrated *in vacuo*. The residue was chromatographed on silica gel with CHCl$_3$-MeOH-AcOH (92 : 3 : 5) to give **7** (110mg, 98%). All the physical data of **7** were identical with those reported previously.[22] **30** : $[\alpha]_D + 65°$ (c 2.3).

25. (a) K. Suzuki, H. Maeta, T. Matsumoto, and G. Tsuchihashi, *Tetrahedron Lett.*, **29**, 3567 (1988); (b) *ibid*, **29**, 3571 (1988); (c) K. Suzuki, H. Maeta, and T. Matsumoto, *ibid*, **30**, 4853 (1989).

26. M. Yoshimura, Y. Matsuzaki, M. Sugimoto, M. Ito, and T. Ogawa, *Japan Kokai Tokkyo Koho*, Heisei 1-228997 (Sept 12, 1989); *Chem. Abstr.*, 112, 119702m (1990).

27. (a) Y. Ito and T. Ogawa, *Tetrahedron Lett.*, **28**, 6221 (1987); (b) *ibid*, **29**, 3987 (1988); (c) *Tetrahedron*, **46**, 89 (1990).

28. H. Iijima, Y. Nakahara, and T. Ogawa, *Tetrahedron Lett.*, **33**, 7907 (1992).

29. B. G. de la Torre, J. L. Torres, E. Bardaji, P. Clapés, N. Xaus, X. Jorba, S. Calvet, F. Albericio, and G. Valentia, *J. Chem. Soc. Chem. Commun.*, 965 (1990).

30. M. Ciommer and H. Kunz, *Synlett,* 593 (1991).

31. T. Rosen, I. M. Lico, and D. T. W. Chu, *J. Org. Chem.*, **53**, 1580 (1988).

32. B. Belleau and G. Malek, *J. Am. Chem. Soc.*, **90,** 1651 (1968).

33. J. F. G. Vliegenthart, L. Dorland, H. van Halbeek, and J. Haverkamp, In *Sialic Acids* (R. Schauer Ed.), Springer-Verlag, New York, p. 127, (1982).

34. T. A. Gerken, *Arch. Biochem. Biophys.*, **247**, 239 (1986).

35. H.-U. Linden, R. A. Klein, H. Egge, J. Peter-Katalinic, J. Dabrowski, and D. Schindler, *Biol. Chem. Hoppe-Seyler*, **370**, 661 (1989).

36. NMR data of the synthetic compounds are available on request to the authors.

37. Y. Nakahara, H. Iijima, and T. Ogawa, *Tetrahedron Lett.*, **35**, 3321 (1994).

Chapter 14

SYNTHESIS OF *C*-GLYCOSIDES; STABLE MIMICS OF *O*-GLYCOSIDIC LINKAGES

CAROLYN BERTOZZI AND MARK BEDNARSKI

Stanford University School of Medicine, Lucas MRS Research Center, Stanford, CA 94305-5488, USA

Abstract The chemical synthesis of *C*-glycoside analogs of *O*-glycosides is reviewed with an emphasis on compounds that can serve to alter biological processes. Strategies for the synthesis of *C*-glycosides are broken down into four main catagories: nucleophilic addition to anomeric carbenium ions and lactones, reactions of glycosyl anions with electrophiles, reactions of glycosyl radicals and transition metal mediated couplings. The physical properties of *C*-glycosides are also discussed and compared to *O*-glycosides. Finally the use of *C*-glycosides to inhibit enzymes such as glycosyl transferases, glycosidases, phosphorylases and their interactions with receptors with an emphasis on cell adhesion is addressed.

1. INTRODUCTION

For many decades, carbohydrate molecules were regarded as energy-storage compounds and support structures (i.e., starch and cellulose) with no significant biological function, despite their ubiquitous presence on the proteins and membranes of all eukaryotic cells. The discovery in 1969 that cell-surface oligosaccharides were profoundly altered in cancer cells, and may be related to cancer cell diffusion, marked the beginning of the third revolution in biology, with carbohydrate molecules joining the ranks of proteins and nucleic acids as determinants of biological activity.[1]   During the next 20 years, the biological functions of carbohydrate molecules came under intense scrutiny, resulting in the discovery of their roles in mediating such diverse cell-cell recognition phenomena as viral and bacterial infection,[2] tumor cell metastasis,[3] and leukocyte adhesion during inflammation.[4] It is now clear that the mystery of carbohydrates, with their enormous structural complexity, is only just beginning to unravel.

As increasing numbers of biologically active oligosaccharides were structurally characterized, organic chemists sought to develop new methods for the synthesis of naturally occurring carbohydrates and their derivatives for use as therapeutic agents. However, like proteins and nucleic acids, carbohydrates are susceptible to biodegradation, thus limiting their therapeutic potential and use in biological studies of structure and function. Synthetic chemists have therefore begun to focus on the design of carbohydrate analogs that can withstand the degradative forces *in vivo*.

The majority of these efforts have been aimed at the synthesis of isosteric carbohydrate analogs in which the reactive acetal functionality of the anomeric center is modified by substitution of an oxygen atom with a sulfur or carbon atom. For example, thioglycosides are carbohydrate analogs that contain a sulfur atom rather than an oxygen atom at the glycosidic linkage as shown in Figure 1.[5] These derivatives are resistant to the action of glycosidases, the enzymes responsible for cleaving glycosides in carbohydrate metabolism. Thioglycosides can be cleaved by chemical hydrolysis, but at a much slower rate than their parent oxygen-linked derivatives.

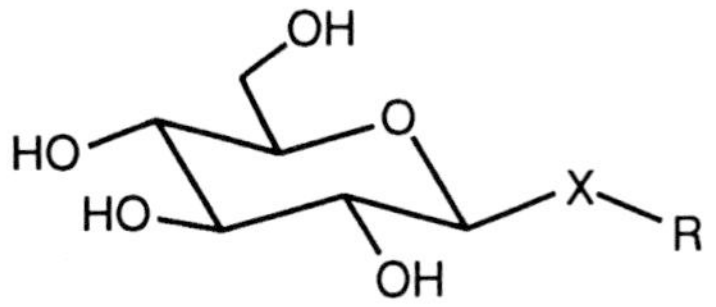

X = O, *O*-glycoside
X = S, thioglycoside

Figure 1. Thioglycosides.

Pseudosugars are another class of stable carbohydrate isosteres. In these analogs, the oxygen atom of the pyranose or furanose ring has been replaced with a methylene group (Figure 2).[6] Pseudosugars are resistant to both chemical and enzymatic hydrolysis at the pseudo-anomeric center, and therefore have an advantage over thioglycosides as biologically active derivatives.   But, pseudosugars are not readily derived from simple carbohydrate substrates, and are therefore more difficult to synthesize than thioglycosides.

Carbon-linked glycosides (C-glycosides) comprise a third class of carbohydrate analogs in which the oxygen atom at the glycosidic linkage has been replaced with a carbon atom (Figure 3).[7] C-Glycosides are resistant to both

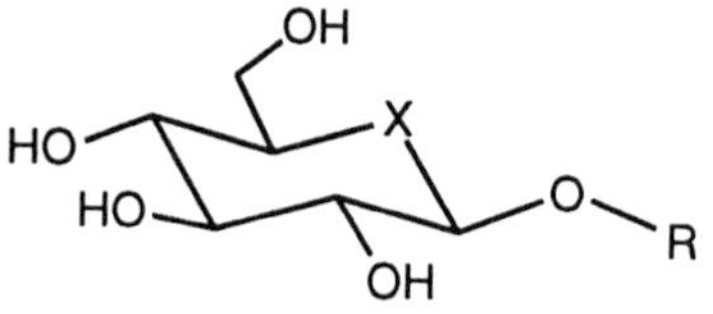

$X = CH_2$, pseudosugar

Figure 2. Pseudosugars.

chemical and enzymatic hydrolysis of the glycosidic bond, and can be readily synthesized from simple sugars. These properties make C-glycosides particularly well suited for use as agents and synthetic tools for studying biological events.

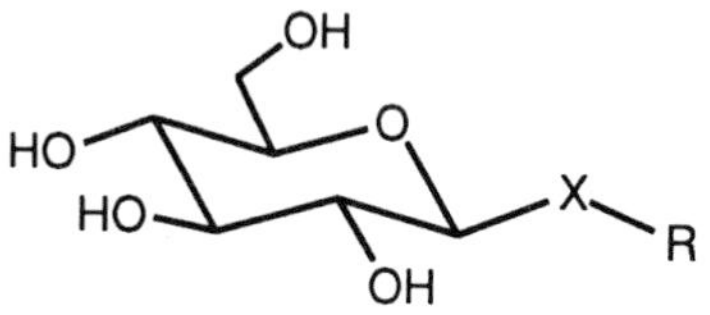

$X = CH_2$, C-glycoside

Figure 3. Carbon-linked glycosides.

This chapter describes synthetic methodology for the construction of C-linked carbohydrate structures, and the evaluation of their biological properties. Specifically we will discuss general methods for the synthesis of C-glycosides, their structural relationship to naturally occurring O-linked glycosides, and describe their interations with carbohydrate binding proteins.

2. SYNTHESIS OF *C*-GLYCOSIDES

Carbon-linked glycosides first captured the interest of synthetic chemists during the 1970's due to the discovery of the *C*-glycosyl functionality in several naturally occurring nucleoside antibiotics such as formycin (**1**) and showdomycin (**2**) (Figure 4).[8] Synthetic efforts aimed at the direct formation of carbon-carbon bonds at the anomeric center intensified during the next decade when a large variety of complex natural products were shown to contain elements derived from *C*-glycosides of naturally occurring sugars.[9] Thus, synthetic *C*-glycosides were initially valued as stereochemically rich starting materials for the synthesis of natural products containing functionalized furan or pyran rings. The most elaborate example of these is the marine natural product palytoxin (**3**) which contains six rings that can be derived from *C*-glycosides, linked together by polyhydroxylated alkyl chains (Figure 5).[10]

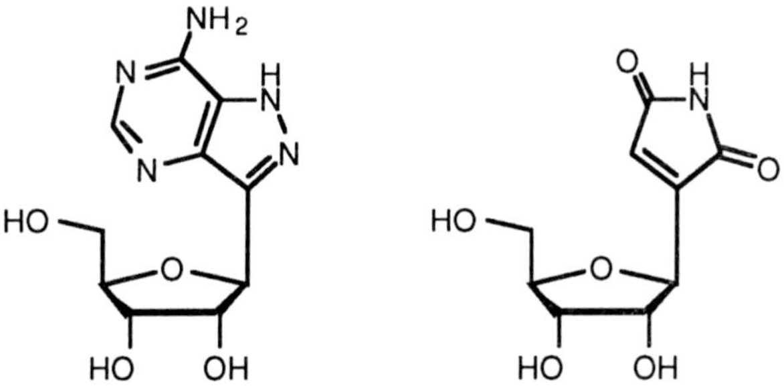

1   2

Figure 4.   Naturally occuring *C*-linked nucleosides, formycin (1) and showdomycin (2).

The two main concerns in forming a carbon-carbon bond at the anomeric position of a carbohydrate molecule are control of stereochemistry (α vs. β) and chemical compatibility of the reaction conditions with the hydroxyl protecting groups. Most methods for the introduction of anomeric alkyl groups take advantage of anomeric stereoselection to achieve a particular configuration, a well known phenomenon in the synthesis of *O*-glycosides and other cyclic acetals.[11]

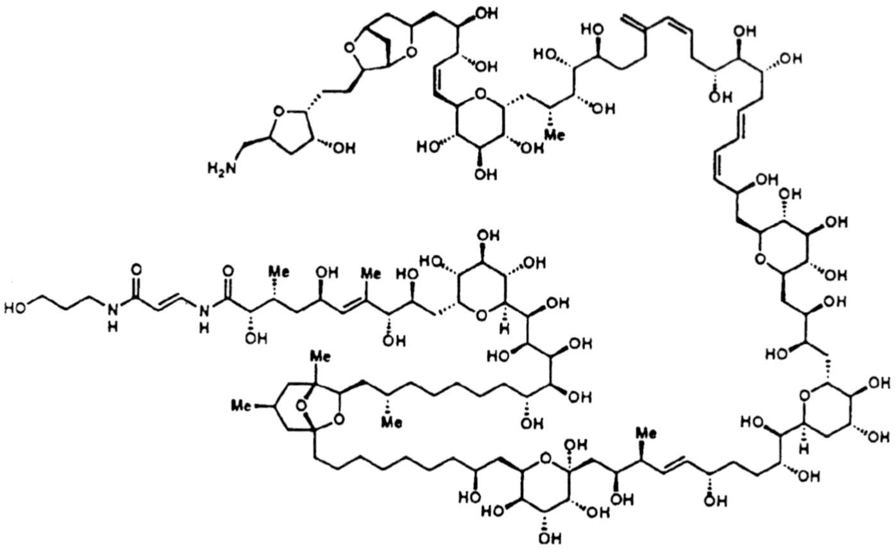

3

Figure 5. Palytoxin.

The chemical reactivity of the anomeric center is multifaceted. The anomeric carbon atom is inherently electrophilic and can be activated for reaction with carbon nucleophiles. The ring oxygen in a carbohydrate molecule promotes the formation of stabilized anomeric carbanions and anomeric radicals as well. This "umpolung" reactivity of the anomeric center has been exploited in the development of diverse C-glycosylation methods such as nucleophilic additions to anomeric oxocarbenium ions[12] and lactones,[13] glycosyl anion additions to electrophiles,[14] glycosyl radical reactions with alkenes and allylstannane,[15] and transition metal-mediated couplings.[16] These methods are discussed individually below.

2. A. Nucleophilic Additions to Anomeric Oxocarbenium Ions

The most widely utilized method for generating C-glycosides involves the Lewis acid catalyzed generation of an anomeric oxocarbenium ion, followed by the addition of a carbon nucleophile.[12] These reactions can be performed in the presence of both ether and ester protecting groups commonly used in carbohydrate chemistry, and proceed with high stereoselectivity to give the α-linked (axial) product. The anomeric centers of the carbohydrate substrates can be activated in the form of glycosyl ethers, esters and halides, *O*-glycosylimidates or thioglycosides, and the nucleophiles include silyl enol ethers, malonate esters and alkylsilanes. Some commonly used Lewis acids in these transformations are boron trifluoride etherate (BF$_3$OEt$_2$), trimethylsilyl trifluoromethanesulfonate (TMSOTf), tin dichloride (SnCl$_2$), tin tetrachloride (SnCl$_4$) and titanium tetrachloride (TiCl$_4$).

The most convenient reaction of this type involves treatment of the methyl glycoside of a protected sugar (**4**) with allyltrimethylsilane and a catalytic amount of TMSOTf as described by Sakurai and coworkers (Figure 6).[12] The products of this reaction are α- and β-linked *C*-allyl glycosides (**5a** and **5b**) that generally form in a ratio between 20:1 and 10:1. The Lewis acid catalyst activates the glycosidic bond for cleavage and the intermediate oxocarbenium ion is trapped from the α-face by the nucleophilic β-silyl alkene (Figure 7). The preference for

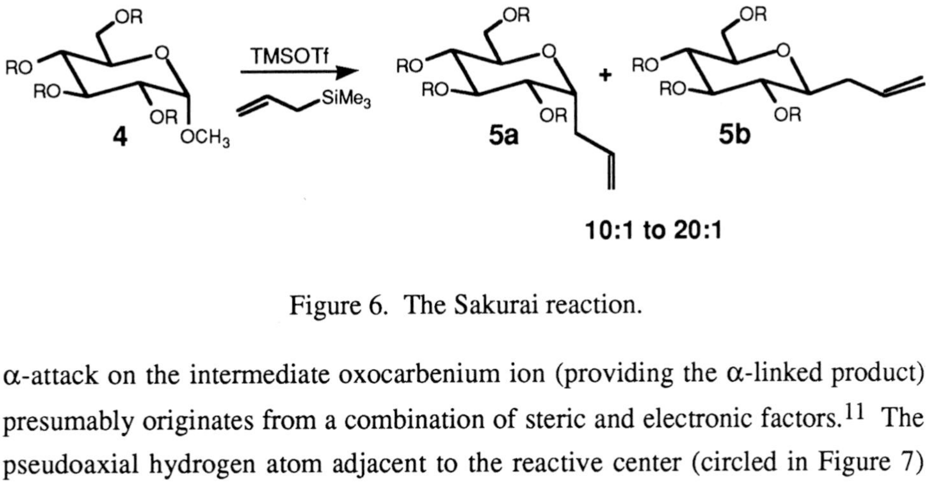

10:1 to 20:1

Figure 6. The Sakurai reaction.

α-attack on the intermediate oxocarbenium ion (providing the α-linked product) presumably originates from a combination of steric and electronic factors.[11] The pseudoaxial hydrogen atom adjacent to the reactive center (circled in Figure 7) blocks the top (β) face of the pyranose ring and destabilizes the transition state for β-attack. Furthermore, the s*-orbital of the incipient carbon-carbon bond in the transition state for α-attack overlaps favorably with the pseudoaxially oriented

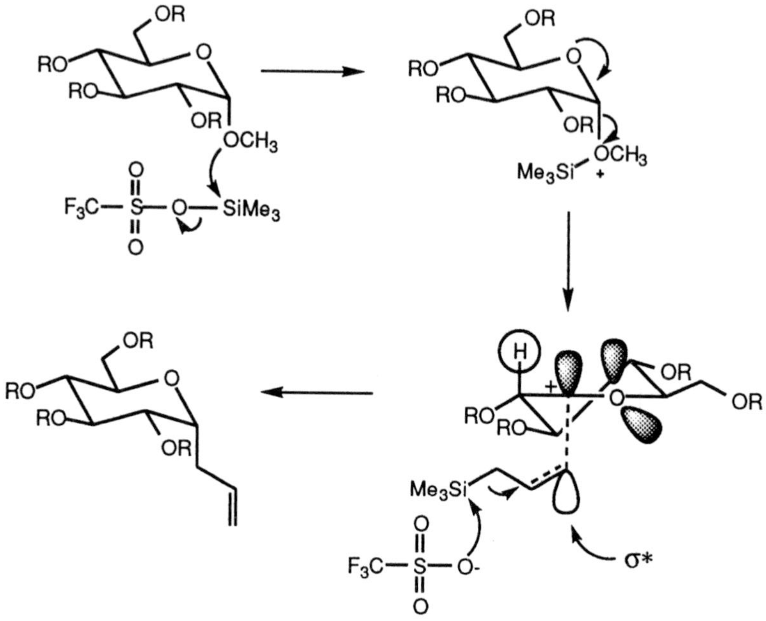

Figure 7. Mechanism of the Sakurai reaction.

lone pair of electrons on the adjacent oxygen atom. The orientation of the same s*-orbital in the transition state for β-attack results in a less favorable interaction.

2. B. Nucleophilic Addition to Anomeric Lactones

A parallel method for constructing β-linked C-glycosyl compounds involves the addition of carbon nucleophiles such as Grignard reagents and alkyllithium derivatives to anomeric lactones of protected sugars (6) (Figure 8).[13] The hemiketal product (7) is then reduced stereoselectively with an alkylsilyl hydride such as triethylsilane in the presence of a Lewis acid catalyst. The reduction proceeds analogously to the previously described Sakurai reaction, with preferential attack of hydride from the α-face, affording the β-linked C-glycosyl product (8) with high stereoselectivity. Unlike Sakurai's method for the synthesis of α-C-glycosides, this method for the synthesis of β-C-glycosides benefits from the structural diversity of readily available Grignard-type nucleophiles.

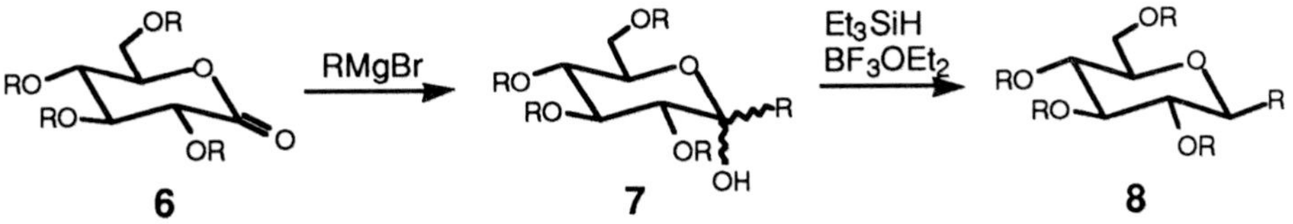

Figure 8. Nucleophilic addition to anomeric lactones followed by stereoselective reduction

2. C. Reaction of Glycosyl Anions with Electrophiles

Another approach to the formation of *C*-glycosyl linkages is the generation of a nucleophilic anomeric center.[14] The majority of these methods utilize a stabilized anionic intermediate generated from anomeric phosphonium salts, nitro and phenylsulfonyl sugars, or 2-deoxy-glycopyranosyllithium derivatives. For example, Vasella and coworkers report the reaction of 1-deoxy-1-nitro sugars (9) with electrophiles such as formaldehyde to afford the tertiary nitro-*C*-glycosides (10) (Figure 9).[14j] Reductive denitration using tributyltin hydride in the presence of a radical initiator affords the β-linked *C*-glycoside (11) with high stereoselectivity.

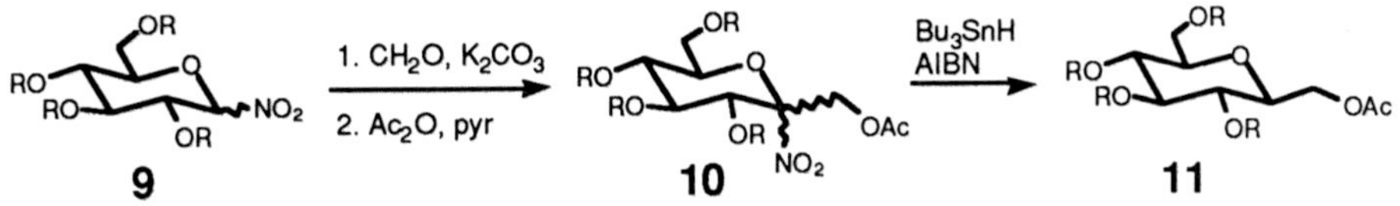

Figure 9. Addition of an anomeric anion to an electrophile.

Although anomeric anions have been utilized in a number of stereoselective *C*-glycosylation methods, these reactions suffer to some degree from the tendency of the anion to undergo β-elimination of the adjacent alkoxy substituent. The electrophilic reactions discussed above are superior in this regard, being fully compatible with the alkoxy functionality of protected sugars.

2.D. Glycosyl Radical Reactions

The great challenge in *C*-glycosylation chemistry is to utilize carbon-carbon bond forming reactions that are compatible with a wide range of protecting groups and functionalities. The previously described reactions involve either Lewis acidic or basic conditions which limit the range of possible protecting groups. Furthermore, many naturally occurring carbohydrate structures contain amine or amide groups, which may preclude the use of these reactions. Radical chemistry is ideally suited for these polyfunctional molecules due to the mild, neutral conditions under which radicals can be generated. In addition, radicals formed at the anomeric center are stabilized by the adjacent ring oxygen atom.[17]

The reaction of anomeric radicals with activated alkenes[15] was first developed by Giese and coworkers.[15a] In one example, the peracetylated glycopyranosyl bromide (**12**) was treated with tributyltin hydride and acrylonitrile under photolysis conditions to afford the α-C-glycosyl product (**13**) exclusively (Figure 10). Mechanistic studies have shown that the glycosyl radical intermediate adopts a boat-like conformation.[17] Subsequent addition of the anomeric radical to acrylonitrile takes place from the pseudoequatorial face of the pyranose ring, providing, after ring-flip to the chair conformation, the α-linked product.

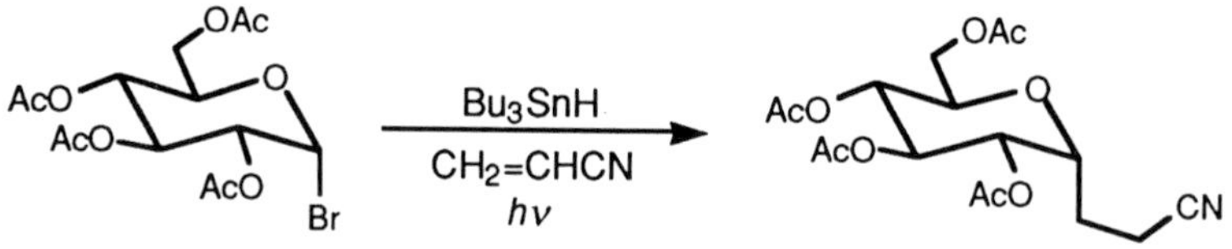

Figure 10.   Addition to an anomeric radical.

Alternatively, radical coupling reactions at the anomeric center have been performed using allyltributylstannane and a radical initiator to give the C-allyl glycoside product.[15c] The development of this radical C-glycosylation method has culminated in the synthesis of C-glycosides of N-acetylneuraminic acid, one of the most difficult carbohydrates to manipulate.[15h,i]

2.E. Transition Metal-Mediated Couplings

The high stereochemical control associated with transition metal-mediated reactions has prompted many researchers to explore such methods for C-glycosylations.[16] Most examples involve the use of palladium(0) catalysts to generate an intermediate palladium-allyl complex which is then susceptible to nucleophilic attack. For example, Sinou and coworkers have described the reaction of 2,3-anhydro sugar (**14**) with 1,2-bis(dibenzylidene acetone) Pd(0) (Pd(dba)$_2$) and a stabilized carbanion nucleophile to obtain the *C*-glycoside product (**15**) with retention of configuration (Figure 11).[161] Addition of the nucleophile opposite to palladium in the p-allylpalladium intermediate results in the observed stereoselectivity.

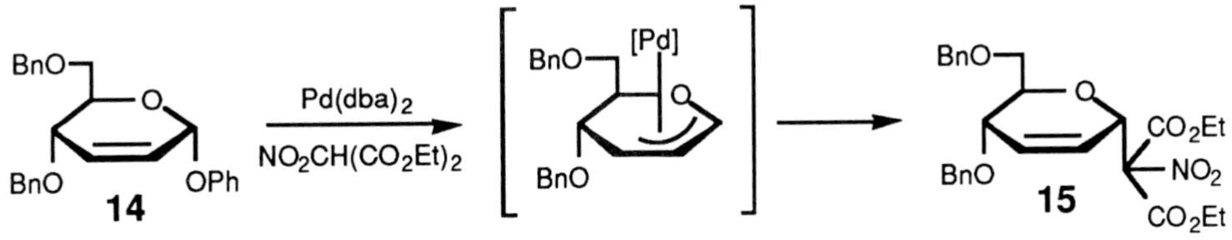

Figure 11. Palladium-mediated *C*-glycosylation.

2.F. Other Synthetic Methods

Other methods for generating *C*-glycosides that are not as widely utilized as those discussed above include enolate ester Claisen rearrangements on glycals,[18] Wittig-type reactions on acyclic sugar aldehydes[19] and anomeric lactones,[20] nucleophilic displacements on glycal methanesulfonates,[21] and carbenoid displacement reactions on phenylthioglycosides.[22] Whitesides and coworkers have reported a combined chemical and enzymatic synthesis of *C*-glycosyl compounds using rabbit muscle aldolase.[23] Recently, Vasella and coworkers have described a method for the generation of anomeric carbenes and their reaction with olefins to afford *C*-glycosyl spirocyclopropanes.[24] A unique application of this methodology is the insertion of anomeric carbenes into C$_{60}$ to generate novel *C*-glycosyl fullerenes.[25]

3. PHYSICAL PROPERTIES OF *C*-GLYCOSIDES

The potential for *C*-glycosides to mediate biological processes involving carbohydrate-protein recognition depends upon their structural similarity to the native *O*-glycosyl derivatives. The substitution of a methylene group for an oxygen atom results in a change in both the size and electronic properties of the glycosyl linkage (Table 1).[26]

Table 1. Physical properties of oxygen and carbon linkages.

Property	*Oxygen*	*Carbon*
Van der Waals Radius	1.52 Å (ether)	2.0 Å (methylene)
Hydrogen Bonding	acceptor	none
Electronegativity (*Pauling values*)	3.51	2.35
Dipole Moment	0.8 D (C-O)	0 D (C-C)

For example, the average Van der Waals radius of an oxygen atom in an ether linkage is 1.52 Å, whereas that of a methylene group is 2.0 Å. This increase in the steric bulk of the glycosyl side chain may interfere with receptor binding in cases involving close protein-ligand contacts. Replacement of the oxygen atom with a methylene group also destroys the hydrogen bonding ability of the glycosyl side chain, which may reduce the binding affinity of the *C*-glycoside if specific hydrogen bonds to this oxygen atom are involved in recognition. Because oxygen is more electronegative than carbon (3.51 and 2.35 on the Pauling scale, respectively), the C-O bond has a strong dipole moment (ca. 0.8 Debye) whereas the C-C bond does not. Therefore, interactions between local dipoles of the receptor and ligand may be affected by substitution of the *C*-glycoside. Finally, the dipole moment and hydrogen bonding ability of the *O*-linkage render it better solvated in an aqueous environment than the *C*-linkage. In this regard, the more hydrophobic *C*-glycoside gains a greater entropic advantage by interacting with a protein receptor and may bind with higher affinity, an effect that is well documented in protein-ligand binding studies.[27]

Although the differences noted above can affect receptor binding activity, perhaps the most important consideration is the conformational similarity of *C*-

and *O*-glycosides in solution. The conformation of *O*-linked glycosides around the glycosidic bond is governed by steric factors and by a stereoelectronic phenomenon termed the "exo-anomeric effect".[28] Figure 12 shows the possible conformers of α- and β-linked *O*- and *C*-glycosides represented by Newman projections of the glycosidic bonds. The *O*-linked sugars α-*O*-Gl and β-*O*-Gl are known to prefer conformer **A**. This preference originates from the steric destabilization of *O*-linked conformers **C** with respect to the other conformers (2 gauche interactions vs. 1), the steric destabilization of *O*-linked conformers **B** compared to **A** (gauche interaction with methylene vs. oxygen), and the stereoelectronic stabilization of conformers **A**. The stereoelectronic contribution derives from the favorable overlap of the glycosidic oxygen lone pair with the low-lying s*-orbital of the ring C-O bond; this stabilizing interaction is absent in *C*-glycosides.

Kishi and coworkers have performed extensive studies on the preferred conformations of *C*-glycosides to determine the importance of stereoelectronic stabilization in governing conformation.[29] By substituting each hydrogen atom on the *C*-glycosyl methylene stereoselectively with a deuterium atom, these researchers were able to measure the coupling constants (*J* values) between the methylene hydrogens and the hydrogen atom on the anomeric carbon. The *J* values were then translated via the Karplus equation into a dihedral angle that represents the weighted average of available conformers. Using this analysis, α- and β-linked *C*-glycosides (α-*C*-Gl and β-*C*-Gl shown in Figure 12) were also shown to prefer conformers **A**, that is the same conformations adopted by *O*-glycosides. Thus, the solution conformations of *O*-linked and *C*-linked sugars are identical due to steric considerations, despite the lack of stereoelectronic stabilization in the latter. This observation suggests that *C*-glycosides will retain the biologically active conformations of their natural carbohydrate counterparts.

4. *C*-LINKED NUCLEOSIDES

The notion that *C*-glycosides can bind to protein receptors similarly to biologically active sugars was first supported by the discovery of naturally occurring *C*-linked nucleosides with anti-viral and anti-tumor activity.[8] Presumably, these *C*-nucleosides bind to and inhibit enzymes involved in the synthesis of nucleic acids or their precursors. In fact, the C-nucleoside pseudouridine (17), a C-linked analog of uridine (16), occurs as a minor component of transfer RNA (Figure 13).[8] Its incorporation into a natural biopolymer implies that pseudouridine is a substrate for the enzymes involved in ribonucleic acid biosynthesis.

5. C-LINKED MONOSACCHARIDES

There are many reports of the synthesis of C-glycosyl isosteres of biologically relevant sugars such as sugar-1-phosphates, sugar nucleotide phosphates, monosaccharides, and disaccharides. However, the biological evaluation of these targets is currently limited to reversible and irreversible enzyme inhibitors and affinity labels for anti-carbohydrate antibodies. C-Glycosyl phosphonate analogs of sugar-1-phosphates, intermediates in the biosynthesis of oligosaccharides, have been synthesized as potential inhibitors of oligosaccharide processing and nucleic acid synthesis.[30] Unlike their chemically labile parent derivatives, C-glycosyl phosphonates are resistant to hydrolysis of both the glycosidic bond and the phosphonate linkage. In this category, C-linked derivatives of ribose-1-phosphate (18),[31] glucose-1-phosphate (19),[32] fructose-1-phosphate (20),[33] KDO-2-phosphate (21) and CMP-KDO (22)[34] have been reported (Figure 14).

Three examples of biological activity have been described from this list of sugar-1-phosphate isosteres. First, Ray et al. report that compound (19) can serve as a substrate for phosphoglucomutase, which phosphorylates glucose-1-phosphate at the C-6 hydroxyl group.[35] The value of k_{cat}/K_m for (19) was 1.7×10^3 $M^{-1}s^{-1}$, substantially lower than that for glucose-1-phosphate (9×10^7 $M^{-1}s^{-1}$). Second, Engel and coworkers have demonstrated the ability of C-linked fructose-1-phosphate (20) to inhibit the hexose phosphate transport system in $E.$ $coli$ at a concentration of 2 mM, although the molecular nature of inhibition was not determined.[33] Finally, Norbeck and coworkers have synthesized C linked analogs of KDO-2-phosphate (21) and (22) as inhibitors of CMP-KDO synthetase, a key enzyme in the synthesis of bacterial lipopolysaccharides.[34] Only compound (22) demonstrated inhibitory activity, with an IC_{50} of 4.1 mM. Compared to the K_m values for native KDO (31 µM) and CTP (11 µM), however, this inhibition value indicates a fairly weak binding interaction between CMP-KDO synthetase and the C-glycoside.

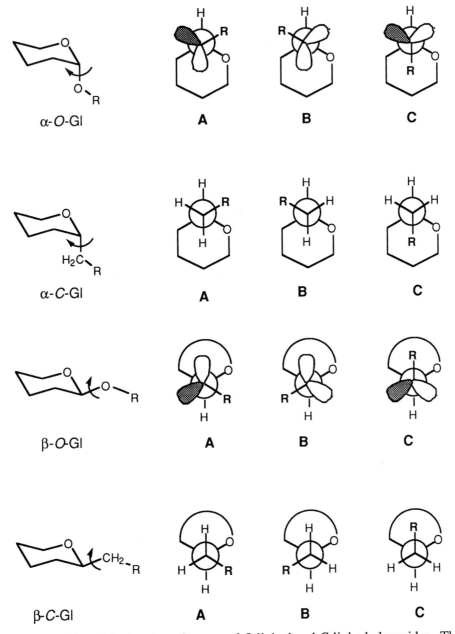

Figure 12. Glycosidic bond conformers of *O*-linked and *C*-linked glycosides. The preferred conformations (**A**) are identical for *O*- and *C*-glycosides. Shaded *n*-orbitals indicate a favorable interaction with the s*-orbital of the ring C-O bond, a stereoelectronic effect that is absent is *C*-glycosides.

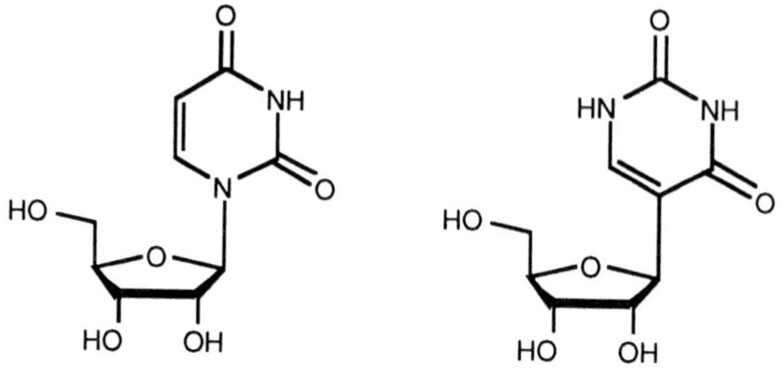

16 17

Figure 13. Uridine and pseudouridine.

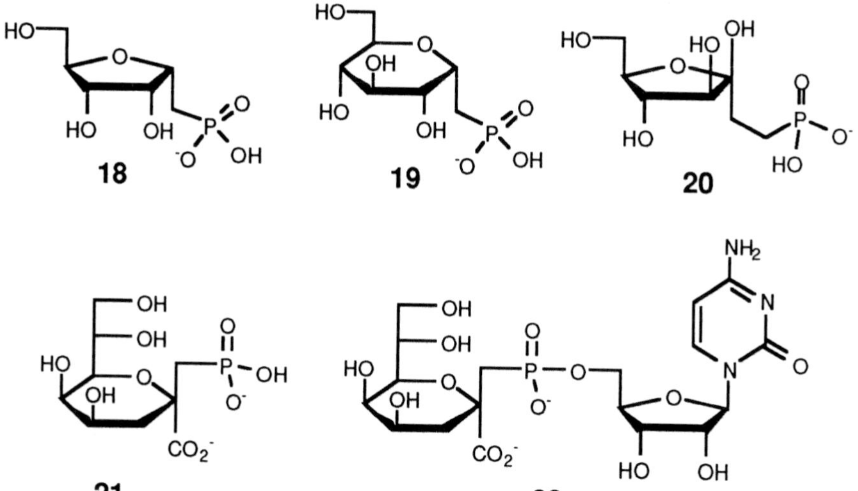

18 19 20 21 22

Figure 14. *C*-Glycosyl sugar phosphonates.

The failure of compounds (**19**), (**21**) and (**22**) to behave more similarly to their parent derivatives may result from the phosphonate substitution rather than the *C*-glycosyl substitution, as phosphonates are markedly different in charge and geometry from natural phosphate esters.[30a] A *C*-linked analog of ribose-1-phosphate (**18**) has been tested as an inhibitor of purine nucleoside phosphorylase (PNP), a therapeutic target for inducing selective T-cell toxicity.[31a] However, this derivative showed no inhibitory activity at a concentration of 6 mM, again perhaps the result of the phosphonate moiety rather than the *C*-glycoside alone.

C-Glycosyl analogs of simple monosaccharides have been synthesized as reversible inhibitors of glycosidases. For example, the *C*-allyl galactoside (**23**) (Fig. 1-15) is a competitive inhibitor of α-galactosidase, but not β-galactosidase, as expected based on the stereochemistry at the anomeric center (a K_i value was not reported).[36] Schmidt and Dietrich have reported the synthesis of a series of *C*-glucosides (**24**)-(**26**), and their inhibitory activity towards sweet almond β-glucosidase.[37] The K_i values of compounds (**24**) and (**25**) are in the millimolar range, whereas the K_i of compound (**26**) is 7.0×10^{-5} M, a value similar to that of the well known inhibitor 1-deoxynojirimycin. The binding advantage of compound (**26**) derives from the positively charged amino group, which is thought to interact with carboxylate groups in the enzyme active site similarly to aza sugar inhibitors. In all three compounds, the hydrophobic phenyl substituent on the *C*-glycosyl side chain is poorly solvated in water, thus enhancing the affinity of these molecules for the binding cleft (hydrophobic effect). Finally, *C*-glycosyl derivatives of fucose have been synthesized and immobilized on sepharose beads.[38] The resulting affinity columns provided a stable matrix on which to purify active porcine liver α-fucosidase.

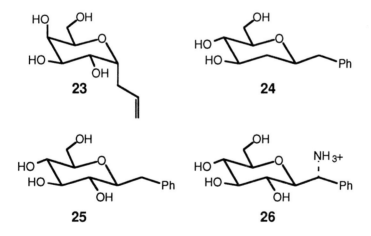

Figure 15. Reversible glycosidase inhibitors.

Because of their chemical stability, *C*-glycosides have also been utilized as affinity labels in cases requiring extremely reactive functionalities such as carbocations and carbenes. Sinnott and Smith report the synthesis of *C*-glycosyl nitrophenyltriazene (**27**) as a suicide inhibitor of *E. coli* β-galactosidase (Figure 16).[39] Protonation of the triazene group by an acidic residue in the enzyme active site results in fragmentation to afford a reactive primary carbocation at the *C*-glycosyl methylene. Compound (**27**) deactivated the enzyme at a rate of 4.02 x 10^{-3} s^{-1}, with an apparent K_i of 73 mM. Lehmann and coworkers have synthesized the disaccharide *C*-glycosyl diazirene (**28**) as a photoaffinity label for the antigalactan antibody IgA X24.[40] Using this molecule, these researchers were able to localize the antigen-combining site on the IgA heavy chain.

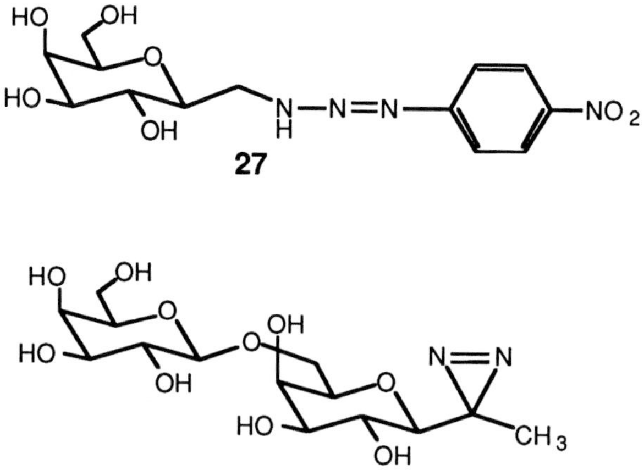

Figure 16. *C*-Glycosyl affinity labels.

6. *C*- LINKED OLIGOSACCHARIDES

Biologically active carbohydrate molecules are often presented in the context of complex oligosaccharides. Within these complex structures, the actual recognition determinants for protein receptors may consist of a single sugar unit, or multiple sugar units linked together in linear or branched arrays. Consequently, synthetic analogs that mimic the activity of multiple-sugar structures must incorporate the essential features of each carbohydrate unit. The construction of stable *C*-glycosyl analogs of multiple-sugar determinants requires the stereoselective formation of a carbon-carbon bond between two pyranose or furanose units, a daunting goal in spite of the availability of methods for simple C-glycosylations. Therefore, it was no small achievement when Sinaÿ and Rouzaud reported in 1983 the first synthesis of a carbon-linked disaccharide (C-disaccharide) analog of glucose β-linked to the 6-position of another glucose unit Glc-C-β-(1,6)-Glc-α-OMe (**29**) (Figure 17).[41]   Since then, several approaches to this challenging class of molecules have been described. These can be divided into two classes: methods in which acyclic precursors are cyclized to form the sugar units, and methods in which pyranose subunits are coupled directly to afford C-disaccharide products.

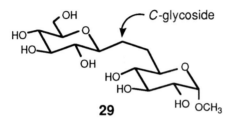

29

Figure 17.   The first C-disaccharide, Glc-C-β-(1,6)-Glc-OMe.

6.A.   Methods for the Synthesis of C-Disaccharides Using Acyclic Precursors
The greatest contribution to the synthesis of C-disaccharides has come from the laboratory of Kishi and coworkers. During the last six years, these researchers have reported syntheses of several C-linked analogs of naturally occurring disaccharides, including maltose (Glc-α-(1,4)-Glc),[42] the repeating unit of starch, cellobiose (Glc-β-(1,4)-Glc),[2] the repeating unit of cellulose, isomaltose (Glc-α-(1,6)-Glc)[43] and gentobiose (Glc-β-(1,6)-Glc).[3] Of particular significance is the report by Kishi and coworkers of a C-linked analog of sucrose (Glc-α-(1,2)-fructose), a molecule that could serve as a non-metabolizable dietary sugar substitute.[44]

All of these syntheses involved the elaboration of a pyranose starting material with a linear polyhydroxylated chain that was later cyclized to form the second carbohydrate unit of the disaccharide. This strategy is illustrated by Kishi's synthesis of C-sucrose as depicted in Figure 18.[44] The initial carbohydrate unit, compound (**30**), was prepared using standard C-glycosylation chemistry described above, and can be further elaborated from the glycosidic carbon chain. Coupling of the vinyl iodide with the carbohydrate derived acetonide-aldehyde resulted in formation of the acyclic precursor (**31**), which upon epoxidation gave (**32**) that was then cyclized to afford the *C*-disaccharide (**33**). In this procedure, the *C*-glycosyl linkage was installed at the glycosidic position of the pyranose precursor, and the second sugar was extended from this position.

An alternative approach is exemplified by Kishi's synthesis of *C*-maltose derivatives as shown in Figure 19 (5-3).[42] The pyranose precursor **34** contains a carbon-linked appendage at C-4 rather than at the glycosidic position, which is installed using standard alkylation chemistry. This appendage was extended by reaction with aldehyde **35** to form the acyclic presursor **36**. After dihydroxylation of the olefin and formation of the terminal epoxide, compound (**37**) was poised for cyclization to the *C*-disaccharide (**38**). Unlike the previous example, the *C*-glycosyl unit ("donor" unit) was formed during the cyclization reaction rather than serving as the pyranose precursor.

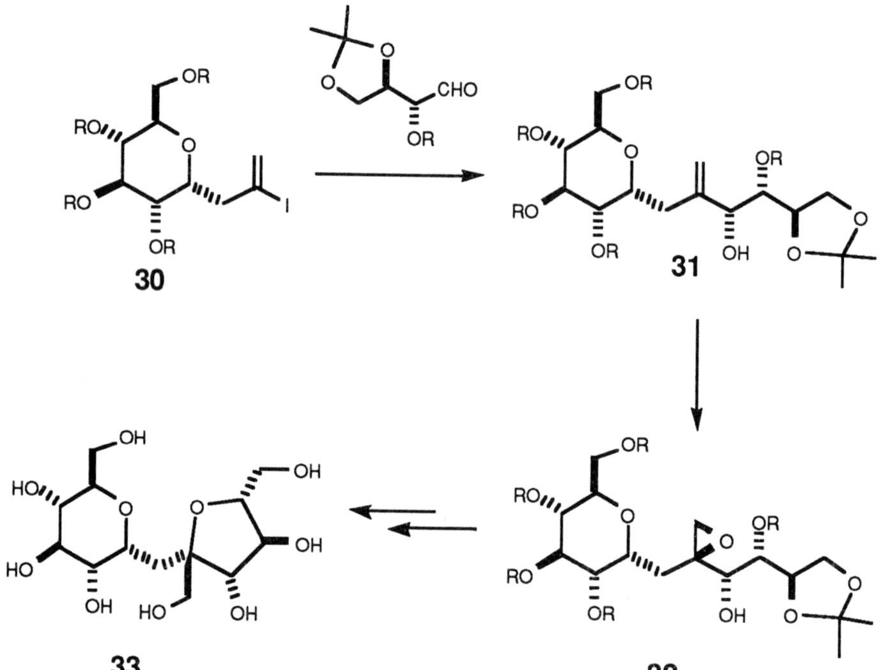

Figure 18.   Kishi's synthesis of *C*-sucrose.

The difficulty in these procedures lies in the need to construct the hydroxylated stereocenters of the acyclic precursor. Each synthesis requires a diastereoselective alkylation, dihydroxylation or epoxidation, resulting in mixtures of diastereomers that must be separated. Each different disaccharide molecule contains a unique array of stereocenters, making this approach cumbersome and difficult to generalize or automate beyond a few examples. Thus, many groups have sought methods to reduce the number of hydroxyl group stereocenter manipulations during the synthesis.

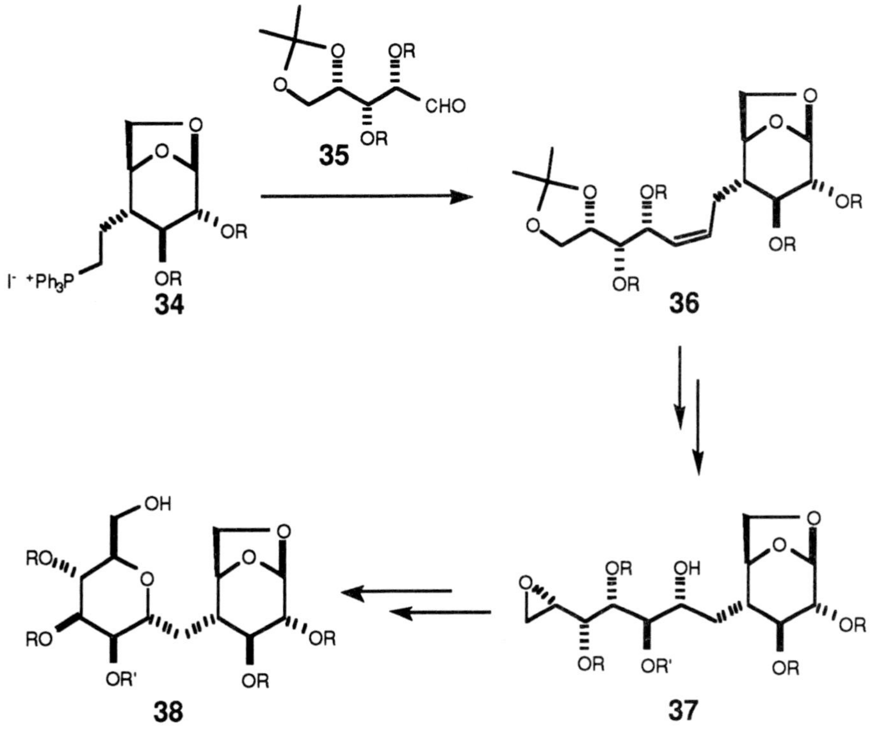

Figure 19. Kishi's synthesis of *C*-maltose derivatives.

6.B. Direct Coupling of Pyranose Subunits

Since carbohydrate starting materials can be obtained with the desired hydroxyl group stereocenters already in place, an intellectually appealing approach to the synthesis of *C*-disaccharides is the direct coupling of pyranose subunits. In one respect, this strategy is more challenging than that utilized by Kishi and coworkers because of the need to functionalize two intact sugar units with reactive centers. However, only two stereocenters need to be controlled, the anomeric position of the C-glycosyl "donor" unit (α or β) and the linkage to the "acceptor" unit.

The direct coupling approach was utilized in Sinaÿ and Rouzaud's[41] pioneering C-disaccharide synthesis as shown in Figure 20. The carbon atom serving as the bridge between the two sugar units was incorporated into the acceptor unit in the form of a vinyl dibromide substituent at the C-6 position (39). Treament of compound (39) with BuLi generated the acetylenic anion (40) *in situ*, which was reacted with protected gluconolactone (41) to form the C-glycosyl bond to the donor unit. Stereoselective reduction of compound (42) with triethylsilane and BF_3OEt_2 gave the β-linked acetylenic C-disaccharide (43), which was reduced by hydrogenolysis to give the desired C-disaccharide (29). Although an elegant approach to the synthesis of a β-(1,6)-linked C-disaccharide, this method cannot be applied to the synthesis of C-disaccharides with other linkages [(1,2), (1,3) and (1,4)] or with α-stereochemistry.

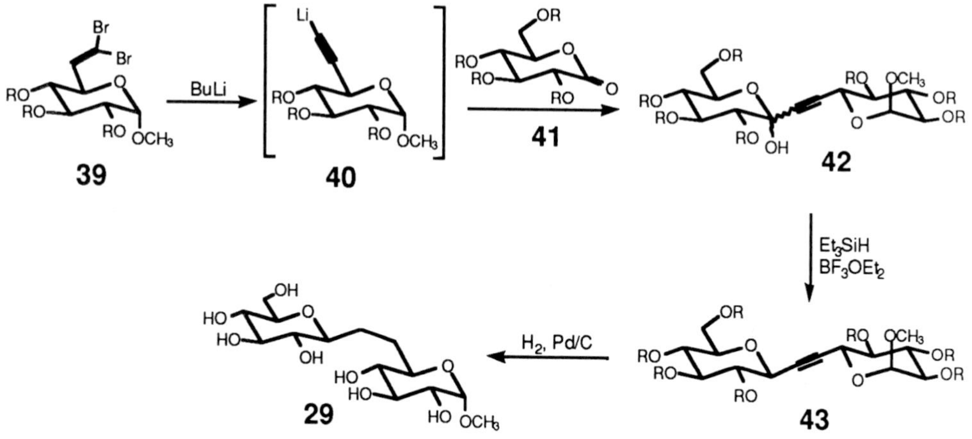

Figure 20. Sinaÿ's synthesis of Glc-C-β-(1,6)-Glc-α-OMe.

Schmidt and Preuss have reported a different approach involving condensation of lithiated glycal sulfoxide (44) with sugar aldehyde (45) (Figure 21).[45] Here, the donor unit is nucleophilic and the acceptor unit serves as the electrophile. The product, (46), was then converted to C-disaccharide (47),

however attempts to remove the hydroxyl group at the methylene bridge were
unsuccessful.

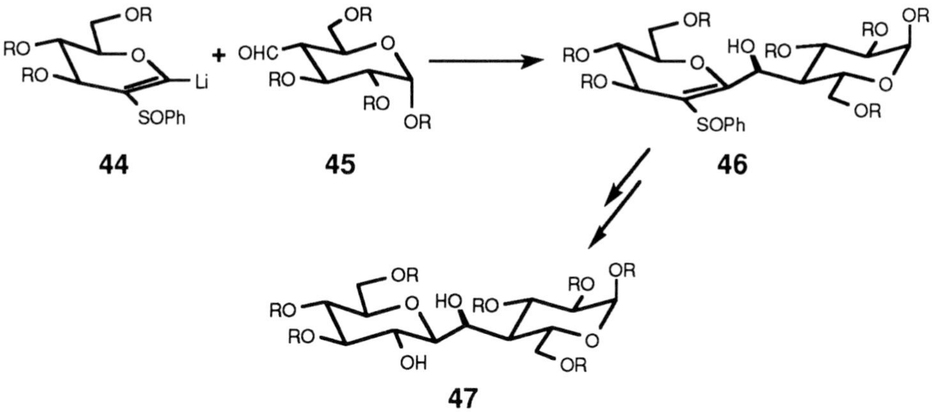

Figure 21.   Schmidt's coupling of sugar aldehydes with glycallithium derivatives.

To overcome the need for deoxygenation at the methylene bridge, Schmidt and
Preuss have recently reported a similar synthesis in which the polarity of the
coupling reaction is reversed.[46] In this synthesis, the acceptor unit (49) has a
nucleophilic lithiated methylene extension, which is condensed with an
electrophilic donor unit (48) to form the C-glycosyl linkage (50) (Figure 22).
Stereoselective reduction at the anomeric position of compound (50) followed by
deprotection gave the desired C-disaccharide (51). Unlike the method described
by Sinaÿ and Rouzaud, Schmidt's procedure can be applied to the synthesis of
(1,3) and (1,4)-linked C-disaccharides with β-stereochemistry.

The polyfunctionalized nature of carbohydrate molecules often precludes the
use of harsh reaction conditions and highly reactive anions or cations. Thus, the
application of radical chemistry to the synthesis of carbohydrate derivatives,
particularly C-glycosides, has met with great success. Giese and coworkers have
described a method for the synthesis of (1,2)-linked C-disaccharides that involves
the addition of anomeric radicals (generated from anomeric bromides such as (52)
to α-methylene lactones (53).[47] The lactone product (54) is then readily
transformed into the corresponding C-disaccharide. This method was used to
generate a number of α-(1,2)-linked structures, but cannot be generalized to C-
disaccharides with other linkages or with β-stereochemistry.

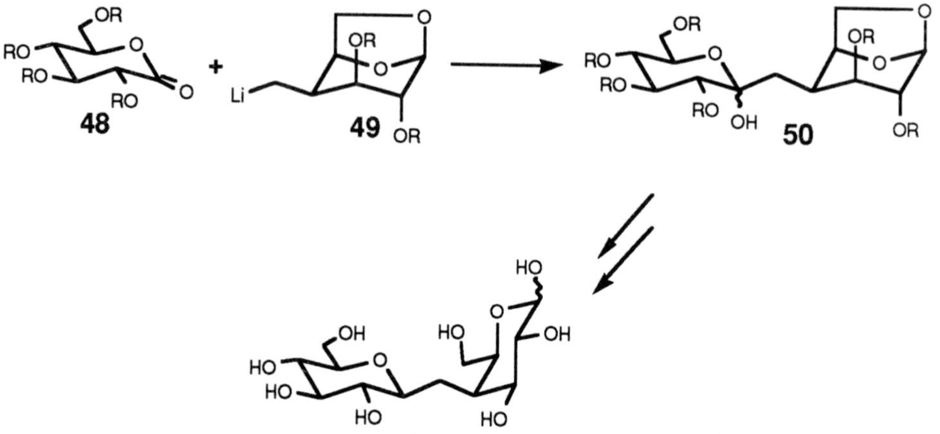

Figure 22.   Schmidt's coupling of sugar alkyllithium derivatives with sugar lactones.

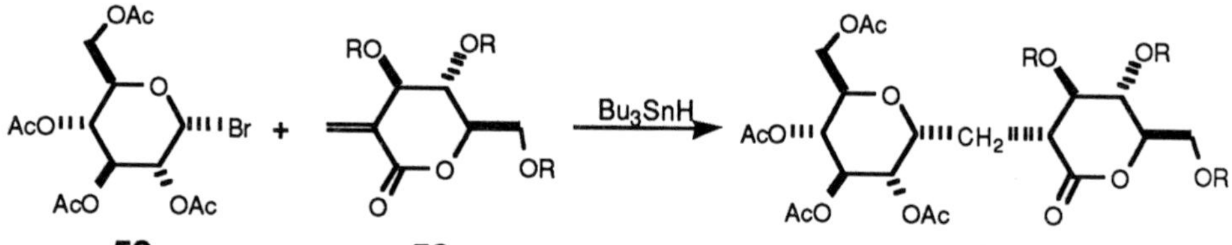

Figure 23.   Giese's synthesis of (1,2)-linked C-disaccharides via a radical coupling reaction.

In a courageous synthetic effort, Kishi and coworkers have described the first synthesis of a C-linked trisaccharide (compound (55), Figure 24).[48] Their synthesis utilized a combination of the previously described approaches. The C-disaccharide comprising pyranose units **B** and **C** was first constructed by cyclization of an acyclic precursor as in the syntheses of C-maltose and C-sucrose. This C-disaccharide was then coupled to the intact pyranose unit **A** to afford C-trisaccharide (55), in which each sugar is linked by a carbon bridge.

were then utilized as agents for receptor-mediated antibody targeting and induction of an immune response to bacterial cells.

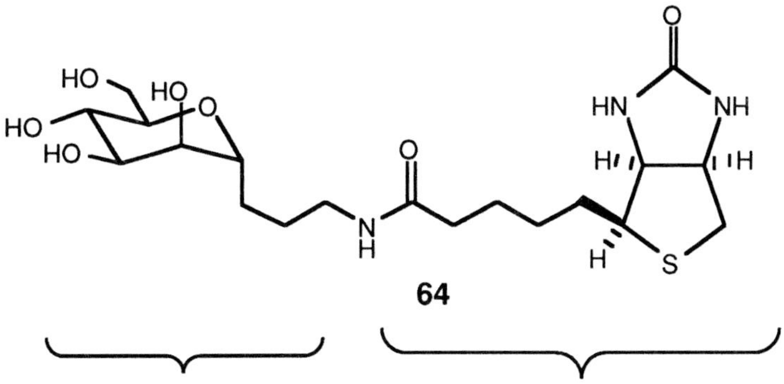

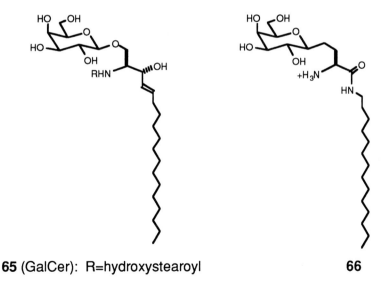

64

C-Mannoside   Biotin

Figure 27.   Mannoside-biotin C-glycosyl analog.

The synthesis of C-linked analogs of the natural glycosphingolipid galactosyl ceramide (**65**) has also been desribed.[72]   Derivatives such as (**66**) bound specifically to the human immunodeficiency virus (HIV) receptor gp120, and are currently being evaluated as antibody targeting agents to neutralize HIV infection.

65 (GalCer): R=hydroxystearoyl   **66**

Figure 28.   GalCer and a synthetic C-glycosyl analog.

Biologically active carbohydrates are often expressed in the context of glycoproteins. Stable analogs of these structures have tremendous therapeutic potential. The synthesis of carbon-linked glycopeptides such as (67) (the analog of O-linked serine (68) and the application of these novel derivatives to glycopeptide structural analysis has been reported.[73]

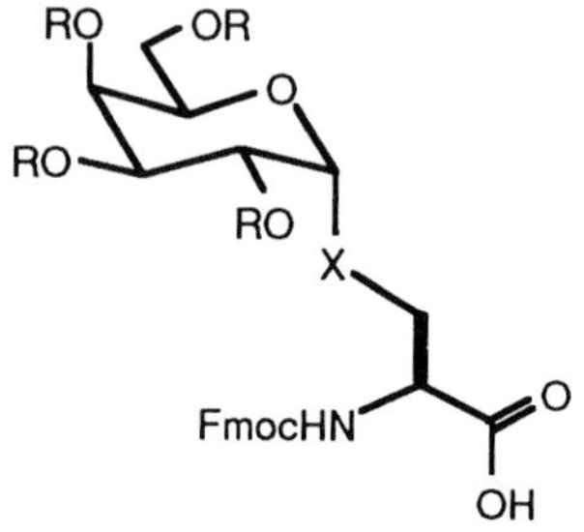

67: X=CH$_2$ (Gal-α-CH$_2$-Ser)
68: X=O (Gal-α-O-Ser)

Figure 29. C-glycosyl amino acid analog.

These C-glycosyl amino acids could be incorporated directly into pollypeptides using commercial peptide synthesizer technology to construct glycopeptides (69) and (70).

Ac-YKAAAAKAA XAKAAAAK-NH$_2$

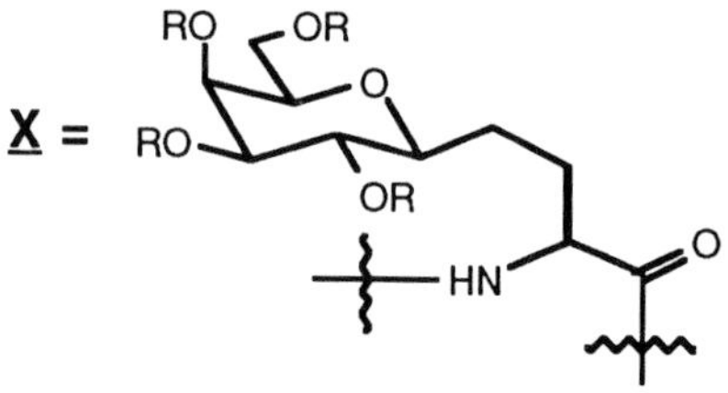

X =

69: R=Bn
70: R=H

Figure 30. Incorporation of a C-glycosyl amino acid into a C-glycopeptide.

8. CONCLUSION

In conclusion, synthetic methodology for C-glycosyl compounds has far outpaced their use in mediating biological processes. The majority of biological investigations have focused on the activity of *C*-glycosides as inhibitors of carbohydrate processing enzymes. However, *C*-glycosides are inherently ground state analogs, being structurally related to enzyme substrates rather than transition states. Consequently, *C*-glycosides generally bind with low affinity to enzymes (as illustrated by the previous examples) and therefore have limited potential as enzyme inhibitors. In contrast, *C*-glycosides are well suited for carbohydrate-binding proteins that have evolved to recognize ground state *O*-linked structures, such as cell-surface receptors and regulatory proteins.

REFERENCES

1. (a) M. Inbar and L. Sachs, *Nature*, **223**, 710 (1969); (b) M. Inbar and L. Sachs, *Proc. Natl. Acad. Sci. U.S.A* ., **63**, 1418 (1969).

2. J. C. Paulson, "Interaction of animal viruses with cell surface receptors." In *The Receptors. Vol. II*. P. M. Conn, Ed. Academic Press: New York, p. 131 (1985).

3. Sell, S. *Human Pathology* , **21**, 1003 (1990) and references cited therein.

4. B. K. Brandley, S. J. Sweidler, and P. W. Robbins, *Cell* , **63**, 861 (1990).

5. (a) D. Horton and D. H. Hutson, *Adv. Carbohydr. Chem. Biochem.*, **18**, 163 (1963); (b) D. Horton and J. D. Wander, "Thio Sugars and Derivatives" in *The Carbohydrates: Chemistry and Biochemistry* , W. Pigman and D. Horton, Eds. Academic Press: New York, (1980); (c) P. Brajeswar and W. Korytnyk, *Carbohydr. Res.* , **126**, 27 (1984).

6. (a) V. E. Marquez and M. Lim, *Med. Res. Rev.* , **6**, 1 (1986); (b) G. E. McCasland, S. Furuta and L. J. Durham, *J. Org. Chem* , **31**, 1516 (1966); (c) T. Suami, *Pure Appl. Chem.*, **59**, 1509 (1987).

7. For reviews of the chemistry and biology of *C*-glycosides see: (a) U. Hacksell, G. and D. Davies, *Prog. Med. Chem.*, **22**, 1 (1985); (b) J. G. Buchanan, *Prog. Chem. Org. Nat. Prod.*, **44**, 243 (1983); (c) J. Goodchild, *Top. Antibiot. Chem.*, **6**, 99 (1982); (d) Special issue, *Carbohydr. Res.*, 171 (1987).

8. S. Hanessian and A. G. Pernet, *Adv. Carbohydr. Chem. Biochem.*, **33**, 111, (1976).

9. S. Hanessian, *Total Synthesis of Natural Products: The Chiron Approach.*
    Pergamon Press: New York, 1983.

10. R. W. Armstrong, J-M. Beau, S. H. Cheon, W. J. Christ, H. Fujioka, W-H.
    Ham, L. D. Hawkins, H. Jin, S. H. Kang, Y. Kishi, M. J. Martinelli, W. W.
    McWhorter, Jr., M. Mizuno, M. Nakata, A. E. Stutz, F. X. Talamas, M.
    Taniguchi, J. A. Tino, K. Ueda, J. Uenishi, J. B. White,, and M. Yonaga, *J.
    Am. Chem. Soc.*, **111**, 7525 (1989).

11. P. Deslongchamps, *Stereoelectronic Effects in Organic Chemistry.*
    Pergamon Press: New York, 1983.

12. (a) G. Grynkiewicz and J. N. BeMiller, *J. Carbohydr. Chem.*, **1**, 121 (1982);
    (b) J. Herscovici, K. Muleka, and K. Antonakis, *Tetrahedron Lett.*, **25**, 5653
    (1984); (c) J. Herscovici, S. Delatre, and K. Antonakis, *J. Org. Chem.*, **52**,
    5691, (1987); (d) M. D. Lewis, J. K. Cha, and Y. Kishi, *J. Am. Chem. Soc*,
    104, 4976 (1982); (e) A. P. Kozikowski and K. L. Sorgi, *Tetrahedron Lett.*,
    25, 2085 (1984). (f) T. L. Cupps, D. S. Wise, and L. B. Townsend, *J. Org.
    Chem.*, **47**, 5115 (1982); (g) A. O. Stewart and R. M. Williams, *J. Am.
    Chem. Soc.*, **107**, 4289 (1985); (h) S. J. Danishefsky, J. F. Kerwin, *J. Org.
    Chem.*, **47**, 3803 (1982); (i) S. J. Danishefsky, S. Deninno, and P. Lartey, *J.
    Am. Chem. Soc.*, **109**, 2082 (1987); (j) G. H. Posner and S. R. Haines,
    Tetrahedron Lett., **26**, 1823 (1985); (k) R. R. Schmidt and H. Hoffman,
    Tetrahedron Lett., **23**, 409 (1982); (l) K. C. Nicolaou, R. E. Dolle, A.
    Chucholowski, and J. L. Randall, *J. Chem. Soc., Chem. Commun.*, 1153
    (1984); (m) S. Hanessian, K. Sato, T. J. Liak, N. Danh, D. Dixit, and B. V.
    Cheney, *J. Am. Chem. Soc.*, **106**, 6114 (1984); (n) Y. Ichikawa, M. Isobe,
    M. Konobe, and T. Goto *Carbohydr. Res.*, **171**, 193 (1987); (o) A. Hosomi,
    Y. Sakata, and H. Sakurai, *Carbohydr. Res.*, **171**, 223 (1987); (p) J. S.
    Panek and M. A. Sparks, *J. Org. Chem.*, **54**, 2034 (1989).

13. (a) Reference 10d. (b) G. A. Kraus and M. T. Molina, *J. Org. Chem.*, **53**,
    752 (1988); (c) S. Czernecki and G. Ville, *J. Org. Chem.*, **54**, 610 (1989).

14. (a) P. Lesimple, J. -M. Beau, and P. Sinay, *J. Chem. Soc., Chem. Commun.*,
    894 (1985); (b) J. -M. Beau, L. M. Allory, and P. Sinay, *J. Chem. Soc.,
    Chem. Commun.*, 355 (1984); (c) J. -M. Beau and P. Sinay, *Tetrahedron
    Lett.*, **26**, 6185 (1985); (d) J. -M. Beau and P. Sinay, *Tetrahedron Lett.*, **26**,

6189 (1985);   (e) J. -M. Beau and Sinay, P. *Tetrahedron Lett.*, **26**, 6193 (1985);   (f) P. Lesimple, J. -M. Beau, and P. Sinay, *Carbohydr. Res.*, **171**, 289 (1987);   (g) S. Valverde, S. Garcia-Ochoa, and M. Martin-Lomas, *J. Chem. Soc., Chem. Commun.*, 383 (1987);   (h) J. B. Ousset, C. Mioskowski, Y. -L. Yang, and J. R. Falck, *Tetrahedron Lett.* , **25**, 5903 (1984);   (i) B. Aebischer, J. H. Bieri, R. Prewo, and A. Vasella, *Helv. Chim. Acta.*, **65**, 2251 (1982); (j) F. Baumberger and A. Vasella, *Helv. Chim. Acta .*, **66**, 2210 (1983).

15.   (a) B. Giese and J. Dupuis, *Angew. Chem., Int. Ed. Engl.*, **22**, 622 (1983); (b) J. Dupuis, B. Giese, J. Hartung, M. Leising, H.-G. Korth, and R. Sustmann, *J. Am. Chem. Soc.*, **107**, 4332 (1985);   (c) G. E. Keck, E. J. Enholm, J. B. Yates, and M. R. Wiley, *Tetrahedron* , **41**, 4079 (1985);   (d) Y. Araki, T. Endo, M. Tanji, J. Nagasawa, and Y. Ishido, *Tetrahedron Lett.*, **28**, 5853 (1987);   (e) F. Baumberger and A. Vasella, *Helv. Chim. Acta,*, **66**, 2210 (1983);   (f) A. De Mesmaeker, P. Hoffmann, B. Ernst, P. Hug, and T. Winkler, *Tetrahedron Lett.*, **30**, 6307 (1989);   (g) A. De Mesmaeker, P. Hoffmann, B. Ernst, P. Hug, and T. Winkler, *Tetrahedron Lett.*, **30**, 6311 (1989);   (h) H. Paulsen and P. Matschulat, *Liebigs Ann. Chem.*, 487 (1991); (i) J. O. Nagy and M. D. Bednarski, *Tetrahedron Lett.*, **32**, 3953 (1991).

16.   (a) G. L. Trainor and B. E. Smart, *J. Org. Chem..*, **48**, 2447 (1983);   (b) P. DeShong, G. A. Slough, V. Elango, G. L. Trainor, *J. Am. Chem. Soc.*, **207**, 7788 (1985);   (c) P. DeShong, G. A. Slough, and V. Elango, *Carbohydr. Res.*, **171**, 343 (1987);   (d) J. C. Y. Cheng and G. D. Daves Jr., *Organometallics*, **5**, 1753 (1986); (e) J. C. Y. Cheng and G. D. Daves Jr., *J. Org. Chem.*, (1987), **52**, 3083. (f) S. Czernecki and F. Gruy, *Tetrahedron lett.*, **22**, 437 (1981);   (g) S. Czernecki and V. B. Dechavanne, *Can. J. Chem.*, **61**, 533 (1983);   (h) H. G. Pandraud, R. Brahmni, V. B. Dechavanne, and S. Czernecki, *Can. J. Chem.*, **63**, 491 (1985);   (i) S. Yougai and T. Miwa, *J. Chem. Soc., Chem. Commun.*, 68 (1983);   (j) L. V. Dunkerton and A. J. Serino, *J. Org. Chem.*, **47**, 2812 (1982);   (k) L. V. Dunkerton, J. M. Enske, and A. J. Serino, *Carbohydr. Res.*, **171**, 89 (1987);   (l) M. Brakta, P. Lhoste, D. Sinou, *J. Org. Chem.*, **54**, 1890 (1989).

17. B. Giese, *Radicals in Organic Synthesis: Formation of Carbon-Carbon Bonds.* Pergamon Press: New York (1986).

18. (a) R. E. Ireland and J. P. Vevert, *Can. J. Chem.*, **59**, 572 (1981); (b) D. P. Curran, and Y.-G. Suh, *Carbohydr. Res.*, **171**, 161 (1987).

19. J. R. Pougny, M. A. M. Nassr, and P. Sinaÿ, *J. Chem. Soc., Chem. Commun.*, 375 (1981).

20. T. V. RajanBabu and G. S. Reddy, *J. Org. Chem.*, **51**, 5458 (1986).

21. T. Ogihara and O. Mitsunobu, *Tetrahedron Lett.*, **24**, 3505 (1983).

22. T. Kametani, K. Kawamura, and T. Honda, *J. Am. Chem. Soc.*, **109**, 3010 (1987).

23. W. Schmid and G. M. Whitesides, *J. Am. Chem. Soc.*, **112**, 9670 (1990).

24. (a) A. Vasella, C. Witzig, and R. Husi, *Helv. Chim. Acta*, **74**, 1362 (1991); (b) A. Vasella and C. A. A. Waldraff, *Helv. Chim. Acta*, **74**, 585 (1991).

25. A. Vasella, P. Uhlmann, C. A. A. Waldraff, F. Diederich, and C. Thilgen, *Angew. Chem., Int. Ed. Engl.*, **31**, 1388 (1992).

26. F. A. Carey and R. J. Sundberg, *Advanced Organic Chemistry, Part A: Structure and Mechanism.* Plenum Press: New York (1984).

27. W. P. Jencks, *Catalysis in Chemistry and Enzymology.* Dover Publications, Inc.: New York (1969).

28. A. J. Kirby, *The Anomeric Effect and Related Stereoelectronic Effects at Oxygen.* Springer-Verlag: Berlin (1983).

29. (a) T.-C. Wu, P. G. Goekjian, and Y. Kishi, *J. Org. Chem.*, **52**, 4819 (1987); (b) P. G. Goekjian, T.-C. Wu, H.-Y. Kang, and Y. Kishi, *J. Org. Chem.*, **52**, 4823 (1987); (c) S. A. Babirad, Y. Wang, P. G. Goekjian, and Y. Kishi, *J. Org. Chem.*, **52**, 4825 (1987); (d) Y. Wang, P. G. Goekjian, D. M. Ryckman, and Y. Kishi, *J. Org. Chem.*, **53**, 4151 (1988); (e) W. H. Miller, D. M. Ryckman, P. G. Goekjian, Y. Wang, and Y. Kishi, *J. Org. Chem.*, **53**, 5580 (1988); (f) P. G. Goekjian, T.-C. Wu, and Y. Kishi *J. Org. Chem.*, **56**, 6412 (1991); (g) P. G. Goekjian, T.-C. Wu, H.-Y. Kang, and Y. Kishi, *J. Org. Chem.*, **56,** 6422 (1991); (h) Y. Wang, S. A. Babirad, and Y. Kishi, *J. Org. Chem.*, **57**, 468 (1992); (i) Y. Wang, P. G. Goekjian, D. M. Ryckman, W. H. Miller, S. A. Babirad, and Y. Kishi, *J. Org. Chem.*, **57**, 482 (1992);

(j) T. Haneda, P. G. Goekjian, S. H. Kim, and Y. Kishi, *J. Org. Chem.*, **57**, 490 (1992).

30. (a) R. Engel, *Chem. Rev.*, **77**, 349 (1977); (b) R. Julina and A. Vasella, *Helv. Chim. Acta.*, **68**, 819 (1985); (c) R. W. McClard, *Tetrahedron Lett.*, **24**, 2631 (1983).

31. (a) R. B. Meyer, T. E. Stone, and P. K. Jesthi, *J. Med. Chem.*, **27**, 1095 (1984); (b) F. Nicotra, L. Panza, F. Ronchetti, and L. Toma, *Tetrahedron lett.*, **25**, 5937 (1984).

32. (a) M. Chmielewski, J. N. BeMiller, and D. P. Cerretti, *Carbohydr. Res.*, **97**, C1 (1981); (b) F. Nicotra, F. Ronchetti, and G. Russo, *J. Org. Chem.*, **47**, 4459 (1982); (c) F. Nicotra, F. Ronchetti, and G. Russo, *J. Chem. Soc., Chem. Commun.*, 470 (1982).

33. J. -C. Tang, B. E.Tropp, and R. Engel, *Tetrahedron Lett.*, **8**, 723 (1978).

34. (a) D. W. Norbeck, J. B. Kramer, and P. A. Lartey, *J. Org. Chem.*, **52**, 2174 (1989); (b) H. Molin, J. -O. Noren, and A. Claesson, *Carbohydr. Res.*, **194**, 209 (1989).

35. W. J. Ray, C. B. Post, and J. M. Puvathingal, *Biochemistry*, **32**, 38 (1993).

36. A. Giannis and K. Sandhoff, *Tetrahedron Lett.*, **26**, 1479 (1985).

37. R. R. Schmidt, H. Dietrich, *Angew. Chem., Int. Ed. Engl.*, **30**, 1328 (1991).

38. S. C. T. Svensson and J. Thiem, *Carbohydr. Res.*, **200**, 391 (1990).

39. M. L. Sinnott and P. J. Smith, *J. Chem. Soc., Chem. Commun.*, 223 (1976).

40. C. -S. Kuhn, C. P. J. Glaudemans, and J. Lehmann, *Liebigs Ann. Chem.*, 357 (1989).

41. D. Rouzaud and P. Sinaÿ, *J. Chem. Soc., Chem. Commun.*, 1353 (1983).

42. Y.Wang, S. A. Babirad, and Y. Kishi, *J. Org. Chem.*, **57**, 468 (1992).

43. P. G. Goekjian, T. -C. Wu, H. -Y. Kang, and Y. Kishi, *J. Org. Chem.*, **56**, 6422 (1991).

44. U. C. Dyer and Y. Kishi, *J. Org. Chem.*, **53**, 3383 (1988).

45. R. R. Schmidt and R. Preuss, *Tetrahedron Lett.*, **30**, 3409 (1989).

46. R. Preuss and R. R. Schmidt, *J. Carbohydr. Chem.*, **10**, 887 (1991).

47. (a) B. Giese and T. Witzel, *Angew. Chem.*, **98**, 459 (1986); (b) B. Giese, M. Hoch, C. Lamberth, and R. R. Schmidt, *Tetrahedron Lett.*, **29**, 1375 (1988).

48. T. Haneda, P. G. Goekjian, S. H. Kim, and Y. Kishi, *J. Org. Chem.*, **57**, 490 (1992).

49. I. M. Dawson, T. Johnson, R. M. Paton, and R. A. C. Rennie, *J. Chem. Soc., Chem. Commun.*, 1339 (1988).

50. A. Boschetti, F. Nicotra, L. Panza, ; G. Russo, ; L. Zucchelli, *J. Chem. Soc., Chem. Commun.*, 1985 (1989).

51. (a) M. Carcano, F. Nicotra, L. Panza, and G. Russo, *J. Chem. Soc., Chem. Commun.*, 642 (1989); (b) S. M. Daly and R. W. Armstrong, *Tetrahedron Lett.*, **30**, 5713 (1989); (c) R. W. Armstrong and B. R. Teegarden, *J. Org. Chem.*, **57**, 915 (1992).

52. R. M. Bimwala and P. Vogel, *Tetrahedron Lett.*, **32**, 1429 (1991).

53. (a) Nielsen, A. T. "Nitronic Acids and Esters" In *The Chemistry of the Nitro and Nitroso Groups*, (H. Feuer, Ed.) Wiley-Interscience: London, Part I (1969); (b) E. W. Colvin, A. K. Beck, and D. Seebach, *Helv. Chim. Acta*, **64**, 2264 (1981).

54. For an historical perspective, see: H. H. Baer, *Adv. Carbohydr. Chem. Biochem.*, **24**, 67 (1969).

55. R. Noyori and H. Takaya, *Acc. Chem. Res.*, **23**, 345 (1990).

56. T. Suami, H. Sasai, and K. Matsuno, *Chem. Lett.*, 819 (1983).

57. S. Kozaki, O. Sakanaka, T. Yasuda, T. Shimizu, S. Ogawa, and T. Suami, *J. Org. Chem.*, **53**, 281 (1988).

58. O. R. Martin and W. Lai, *J. Org. Chem.*, **55**, 5188 (1990).

59. A. Jobe and S. Bourgeois, *J. Mol. Biol.*, **69**, 397 (1972).

60. A. E. Chakerian, J. S. Olson, and K. S. Matthews, *Biochemistry*, **26**, 7250 (1987).

61. F. Jacob and J. Monod, *J. Mol. Biol.*, **3**, 318 (1961).

62. M. Pfahl, *J. Mol. Biol.*, **147**, 1 (1981).

63. J. H. Miller and W. S. Reznikoff, *The Operon*, Cold Spring Harbor Laboratory (1980).

64. P. Angibeaud and J. -P. Utille, *Synthesis*, 737 (1991).

65. A. Brandstrøm, B. Lamm, and I. Palmertz, *Acta Chem. Scand. B*, **28**, 699 (1974).

66. K. E. Gilbert and W. T. Borden, *J. Org. Chem.*, **44**, 659 (1979).

67. O. Sakanaka, T. Ohmori, S. Kozaki, T. Suami, T. Ishii, S. Ohba, and Y. Saito, *Bull. Chem. Soc. Jpn.*, **59**, 1753 (1986).
68. N. Ono and A. Kaji, *Synthesis*, 693 (1986).
69. M. J. Robins, J. S. Wilson, and F. Hansske, *J. Am. Chem. Soc.*, **105**, 4059 (1983).
70. W. R. Kobertz, C. R. Bertozzi, and M. D. Bednarski, *Tetrahedron Lett.*, **33**, 737 (1992).
71. C. R. Bertozzi and M. D. Bednarski, *J. Am. Chem. Soc.*, **114**, 2242 (1992).
72. C. R. Bertozzi and M. D. Bednarski, *J. Am. Chem. Soc.*, **114**, 1014 (1992).
73. C. R. Bertozzi, P. D. Hoeprich, and M. D. Bednarski, *J. Org. Chem.*, **57**, 6092 (1992).

Chapter 15

RECENT DEVELOPMENTS IN GLYCOPEPTIDE SYNTHESIS

STEFAN PETERS, MORTEN MELDAL AND KLAUS BOCK

Carlsberg Laboratory, Department of Chemistry, Gamle Carlsberg Vej 10, DK-2500 Valby, Copenhagen, Denmark

Abstract   An overview of convenient methodology and techniques used in solid phase glycopeptide synthesis by the building block approach is given. The practical details of importance in aliphatic O-linked, aromatic O-linked and N-linked glycopeptide synthesis are emphasized. The description is centred around the use of pentafluorophenyl esters as a protecting group during glycosylation and an activating group during glycopeptide assembly. Different kinds of orthogonal protection schemes are discussed dependent on the type of glycopeptide synthesized. An experimental part gives the typical procedures necessary for a successful glycopeptide assembly.

1.   INTRODUCTION

The importance of carbohydrate recognition in biological events[1] and the growing interest in the use of synthetic glycopeptides as model compounds, has brought about some remarkable advances in methodology in the area of glycopeptide synthesis. For example, in the past decade, various kinds of O- and N-linked glycopeptides and oligosaccharides became accessible as a result of the development of new techniques in both peptide and carbohydrate chemistry especially in the solid phase peptide synthesis.

The large number of recent publications dealing with glycopeptide synthesis certainly substantiates the great interest in this topic. This review will discuss the most recent developments in the synthesis of glycosylated amino acids used as building blocks in solid phase synthesis as well as the general methodology on glycopeptide synthesis. The solid phase synthesis of glycopeptides will be exemplified by the assembly of aliphatic O-linked, aromatic O-linked and N-linked structures, which were synthesized by the application of different techniques as for instance a) multiple column, b) plastic syringe and c) continuous flow solid phase synthesis. The present review will also include an experimental description of how to synthesize glycosylated amino acids and how to carry out the new techniques of solid phase glycopeptide synthesis. Detailed reviews on glycopeptide synthesis were previously published.[2-6]

2. GLYCOPEPTIDE ASSEMBLY

The synthesis of glycopeptides can be realized by different strategies. The currently most commonly applied methodology for the insertion of a carbohydrate into a peptide is the use of a glycosylated amino acid as a building block in a stepwise peptide assembly by either solution or solid phase techniques.[4] Solid phase synthesis has become particularly attractive for glycopeptide syntheses due to fast assembly, the possibility of automation and mild conditions used during deprotection steps. Furthermore, multiple column[7] or library techniques[8] have opened the possibility of synthesizing a large number of compounds in parallel. Solution phase glycopeptide synthesis is frequently applied for sequences with a small peptide part, large scale synthesis or in those cases where the saccharide part contains extremely labile residues which might be lost under the conditions of solid phase synthesis.

An alternative strategy for glycopeptide assembly is the connection of the carbohydrate to a peptide after completion of the peptide synthesis by a selective glycosylation reaction. The approach to *O*-glycosylate peptides *en block* with an activated saccharide donor has often failed due to a low solubility of the peptide acceptor in organic solvents. However, it was demonstrated recently that protected peptides can be *N*-glycosylated in solution[9] and that polymer bound peptides can be both *N*- and even *O*-glycosylated.[10-12]

2.A.   Synthesis of Building Blocks Suitable for Glycopeptide Assembly

During the synthesis of building blocks suitable for *O*-glycopeptide synthesis an activated glycosyl donor is coupled to an *N*- and *C*-terminally protected hydroxy amino acid. The *O*-glycosidic bond formed is labile to both strong acid and base. This is why the functional groups of the carbohydrate are frequently protected as acetates or benzoates which can be removed afterwards under mild basic conditions e.g. with hydrazine hydrate in methanol[13-19] or a dilute solution of sodium methoxide in methanol.[20,21] Protection of the saccharide with benzyl groups is not generally applicable due to the possible presence of amino acid components which are unstable under hydrogenolytical conditions. Protection of the N^{α}-amino group has usually been performed with the fluoren-9-ylmethoxycarbonyl (Fmoc) group particularly if the building block was subsequently used in solid phase synthesis. The Fmoc group can be removed under mild conditions with morpholine,[22] or the more basic piperidine or 1,8-diazabicyclo[5.4.0]undec-7-en (DBU), without any β-elimination of *O*-linked sugars like *N*-acetyl-galactosamine (GalNAc).[23,24] During the synthesis of larger glycopeptides Fmoc deprotection with morpholine was found to be incomplete and use of piperidine as deprotection reagent gave more pure products.[25] As *C*-terminal protecting groups the tert.-butyl,[26,27] allyl[19,28-30] or phenacyl esters,[31,32] which can be cleaved under mild conditions have frequently been applied. However, when one of the above mentioned *C*-terminal protecting groups had been selected, it was removed selectively after glycosylation and the free carboxylic acid needed subsequently to be activated before use in solid phase synthesis. These two steps were avoided by choosing e.g. the pentafluorophenyl (Pfp) ester as the C-terminal protecting group.[33] It was demonstrated several times on both O- and N-glycosylations that the Pfp ester is stable under conditions of glycoside synthesis and chromatographic purification.[17,18,21,25,34-41] Furthermore the Pfp ester activates the carboxylic acid for a nucleophilic attack of an amine during peptide bond formation. Thus glycosylation of Fmoc/Pfp protected amino acids affords building blocks which can directly be used for solid phase glycopeptide synthesis.

Fmoc/Pfp protected amino acids, such as serine, threonine and tyrosine containing an unprotected hydroxyl group in the side chain, can be synthesized either by esterification of the N^α-Fmoc protected acids **1**, **2** or **7** with pentafluorophenol (Pfp-OH) under *in situ* activation of dicyclohexyl carbodiimide (DCC)[42] or by trifluoroacetic acid (TFA)-cleavage of the ^{t}Bu-ether in the commercially available compounds **5**, **6** and **9**.[41]

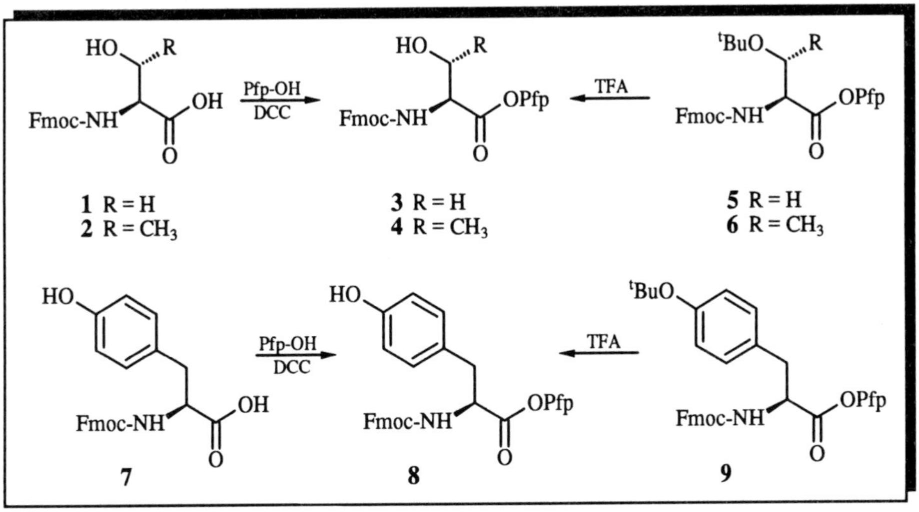

Figure 1.   Synthesis of Fmoc/Pfp protected hydroxy amino acids.

2.A.i.   Aliphatic O-linked glycosyl amino acid

The synthesis of aliphatic O-linked glycosyl amino acids will be discussed first using mucin glycopeptides containing GalNAc-sugars α-linked to the hydroxy groups of serine and threonine as an example. Protected GalNAc-glycosyl donor cannot be used for the glycosylation reaction because the acetamido group in position-2 favours the formation of either a β-glycosidic bond or an oxazoline. Reasonable α-stereoselectivities have only been obtained by application of glycosyl donors with a non participating substituent at C-2 in the carbohydrate[43]. The azido group has been found to be the best choice of amino protective group in this position and can easily be converted into the naturally occurring acetamido derivative. The 2-azido galactose derivatives are available by azidonitratization of 3,4,6-tri-O-acetyl-D-galactal.[44,45] Glycosylation of the Fmoc/Pfp protected serine and threonine derivatives **3** and **4** with the β-chloride **10**[44,46,47] was performed at room temperature with silver perchlorate/silver carbonate catalyst. The α-linked glycosyl amino acids **11** and **12** were isolated in yields of 70-78% after chromatography on silica gel or reversed phase HPLC.[17,18]

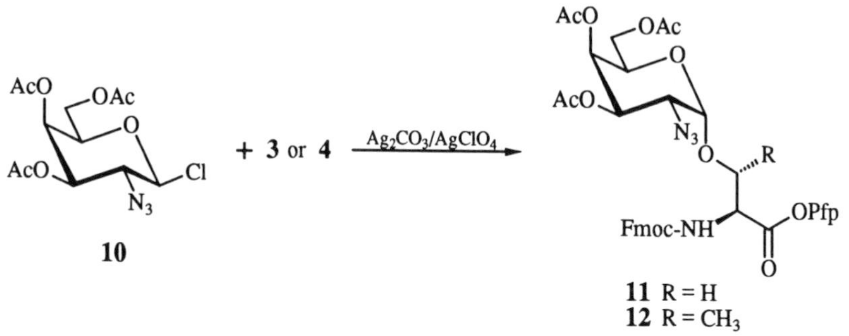

13 Ac-Pro-Thr-Xxx-Thr-Pro-Ile-Ser-Thr-NH₂   (α-D-GalNAc)

14 Ac-Pro-Thr-Thr-Thr-Xxx-Ile-Ser-Thr-NH₂   (α-D-GalNAc)

Xxx = Pro, Gly, Ser, Thr, Ala, Arg, Asn, Asp, Cys(Acm), Gln, Glu, His, Ile, Leu, Lys, Met, Phe, Trp, Tyr, Val

α-D-GalNAc

Ac-Pro-Thr-Pro-Thr-Gly-Thr-Gln-Thr-Pro-Thr-Thr-Thr-Pro-Ile-Thr-Thr-Thr-Thr-Thr-Val-Thr-Pro-Thr-NH₂   **15**

Figure 2.   Solid phase synthesis of mucin glycopeptides.

Building block **12** has been used for a multiple column glycopeptide synthesis of 40 different mucin glycopeptides based on the parent sequences **13** and **14** in which the flanking amino acids on both sides of Thr-5 were replaced with the 20 naturally occurring amino acids.[17,48]

Furthermore, glycosyl amino acid **12** was applied in the automated synthesis of the triicosapeptide **15** using deprotection with DBU in DMF. The 2-azido group was quantitatively transformed into the acetamido derivative after the peptide synthesis using thioacetic acid while the peptide was linked to the solid support.[23]

The second example shows the synthesis of glycopeptides carrying a mannose-6'-phosphate disaccharide α-linked to threonine. The 6'-*O*-phosphorylated glycosyl donor **16**, benzoylated and protected at the phosphate with the 2,2,2-trichloroethyl (TCE) group, was reacted with N^{α}-Fmoc-Thr-OPfp **4** under catalysis with silver triflate (AgOTf) to afford building block **17** in 74% yield after chromatography on silica gel.[38] Compound **17** was used for the solid phase synthesis of the di-glycosylated tripeptide **18** by the plastic syringe technique. The TCE group was

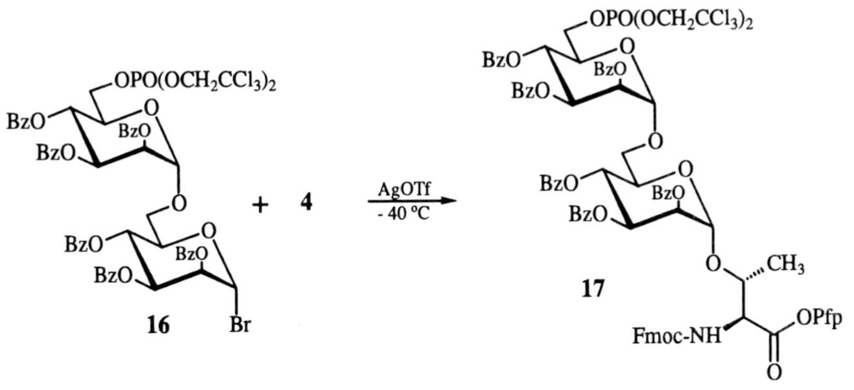

Ac-Thr(R)-Lys(ABz)-Thr(R)-NH$_2$

18

R = α-D-Man-6-P-(1→6)-α-D-Man-(1→O)-

ABz = 2-aminobenzoic acid

Figure 3. Synthesis of *O*-linked mannose-6'-phosphate glycopeptides.

removed at the end of the synthesis with a mixture of zinc and silver carbonate.[40]

2.A.ii. Aromatic *O*-linked glycosyl amino acids

The glycosylation of aromatic hydroxyl groups is more difficult than that of aliphatic ones due to the low nucleophilicity of phenols. However Fmoc/Pfp protected tyrosine derivatives have been glycosylated under Koenigs Knorr conditions with various glucose and maltose derivatives to provide building blocks suitable for solid phase glycopeptide synthesis.[35] For instance, Fmoc-Tyr(ᵗBu)-OPfp **9** was reacted with the peracetylated maltosyl bromide **19** in dichloromethane promoted by silver triflate to afford the β-linked glycoside **20** in 81% isolated yield. Under the same conditions the tyrosine acceptor **8** was glycosylated in only 42% yield. Further studies on glycosylations of tyrosine derivatives have shown that the ᵗBu-protected aglycone **9** was always glycosylated faster and in better yields than the unprotected compound **8** due to increased accessibility of the electron lone pairs of the phenol.[49] The synthesis of tyrosine derivatives with an α-glycosidic linkage was performed in a similar manner by using per-*O*-benzoylated glycosyl donors and acetonitrile as the solvent.[35]

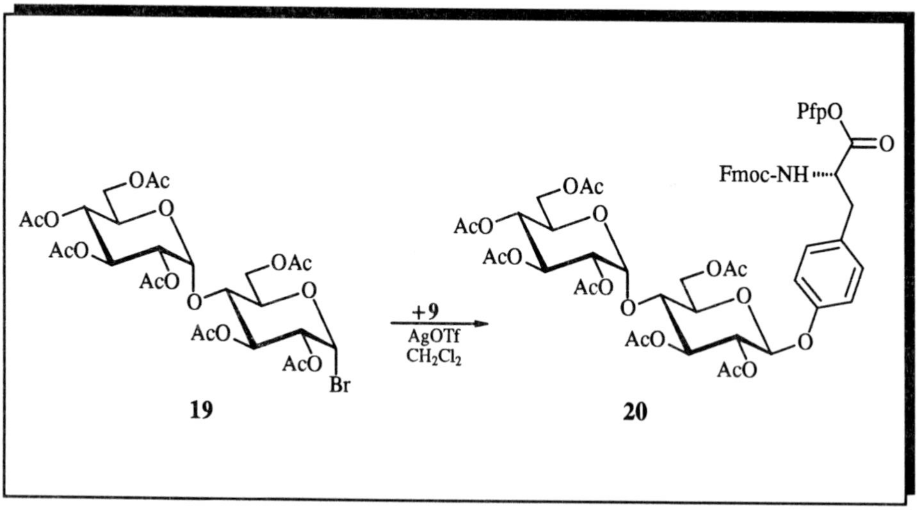

Figure 4.   Synthesis of aromatic *O*-linked glycosyl amino acids.

2.A.iii.   N-linked glycosyl amino acids

Glycosyl amino acids containing N-linked carbohydrates are available by condensation of an O-protected or O-unprotected glycosyl amine with the C^β-carboxylic group of an aspartic acid protected at the N^α- and C^α- functional groups. The 1-amino sugars are readily available by reacting the unprotected sugar with either methanolic ammonia[50,51] or aqueous ammonium bicarbonate.[52] The former yields crystalline glycosyl amines of monosaccharides, whereas the latter yields glycosyl amines of oligosaccharides often contaminated with bisglycosyl amines and unreacted starting material. However, the purification of the crude glycosyl amines by ion exchange chromatography is only efficient with monosaccharides. Different approaches to the removal of these byproducts have been reported. Crude glycosyl amines of different maltose oligosaccharides were used in the unprotected mode with coupling of *in situ* generated N^α-Fmoc-Asp(OPfp)-O^{t}Bu, and the condensation products were purified by HPLC to yield 13-35% of glycosylated material.[53]

Alternatively the crude glycosyl amines, for instance compound **22**, was transformed into the Fmoc protected, peracetylated derivative **23** which was purified by crystallization or by chromatography. Cleavage of the Fmoc group with piperidine/DMF afforded the pure, per-O-acetylated glycosyl amine **24** to be used for coupling with an aspartic acid residue.[39]

The versatility of the Fmoc/Pfp active ester approach was further demonstrated by the synthesis of N-linked glycosyl amino acids. The glycosyl amine **24** was reacted with the acid chloride N^α-Fmoc-Asp(Cl)-OPfp **26** in the presence of N-ethylmorpholine (NEM) to afford the glycosyl amino acid **27** in 76% yield.[39] The acid chloride **26** was synthesized in a quantitative "one pot"[39] reaction from commercially available N^α-Fmoc-Asp(O^{t}Bu)-OPfp **25**. The C^β-carboxylic acid in **26** is highly activated as an acid chloride which undergoes selective nucleophilic attack by the amine at the C^β-carboxylic acid.

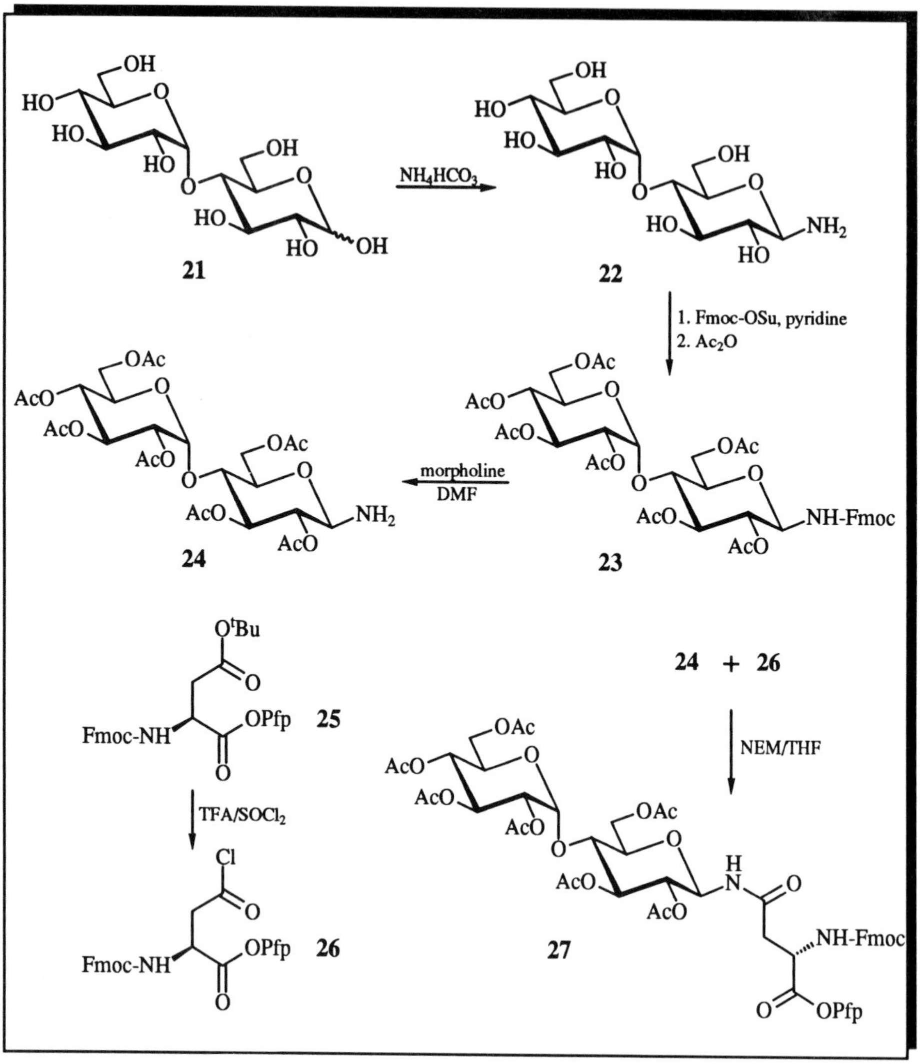

Figure 5.   Synthesis of N-linked glycosyl amino acids.

Glycopeptides containing N-linked 1-amino-1-deoxy alditols[54] are interesting
model compounds for investigations requiring a more flexible carbohydrate side
chain. Attempts to protect OH-groups of 1-amino-1-deoxy alditols by acetates
failed due to an O→N acyl migration to the more basic primary amino group.
However, it was demonstrated that the carbohydrate hydroxyl groups of 1-amino-

alditols as well as of glycosyl amines can be protected by trimethylsilyl (TMS) groups.[55,56] For instance the 1-amino-1-deoxy-lactitol[54] 28 was reacted with TMS-Cl in pyridine to provide the per-*O*-silylated derivative which was condensed with N^{α}-Fmoc-Asp(Cl)-OPfp 26 affording the glycosyl amino acid 29. This building block was used for the parallel assembly of glycosylated and non-glycosylated internally quenched fluorescent substrates as for instance compound 30.[56]

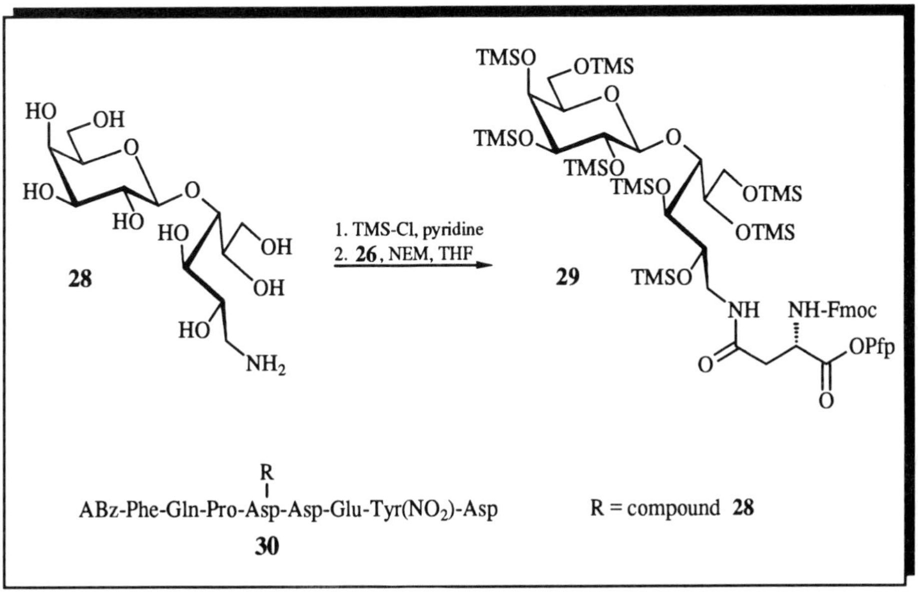

Figure 6.   Preparation of 1-amino-1-deoxy-lactitol aspartic acid derivatives.

For glycosyl amines, bis-trimethylsilylacetamide (BSA) has proven to be the best silylation reagent.[56] Crude preparations of glycosyl amines can be used for this reaction, e.g. maltoheptosyl amine, which was converted into the per-*O*-silylated derivative in 86% yield, and then reacted with the amino acid 26 to provide an aspartic acid derivative linked to a heptasaccharide.[56] Extended sugar chains have also been incorporated in the unprotected mode.[53]

A different approach to the synthesis of *N*-linked glycosyl amino acids has been reported by von dem Bruch and Kunz[30] using the 1-azido-*N*-acetylglucosamine derivative **31**[57-59] as starting material. Glycosyl azides are suitable anomeric amine precursors stable toward acidic and basic reaction conditions which allow the assembly of oligosaccharides as for instance the branched trisaccharide **32**.[30] Reduction of the azide afforded the glycosyl amine **33** which was condensed in the presence of the coupling reagent isobutyl-2-isobutoxy-1,2-dihydrochinolin-1-carboxylate (IIDQ) with N^α-Boc-Asp-OAll. Building block **34** was used in solution phase synthesis of glycopeptides containing clustered Lewis[x] side chains.[30]

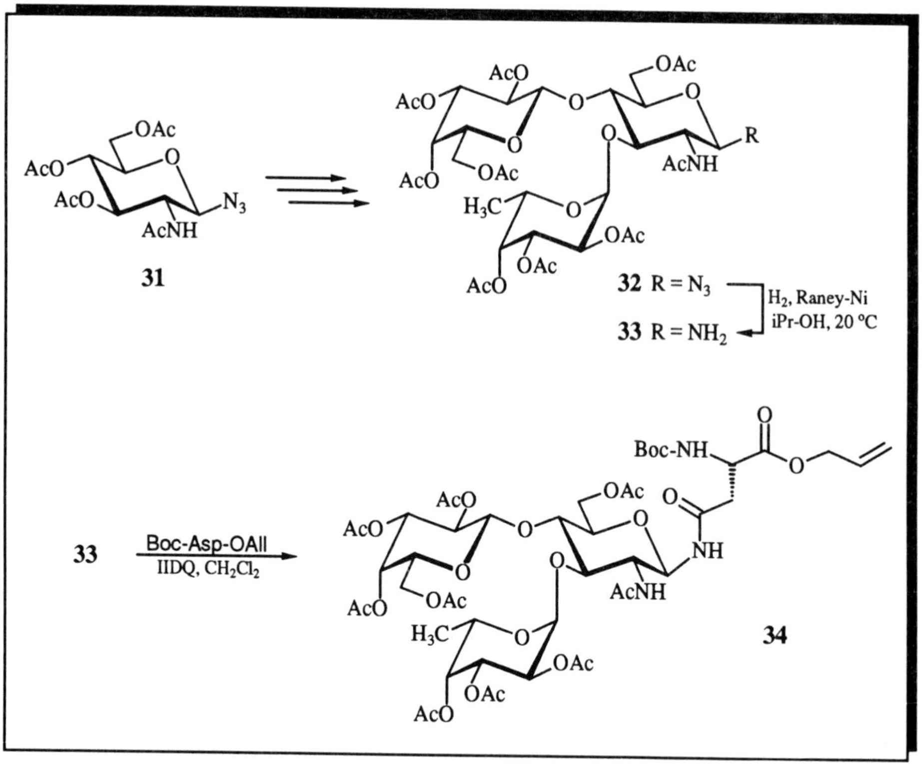

Figure 7.   Synthesis of a building block with the Lewis[x] antigen.

2.B.   Solid Phase Glycopeptide Synthesis

The methodology of solid phase peptide synthesis has been well established and reviewed.[60] However in solid phase glycopeptide synthesis valuable glycosyl amino acids are used and therefore the selected coupling methods should secure high yields, quantitative deprotection reactions and pure glycopeptide products. In the first step of the solid phase synthesis the polymer matrix has to be derivatized with a linker. Depending on the type of linker used the peptide may be cleaved under acidic, basic or almost neutral conditions. Considering the sensitivity of O-linked glycopeptides base labile linkers e.g. hydroxymethylbenzoic acid should be used with great care. The [4-(Fmoc-amino)-2,4-dimethoxybenzyl-phenoxyacetic acid] (Rink linker)[61] and the 5-(4-aminomethyl-3,5-dimethoxy-phenoxy)valeric acid (PAL-linker),[62] which are both acid labile, provide C-terminal amides. Theses and the hydroxymethylphenoxyacetic acid[63] which afford C-terminal acids, have been extensively used for glycopeptide synthesis. Kunz et al. have investigated a resin derivatized with an allyl type linker which allows the cleavage of peptides and glycopeptides under neutral conditions.[64]

The choice of resin does not in principle influence the outcome of the synthesis significantly when the peptide sequence is less than 10 amino acids long, but with longer sequences the reaction rates may become much reduced due to sterical interactions or aggregation of the peptide caused by β-sheet or α-helix formations. Therefore a resin which provides fast reaction conditions is preferred. One such efficient polymer, the PEGA-resin (polyethylene glycol poly-acrylamide copolymer), has recently been described.[65-67] This polymer has been used extensively for glycopeptide synthesis with good results. The high swelling potential and deaggregating properties of the flexible PEG-polymer network allows a high rate of mass transfer resulting in fast acylation reactions.[67] Even enzymes e.g. glycosyltransferases have access into the open interior of the polymer.[66] Furthermore, the swelled polymer is transparent and completely flow stable providing compatibility with the continuous flow version of solid phase peptide synthesis. The polystyrene polyethyleneglycol (PS-PEG)[68,69] and the kieselguhr supported polyamide types of resins[70] may also be used with excellent results.

For a successful solid phase glycopeptide synthesis it is essential to have quantitative, racemization-free coupling reactions throughout the synthesis. In this review it has not been considered relevant to discuss whether *in situ* activation or the use of activated ester is the method of choice. In our laboratory the best results have been obtained by application of N^α-Fmoc amino acid Pfp- or 3,4-dihydro-4-oxo-1,2,3-benzotriazine-3-yl (Dhbt) esters. The reactivity of Pfp esters can be increased by addition of 1-hydroxybenzotriazole (HOBt) or Dhbt-OH as auxiliary nucleophiles.[71-73] The application of Dhbt esters or Pfp esters with addition of Dhbt-OH makes it possible to follow the progress of the peptide coupling by a colour reaction.[73,74]

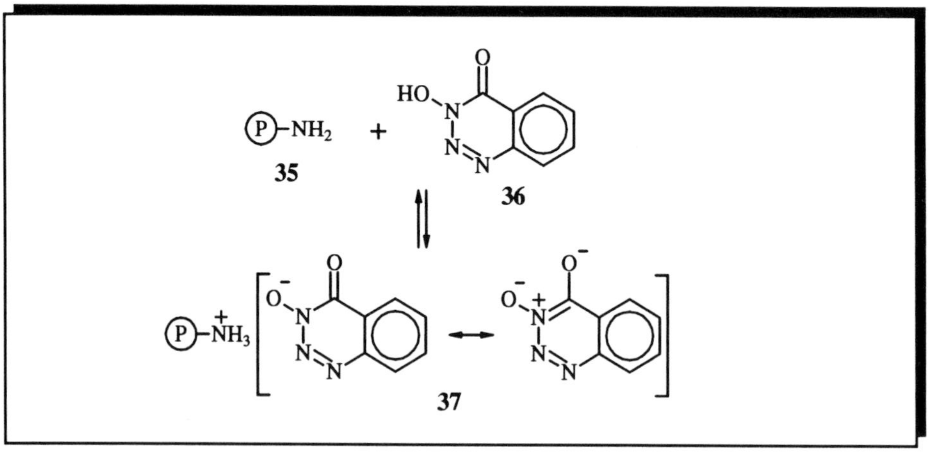

Figure 8.   Acylation control by the yellow Dhbt-colour reaction.

The Dhbt-OH molecule **36** will be deprotonated by the resin bound terminal amino groups **35** to form an intensively yellow coloured ion-pair **37**. As soon as all amino groups are acylated the yellow colour of the resin will disappear. The ability to follow the coupling reaction either visually or by a solid phase spectrometer is particularly valuable for multiple column and automated synthesis. The same kind of colour reaction can be induced with other reagents e.g. bromophenol blue.[75,76]

2.C.   Techniques in solid phase glycopeptide synthesis

Depending on the required number of glycopeptides and the length of their sequence different kinds of peptide synthesizers may be used for the solid phase assembly. Small glycopeptides like **18** have conveniently been synthesized by application of the "plastic syringe technique".[38,39] This kind of low cost technology makes peptide synthesis possible even for those laboratories which do not have access to any peptide synthesizer. A disposable plastic syringe A (without piston) fitted with a sintered Teflon filter (pore size 70 µm), is used as the reactor. The outlet of A is connected to the outlet of a 50 ml plastic syringe B via a Teflon tube with female Luer adapters. Syringe B is used as a waste syringe to remove solvents and used reagents. Solvents and reagents are added manually, and mixing of the resin with reagents is performed with a small spatula.

The synthesis of glycopeptides containing more than 15 amino acids is most conveniently carried out in an automated peptide synthesizer. The synthesis of longer peptides by commercially available peptide synthesizers is presently considered routine work. In our laboratory an automated peptide synthesizer has been constructed which allows the on line monitoring of the coupling reaction based on the above mentioned Dhbt colour reaction.[73,74] The continuous flow version of Fmoc solid phase peptide synthesis is used.[77] Solvents and reagents are pumped and recirculated through the resin packed into a glass column. Activated amino acids or glycosyl amino acids are delivered by recirculation through an in-line sampler.[78] The glass column containing the resin is flatted in the middle to form a flow cell through which the colour reaction can be monitored by a spectrophotometer.

If it is necessary to synthesize more than a single glycopeptide sequence multiple column techniques are particularly useful. Multiple peptide synthesis has recently been reviewed by Jung.[79] The 20-column or 96-column synthesizers built in our laboratory consists of a Teflon block with either 20 or 96 parallel columns for peptide synthesis.[7,80] Each column is equipped with a Teflon filter at the bottom and an outlet. All outlets are connected to a vacuum-chamber for removing solvents or reagents by suction. Activated amino acids, reagents or solvents are delivered manually by pipette in the case of the 20-column synthesizer. The 96-column apparatus is equipped with two parallel 96-channel washers (one for DMF, the other for deprotection reagent) which are connected to dispensing bottles. Thus all 96 columns can be washed simultaneously with the required solvent by mounting the block under the respective washer and dispensing once from the bottle. The delivery of activated amino- or glycosyl amino acids is performed from a 96-well reagent tray using an 8-channel multi pipette. During the coupling reaction the Teflon block is placed on a shaker in order to provide mixing of resin and reagents. The progress of coupling reactions can be followed visually by the Dhbt-colour reaction. Thus 20-96 different sequences can be synthesized in parallel. A detailed description of the synthesis of 40 different mucin glycopeptides will be given in this chapter.

3. EXPERIMENTAL

Synthesis of N^α-Fmoc-Ser-OPfp 3, N^α-Fmoc-Thr-OPfp 4 and N^α-Fmoc-Tyr-OPfp 8. Method A - The Pfp esters were synthesized according to the method of Kisfaludy[42] by esterification of the respective N^α-Fmoc protected free acids with pentafluorophenol (1 equiv.) and DCC (1 equiv.) at 0-5 °C in dry dioxane (for 1 and 2) or at -20 °C in tetrahydrofuran (THF) (for 7). The dicyclohexyl urea is removed by filtration and the concentrated product is recrystallized from ethyl acetate/n-hexane. However, this procedure may not remove all urea byproducts which can not be tolerated in glycosylation reactions and therefore it is recommended to purify the product by silica gel chromatography using dry solvents (e.g. ethyl acetate/light petroleum 1:5 for compound 8) and dried silica gel.[35]

Method B - Fmoc-Thr(^{t}Bu)-OPfp 6 (12.1 g, 21 mmol) was reacted with TFA

(100 ml) for 45 min at r.t. The solution was concentrated to dryness and coevaporated twice with toluene to yield compound **4** (10.8 g, 99%), m.p. 122 °C (Et$_2$O). Similarly, Fmoc-Ser-OPfp **3** (yield 94%, m.p. 139-140 °C) and Fmoc-Tyr-OPfp **8** (yield 94%, m.p. 170.5-172.5 °C) were prepared.[41]

Synthesis of N^α-Fmoc-Thr(Ac$_3$-α-D-GalN$_3$)-OPfp **12** - N^α-Fmoc-Thr-OPfp **4** (290 mg, 0.57 mmol), Ag$_2$CO$_3$ (355 mg, 1.29 mmol) and powdered molecular sieves 4 Å (1 g), dried *in vacuo* overnight, were suspended in dry toluene/dichloromethane 1:1 (30 ml) and stirred for 1 h under nitrogen atmosphere. Crystalline AgClO$_4$ (43 mg, 0.21 mmol) and β-chloride **10**[44,46,47] (300 mg, 0.86 mmol) were added and the suspension was stirred at r.t. for 1-3 days. The mixture was diluted with dichloromethane, filtered, washed with NaHCO$_3$-solution and water and dried over Na$_2$SO$_4$. The product was purified by either RP-HPLC or by medium pressure liquid chromatography (light petroleum/ethyl acetate 4:1). The pure product **12** was precipitated from diethyl ether/n-pentane 1:10. Yield 342 mg (73%), m.p. 78 °C, $[\alpha]_D^{22}$ 39.8 (c 1, CHCl$_3$).[17,18]

Synthesis of N^α-Fmoc-Thr[Bz$_3$-α-D-Man-6P(TCE)$_2$-(1→6)-α-D-Bz$_3$-Man]-OPfp **17** - The bromide **16** (1.3 g, 0.95 mmol) and N^α-Fmoc-Thr-OPfp **4** (0.48 g, 0.95 mmol) and molecular sieves 3 Å (0.5 g) were suspended in dry dichloromethane (90 ml) and stirred for 0.5 h under argon atmosphere at -40 °C. AgOTf (0.3 g, 1.14 mmol) was added quickly and the mixture was stirred 1 h. 2,4,6-collidine (2.5 ml, 1.89 mmol) was then added, and the temperature was slowly raised to 20 °C. The mixture was diluted with dichloromethane, filtered through Celite, extracted with cold 10% aq. TFA and water. After drying (MgSO$_4$) and concentration title compound **17**[38] was purified by vacuum liquid chromatography (light petroleum/ethyl acetate 3:1, dry solvents). Yield 1.26 g (74%), $[\alpha]_D^{25}$ -46.6 (c 1.3, CH$_2$Cl$_2$).

Synthesis of N^α-Fmoc-Asn[(TMS)$_8$-1-deoxy-lactitol-1-yl-]-OPfp **29** - 1-Amino-1-deoxy-lactitol **28**[54] (1.5 g, 4.4 mmol) was stirred with dry diethyl ether (50 ml) for one week, then the diethyl ether was decanted leaving the amorphous amino-lactitol. Pyridine (80 ml, 992 mmol) and TMS-Cl (55 ml, 435 mmol) were added and the reaction mixture was heated to reflux for 24 h. The reaction mixture was concentrated, suspended in dry n-pentane, filtered through dry Celite and the filtrate concentrated to give the per-O-silylated 1-amino-1-deoxy-lactitol hydrochloride (2.33 g, 56%, syrup) which was more than 90% pure according to ^{1}H NMR. The crude compound (1.6 g, 1.67 mmol) was dissolved in THF (15 ml) and NEM (442 µl, 3.5 mmol) was added. The mixture was added dropwise to a solution of the acid chloride **26** (1.73 mmol) in THF (20 ml) at -40 °C. NEM·HCl precipitated instantaneously and the mixture was allowed to warm to 20 °C. The hydrochloride was removed by filtration. The crude product was passed through a short column of dry silica gel (light petroleum/ethyl acetate 12:1). Yield 1.51 g (68%), compound **29** was characterized by NMR spectroscopy and MS.[55,56]

3.A. Solid Phase Glycopeptide Synthesis - General Procedures

DMF was distilled under reduced pressure and checked for free amine by addition of Dhbt-OH. The resins were derivatized with the PAL-[62] or the Rink-linker[61] by the O-(1H-benzotriazol-1-yl)-N,N,N',N'-tetramethyluronium tetrafluoroborate (TBTU) method.[81] In case of multiple column syntheses it is recommended to introduce norleucine as an internal standard before the linker in order to determine exactly the loading of the resin by amino acid analysis. The solid phase cycle (Fmoc deprotection, washing, coupling, washing) starting with the C-terminal amino acid was repeated until completion of the respective glycopeptide sequence. Acetylation in cases where N-terminal protected amino groups are wanted, was performed with Ac$_2$O in DMF. The amount of solvent for Fmoc-deprotections and for each DMF-washing is about twice the volume of the swollen resin and coupling reactions were carried out in the minimum volume of DMF that keeps the swollen resin covered. Usually 5-6 DMF washings for 2 min were enough to remove all reagents. Excess solvent was removed by applying vacuum for 2 min.

Side chain protection of activated N^{α}-Fmoc amino acids was as follows: ^{t}Bu-ethers for Ser, Thr, Tyr, O^{t}Bu-esters for Asp and Glu, Boc for Lys and His, 2,2,5,7,8-pentamethylchroman-6-sulfonyl (Pmc) for Arg, acetamidomethyl (Acm) for Cys, Asn and Gln were used without side chain protection. During coupling of N^{α}-Fmoc amino or glycosyl amino acids Pfp esters Dhbt-OH (1 equiv.) was added. 3-4.5 equiv. of activated esters were used for acylations, usually less (1.5-2 equiv) of the valuable glycosylated amino acids. Cleavages of glycopeptides from acid labile linkers was performed with 95% aq. TFA for 2 h. The resins were then carefully rinsed with TFA and the TFA fractions were concentrated. All glycopeptides were purified by RP-HPLC and were pure according to analytical HPLC, amino acid analysis, ^{1}H NMR spectroscopy and mass spectroscopy.

3.A.i.   Plastic syringe technique

The plastic syringe technique will be exemplified by the solid phase synthesis of the mannose-6-phosphate containing glycopeptide 18.[38,40] PEGA-resin (0.3 g, 0.045 mmol) was packed into a 10 ml disposable syringe (Discardit II, Beckton Dickinson), fitted with a sintered Teflon filter (Omnifit) and connected to a 50 ml waste syringe via a Teflon tube (Omnifit). Excess of solvents and reagents were removed with the second syringe. Fmoc deprotections were effected with 20% piperidine in DMF (3 ml) for 50 min. The first glycosyl amino acid 17 (120 mg, 1.5 equiv.) was dissolved in DMF (1.5 ml) and the solution was added to the resin. The suspension was agitated twice and then left for 24 h. After Fmoc deprotection the resin was reacted with N^{α}-Fmoc-Lys(ABz)-OPfp (68 mg, 2 equiv.) and Dhbt-OH (11 mg, 1.5 equiv.) in DMF (1.5 ml) for 24 h. After Fmoc cleavage the last glycosyl amino acid was coupled as described above followed by Fmoc removal and acetylation with Ac$_2$O/DMF 1:7 (20 min). The resin was transferred to a glass flask and dried. The glycopeptide was cleaved off the polymer with TFA and purified by RP-HPLC (yield 85 mg, 92%). For the deprotection of the TCE groups the glycopeptide (10 mg) was dissolved in acetic acid/pyridine 1:10 (3 ml) and reacted with zinc (170 mg) and Ag_2CO_3 (60 mg) at 50-60 °C for 18 h. The suspension was filtered and purified by RP-HPLC to yield 6.8 mg (90%) product, which was de-O-benzoylated with a mixture of chloroform/methanol/hydrazine hydrate 1:4:1 (1.2 ml) for 2 h. The product was purified by gel filtration and RP-HPLC to yield 1.5 mg (90%) of the glycopeptide 18.[40]

3.A.ii.   Multiple column glycopeptide synthesis in 20-well synthesizer

The assembly of glycopeptides containing N-linked lactitol disaccharides based on the parent sequence 30 walking through the sequence with a glycosylated and non glycosylated asparagine and using a 20 column synthesizer will be presented.[56] The PEGA resin was derivatized outside the reactor with the HMBA-linker by the usual procedure and then reacted with N^{α}-Fmoc-Asp(O^{t}Bu)-OH by activation with 2,4,6-mesitylenesulfonyl-3-nitro-1,2,4-triazolide and 1-methylimidazole.[82] The derivatized resin was distributed into 16 of the 20 wells (60 mg each, 0.087 mmol) of the synthesizer. Addition of solvents (DMF or deprotection reagent) was performed manually using an Eppendorf Multipette. Fmoc deprotections were performed with 2% DBU/2% piperidine in DMF (1 min + 30 min) for peptides and 50% morpholine in DMF (1 min + 30 min) for glycopeptides after coupling the glycosylated amino acid 29. During coupling of the respective amino- or glycosyl amino acids (17-20 h) the block was placed on a shaker table. The progress of the coupling reactions was followed visually by the Dhbt-colour reaction. After completion of the peptide assembly the resins were treated with 95% aq. TFA to remove all ^{t}Bu and TMS protecting groups. Cleavage off the polymer was performed with 0.1 M NaOH. The synthesis block was taken apart and the outlet of each column was connected via a silicon tube with a 5 ml glass vial. The resins were treated with base for 2 h and the solutions were eluted by supplying pressure to the column with a rubber bulb. The resins were rinsed extensively with water. After neutralization with 0.1 M HCl and purification the peptides and glycopeptides were isolated in 25-71% yield.

3.A.iii.   Multiple column synthesis of 40 mucin glycopeptides

The multiple column synthesis of 40 different mucin glycopeptides[17,18] based on the parent sequences **13** and **14** in Figure 2 was performed in the manual 96-well synthesizer. Each sequence was synthesized in two columns, the remaining columns were used for other synthesis. The kieselguhr supported poly(dimethylacrylamide) resin[70] was derivatized with norleucine, the PAL-linker[62] and the first amino acid Fmoc-Thr(^{t}Bu) was added outside the synthesizer. The dried resin (30 mg) was weighed into the wells. Fmoc deprotections were effected with 50% morpholine in DMF (1 x 2 min, 2 x 20 min), coupling reactions were performed using 4.61 equiv. of activated amino or glycosyl amino acids in DMF (0.2 ml) for 5-22.5 h. Fmoc amino acids Dhbt esters were used except for His, Cys and the glycosyl amino acid **12** which were coupled as Pfp esters. The progress of each acylation reaction was followed visually by the disappearance of the yellow colour in the column. After completion of the glycopeptide sequences all azido groups were converted into the acetamido derivatives by reacting the resins with thioacetic acid (5 times distilled under vacuum into a cold trap) for 90 h. The reduction was monitored by IR spectroscopy observing the disappearance of the azide absorption band. Deacetylation of the carbohydrates was also performed on the polymer bound glycopeptides by reacting the resin in each column with 15% hydrazine hydrate in methanol (0.4 ml) for 6 h. The resins were transferred into glass vials and treated with TFA-scavenger mixtures. After RP-HPLC, the pure glycopeptides were obtained in yields of 1.7-5.7 mg (12-70%).

3.A.iv.   Automated synthesis of the glycosylated triicosapeptide 15

Glycopeptide **15** was synthesized on the PEGA-resin (250 mg), derivatized with the Rink linker, in the automated peptide synthesizer described above.[23] Deprotection of the Fmoc group was achieved by a 13 min flow of 2% DBU/DMF through the column. Fmoc amino acids (4 equiv.) were coupled for 1.5 h recirculation after *in situ* preactivation (1.5 min) with TBTU (3.9 equiv.) and NEM (4 equiv.). The glycosylated Fmoc amino acid Pfp ester **12** was coupled 3 h in the presence of Dhbt-OH. The acylation curves measured by the solid phase spectrophotometer indicated a quantitative coupling of all amino acids. Reduction of the azido group was performed on the polymer bound glycopeptide with thioacetic acid. The glycopeptide was cleaved off with 95% aq. TFA, filtered through a SEP-PAK C_{18} cartridge and purified by RP-HPLC. The product was dissolved in methanol water 1:1 (4 ml) and treated with sodium methoxide solution (1%, 22 µl) for 3 h. After RP-HPLC the pure glycopeptide **15** was isolated in 61% yield.

ACKNOWLEDGEMENT

S.P. thanks the Danish Cancer Association and the Alexander von Humboldt-Stiftung for financial support.

REFERENCES

1.   A. Varki, *Glycobiology*, **3**, 97 (1993).

2.   H. Kunz, *Angew. Chem.*, **99**, 297 (1987).

3.   H. Kunz, *Pure & Appl. Chem.*, **65**, 1223 (1993).

4.   M. Meldal, In *Neoglycoconjugates: Preparation and application* (Y.C. Lee and R.T. Lee, Eds.) Academic Press, San Diego, p. 145 (1994).

5.   J. Tamura, *TIGG*, **6**, 29 (1994).

6.   H. Paulsen, S. Peters, and T. Bielfeldt, In *Glycoproteins* (J. Montreuil, H. Schacter, and J.F.G. Vliegenthart, Eds.) Elsevier, Amsterdam, In Press (1995).

7.   M. Meldal, C.B. Holm, G. Bojesen, M.H. Jacobsen, and A. Holm, *Int. J. Peptide Protein Res.*, **41**, 250 (1993).

8.   K.S. Lam, S.E. Salmon, E.M. Hersh, V.J. Hruby, W.M. Kazmierski, and R.J. Knapp, *Nature*, **354**, 82 (1991).

9.   S.T. Cohen-Anisfeld and P.T. Lansbury Jr., *J. Am. Chem. Soc.*, **115**, 10531 (1993).

10.   S.A. Kates, B.G. de la Torre, R. Eritja, and F. Albericio, *Tetrahedron Lett.*, **35**, 1033 (1994).

11.   D.M. Andrews and P.W. Seale, *Int. J. Peptide Protein Res.*, **42**, 165 (1993).

12.   M. Hollosi, E. Kollat, I. Laczko, K.F. Medzihradszky, J. Thurin, and L. Otvos Jr., *Tetrahedron Lett.*, **32**, 1531 (1991).

13.   H. Kunz, S. Birnbach, and P. Wernig, *Carbohydr. Res.*, **202**, 207 (1990).

14.   F. Filira, L. Biondi, B. Scolaro, M.T. Foffani, S. Mammi, E. Peggion, and R. Rocchi, *Int. J. Biol. Macromol.*, **12**, 41 (1990).

15.   F. Filira, L. Biondi, B. Cavaggion, B. Scolaro, and R. Rocchi, *Int. J. Peptide Protein Res.*, **36**, 86 (1990).

16.   E. Bardaji, J.L. Torres, P. Clapes, F. Albericio, G. Barany, R.E. Rodriguez, M.P. Sacristan, and G. Valencia, *J. Chem. Soc., Perkin Trans. 1*, 1755 (1991).

17.   T. Bielfeldt, S. Peters, M. Meldal, K. Bock, and H. Paulsen, *Angew. Chem.*, **104**, 881 (1992).

18.   H. Paulsen, T. Bielfeldt, S. Peters, M. Meldal, and K. Bock, *Liebigs Ann. Chem.*, 369 (1994).

19.   S. Rio, J.-M. Beau, and J.-C. Jacquinet, *Carbohydr. Res.*, **255**, 103 (1994).

20.   S. Peters, T. Bielfeldt, M. Meldal, K. Bock, and H. Paulsen, *J. Chem. Soc., Perkin Trans. 1*, 1163 (1992).

21.   A.M. Jansson, M. Meldal, and K. Bock, *J. Chem. Soc., Perkin Trans. 1*, 1699 (1992).

22.   P. Schultheiβ-Reimann and H. Kunz, *Angew. Chem.*, **95**, 64 (1983).

23.   M. Meldal, T. Bielfeldt, S. Peters, K.J. Jensen, H. Paulsen, and K. Bock, *Int. J. Peptide Protein Res.*, **43**, 529 (1994).

24.   J. Kihlberg and T. Vuljanic, *Tetrahedron Lett.*, **34**, 6135 (1993).

25.   M. Meldal, S. Mouritsen, and K. Bock, In *Carbohydrate Antigens*, A.C.S. Symp. Ser. 519 (P.J. Garegg and A.A. Lindberg, Eds.) Academic Press, New York, p. 19 (1993).

26.   M. Schultz, P. Hermann, and H. Kunz, *Synlett*, 37 (1992).

27.   H. Paulsen, K. Adermann, G. Merz, M. Schultz, and U. Weichert, *Starch/Stärke*, **40**, 465 (1988).

28.   S. Friedrich-Bochnitschek, H. Waldmann, and H. Kunz, *J. Org. Chem.*, **54**, 751 (1989).

29.   W.A. Macindoe, H. Iijima, Y. Nakahara, and T. Ogawa, *Tetrahedron Lett.*, **35**, 1735 (1994).

30.   K. von dem Bruch and H. Kunz, *Angew. Chem.*, **106**, 87 (1994).

31.   B. Lüning, T. Norberg, C. Rivera-Baeza, and J. Tejbrant, *Glycoconjugate J.*, **8**, 450 (1991).

32.   Y. Nakahara, H. Iijima, S. Shibayama, and T. Ogawa, *Carbohydr. Res.*, **216**, 211 (1991).

33.   M. Meldal and K.J. Jensen, *J. Chem. Soc., Chem. Commun.*, 483 (1990).

34.   K.R. Reimer, M. Meldal, S. Kusomoto, K. Fukase, and K.Bock, *J. Chem. Soc., Perkin Trans. 1*, 925 (1993).

35.   K.J. Jensen, M. Meldal, and K. Bock, *J. Chem. Soc., Perkin Trans. 1*, 2119 (1993).

36.   M. Meldal, K.J. Jensen, A.M. Jansson, and K. Bock, In *Innov. Perspec. S.P.S.* (R. Epton, Ed.) p. 179 (1992).

37.   S. Peters, T. Bielfeldt, M. Meldal, K. Bock, and H. Paulsen, *Tetrahedron Lett.*, **33**, 6445 (1992).

38.   M.K. Christensen, M. Meldal, and K. Bock, *J. Chem. Soc., Perkin Trans. 1*, 1453 (1993).

39.   I. Christiansen-Brams, M. Meldal, and K. Bock, *J. Chem. Soc., Perkin Trans. 1*, 1461 (1993).

40. M.K. Christensen, M. Meldal, K. Bock, H. Cordes, S. Mouritsen, and H. Elsner, *J. Chem. Soc., Perkin Trans. 1*, 1299 (1994).

41. A. Vargas-Berenguel, M. Meldal, H. Paulsen, and K. Bock, *J. Chem. Soc., Perkin Trans 1*, 2615 (1994).

42. L. Kisfaludy and I. Schön, *J. Chem. Soc., Chem. Commun.*, 325 (1983).

43. H. Paulsen, *Angew. Chem.*, **94**, 184 (1982).

44. R.U. Lemieux and R.M. Ratcliffe, *Can. J. Chem.*, **57**, 1244 (1979).

45. J. Broddefalk, U. Nilsson, and J. Kihlberg, *J. Carbohydr. Chem.*, **13**, 129 (1994).

46. H. Paulsen and J.-P. Hölck, *Carbohydr. Res.*, **109**, 89 (1982).

47. B. Ferrari and A.A. Pavia, *Carbohydr. Res.*, **79**, C1 (1980).

48. H. Paulsen, T. Bielfeldt, S. Peters, M. Meldal, and K. Bock, *Liebigs Ann. Chem.*, 381 (1994).

49. A. Vargas-Berenguel, M. Meldal, H. Paulsen, K.J. Jensen, and K. Bock, *J. Chem. Soc., Perkin Trans 1*, 3287 (1994).

50. H.S. Isbell and H.L. Frush, *Methods Carbohydr. Chem.*, **8**, 255 (1980).

51. H.S. Isbell and H.L. Frush, *J. Org. Chem.*, **23**, 1309 (1958).

52. L.M. Likhosherstov, O.S. Novikova, V.A. Derevitskaya, and N.K. Kochetkov, *Carbohydr. Res.*, **146**, C1 (1986).

53. L. Urge, D.C. Jackson, L. Gorbics, K. Wroblewski, G. Graczyk, and L. Otvos Jr., *Tetrahedron*, **50**, 2373 (1994).

54. I. Christiansen-Brams, M. Meldal, and K. Bock, *J. Carbohydr. Chem.*, **11**, 813 (1992).

55. I. Christiansen-Brams, M. Meldal, and K. Bock, *Tetrahedron Lett.*, **34**, 3315 (1993).

56. I. Christiansen-Brams, A.M. Jansson, M. Meldal, K. Breddam, and K. Bock, *Bioorg. Med. Chem. Lett.*, In Press (1995).

57. Z. Györgydeak, L. Szilagyi, and H. Paulsen, *J. Carbohydr. Chem.*, **12**, 139

(1993).

58.   H. Kunz, H. Waldmann, and J. März, *Liebigs Ann. Chem.*, 45 (1989).

59.   J. Thiem and T. Wiemann, *Angew. Chem.*, **102**, 78 (1990).

60.   G.B. Fields and R.L. Noble, *Int. J. Peptide Protein Res.*, **35**, 161 (1990).

61.   H. Rink, *Terahedron Lett.*, **28**, 3787 (1987).

62.   F. Albericio, N. Kneib-Cordonier, S. Biancalana, L. Gera, R.I. Masada, D.
       Hudson, and G. Barany, *J. Org. Chem.*, **55**, 3730 (1990).

63.   E. Atherton, C.J. Logan, and R.C. Sheppard, *J. Chem. Soc., Perkin
       Trans. 1*, 538 (1981).

64.   H. Kunz and B. Dombo, *Angew. Chem.*, **100**, 732 (1988).

65.   M. Meldal, *Tetrahedron Lett.*, **33**, 3077 (1992).

66.   M. Meldal, F.-I. Auzanneau, O. Hindsgaul, and M.M. Palcic, submitted for
       publication.

67.   F.I. Auzanneau, M. Meldal, and K. Bock, *J. Peptide Sci.*, **1**, 31 (1995).

68.   W. Rapp, L. Zhang, R. Häbish, and E. Bayer, In *Peptides 1988*, Proc. Eur.
       Pept. Symp., 20'th (G. Jung and E. Bayer, Eds.) Walter de Gruyter, Berlin,
       p. 199 (1989).

69.   S. Zalipsky, F. Albericio, and G. Barany, In *Peptides 1985*, Proc. Am. Pept.
       Symp., 9'th (C.M. Deber, V.J. Hruby, and K.D. Kopple, Eds.) Pierce
       Chemical Company, Rockford, Illinois, p. 257 (1986).

70.   E. Atherton, E. Brown, R.C. Sheppard, and A. Rosevear, *J. Chem. Soc.,
       Chem. Commun.*, 1151 (1981).

71.   W. König and R. Geiger, *Chem. Ber.*, **103**, 788 (1970).

72.   W. König and R. Geiger, *Chem. Ber.*, **103**, 2034 (1970).

73.   E. Atherton, J.L. Holder, M. Meldal, R.C. Sheppard, and R.M. Valerio, *J.
       Chem. Soc., Perkin Trans. 1*, 2887 (1988).

74.   L.R. Cameron, J.L. Holder, M. Meldal, and R.C. Sheppard, *J. Chem. Soc.,
       Perkin Trans. 1*, 2895 (1988).

75.   M. Flegel and R.C. Sheppard, *J. Chem. Soc., Chem. Commun.*, 536 (1990).

76.   V. Krchnak, J. Vagner, P.Safar, and M. Lebl, *Coll. Czech. Chem. Commun.*, **53**, 2542 (1988).

77.   A. Dryland and R.C. Sheppard, *Tetrahedron*, **44**, 859 (1988).

78.   M. Meldal, A. Holm, and O. Buchardt, PCT Int. Appl. WO 90 07,975 (Cl. B01F1/00) 26 Jul 1990, DK Appl. 89/213,18 Jan 19, (1990).

79.   G. Jung and A.G. Beck-Sickinger, *Angew. Chem.*, **104**, 375 (1992).

80.   A. Holm and M. Meldal, In *Peptides 1988*, Proc. Eur. Pept. Symp., 20'th (G. Jung and E. Bayer, Eds.) Walter de Gruyter, Berlin, p. 208 (1989).

81.   R. Knorr, A. Trzeciak, W. Bannwarth, and D. Gillessen, *Tetrahedron Lett.*, **30**, 1927 (1989).

82.   B. Blankemeyer-Menge, M. Nimtz, and R. Frank, *Tetrahedron Lett.*, **31**, 1701 (1990).

Chapter 16

DESIGN AND SYNTHESIS OF GLYCOCONJUGATES

RENÉ ROY

Department of Chemistry, University of Ottawa, Ottawa, Ontario,
Canada K1N 6N5

Abstract   The design and syntheses of glycoconjugates and their precursors
are reviewed.   Particular emphasis was given to sialic acid containing
conjugates. Phase transfer catalysis is described as a useful and stereospecific
strategy to suitably functionalized glycoconjugate precursors. N-Acryloylated
glycosyl derivatives were used as common precursors to both neoglycoproteins and glycopolymers.   A few examples of glycopolymers having
custom-designed properties are discussed. Finally, new generations of multibranched dendritic glycoconjugates are described.

1. INTRODUCTION

Cell surface glycoproteins, glycolipids and proteoglycans constitute forefront
molecules having well-documented implications in cell-cell recognitions, growth,
differentiation and infections. Glycobiology has emerged into an active field of
research dedicated to studies, at the molecular level, of the exact role played by
carbohydrates in these phenomena. Fundamental insights into these carbohydrate-
based interactions are critically dependent on appropriate synthetic models that
ought to take into considerations the multivalent nature of cell surface
carbohydrates. Multivalent carbohydrate structures or macromolecules should be
well defined and easy to prepare in various shapes and orientations. Furthermore,
their design should also result in forms suitable for both *in vitro* and *in vivo*

being valuable immunodiagnostics for high concentrations of sialic acid found in cancer patients.[22]

An excellent review by Stowell and Lee[9] describes a wide variety of chemical methods for the syntheses of neoglycoproteins.[1,3,4] Because reductive amination offers the desired efficiency and chemoselectivity, it was adopted for the early syntheses of neoglycoproteins containing sialosides.[18,22-25] Sialyloligosaccharides can be directly coupled to proteins' lysine ε-amino groups via their reducing end with sodium cyanoborohydride (NaBH$_3$CN) in various buffers.[24,25] They can also be conjugated with added spacer arms such as phenethylamine.[26] Borate buffer (0.2M, pH 9) appeared to be better suited for the conjugations of reducing sugars,[27] while phosphate buffer (pH 7) appeared to be better for aldehyde-containing aglycons.[23] As opposed to neutral oligosaccharides, sialosides tend to accumulate anionic repulsive properties on the conjugates and, as a result, complete substitutions of amino groups require more stringent conditions.[23]

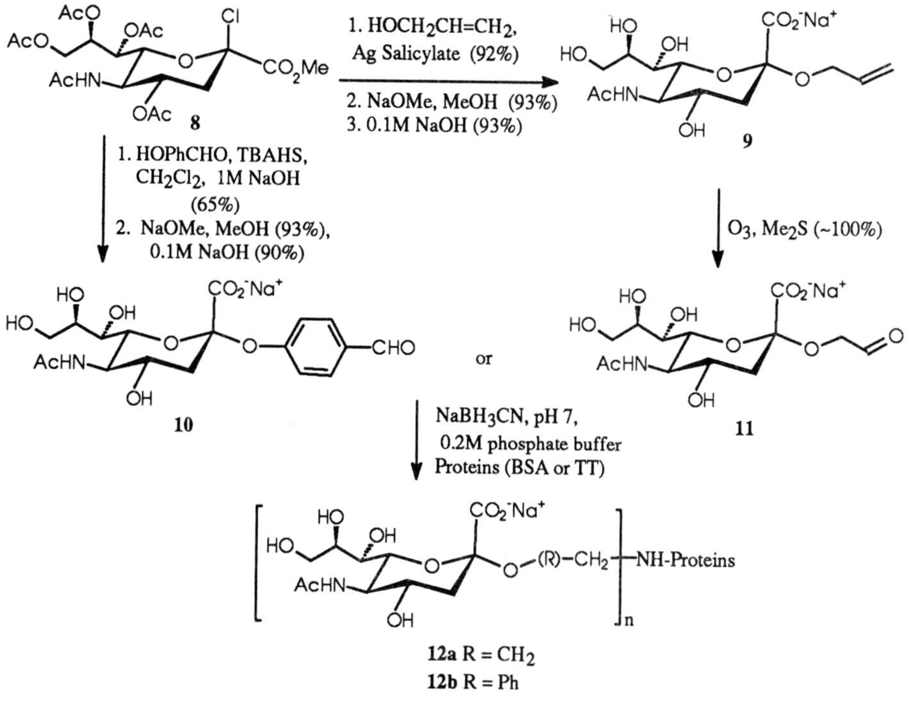

Scheme 3.

In a preliminary work, allyl α-sialoside (**9**) was ozonolyzed and the resulting aldehyde was conjugated to both BSA and TT.[22] The extents of conjugations were followed as a function of time, buffers, pH and temperatures. Conjugates of less than ~18 and ~8 Neu5Ac residues per BSA and TT respectively were poorly antigenic. Similarly, *p*-formylphenyl sialoside **10** was also conjugated to BSA (Scheme 3).[18] It was concluded that threshold carbohydrate densities were required for efficient binding to model lectins. Both conjugates **12a** (BSA and TT) were only moderate inhibitors of hemagglutinations of influenza virus[24] and conjugates with high Neu5Ac contents were highly immunogenic in rabbits, triggering IgG antibodies with high specificities against α-sialosides.[28]

2.C. Glycopolymers

The observed immunogenicities of sialylated neoglycoproteins preclude their use as therapeutic antagonists in viral infections. Synthesis of potentially non-immunogenic glycopolymers was therefore considered as an alternative. With few exceptions, there are essentially four different approaches to glycopolymer syntheses.[13,29-32] Carbohydrate monomers having (1) alkenyl[29] or styryl[30] glycosides and (2) N-[13] or O-acryloyl[31] residues have thus been prepared. These monomers can be copolymerized with other monomers such as acrylamide to confer water-solubility to the glycopolymer. Alternatively, carbohydrate precursors can be synthesized with (3) amino or (4) reactive carboxyl functions on the aglycon moieties. These derivatives can be efficiently grafted to pre-formed polymers having either amino or active ester functionality.[32]

There are numerous advantages to glycopolymer antigens over their neoglycoprotein counterparts that have yet to be fully recognized. Glycopolymers can possess uniform and stable structures having a wide range of molecular weights, carbohydrate densities and functionalities. Their purifications and characterizations are easier. They can be produced inexpensively and in large scale. They can also be used advantageously as multivalent inhibitors or for cell targeting since they are non or poorly immunogenic. In addition, they can be used directly in quantitative immunoprecipitation and in enzyme-linked immunosorbent assays (ELISA).[13,33,34] In that respect, it has been demonstrated that they are usually more sensitive in microtiter plate ELISA assays. In the following sections, glycopolymer syntheses using pre-formed *N*-acryloylated carbohydrates as monomer precursors are described.

Copolymerization studies with allyl α-sialoside **9** and acrylamide using ammonium persulfate $(NH_4)_2S_2O_8$ at 100°C (15 min) or with TMEDA at room temperature (16 h) provided copolymer **13** (Scheme 4).[22] The yields of these copolymerizations were usually low (<60%) and the sialoside contents of the polymers were also low due to a large reactivity discrepancy between the alkenyl and acryloyl functionalities.

To improve its copolymerizing property relative to that of acrylamide, allyl sialoside **9** was extended by light-induced free radical addition with cysteamine (2-aminoethanethiol) (84%).[35] Subsequent *N*-acryloylation with acryloyl chloride gave monomer **14** (88%). Copolymerization of **14** afforded copolymer **15** in which the molar ratios of the two monomers (^{1}H-NMR) coincided to those used in the initial reaction mixtures. This strategy (Scheme 4) was applied to the

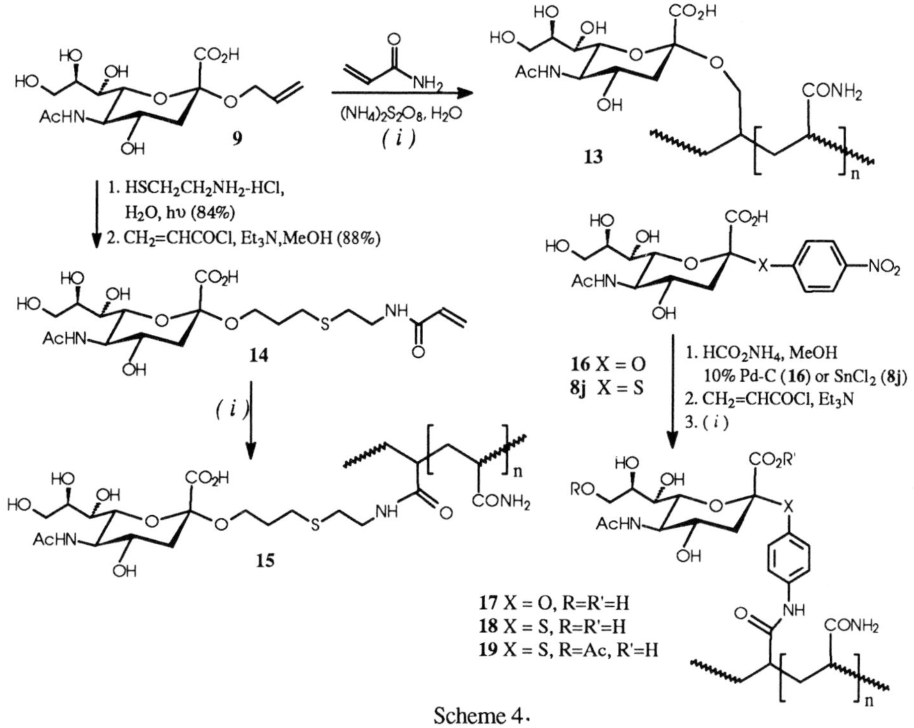

Scheme 4.

syntheses of other copolymers containing rigid aromatic spacers, including thiosialosides **18-19** designed to be resistant to viral neuraminidases.[36]

Sialopolymer **15** has been used for the inhibition of hemagglutination of human erythrocytes by influenza A virus (IC_{50} 2.3 μM).[24] While polymer **18** appeared to be inactive against the A strain, its homologous 9-O-acetylated derivative **19** showed good inhibition of HA of influenza C virus (IC_{50} 3.3 μM).[36] Following our first reports, other groups have made similar observations.[32,37] The above strategy was then used to prepare more complex glycopolymers containing Thomsen-Friedenreich (T) antigen (**20**),[38] lactose (**21**),[13] N-acetyllactosamine (**22**), blood group ABH type 1 and type 2 antigens, lacto-N-tetraose (**23**),[38] GM_3 (**24**) and GD_3 (**25**)[39] oligosaccharides (Scheme 5).

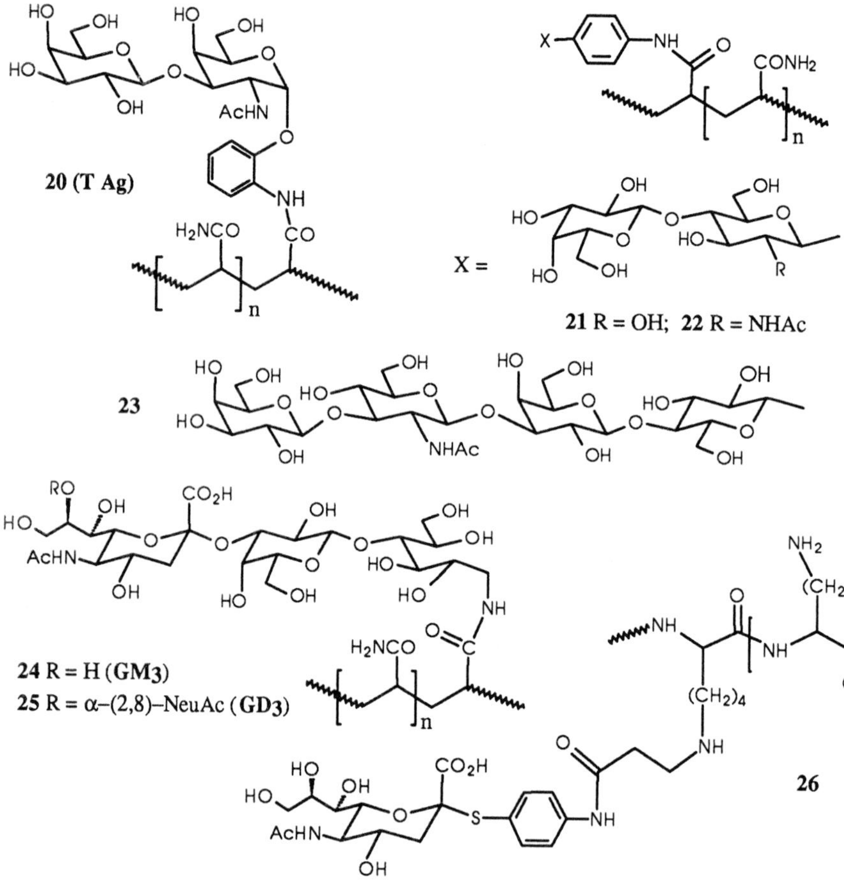

20 (T Ag)

21 R = OH; 22 R = NHAc

23

24 R = H (GM3)
25 R = α–(2,8)–NeuAc (GD3)

26

Scheme 5.

2.D.   *N*-Acryloylated Derivatives as Common Precursors to Both Neoglyco-proteins and Glycopolymers

Neoglycoproteins were previously designed to trigger anti-carbohydrate antibodies.[4] However, ensuing serological investigations with the same conjugates were usually hampered by high background and cross-reacting antibodies against the protein carriers. To simplify immunoassays, homologous glycopolymers, having only the carbohydrate haptens in common with the neoglycoproteins used as immunogens, were highly needed. To achieve this goal, a simplified strategy was suggested based on the rational design of one single precursor to both glycoconjugates. *N*-Acryloylated derivatives such as **14** were used for such purposes (Scheme 6).[38-41] Neoglycoproteins generated by Michael addition of lysine ε-amino groups onto *N*-acryloylated monomers were used as immunogens while glycopolymers were used as screening antigens.

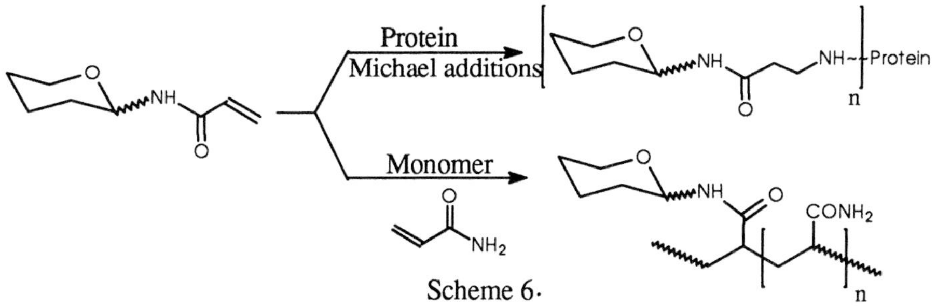

Scheme 6.

In immunochemical investigations of sialylated neoglycoproteins **12a,b**, glycopolymer **15** was required for antibody titrations and combining site mapping. Glycopolymer **15** was directly used as coating antigen in ELISA or in quantitative precipitation assays. In that work, two different sialoside precursors were used (**11** and **14**). Using the above strategy, *N*-acryloylated derivative **14** was directly conjugated to BSA and TT using base catalyzed Michael additions. Optimum coupling conditions were identified (0.2M carbonate buffer, pH 10.5, 37°C, 3 days) without noticeable denaturation of the protein carriers.[42] Most glycopolymers illustrated in Scheme 5 were also simultaneously prepared as neoglycoproteins.[38-42] The homologous *ortho*- or *para*-*N*-acryloylated phenyl glycoside precursors were conjugated to BSA or TT by base catalyzed Michael additions. In some cases, poly-L-lysine was used as a protein mimic to facilitate structural determinations (**26**, Scheme 5).[41]

3. CUSTOM-DESIGNED GLYCOPOLYMERS

Glycopolymers are now receiving attention because of the many advantageous physico- and immuno-chemical properties with which they can be constructed. To further expand the usefulness of glycopolymers in various immunoassays, this section will briefly describe custom-designed three-component copolymerizations (terpolymerizations).

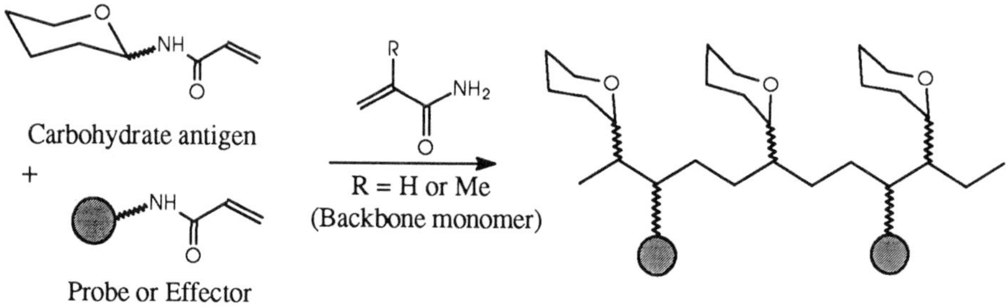

Carbohydrate antigen

+

Probe or Effector

R = H or Me
(Backbone monomer)

Scheme 7.

The strategy described herein (Scheme 7) is very simple and relies on readily available *N*-acryloylated precursors that can be used as monomers in terpolymerization reactions. Reaction mixtures are composed of the carbohydrate haptens, probes (or effector molecules) and monomers such as acrylamide or methacrylamide, to be incorporated as polymer backbone.

Biotin, stearylamine or 2-(p-hydroxyphenyl)ethylamine (tyramine) as their *N*-acryloylated derivatives were first used as probes or effectors.[43,44] The choice of biotin was obvious in light of its universal utilization in biochemical probes through commercially available avidin or streptavidin kits. Stearylamine was chosen to confer glycopolymers improved hydrophobic properties as coating antigens in microtiter plate ELISA assays, while tyramine was used for radiolabeling experiments with ^{125}I.

Hetero-bispecific glycopolymer antigens containing L-rhamnose and *N*-acetylglucosamine **30** were also synthesized and used in model serological studies

(Scheme 8). All terpolymers retained their specific avidin/streptavidin, lectin and antibody binding properties as determined by agar gel diffusions, quantitative precipitations and ELISA. Noteworthy was the observation that commercially available rabbit anti-capsular polysaccharide (CPS) antibodies used to diagnose *Streptococcus pneumoniae* type 23F and β-hemolytic *Streptococci* Group A infections was found to bind independently to the bispecific polymer **30**. This observation resulted from the fact that the antibodies recognize α-L-rhamnose (type 23F) and β-D-*N*-acetylglucosamine as antigenic determinants present as branches on the repeating units of the CPSs.

It is also worth mentioning that all glycopolymers described in this section are highly antigenic with both lectins and antibodies. The optimum binding efficiency was achieved with an acrylamide to carbohydrate ratio of 10 to 1, while average molecular weights were in the range 50-100 kDa.[45] Interestingly, the

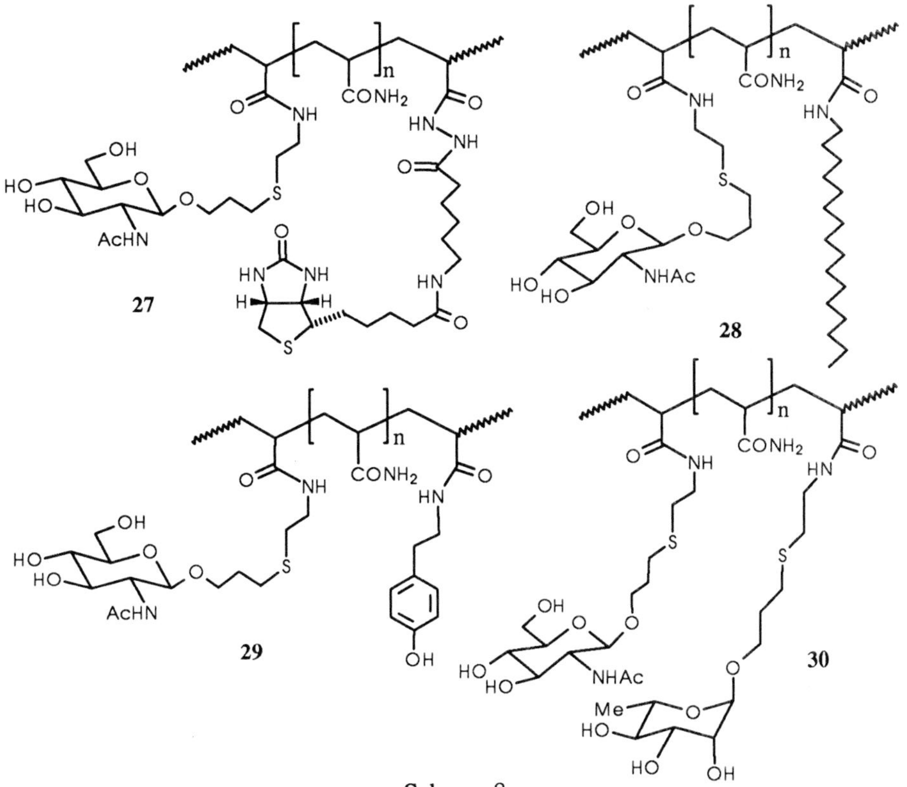

Scheme 8.

molecular   weight appeared to have no influence on the antigenic properties. Recent molecular dynamic studies using ^{13}C-NMR relaxation data and viscometry have shown that glycopolymers behave as dynamic random coils.[46]

4. NEW GLYCOCONJUGATE STRUCTURES

Growing interest for glycoconjugates and glycomimetics in drug research has stressed the need for specially designed carbohydrate ligand structures. Rational design of well-defined carbohydrate supramolecules can offer effective tools for inhibition studies.   It is with this strategy in mind that we initiated a program aimed at investigating supramolecular recognition phenomena.

4.A. Carbohydrate Telomers
The study of cluster effects suggests that the number of sugar residues and their respective propinquity confer to glycosylated clusters their improved overall binding affinity. Low valency lactosylated glycoconjugates (clusters) have been found useful as inhibitors or targeting reagents for endogenous lectins such as those present on mammalian hepatic asialoglycoprotein receptors (ASGP-R).[47] More recently, lactosylated clusters have been shown to be potent inhibitors of lung colonizations by metastatic cell lines in mice.[10]   As previous lactosylated cluster syntheses were tedious,[10,11,47] a straightforward   approach has been designed to circumvent this difficulty. The strategy described herein is based on a single step synthesis of families of lactosylated clusters by telomerization using N-acryloylated lactosyl derivatives as monomeric units.[48]

Glycopolymers were originally designed as sources of optimum multivalent clusters, but such high molecular weight conjugates may represent chemically inappropriate structures.   It was therefore anticipated that oligomers with controlled hapten distributions along the backbone would offer some advantages. Quenching the propagation process with chain transfer reagents (telogens) such as thiols during polymerization provides oligomers (telomers) of few repeating units. The number of repeating units within the telomers can be controlled by the thiol concentrations incorporated into the initial reactions. In this section, single step syntheses of families of lactosylated telomers composed of one to six disaccharide residues are described. Two series of telomers were synthesized having short and long spacer arms between the telomer backbones (Scheme 9).[48,49]

Reduction (Ra-Ni or H_2, Pd-C) of lactosyl azide **4a**, obtained under PTC conditions, afforded anomerically unstable lactosylamine **31** which was immediately treated with either acryloyl chloride (EtOAc, pyridine, 0°C, 45 min, 85%) or 6-*N*-acrylamidohexanoic acid (EtOH, EEDQ, 25°C, 16 h, 73%) to provide monomer precursors **32** and **33** as taxogens. After de-O-acetylation, the resulting monomers of **32a** and **33a** were telomerized in refluxing methanol under nitrogen with 2,2'-azobisisobutyronitrile (AIBN) as initiator and *tert*-butyl mercaptan in various molar ratios as telogen. The reactions provided, in one single step, families of low molecular weight lactosylated clusters. Separation of each telomer was accomplished by size exclusion chromatography on a Bio-Gel P-2 column. Results from the telomerizations of **32a** are shown in Table 1.

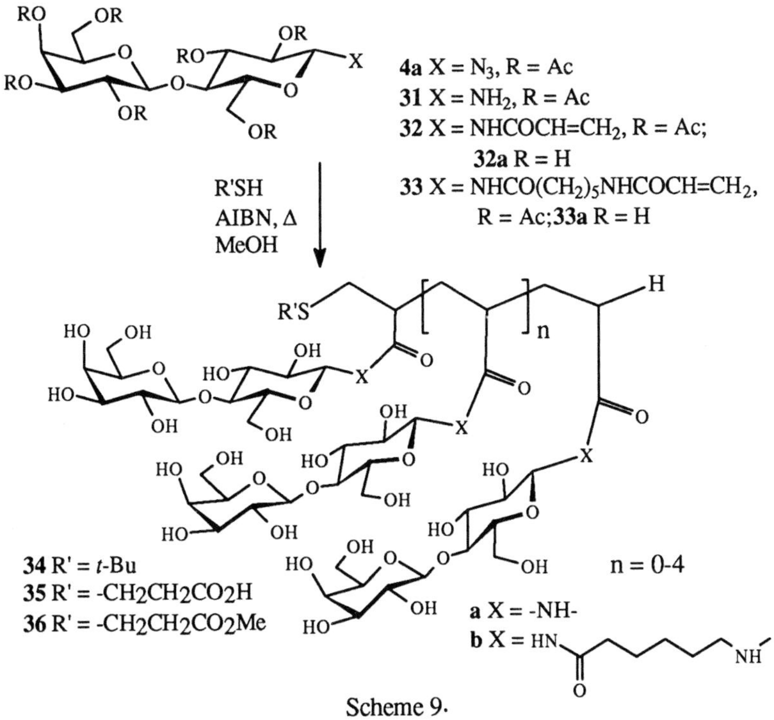

4a X = N_3, R = Ac
31 X = NH_2, R = Ac
32 X = $NHCOCH=CH_2$, R = Ac;
    32a R = H
33 X = $NHCO(CH_2)_5NHCOCH=CH_2$,
    R = Ac;33a R = H

R'SH
AIBN, Δ
MeOH

34 R' = *t*-Bu
35 R' = $-CH_2CH_2CO_2H$
36 R' = $-CH_2CH_2CO_2Me$

n = 0-4

a X = -NH-
b X = HN—...—NH

Scheme 9.

Other chain transfer reagents such as 3-mercaptopropionic acid and its methyl ester were also successfully incorporated onto 6-*N*-acrylamidohexanoic acid containing telomers to provide compounds **35b** and **36b** appropriately functionalized for further manipulations.

Table 1. Effect of *t*-BuSH concentrations on the average telomer size for **34a**.

Telogen (equiv.)	Average number of lactose residues in the telomer mixture[a]
0.1	190
0.2	85
1.0	43
5.0	6
10.0	2[b]

[a] Based on ^{1}H NMR spectral data

[b] Telomer distribution was: 29.4% monomer, 15.8% dimer, 8.4% trimer, 5.8% tetramer and 41% >5.

4.B. Dendritic Glycosides

Among the various structural units on which multivalent carbohydrate clusters could be built, dendritic molecules[50] offer distinctive advantages over glycopolymers. Dendrimers are macromolecules of well defined molecular weights and structures that can be scaffolded in desired valencies. Furthermore, due to their shapes, they can effectively mimic multi-antennary glycoproteins present at cell surfaces. Therefore, it was thought that dendritic sialylated clusters should yield improved therapeutic potential over glycopolymers.

Existing starburst or convergent[50] approaches to dendritic macromolecules suffer some technical drawbacks and to circumvent them, solid-phase synthesis of dendritic sialosides has been proposed.[20,51] To provide a general strategy applicable to different oligosaccharides, thiolated glycosides were chosen for efficient conjugation to a dendritic backbone at a late stage of the syntheses. Dendrimers with *N*-chloroacetyl end groups were conjugated to thiolated carbohydrate residues by the following reaction:

n Sugar-SH + [ClCH$_2$C(O)-NH]$_n$-Dendrimer →

[Sugar-S-CH$_2$C(O)-NH]$_n$-Dendrimer

As mentioned above, S-acetylated glycosyl derivatives were readily accessible under PTC conditions using glycosyl halides and thiolacetic acid as nucleophile. Other thiolated derivatives containing extended spacer arms were also required and were made available via Michael additions of thiolacetic acid onto N-acryloylated precursors formed by reduction of glycosyl azide, followed by treatment with acryloyl chloride. This further exemplifies the wide application and usefulness of N-acryloylated derivatives.

4.B.i. Chemoselective removal of thiolacetates with hydrazinium acetate

Deprotection of thiolacetates has been known to proceed with some chemoselectivity under Zemplén conditions (NaOMe/MeOH), however partial or complete de-O-acetylation has been frequently observed. Selective removal of anomeric S-acetates has been shown to occur with dialkylamine,[52] cysteamine in hexamethylphosphoramide (HMPA) in the presence of 1,4-dithioerythritol (DTE)[53] and under low temperature (-40°C) Zemplén conditions.[54]

$$RSAc \xrightarrow[\substack{DMF, N_2 \\ r.t., < 30\ min}]{H_2NNH_2\text{-HOAc}} RSH \xrightarrow[\substack{Et_3N \\ DMF}]{R'X} R\text{-}S\text{-}R'$$

Scheme 10.

Although hydrazinium acetate (H$_2$NNH$_2$-HOAc) has been widely used to accomplish anomeric de-O-acetylations,[55] this reagent has not yet been exploited for de-S-acetylations.[56] Treatment of thiolacetates 2d, 3d, 8e and 41 with hydrazinium acetate exhibited complete chemoselectivity not only at the anomeric positions but also at primary thioesters (Scheme 10). Furthermore, even after prolonged reaction time at ambient temperature, complete chemoselectivity was still observed. Freshly deprotected thiolacetate derivatives were treated with various electrophiles to give thioethers 37, 38, 40 and 42-45 in good to excellent yields. Performing the reactions under strict nitrogen atmosphere allows disulfide formation to be minimized (Scheme 11).

4.B.ii. Solid-phase synthesis of dendrimers using divalent L-lysine

As mentioned above, polymers containing multivalent α-sialosides have shown improved inhibitory properties against many viral hemagglutinins.[5] It was thus postulated that dendritic sialosides should be equally efficient. To this end, multibranched divalent L-lysine repeating units having terminal N-chloro-acetylated groups with 2^n valencies per generation appeared to represent interesting chemical targets. Dendritic L-lysine cores were constructed on p-benzyloxybenzyl alcohol resin (Wang resin) to which was anchored a β-alanyl spacer using well established and high yielding solid phase peptide chemistry with 9-fluorenylmethoxycarbonyl (Fmoc) amino-protecting groups and benzotriazolyl esters as coupling reagents. After Fmoc-deprotection (20% piperidine), freshly prepared N^α,N^ε-di-Fmoc-L-lysine-OBt ester was added to the Wang resin. Deprotection-coupling cycles were repeated until di-, tetra-, octa- and hexadeca-valent dendrimers were obtained.

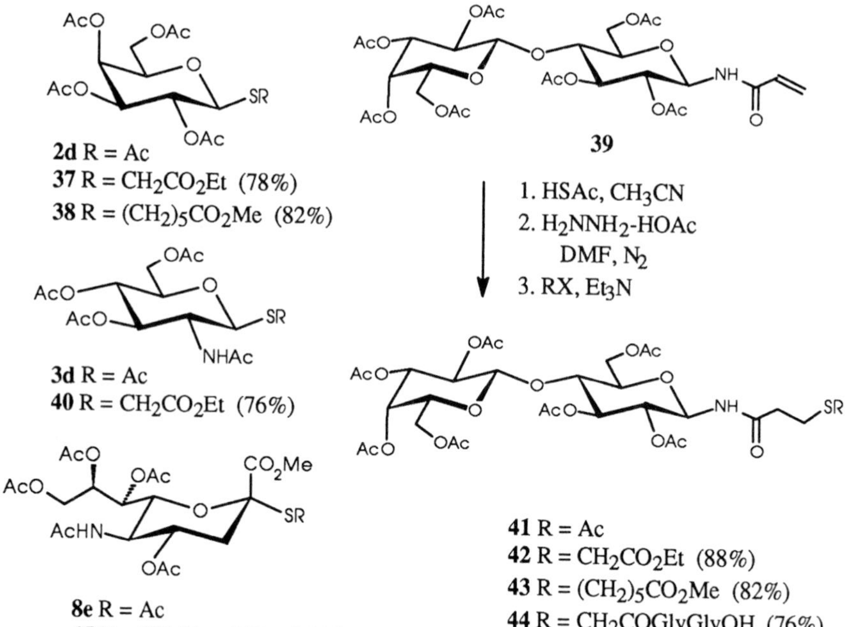

2d R = Ac
37 R = CH$_2$CO$_2$Et (78%)
38 R = (CH$_2$)$_5$CO$_2$Me (82%)

3d R = Ac
40 R = CH$_2$CO$_2$Et (76%)

8e R = Ac
45 R = CH$_2$Ph-p-NO$_2$ (74%)

39

1. HSAc, CH$_3$CN
2. H$_2$NNH$_2$-HOAc
   DMF, N$_2$
3. RX, Et$_3$N

41 R = Ac
42 R = CH$_2$CO$_2$Et (88%)
43 R = (CH$_2$)$_5$CO$_2$Me (82%)
44 R = CH$_2$COGlyGlyOH (76%)

Scheme 11.

Aminated monodendrons resulting from each sequential generation were then treated with pre-formed N-chloroacetylglycylglycine benzotriazolyl ester. Resulting dendrimers were generally obtained in >90% yields and 90-95% purity after hydrolysis from the polymer support (95% aq. TFA, 1.5 h) (Scheme 12). While still anchored to the resin, each dendrimer generation was treated (1% Et$_3$N, DMF, 16 h, 25°C) with an excess of thiosialoside **8e** (R=H) prepared using the procedure described above. Peracetylated sialylated dendrimers were released from the polymer support as described above and were obtained in 66-99% yields. Deprotected dendrimers such as **46** (NaOMe/MeOH, 25°C, 1 h, then 0.05M NaOH, 2 hr) were obtained as freely water-soluble materials in essentially quantitative yields.

The antigenic properties of dendritic thiosialosides were evaluated in model plant lectin studies using wheat germ agglutinin (WGA) in solid phase lectin assays (ELLA).[20] Poly(acrylamide-co-*p*-*N*-acrylamidophenyl-thio sialoside) (**18**) used as capture antigen, was coated on the surface of polystyrene plates and each of the above dendrimers was used to inhibit the binding of horseradish peroxidase labeled WGA. The concentrations of dendrimers required for 50% inhibitions were then plotted as a function of dendrimer valency. The results indicated that the inhibition potential of these dendrimers follows an exponential growth for a linear increase in dendrimer valency.[20,49] Inhibition of hemagglutination of human erythrocytes with influenza A virus (X-31) by hexadecavalent dendrimer showed 50% inhibition at 91 mM, a value similar to that of previous sialopolymers.[51]

The strategy was then applied to *N*-acryloylated lactose (**31**) and *N*-acetyllactosamine derivatives which were previously thiolated (AcSH, CH$_3$CN, then H$_2$NNH$_2$-AcOH, DMF) to provide dendrimers such as **47** and **48**.[57] These dendritic structures constitute useful precursors to sialylated Lewis[X] multivalent conjugates which might be potential inhibitors of the selectin families of receptors involved in inflammation processes.

4.B.iii. Dendrimers with gallic acid as trivalent repeating unit

Lysine-based dendrimers present scaffolding valencies of 2^n at each n'th generation. Dendrimers with 3^n valencies were also desired for structure-activity relationship inhibition experiments. To this end, gallic acid was chosen as the core repeating unit with hydrophilic oligoethylene glycol as spacer arms. Gallic acid methyl ester (**50**) was alkylated with azido tri- and tetra- ethylene-glycol

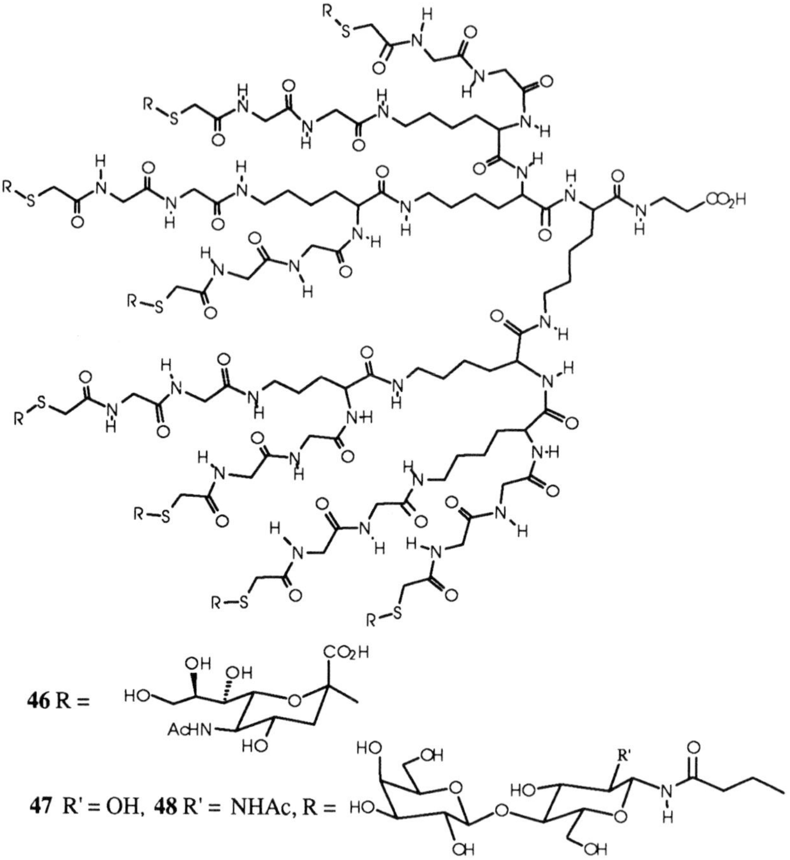

Scheme 12.

tosylates (**49**). Azide reduction followed by *N*-chloroacetylation provided first generation dendritic precursors. Subsequent treatment with thiolated sialoside **8e** (R=H) gave trivalent sialylate dendrimers (**51**) (Scheme 13).

Structures containing lactose and *N*-acetyllactosamine were similarly prepared. Second generation dendrimers (nonavalent) were also synthesized using azido-acid and amino-methyl ester derivatives with carbodiimide (EDC) and 1-hydroxybenzotriazole (HOBt) coupling methodology.[58]

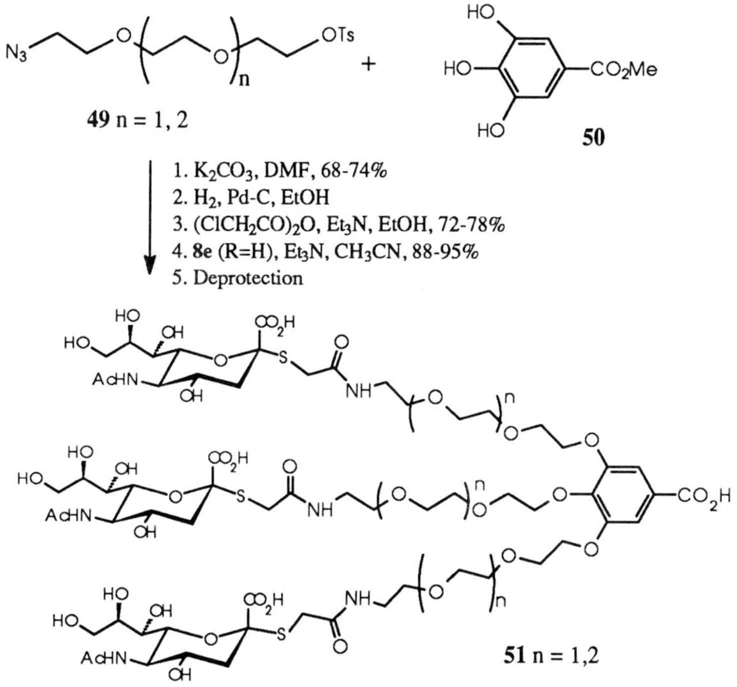

Scheme 13.

5. CONCLUSIONS

Efficient and widely applicable synthetic strategies in glycoconjugate chemistry have given access to new types of glycoforms containing important haptenic carbohydrate ligands. Of particular interest were sialylated glycopolymers and glycodendrimers having good inhibition properties against influenza virus hemagglutinins. The strategies now in hand are being currently applied to more complex oligosaccharides including sialyl Lewis[X] antigens. It is also worth mentioning that most glycopolymers described in this chapter were highly antigenic with both lectins and antibodies. Glycopolymers and glycodendrimers were useful in various immunoassays. Terpolymers of modulated physical and immunochemical properties including probes or effector molecules have also been successfully used in microtiter plate ELLA and ELISA. Of particular interest are the newly described glycodendrimers having carbohydrate densities and orientations that can effectively mimic multi-antennary glycoproteins.

ACKNOWLEDGMENTS

This work was supported by the Natural Sciences and Engineering Research Council of Canada (NSERC). I am truly indebted to my coworkers who have made this contribution possible. I wish to address my sincere thanks to: F.O. Andersson, S. Aravind, S. Cao, C.A. Laferrière, S.J. Meunier, W.K.C. Park, R.A. Pon, Q.Q. Wu, A. Romanowska, F.D. Tropper, S.-N. Wang and D. Zanini. I am also thankful to Dr. R. Schauer and his collaborators Dr. S. Kelm and G. Harms (Kiel, Germany) for the influenza virus experiments. The contribution of Dr. K.L. Matta and his colleagues (Roswell Park Cancer Institute, Buffalo, NY) is also acknowledged.

Space limitation has prevented the listing of all relevant publications related to this chapter. I apologize to all authors whose excellent contributions were omitted.

REFERENCES

1.   Y.C. Lee and R.T. Lee (Eds.), *Neoglycoconjugates: Preparation and Applications*, Academic Press, San Diego, CA (1994).

2.   I.J. Goldstein and R.O. Poretz, In *The Lectins, Properties, Functions, and Applications in Biology and Medicine* (I.E. Liener, N. Sharon, and I.J. Goldstein, Eds.) Academic Press, Orlando, FL, p. 35 (1986).

3.   R.U. Lemieux, *Chem. Soc. Rev.*, **18**, 347 (1989).

4.   W.E. Dick and M. Beurret, In *Contribution to Microbiology and Immunology*, Vol. **10**, *Conjugate Vaccines* (J.M. Cruse and R.E. Lewis, Jr., Eds.), Karger, Basel, p. 48 (1989).

5.   J.C. Paulson, In *The Receptors* (M. Conn, Ed.) Academic Press, New York, Vol. **2**, p. 131 (1985).

6.   J.H. Pazur, *Adv. Carbohydr. Chem. Biochem.*, **39**, 405 (1981).

7.   R. L. Schnaar, *Anal. Biochem.*, **143**, 1 (1984).

8.   J. Kopecek, *J. Controlled Release*, **11**, 279 (1990).

9.   C. P. Stowell and Y.C. Lee, *Adv. Carbohydr. Chem. Biochem.*, **37**, 225 (1980).

10.   B. Dean, H. Oguchi, S. Cai, E. Otsuji, K. Tashiro, S.-i. Hakomori, and T. Toyokuni, *Carbohydr. Res.*, **245**, 175 (1993).

11.   R.T. Lee, P. Lin, and Y.C. Lee, *Biochemistry*, **23**, 4255 (1984).

12.   C. Lancelon-Pin and H. Driguez, *Tetrahedron Lett.*, **33**, 3125 (1992).

13.   R. Roy, F.D. Tropper, and A. Romanowska, *Bioconjugate Chem.*, **3**, 256 (1992).

14.   R. Roy, F.O. Andersson, and M. Letellier, *Tetrahedron Lett.*, **33**, 6053 (1992).

15.   R. Roy and F.D. Tropper, *Can. J. Chem.*, **69**, 817 (1991).

16.   F.D. Tropper, F.O. Andersson, C. Grand-Maître, and R. Roy, *Carbohydr. Res.*, **229**, 149 (1992).

17.   F.D. Tropper, F.O. Andersson, S. Braun, and R. Roy, *Synthesis*, 618 (1992).

18.   R. Roy, F.D. Tropper, A. Romanowska, M. Letellier, L. Cousineau, S.J. Meunier, and J. Boratyñski, *Glycoconjugate J.*, **8**, 75 (1991).

19.   R. Roy, F.D. Tropper, and C. Grand-Maître, *Can. J. Chem.*, **69**, 1462 (1991).

20.   R. Roy, D. Zanini, S.J. Meunier, and A. Romanowska, *J. Chem. Soc., Chem. Commun.*, 1869 (1993).

21.   R.S. Becker, *Springer Semin. Immunopathol.*, **15**, 217 (1993).

22.   (a) R. Roy, C.A. Laferrière, A. Gamian, and H.J. Jennings, *J. Carbohydr. Chem.*, **6**, 161 (1987); (b) C.A. Laferrière, *Ph. D. Dissertation*, University of Ottawa, Ottawa, Canada (1990).

23.   R. Roy and C.A. Laferrière, *Can. J. Chem.*, **68**, 2045 (1990).

24.   A. Gamian, M. Chomik, C.A. Laferrière, and R. Roy, *Can. J. Microbiol.*, **37**, 233 (1991).

25.   R. Roy, C.A. Laferrière, and H. Dettman, *Carbohydr. Chem.*, **186**, C1 (1989).

26.   D.F. Smith, D.A. Zopf, and V. Ginsburg, *Methods Enzymol.*, **50**, 169 (1978).

27.   R. Roy, E. Katzenellenbogen, and H.J. Jennings, *Can. J. Biochem. Cell Biol.*, **62**, 270 (1983).

28. R. Roy, C.A. Laferrière, R.A. Pon, and A. Gamian, *Methods Enzymol.*, **247**, 351 (1994).

29. See for examples: (a) V. Horeksí, P. Smolek, and J. Kocourek, *Biochim. Biophys. Acta*, **538**, 293 (1978); (b) N.K. Kochetkov, *Pure Appl. Chem.*, **56**, 923 (1984); (c) P. Kosma, P. Waldstätten, L. Daoud, G. Schulz, and F.M. Unger, *Carbohydr. Res.*, **194**, 145 (1989); (d) S.-I. Nishimura, K. Matsuoka, T. Furuike, N. Nishi, S. Tokura, K. Nagani, S. Murayama, and K. Kurita, *Macromolecules*, **27**, 157 (1994).

30. K. Kobayashi, A. Kobayashi, S. Tobe, and T. Akaike, In *Neoglyco-conjugates: Preparation and Applications* (Y.C. Lee and R.T. Lee, Eds.) Academic Press, San Diego, CA, p.262 (1994).

31. V. Chytrý and H. Driguez, *Makromol. Chem., Rapid Commun.*, **13**, 499 (1992).

32. N.E. Byramova, L.V. Mochalova, I.M. Belyanchikov, M.N. Matrosovich, and N.V. Bovin, *J. Carbohydr. Chem.*, **10**, 691 (1991).

33. R. Roy and F.D. Tropper, *Glycoconjugate J.*, **5**, 203 (1988).

34. R. Roy and F.D. Tropper, *J. Chem. Soc., Chem. Commun.*, 1058 (1988).

35. R. Roy and C.A. Laferrière, *Carbohydr. Res.*, **177**, C1 (1988).

36. R. Roy, F.O. Andersson, G. Harms, S. Kelm, and R. Schauer, *Angew. Chem. Int. Ed. Engl.*, **31**, 1478 (1992).

37. A. Spaltenstein and G.M. Whitesides, *J. Am. Chem. Soc.*, **113**, 686 (1991).

38. R. Roy, F.D. Tropper, A. Romanowska, R.K. Jain, C.F. Piskorz, and K.L. Matta, *Bioorg Med. Chem. Lett.*, **2**, 911 (1992).

39. R. Roy and C.A. Laferrière, *J. Chem. Soc., Chem. Commun.*, 1709 (1990).

40. R. Roy, F.D. Tropper, T. Morrison, and J. Boratyñski, *J. Chem. Soc., Chem. Commun.*, 536 (1991).

41. R. Roy, R.A. Pon, F.D. Tropper, and F.O. Andersson, *J. Chem. Soc., Chem. Commun.*, 264 (1993).

42. A. Romanowska, S.J. Meunier, F.D. Tropper, C.A. Laferrière, and R. Roy, *Methods Enzymol.*, **242**, 90 (1994).

43. R. Roy, F.D. Tropper, and A. Romanowska, *J. Chem. Soc., Chem. Commun.*, 1611 (1992).

44.   F.D. Tropper, A. Romanowska, and R. Roy, *Methods Enzymol.*, **242**, 255
      (1994).

45.   F.D. Tropper, *Ph.D. Dissertation*, University of Ottawa, Ottawa, Canada
      (1992).

46.   R. Roy, F.D. Tropper, A.J. Williams, and J.-R. Brisson, *Can. J. Chem.*, **71**,
      1995 (1993).

47.   J. Haensler and F. Schuber, *Glycoconjugate J.*, **8**, 116 (1991).

48.   W.K.C. Park, S. Aravind, A. Romanowska, J. Renaud, and R. Roy,
      Methods Enzymol., **242**, 292 (1994).

49.   (a) W.K.C. Park, S. Aravind, S. Brochu, and R. Roy,   *Abstracts of the
      XVIIth Int. Carbohydr. Symp.*, Ottawa, Canada, p.303 (1994); S. Aravind,
      W.K.C. Park, S. Brochu, and R. Roy, *Tetrahedron Lett.*, **35**, 7739 (1994).

50.   See for examples: (a) D.A. Tomalia, A.M. Naylor, and W.A. Goddard, III,
      Angew. Chem. Int. Ed. Engl., **29**, 138 (1990); (b) G.R. Newkome, C.N.
      Moorefield, and G.R. Baker, *Aldrichimica Acta*, **25**, 31 (1992); (c) H.-B.
      Mekelburger, W. Jaworek, and F. Vögtle, *Angew. Chem. Int. Ed. Engl.*,
      31, 1571 (1992); (d) C.J. Hawker and J.M.J. Fréchet, *J. Am. Chem. Soc.*,
      112, 7638 (1990).

51.   R. Roy, D. Zanini, S.J. Meunier, and A. Romanowska, *ACS Symp. Ser.*,
      560, 104 (1994).

52.   S. B. Bennett, M. von Itzstein, and M.J. Kiefel, *Abstracts of the XVIIth Int.
      Carbohydr. Symp.*, Ottawa, Canada, p. 249 (1994).

53.   C. Chou, G. Rasmussen, M. Schulein, B. Henrissac, and H. Driguez, *J.
      Carbohydr. Chem.*, **12**, 743 (1993).

54.   A. Hasegawa, J. Nakamura, and M. Kiso, *J. Carbohydr. Chem.*, **5**, 11
      (1986).

55.   G. Excoffier, D. Gagnaire, and J.-P. Utille, *Carbohydr. Res.*, **39**, 368
      (1975).

56.   W.K.C. Park, S.J. Meunier, D. Zanini, and R. Roy, *Abstracts of the XVIIth
      Int. Carbohydr. Symp.*, Ottawa, Canada, p. 306 (1994).

57.   W.K.C. Park, D. Zanini, S.-N. Wang, and R. Roy, *Abstracts of the XVIIth
      Int. Carbohydr. Symp.*, Ottawa, Canada, p.307 (1994).

58.   S.-N. Wang, S.J. Meunier, Q.Q. Wu, and R. Roy, *Abstracts of the XVIIth Int. Carbohydr. Symp.*, Ottawa, Canada, p.304 (1994).

Chapter 17

SYNTHESIS OF BIOLOGICALLY ACTIVE SULFATED AND PHOSPHORYLATED OLIGOSACCHARIDES

WASIMUL HAQUE AND ROBERT M. IPPOLITO

Department of Biotechnology, Alberta Research Council
Karl Clark Road, Edmonton, Alberta, Canada T5K 2E5

Abstract This review will highlight selected syntheses of biologically active charged carbohydrate structures and includes examples of glycosaminoglycans, *i.e.*, heparin, and chondroitin 4 and 6 sulfates; selectin binding structures such as fucoidan repeating units, $3'$-SO_3-Lewis^x and $3'$-SO_3-Lewis^a; bacterial sulpholipids; phosphorylated high mannose structures, polyphosphorylated galactose compounds and inositol phosphates. Synthetic strategies and the methods utilized will be emphasized. The chemistry of carbohydrate structures having phosphodiester bonds between sugar moieties, *i.e.*, nucleic acid synthesis, sugar nucleotide synthesis and the synthesis of glycosylphosphatidylinositide hydrolysis products will not be reviewed here due to the sheer volume of work in the respective areas. The biological activity of the compounds will only be discussed briefly in the introduction. The reader is encouraged to seek out the appropriate references for more in depth explanations of the biological significance of the targeted compounds.

1.   INTRODUCTION

A structurally diverse group of sulphated and phosphorylated carbohydrates have biological activity. The selected examples examine the synthetic strategies employed and compare the methods utilized.

Glycosaminoglycans are a large family of polydisperse anionic polysaccharides that bind numerous proteins of biological interest including antithrombin III, fibroblast growth factor and perhaps even the selectins. The polydispersity of these materials makes it difficult to determine the precise structure-activity relationships responsible for their biological functions and hence there is a need to synthesize deliberate portions of their structures. In this light the synthesis of heparin and chondroitin fragments will be reviewed. These materials have numerous glycosidic linkages and many charged moieties and represent some of the most challenging synthetic efforts ever attempted.

The discovery of the selectins in the late 1980's has generated intense interest in the therapeutic potential of carbohydrates. The three selectins E (endothelial), P (platelet and endothelial), and L (lymphocyte) share significant structural features and DNA homology, in particular C type lectin domains thought to be responsible for their involvement with cells of the immune system.

E selectin is synthesized and expressed at the surface of endothelial cells some 3 to 5 hours after activation of the endothelium by inflammatory mediators such as the cytokines IL-1β or TNFα. P selectin is contained in the hard granules of platelets and Wiebel-Palade bodies of endothelial cells. It is mobilized to the surface of these cells rather quickly after cytokine activation and likely plays a role both in the inflammatory and hemostatic cascades. L selectin is found constitutively expressed on the surface of neutrophils and lymphocytes and mediates lymphocyte traffic from the circulatory system to the lymph system through high endothelial venules.

Carbohydrates displayed on the surface of neutrophils allow these cells to roll on the pro-inflammatory endothelial surface perhaps through multiple carbohydrate-selectin interactions along their respective surfaces and initiating a cascade of events that leads to flattening and extravasation of these cells into sites of inflammation. Potential ligands for these selectins include sialyl Lewis[x], [1] sialyl Lewis[a], [2] 3'-sulfo Lewis[x], [3] 3'-sulfo Lewis[a], [4] sulfatide, fucoidan and heparin fragments to some extent. The natural ligands for P and L selectin[5,6] have been isolated and cloned, and appear to be O-linked mucin-like carbohydrate structures bearing multiple copies of the as yet unidentified natural ligands. The natural ligand for E selectin may be *N*-linked tetra-antennary sialyl Lewis[x] (and sialyl di-Lewis[x]) containing structures.[7]

Different approaches to the synthesis of 3'-SO$_3$-Lewis[x], 3'-SO$_3$-Lewis[a] as reducing sugars, with other aglycons and as extended structures will be examined in more detail.

Hormones such as lutropin require sulfated carbohydrate structures for expression of their biological function.[8] The synthesis of these structures will be reviewed here. Cyanobacterial sulfolipids have anti-HIV properties. Synthesis of this sulfonic acid containing glycolipid will also be reviewed.

D-Myo-inositol 1,4,5-triphosphate and other hydrolysis products[9] from glycosylphosphatidylinositides that serve as intracellular second messengers responsible for Ca[++] flux have seen an explosion of synthetic efforts directed at them and their analogs in an effort to better elucidate the structure-activity relationships responsible for their mechanisms of action. One example of the synthesis of D-myo-inositol 1,4,5-triphosphate[9a] has been reviewed to demonstrate more modern methods of phosphorylation.

Lysosomal degradation of glycoproteins utilizes the mannose-6-phosphate receptor. The synthesis of several phosphorylated high mannose structures will also be reviewed.

The review of sugar phosphorylation methods will be brief and directed primarily at more recent methods of phosphorylation.

2.   GLYCOSAMINOGLYCANS AND RELATED STRUCURES

Heparin, heparan sulfate, chondroitin sulfate, dermatan sulfate, hyaluronic acid and keratin sulfate constitute the most abundant structures of glycosylaminoglycans. Chondroitin and dermatan sulfates are essentially polydisperse disaccharides with alternating 1→3 and 1→4 glycosidic linkages. Heparan sulfate and heparin also have a polydisperse disaccharide backbone but the glycosidic linkages are all 1→4.

2.A.   Heparin Fragments

Heparan sulfate consists of repeating disaccharide units of glucosamine and uronic acid (either D-glucuronic acid or its C(5) epimer L-iduronic acid). The glucosamine is partly *N*-sulfated and partly *N*-acetylated.

The glycosaminoglycans have been the subject of intense synthetic efforts over the last ten years. In particular the elucidation of the binding site for the anticoagulant serine protease inhibitor anthrombin III to a common pentasaccharide binding sequence (sugars DEFGH, Scheme 1) in various anticoagulant heparin fractions has spawned a monstrous synthetic effort[11,12] aimed at defining a more

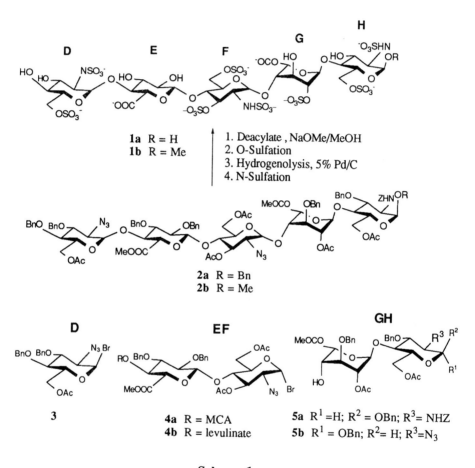

1a R = H
1b R = Me

1. Deacylate , NaOMe/MeOH
2. O-Sulfation
3. Hydrogenolysis, 5% Pd/C
4. N-Sulfation

2a R = Bn
2b R = Me

3

4a R = MCA
4b R = levulinate

5a R^1 =H; R^2 = OBn; R^3= NHZ
5b R^1 = OBn; R^2= H; R^3=N$_3$

Scheme 1.

complete structure activity profile for each of the sugars in this sequence and their charged moieties and in particular the conformational preferences induced in each of the rings by the presence of the sulfate groups. The work should serve as the paradigm for future synthetic efforts aimed at defining the individual contributions of the charged moieties of other sulfated and phosphorylated carbohydrates with

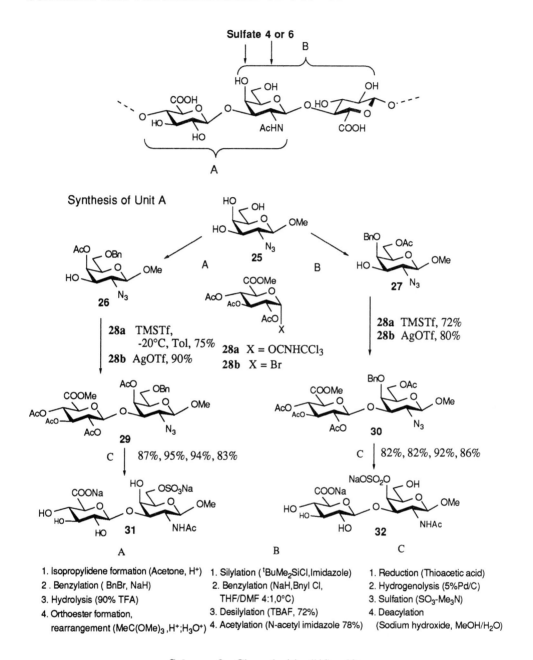

Scheme 3. Chrondroitin 4'/6'-sulfate.

Synthesis of Unit B

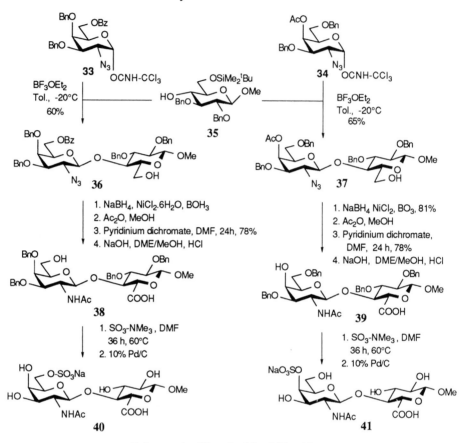

Scheme 4. Chondroitin 4'/6' sulfate.

similar to those synthesized by Jacquinet[16] except that galactose replaces N-acetylgalactosamine.

Benzyl lactose was benzylidenated with benzaldehyde/ZnCl$_2$ (66%, Scheme 5) and the 6-OH of the glucose was blocked with *t*-butyl dimethylsilylchloride/imidazole to **42b**. After perbenzoylation, the silyl blocking group is removed with tetrabutylammonium fluoride and the 6-OH oxidized under neutral conditions with pyridinium dichromate (46%) or Corey-Samuelsson conditions and esterified with diazomethane to **43**. After TFA (90% aqueous) hydrolysis of the benzylidene group and selective benzoylation of 6-OH to **44**, the 4-OH is sulfated with

4 or 6 SO₃-β-D-Gal-(1→4)-β-D-GlcA Fragments

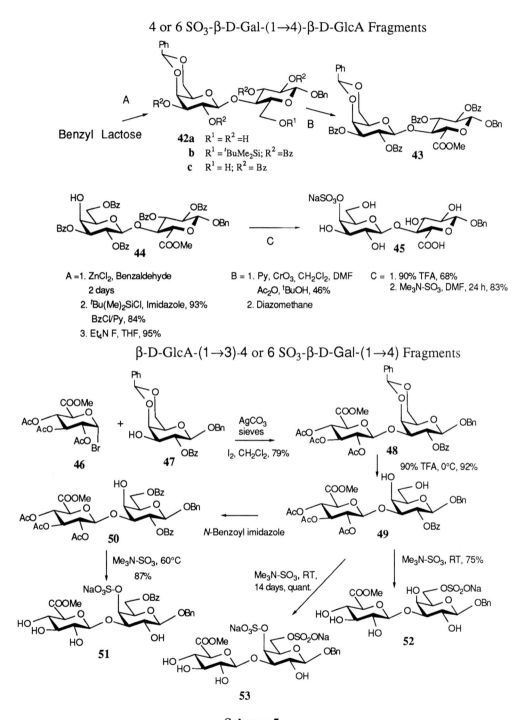

A = 1. ZnCl₂, Benzaldehyde
     2 days
   2. ᵗBu(Me)₂SiCl, Imidazole, 93%
      BzCl/Py, 84%
   3. Et₄N F, THF, 95%

B = 1. Py, CrO₃, CH₂Cl₂, DMF
      Ac₂O, ᵗBuOH, 46%
   2. Diazomethane

C = 1. 90% TFA, 68%
   2. Me₃N-SO₃, DMF, 24 h, 83%

β-D-GlcA-(1→3)-4 or 6 SO₃-β-D-Gal-(1→4) Fragments

Scheme 5.

SO$_3$-NMe$_3$ (83%) at room temperature for 24 h and debenzoylated using Zemplén conditions to give **45**.

Zsiska and Meyer also synthesized β-D-GlcA-(1→3)-β-D-Gal fragments (Scheme 5).[17b] Glycosylation of acceptor **47** with methylbromoglucuronate **46** using silver carbonate/ I$_2$ /molecular sieves gives disaccharide **48** (79%), then the benzylidene is hydrolyzed to **49** and the diol benzoylated on 6-OH with N-benzoylimidazole to give **50**, followed by sulfation and deblocking to **51**. The diol **49** is sulfated selectively on 6-OH to **52** or persulfated to disulfate **53**.

Goto and Ogawa[18] have synthesized the hexasaccharide from the linkage region of Swarm rat chrondrosarcoma proteoglycan using a 3 + 3 block approach. The synthesis is detailed in Scheme 6. The β-D-GlcA-(1→3)-β-D-GalNAc-(1→4)-β-D-GlcA donor is synthesized first. A suitably blocked glucose acceptor **55** and selectively blocked azidogalactose imidate **56** are reacted using TMSOTf/molecular sieves at -78°C to give 62% of desired β disaccharide **57**. Conversion of **57** into acceptor **58**, requires ceric ammonium nitrate (CAN) removal of the p-methoxyphenol blocking group at 6-OH, then two step oxidation with oxalyl chloride/DMSO/(i-Pr)$_2$EtN at -78°C and NaClO$_2$/2-methyl-2-butene in phosphate buffer. After diazomethane treatment and Zemplén deacylation **58b** is obtained in 73% yield. The glucouronate imidate donor **59** is then reacted with **58b** using BF$_3$OEt$_2$ at -20°C to give **60** in 76% yield. To complete synthesis of donor **62**, compound **60** is hydrolyzed to remove the benzylidene, the 6-OH blocked by acetylation and the 4-OH blocked by levulination. The azido group is reduced with Ph$_3$P then acetylated with Ac$_2$O-DMAP in pyridine. The allyl group is isomerized with the iridium catalyst [(Ph$_2$MeP)$_2$Ir(COD)]PF$_6$, H$_2$ in THF and hydrolyzed to the reducing sugar and acetylated. After hydrogenolysis of the benzyl groups and p-methoxybenzoylation the anomeric acetate is hydrolyzed with piperidine acetate in THF and converted to imidate **62** with CCl$_3$CN and 1,8-diaza bicyclo[5.4.0]undec-7-ene (DBU). Overall yield for the ten steps is 30%.

Similarly the other trisaccharide fragment β-D-Gal-(1→3)-β-D-Gal-(1→3)-β-D-Xyl (unit B) is prepared as an acceptor and reacted with imidate **62** (BF$_3$OEt$_2$, Toluene/CH$_2$Cl$_2$, -20°C) to give the blocked hexasaccharide **63** (47%).

Chondroitin Sulfate Linkage Region

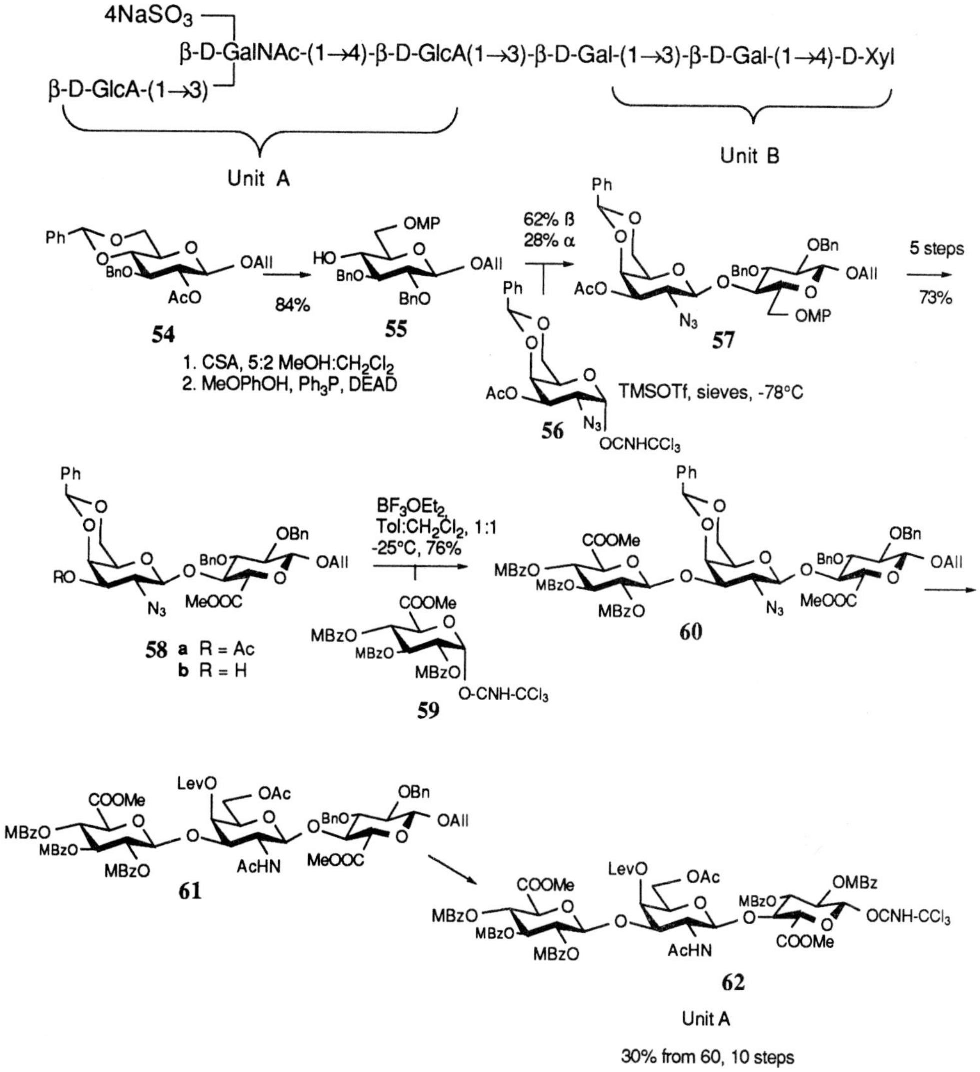

β-D-GalNAc-(1→4)-β-D-GlcA(1→3)-β-D-Gal-(1→3)-β-D-Gal-(1→4)-D-Xyl

4NaSO₃

β-D-GlcA-(1→3)

Unit A   Unit B

Scheme 6.

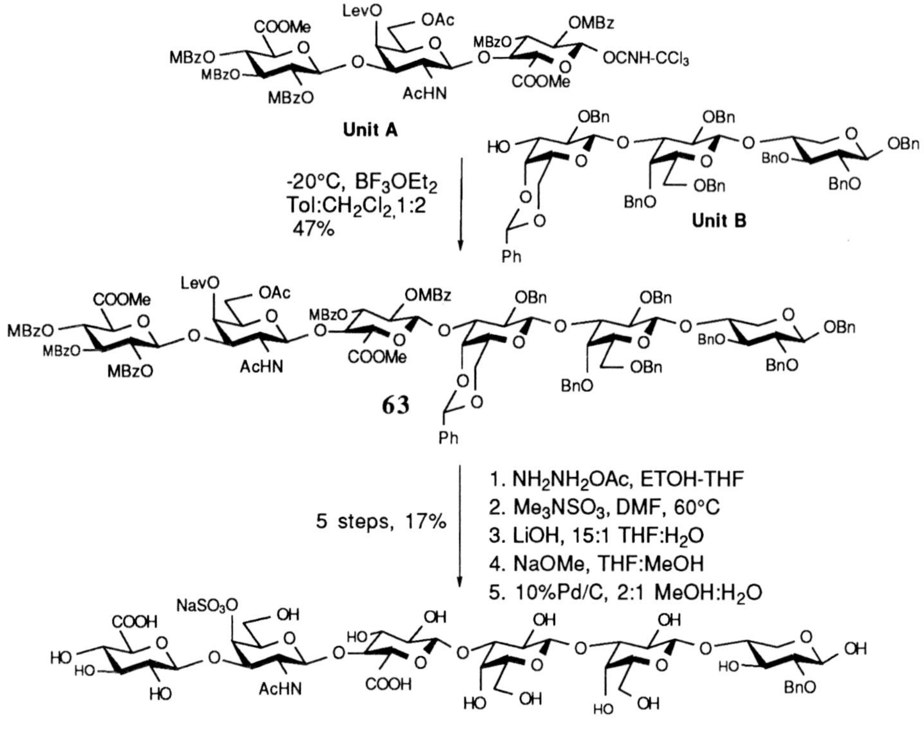

64

CSA = Camphorsulphonic acid
Lev = Levulinate
MBz = p-Methoxybenzoyl
DEAD = Diethylazodicarboxylate

Scheme 6 (continued).

Conversion of **63** to the desired sulfated hexasaccharide **64** requires removal of the levulinate with hydrazine acetate, sulfation with SO_3-NMe_3 in DMF at 60°C, hydrolysis of the uronic methyl esters with LiOH in 15:1 THF:H_2O, Zemplén removal of the remaining esters and hydrogenolysis of the benzyl groups.

Jacquinet *et al.*[19] has published the synthesis of a similar chondroitin 4-sulfate fragment. Sinay and co-workers[20] have published the synthesis of dermatan sulfate fragments and Ogawa *et al.*[21] the synthesis of keratin sulfate fragments.

3. SULFATED OLIGOSACCHARIDES FROM HORMONES AND BACTERIA

Baenziger *et al.*[8] reported that Asn-linked oligosaccharides found on the bovine pituitary hormone lutropin (LH) terminates in 4-SO_3-β-D-GalNAc-(1→4)-β-D-GlcNAc-(1→2)-α-D-Man and that the structures are essential for aspects of its biological activity.

Hindsgaul *et al.*[22] have synthesized the β-D-GalNAc-(1→4)-β-D-GlcNAc-(1→2)-α-D-Man trisaccharide sulfated at both the 3 and 4 positions of the terminal galactose. This process (Scheme 7) involves the reaction of the selectively protected *N*-phthalamidoglucosamine chloride **67** with the suitably protected 8-methoxycarbonyloctyl glycoside of **68** to give after silver trifate/collidine glycosylation and deacetylation disaccharide **69**. A suitably protected *N*-phthalamido galactose derivative was prepared from *N*-phthalamido blocked glucosamine **66** by triflate formation at 4-OH, inversion with tetrabutylammonium acetate in DMF, deallylation to reducing sugar and treatment with oxalyl chloride/DMF to chloride **70**. Silver trifate catalyzed glycosylation of **69** with **70** gave the suitably blocked trisaccharide **71**. The hydrazine hydrate sensitive aglycon requires that **71** be treated with hydrazine acetate in refluxing methanol followed by acetylation of the amino group, deacylation of 4'-OAc, sulfation with SO_3-pyridine in DMF and hydrogenolysis to produce **65a**.

The 3'-SO_3 structure **65b** requires reaction of *N*-phthalamidogalactosamine derivative **72** with **69** (AgOTf, collidine), removal of phthalamido protecting groups and N-acetylation, de-O-acetylation and benzylidene formation to **73** followed by sulfation on 3'-OH with SO_3-pyridine in DMF and hydrogenolysis over 5% Pd on C .

Cyanobacterial sulfolipids appear to have good anti-HIV activity. Conant *et al.*[23] synthesized the dipalmitate glyceride analogs while a second synthesis by Halcomb and Danishefsky[24] (Scheme 8) has produced the natural glycerolipid. Schmidt *et al.*[25] have published a synthesis of the 1-phosphate of 6-deoxy-6-sulfo-D-glucose, the precursor of the nucleotide likely responsible for the biosynthesis of the natural glycolipid.

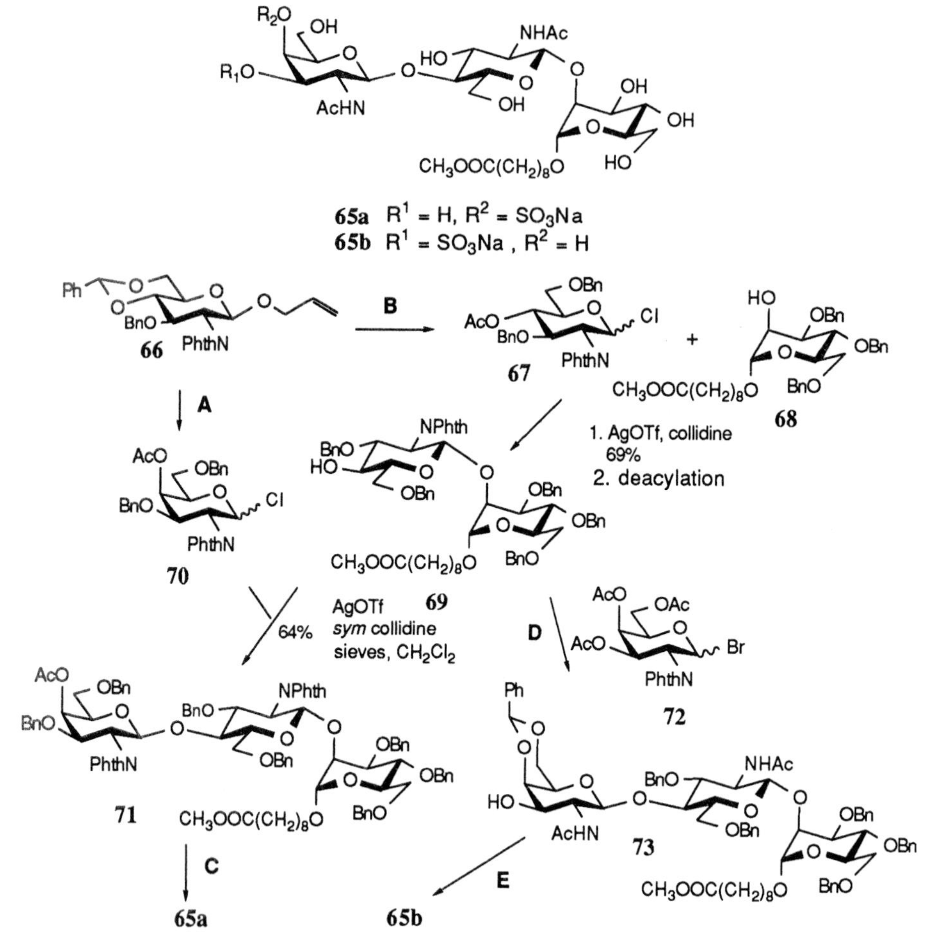

65a R¹ = H, R² = SO₃Na
65b R¹ = SO₃Na , R² = H

A

1. NaCNBH₃, HCl 89%
2. Triflic anhydride, CH₂Cl₂/Py
   Et₄NOAc, DMF, 83%
3. Wilkinson's cat, Hg⁺ hyd, 86%
4. Oxalyl chloride, DMF

B

1. NaCNBH₃, HCl, 89%
2. Ac₂O-Py
3. Ph₃PRhCl, Hg⁺ hyd, 86%
4. Oxalyl chloride, DMF

E

1. SO₃-Py/DMF, 93%
2. 5%Pd/C, Na⁺ resin, 84%

C

1. NH₂NH₂OAc, MeOH, Rflx, 69%
2. NaOMe
3. SO₃-Py/DMF, 93%
4. 5%Pd/C, Na⁺ resin, 84%

D

1. AgOTf, 85%
2. Hydrazine acetate, MeOH, Rflx, Ac₂O, 79%
3. NaOMe
4. Benylidenation, 80%

Scheme 7.

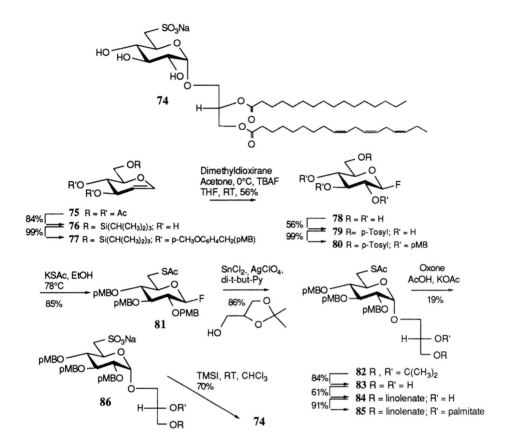

Scheme 8.   Cyanobacterial sulfolipid.

The synthesis by Halcomb and Danishefsky[24] (Scheme 8) has several interesting features. The glucal derivative **77** protected at 6-OH with tri-isopropylsilyl (TIPS) and *p*-methoxybenzyl groups at 3-OH and 4-OH (respectively) is reacted with 2,2-dimethyldioxirane to give an epoxide which is then opened with tetrabutylammonium fluoride (TBAF) to give **78** in 56% yield. After tosylation of 6-OH and *p*-methoxybenzylation of 2-OH to give **80**, the tosylate is displaced with KSAc to give **81** in 85% yield.

Mukaiyama glycosylation using $SnCl_2$ and $AgClO_4$/ether (86%) and (S)-isopropylidene glycerol proceeded in the presence of the thioacetate to give a single diastereomer. The glycerol acetonide **82** is then hydrolyzed and the primary and secondary hydroxyls acylated sequentially with linolenic acid and palmitic acids respectively. The thioacetate function was oxidized to the sulfonic acid in 19% yield using oxone, the triene remaining intact during this oxidation. The *p*-methoxybenzyl groups were removed with iodotrimethylsilane to give the desired **74**. In contrast both Conant's original synthesis[23] and Schmidt's synthesis[25] of the 1-phosphate compound utilized oxidation of the 6 disulfide structure with *m*-chloroperbenzoic acid to give the sulfonic acid in very good yields.

4.   SELECTIN LIGANDS

Natural ligands for the selectins include sialyl Lewis[x], [1] sialyl Lewis[a], [2] and a variety of sulphated structures including sulfatide,[26] fucoidan,[27c] 3'-sulfo Lewis[x], [3] 3'-sulfo Lewis[a], [4] and others.[28] Sulfatide is a simple 3-SO_3 galactose glycolipid. The glycerolipid form of this material has been synthesized by Flowers.[29]

4.A.   Fucoidan Structures

Fucoidan is a polysulfated fucose containing polymer with the ability to inhibit cell adhesion events including sperm egg binding[27a] in diverse species, infection of human cell lines with enveloped viruses,[27b] and cell-cell binding mediated by P or L selectin.[27c] The structure of fucoidan has recently been revised.[30] Matta *et al.*[31] have reported the synthesis of the sulfated disaccharides α-L-Fuc-(1→2)-3 and 4-SO_3-α-L-Fuc (termed unit A in Scheme 9) and 3 and 4 SO_3-α-L-Fuc-(1→2)-α-L-Fuc based on the Percival model of fucoidan. This second fragment may not be part of fucoidan based on the revised model. No data as to the activity of these fucoidan fragments in inhibiting selectin mediated cell adhesion events has appeared.

In Matta's[31] synthesis of unit A (Scheme 9) the methyl 3,4-isopropylidene fucoside **88** is allylated on 2-OH and the isopropylidene then hydrolyzed. Orthoacetate formation and rearrangement to the 4-OAc, benzylation on 3-OH and deallylation gives donor **91**. Fucosylation of **91** with perbenzylated thiofucoside **87** using $CuBr_2$/DMF/Et_4NBr gives the desired disaccharide **92**. Deacetylation and sulfation with SO_3-pyridine followed by hydrogenolysis gives the desired α-L-Fuc-(1→2)-4-SO_3-α-L-Fuc disaccharide **93**. Similarly the 3-SO_3 disaccharide is prepared by reaction of **87** and **88** followed by a similar sequence of reactions to give the desired product **90**.

Revised Structure of Fucoidan

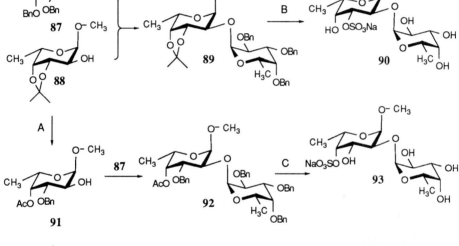

Unit A

Synthesis of Unit A

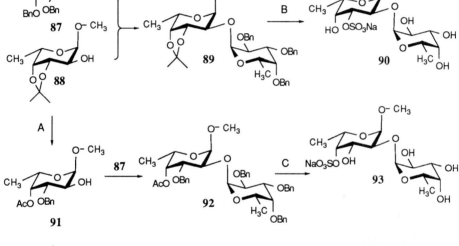

| A | 1. 2 O Allylation (allyl bromide, NaH)
2. Deisopropylidenation
3. Orthoacetate, rearrangement
4. Benzylation
5. Deallylation | B | 1. Deisopropylidenation
2. Orthoacetate formation
3. Rearrangement of orthoacetate
4. Sulfation
5. Deacylation
6. Hydrogenolysis | C | 1. Deacylation
2. Sulfation
3. Hydrogenation |

Scheme 9.

4.B.   3'-Sulfo Lewis^x and 3'-Sulfo Lewis^a

Matta *et al.*[32] have reported a synthesis of the 3'-sulfo Lewis[x] **94a** as its reducing sugar (Scheme 10) . An appropriately blocked *N*-acetylglucosamine **96** is reacted with acetobromogalactose **95** to give disaccharide **98**, the 3-MeOBn group was removed by 2,3-dichloro-5,6-dicyano-1,4-benzoquinone (DDQ) in aqueous

3'-Sulfo Lewis^x

94a R = H
  b R = 8-Methoxycarbonyloctyl
  c R = 3-Galactose-OH

CH_2Cl_2 to give **99**, which was subsquently fucosylated with protected fucose thioglycoside **97** using $CuBr_2$/DMF/Et_4Br to give **100**. Zemplén deacetylation followed by isopropylidenation and acetylation gave **101**. After hydrolysis of the isopropylidene, orthoacetate formation and rearrangement to the 4'-OAc **102**, the compound is sulfated with SO_3-pyridine and deblocked by Zemplén deacylation and hydrogenolysis to give **94a**.

Ippolito et al.[33] have reported the synthesis of sulfated Lewis^x and Lewis^a oligosaccharides as immunomodulating and tolerogenic compounds in the mouse DTH model. The synthesis of 3'-sulfo Lewis^x as its 8-methoxycarbonyloctyl glycoside **94b** is described (Scheme 11). The appropriate galactose thioglycoside **103** is converted into its halide and reacted with the N-acetylglucosamine **104** derivative according to literature conditions.[34] The removal of p-methoxybenzyl by treatment with DDQ in aqueous CH_2Cl_2 gives disaccharide **106** which is then reacted with thiofucoside **105** to give the blocked Lewis^x derivative **107**. After debenzoylation with sodium methoxide in methanol the partially blocked Lewis^x diol is then sulfated selectively on 3-OH with SO_3-pyridine in DMF at -30°C for 16 h and hydrogenolyzed to give **94b**. Brandley et al.[35] and Bevilacqua[36] have confirmed P selectin binding activity for this compound.

Nicoloau et al.[37](Schemes 12, 13) have synthesized both trisaccharides 3'-sulfo-Lewis^x **94a** and 3'-sulfo-Lewis^a **137** as well as the same structures bearing the internal galactose moieties, tetrasaccharides **94c** and **136**. His approach to these difficult structures utilizes the glycosyl fluoride donors **110**, **111** and **114** and acceptors **109**, **112**, and **113**.

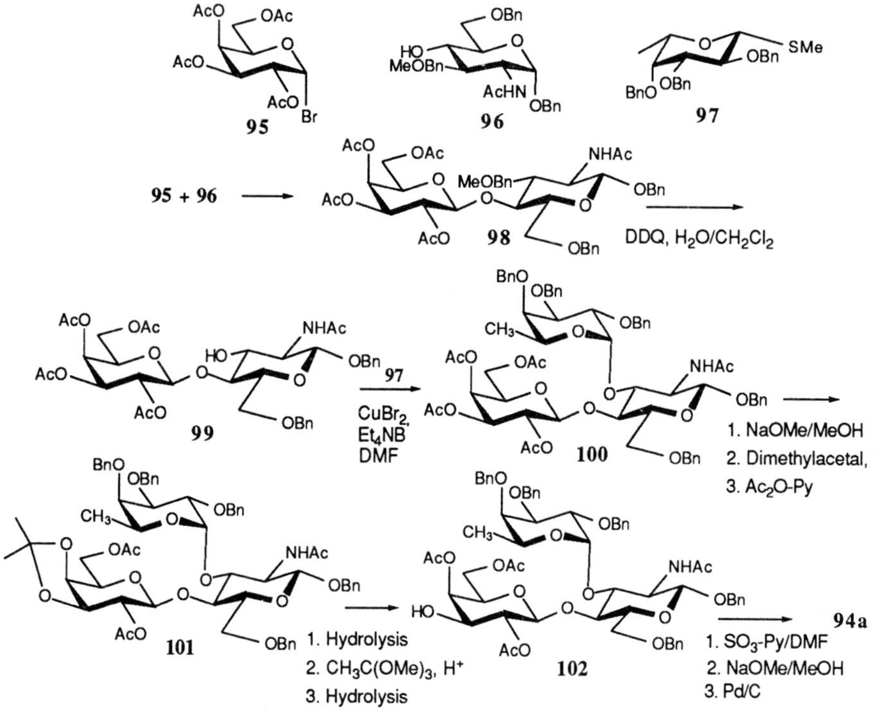

Scheme 10.

Most of the glycosylations employ Mukaiyama conditions,[38] ($AgClO_4$-$SnCl_2$) for activation of the respective fluorides and give excellent yields. The Lewis[x] tetrasaccharide **94c** (Scheme 12) is built from the internal galactose to the terminal galactose **119** followed by deallylation to **120**, fucosylation to **121** and dechloroacetyation using thiourea/2,6 lutidine for 5 h at 65°C in EtOH to **122** and sulfation on 3-OH using SO_3-NMe_3 in pyridine for 24 h at 25°C. Deacylation and hydrogenolysis provides the desired **94c** in excellent yield.

In the synthesis of the 3'-SO_3-Lewis[a] tetrasaccharide (Scheme 13) the molecule is assembled as the Lewis[a] trisaccharide donor and the internal galactose added last. In the initial glycosylation in this sequence donor **110** and acceptor **112** were reacted under Mukaiyama-Suzuki glycosylation conditions,[38] i.e., bis-(cyclopentadienyl)hafnium dichloride-silver triflate ($CpHfCl_2$-AgOTf),

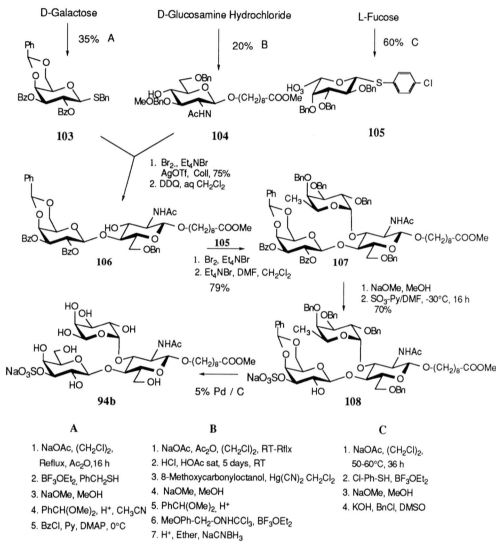

Scheme 11.

di-t-butyl-4-methylpyridine to give **124** in 63% yield. The acceptor **125** was produced in 76% yield by treatment of **124** with ethereal HCl/NaCNBH₃ (10 equiv.) in THF. Reaction of **125** with donor **111** under Mukaiyama conditions in ether/THF (5:1) gave **126** in 95% yield. Thioglycoside **126** was treated with 3

equiv. of DAST, and 1.25 equiv. of NBS at -78°C (CH_2Cl_2 , 2 h) to give β anomer
fluoride **127** and this was reacted under Mukaiyama-Suzuki conditions with donor

109 to give **128** (58%) or with benzyl alcohol to give **132** (95%). The respective
compounds were dechloroacetylated and then sulfated as described above to give
130 (50%) and **134** (76%) respectively and the phthalamido groups removed by
hydrazine hydrate/EtOH treatment at 100°C for 3 h and then converted to NHAc
with Ac_2O-Et_3N-MeOH in 10 min. Finally the compounds were hydrogenolyzed to
give **136** (95%) and **137** (82%).

5.   PHOSPHORYLATED CARBOHYDRATE STRUCTURES

Phosphorylated carbohydrates show a greater diversity of structure than their
sulfated counterparts. The chemistry associated with forming phosphodiester bonds
between carbohydrate structures, *i.e.*, in nucleic acids, in the sugar nucleotides that
serve as co-factors for the post-translational glycosylation[39] of proteins, in the
teichoic   acid   outer   cell   wall   polymers   of   bacteria[40]   and   in
phosphatidylinositides[41,42] is voluminous and not able to be adequately reviewed
here. Theim and Franzkowiak[43] have published a good review of the chemistry of
carbohydrate phosphodiester linkages.

5.A. Phosphorylated Mannose Structures

Srivastava and Hindsgaul[10] have synthesized a large variety of phosphorylated
mannose structures (Scheme 14) bearing terminal mannose 6-phosphate. These
structures should be typical of those found on glycoproteins targeted for destruction
by lysosomal degradation through the mannose 6-phosphate receptor. Compounds

138 to **144** are utilized along with **146** to produce disaccharides **148** to **150**. These disaccharides are then each reacted with donor **147b** bearing the 6-blocked phosphate moiety. This interesting approach gave acceptable glycosylation yields and thus avoided functional group manipulations at the trisaccharide stage. These compounds were easily deblocked to the desired **151** to **153**.

In another approach, Matta *et al.*[44] (Scheme 15) synthesized the 6 and 6' mono phosphorylated α-D-Man-(1→2)-α-D-Man disaccharides **156**, **159**, and

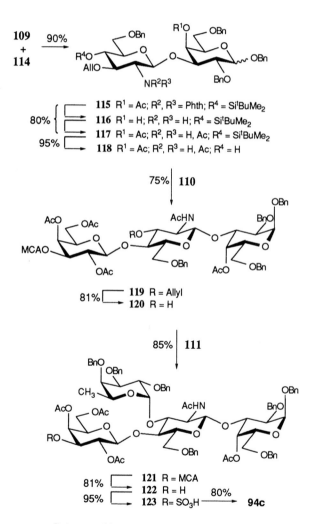

Scheme 12. Extended 3'-sulfoLewisx structure.

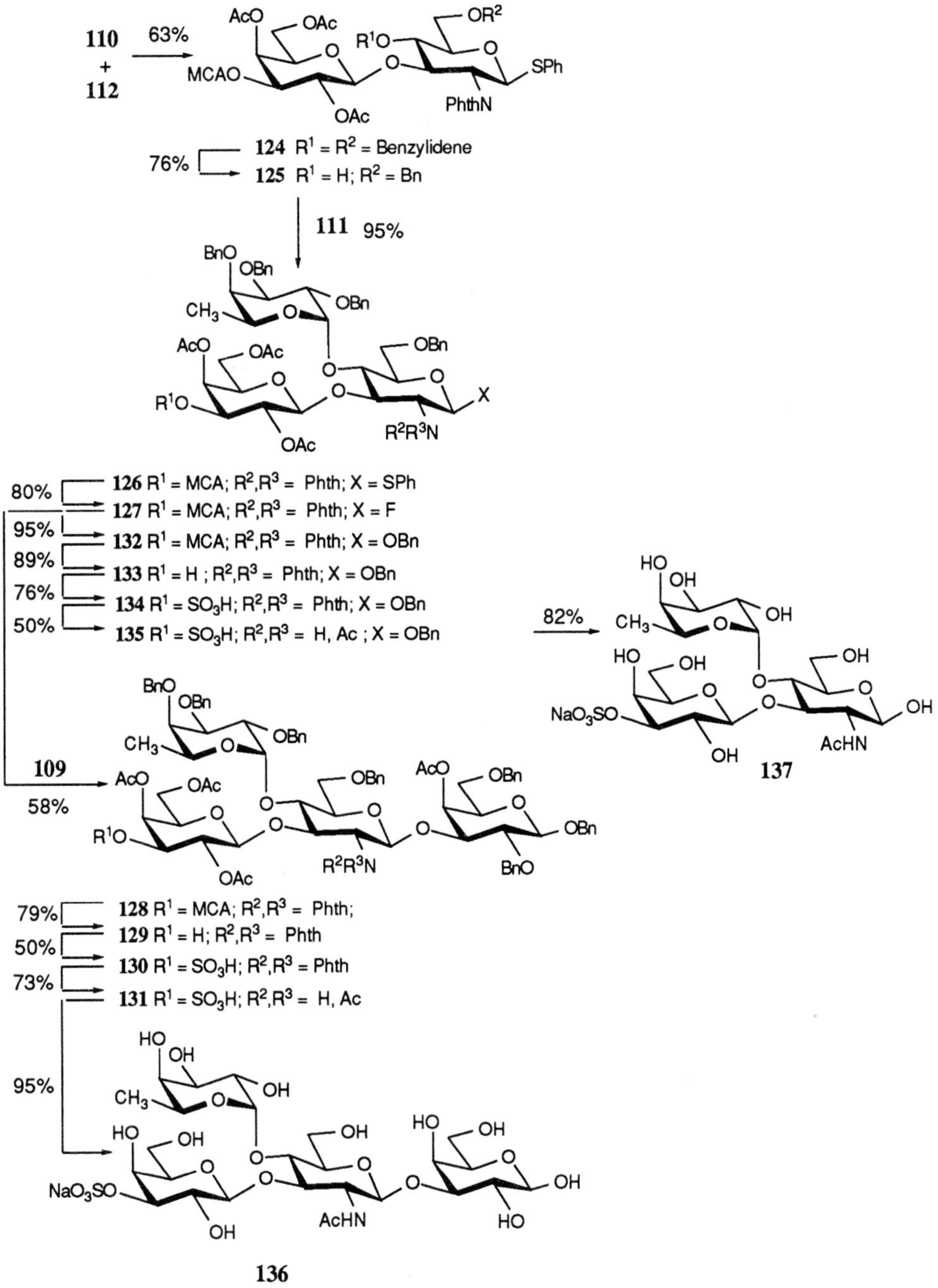

Scheme 13. Nicolaou synthesis of 3'-sulfoLewis[a] structures.

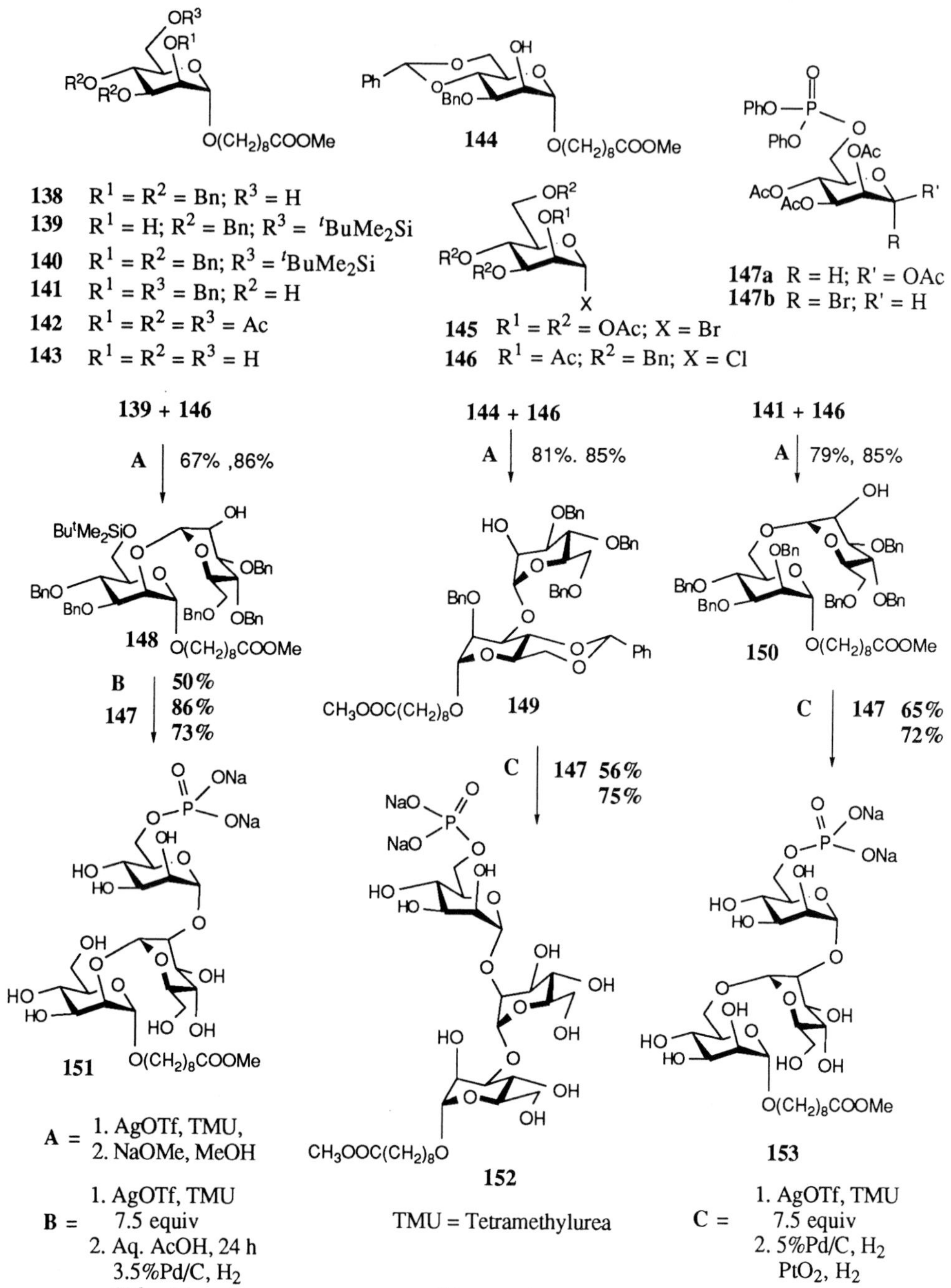

138 $R^1 = R^2 = Bn; R^3 = H$
139 $R^1 = H; R^2 = Bn; R^3 = {}^tBuMe_2Si$
140 $R^1 = R^2 = Bn; R^3 = {}^tBuMe_2Si$
141 $R^1 = R^3 = Bn; R^2 = H$
142 $R^1 = R^2 = R^3 = Ac$
143 $R^1 = R^2 = R^3 = H$

144

145 $R^1 = R^2 = OAc; X = Br$
146 $R^1 = Ac; R^2 = Bn; X = Cl$

147a $R = H; R' = OAc$
147b $R = Br; R' = H$

139 + 146
A | 67% ,86%

148

B | 50%
147 | 86%
| 73%

151

144 + 146
A | 81%. 85%

149

C | 147 56%
| 75%

152

TMU = Tetramethylurea

141 + 146
A | 79%, 85%

150

C | 147 65%
| 72%

153

A = 1. AgOTf, TMU,
2. NaOMe, MeOH

B = 1. AgOTf, TMU
7.5 equiv
2. Aq. AcOH, 24 h
3.5%Pd/C, H$_2$
PtO$_2$, H$_2$

C = 1. AgOTf, TMU
7.5 equiv
2. 5%Pd/C, H$_2$
PtO$_2$, H$_2$

Scheme 14.

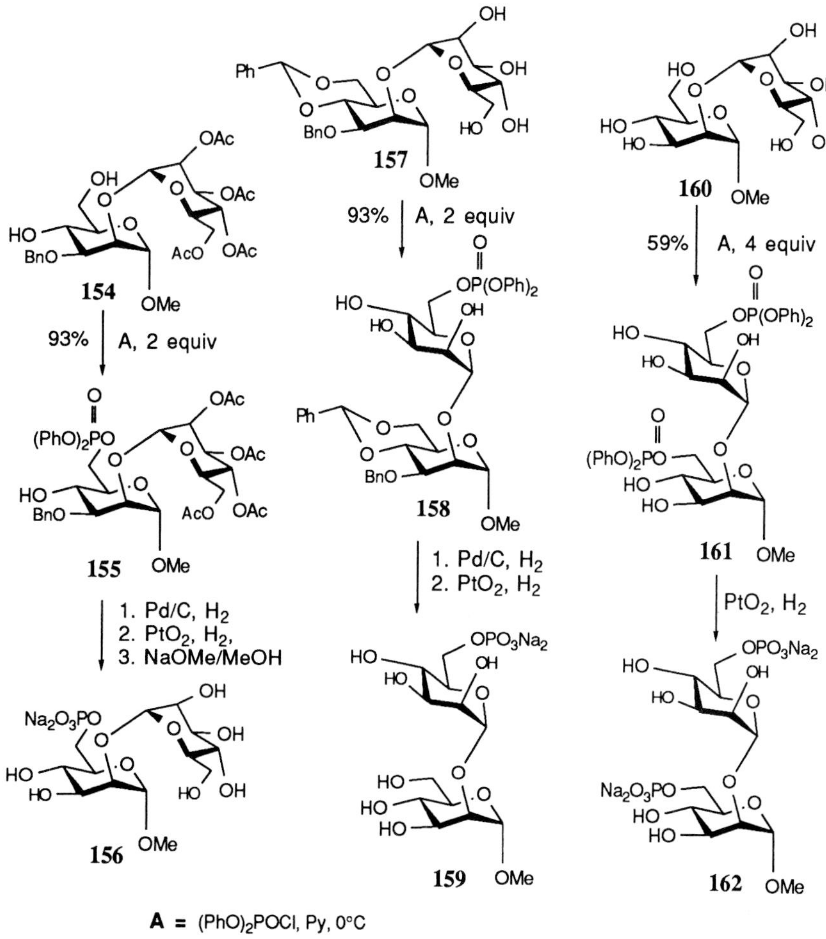

Scheme 15.

as well as 6,6'diphosphorylated α-D-Man-(1→2)-α-D-Man disaccharide **161** utilizing selective phosphorylation on the primary hydroxyl positions of **154**, **157** and **160**, respectively, using diphenylphospho chloridate (2-4 equiv, 0°C) in pyridine. Interestingly the unprotected disaccharide **160** is phosphorylated selectively on the 6 and 6' positions in reasonable yield. Diphenylphosphochloridate is a highly selective phosphorylating reagent but requires hydrogenation over Adam's catalyst (PtO₂) to remove the phenyl groups.

5.B.   Inositol   and   Galactose   Phosphates

D-Myo-inositiol-1,4,5-triphosphate is believed to be an intracellular second messenger through its effects on cellular calcium flux. The synthesis of it and its analogs have been an area of frantic synthetic activity. Yu and Fraser-Reid[9a] have synthesized D-myo-inositiol-1,4,5-triphosphate and several analogs **170, 174,** and **176** (Scheme 16) utilizing *N,N*-diisopropyl dibenzylphosphoramidite/1H-tetrazole

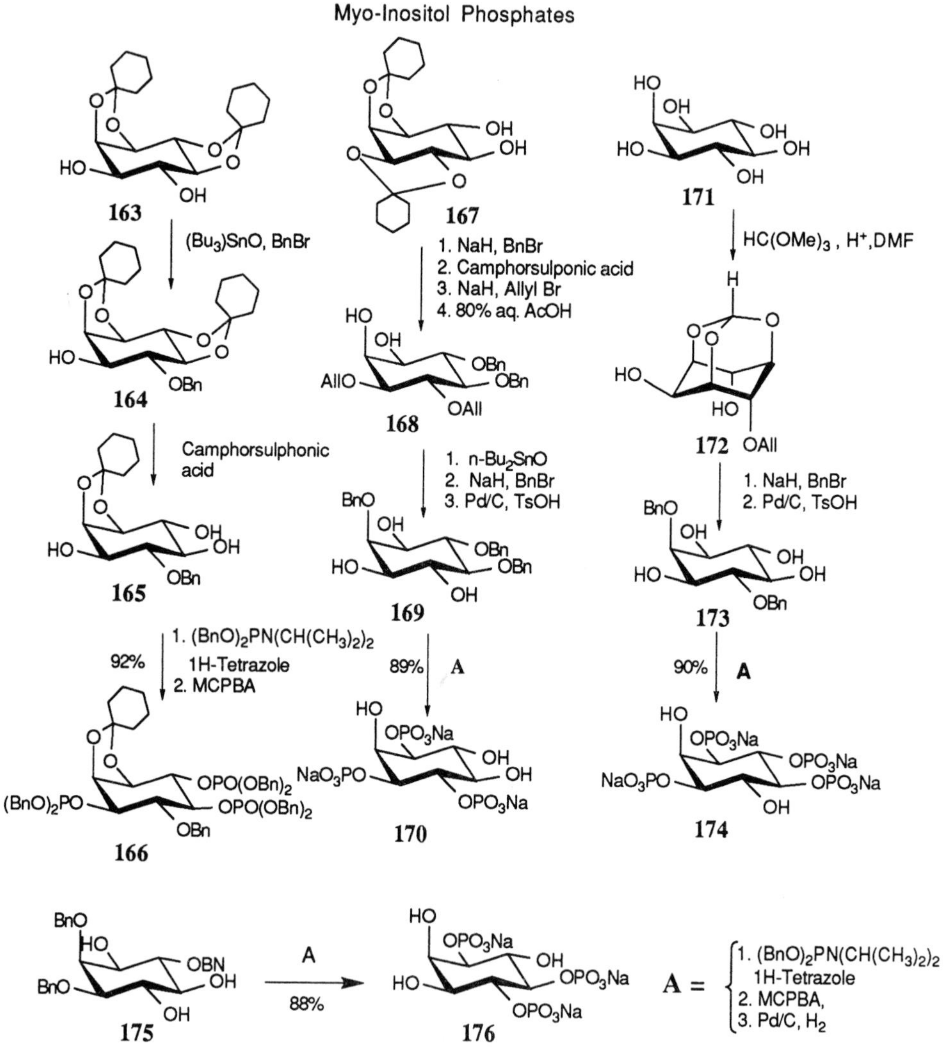

Myo-Inositol Phosphates

Scheme 16.

as the phosphititylating reagent followed by oxidation at -40°C with *m*-chloroperbenzoic acid (MCPBA) to the dibenzylphosphate and hydrogenolysis to the desired materials. The use of *t*-butylhydroperoxide (*t*-BuOOH) as the oxidant resulted in mixtures of products that were not separable.

Bernabe *et al.*[45] have synthesized the polyphosphorylated galactose derivatives **178** and **180** (Scheme 17) as inhibitors of the cyclic AMP-dependent protein kinase A responsible for insulin metabolism. His method is similar to that employed by Yu and Fraser-Reid,[9a] employing the same phosphititylating system but using catalytic $RuCl_3/NaIO_4$ as the oxidant. His yields of tetraphosphate **178** and **180** from galactose derivatives **177** and **179** were 61% and 82%, respectively.

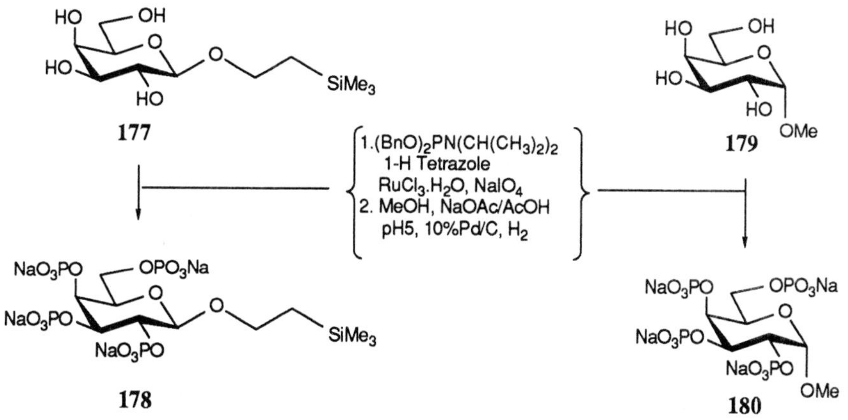

Scheme 17.

Other methods of phosphorylation include reagents such as bis (2,2,2-trichloroethyl) phosphorochloridate $(Cl_3CCH_2O)_2POCl)$[46] and 2-cyanoethyl *N,N*-diisopropylchloro phosphoroamidite (NC-CH_2-CH_2-O)PClN $(CH(CH_3)_2)_2$.[47]

One of the complexities encountered in deprotection is the migration of phosphate groups under acidic or basic conditions. It is most often observed that the Zemplén conditions for de-O-esterification of phosphorylated sugars may result in partial or total migration of phosphate groups. The use of triethylamine in methanol has avoided these difficulties.

6.   CONCLUSIONS

This review demonstrates the complexity associated with the synthesis of mono or poly-sulphated/phosphorylated carbohydrate structures. Methods of sulfation are mostly limited to blocked structures having one or two hydroxyl groups available for sulfation. Sulfation is generally accomplished using sulfur trioxide complexes with pyridine or trimethylamine, the trimethylamine complex being less reactive and often requiring elevated temperatures and longer reaction times. DMF is a common solvent although pyridine has also been used. The charged structures are often isolated as triethylamine or pyridinium salts and converted to sodium or potassium salts using ion exchange resins.

The range of phosphorylated carbohydrate structures is diverse. Phosphorylating reagents have a wider range of activities and thus allow a wider diveristy of structures to be accessed.

The diphenyl and dibenzylphosphodichloridate methods are still very useful and generally applicable. The newer phosphitylating agents are also proving to be very useful in the field of carbohydrates and will likely continue to be the reagents of choice for introducing a phosphate group.

REFERENCES

1. (a) M. L. Philips E. Nudelman, F. C. A. Gaeta, M. Perez, A. K. Singhal, S. Hakomori, and J. C. Paulson, *Science*, **250**, 1130 (1990); (b) W. Walz, A. Aruffo W. Kolanus, M. Bevilacqua, and B. Seed, *Science*, **250**, 1132 (1990).

2. E. L. Berg, M. K. Robinson, O. Mannson, E. C. Butcher, and J. L. Mangnani, *J. Biol. Chem.,* **266**, 14869 (1991).

3. C. T. Yuen A. M. Lawson, W. Chai, M. Larkin, M. S. Stoll, A.Stuart, F. S. Sulllivan, T. J. Ahern, and T. Feizi, *Biochem*istry, **31**, 9126 (1992).

4. C. T. Yuen, K. Bezouska, J. O'Brien, M. Stoll, R. Lemoine, A. Lubineau, M. Kiso, A. Hasegawa, N. J.Bokovich, K.C. Nicalaou, and T. Fiezi, *J. Biol. Chem.*, **269**, 1595 (1994)

5. D. Sako, X.J.Chang, K. M. Barone, G. Vachino, H. M. White, G. Shaw, G. M. Valedman, K. M. Bean, T. J. Ahem, B. Furie, D.A. Cumming, and G. R. Lawson, *Cell*, **75,** 1179 (1993)

6. (a) L. A. Lasky, M.S. Singer, D. Dowbenko, Y. Imai, W. J. Henzel, C. Grimley, C. Fennie, N. Gillett, S. R. Watson, and S. D. Rosen, *Cell*, **69,** 927 (1992); (b) Y. Imai, L. A. Lasky, and S. D. Rosen, *Nature,* **361,** 555 (1993)

7. R. Parekh, "*Carbohydrates in Cell Adhesion, Inflammation and Cancer,*" CHI Meeting, San Diego, December 6-8 (1993).

8. T. P. Skeleton, L. V. Hooper, V. Srivastava, O. Hindsgaul, and J. V. Baenziger, *J. Biol. Chem.*, **266,** 17142 (1991).

9. (a) K-L. Yu and B. Fraser-Reid, *Tetrahedron Lett.*, **29,** 979 (1988); (b) G. Salamonczyk and K. M. Pietrusiewicz, *Tetrahedron Lett.*, **43,** 6167 (1991); (b) L.Ling and S. Ozaki, *Carbohydr. Res.*, **256,** 49 (1994).

10. (a) O. P. Srivastava and O. Hindsgaul, *Carbohydr. Res.*, **161,** 195 (1987); (b) O. P. Srivastava and O. Hindsgaul, *Carbohydr. Res.*, **155,** 57 (1986); (c) O. P. Srivastava and O. Hindsgaul, *J. Org. Chem.*, **52,** 2869 (1987).

11. (a) P. Sinay, J. C. Jacquinet, M. Petitou, P. Duchaussoy, I. Ledermam, J. Choay, and G Torri, *Carbohydr. Res.*, **132,** C5 (1984); (b) M. Petitou, P. Duchaussoy, I. Lederman, J. Choay, P.Sinay, J. C. Jacquinet, and G. Torri, *Carbohydr. Res.,* **147,** 221 (1986); (c) T. Chiba, J. C.Jacquinet, P. Sinay, M. Petitou, and J. Choay, *Carbohydr. Res.*, **174,** 253 (1988).

12. C. A. A. van Boeckel, T. Beetz, J. N. Vos, J. M. de Jong, S. F. van Aelst, R.H. van den Bosch, J. M. R. Mertens, and F. A. van der Vlugt, *J. Carbohydr. Chem.*, **4,** 293 (1985).

13. C. A. van Boeckel and M. Petitou, *Angew. Chem Int. Ed. Engl.*, **32,** 1671 (1993).

14. (a) M. Petitou, P. Dauchaussoy, I. Lederman, J. Choay, J.C. Jacquinet, P. Sinay, and G. Torri, *Carbohydr. Res*, **167,** 67 (1987); (b) M. Petitou, P. G. Jaurand, M. Derrien, and J. Choay, *Bioorg. Med. Chem. Lett.*, **1,** 95 (1991).

15. J. Ruggiero, R. P. Vieira, and P. A. S. Mourao, *Carbohydr. Res*, **256,** 275 (1994).

16. J-C. Jacquinet, *Carbohydr. Res.*, **199,** 153 (1990).

17. (a) M. Zsiska and B. Meyer, *Carbohydr. Res.*, **215,** 261 (1991); (b) *ibid.* p. 279.

18. (a) F. Goto and T. Ogawa, *Tetrahedron Lett.*, **33,** 6841 (1992); (b) *ibid.* p. 5099.

19. S. Rio, J-M. Beau, and J-C. Jacquinet, *Carbohydr. Res.*, **255,** 103 (1994).

20. (a) A. Marra, X. Dong, M. Petitou, and P. Sinay, *Carbohydr. Res.*, **195,** 39 (1989); (b) J-C. Jacquinet and P. Sinay, *Carbohydr. Res.*, **159,** 229 (1987).

21. M. Kobayashi, F. Yamazaki, Y. Ito, and T. Ogawa, *Carbohydr. Res*, **201,** 51 (1990).

22. (a) V. Srivastava, O. Hindsgaul, and J. U. Baenziger, *Can. J. Chem.*, **65,** 1643 (1987); (b) V. Srivastava and O. Hindsgaul, *Carbohydr. Res.*, **185,** 163 (1989).

23. R. Gigg, A. A. E. Penglis, and R. Conant, *J. Chem. Soc. Perkin. Trans 1*, 2490 (1980).

24. R. L. Halcomb and S. J. Danishefsky, *J. Am. Chem. Soc.*, **111,** 6661 (1989).

25. M. Hoch, E. Heinz and, R. R. Schmidt, *Carbohydr. Res.*, **191,** 21 (1989).

26. A. Aruffo, W Kolanus, G. Walz, P. Fredman, and B. Seed, *Cell*, **67,** 35 (1991).

27. (a) S. Oehninger, G. F. Clark, A. A. Acosta, and G. D. Hodgen, *Fertil. Steril.*, **55,** 165 (1991); (b) G. F. Clark, S. C. Oehninger, and L. K. Moen, *FASEB J.*, **6,** 233 (1992); (c) C. Foxall, S. R. Watson, D. Dowbenko, L. A. Lasky, M. Kiso, A. Hasegawa, D. Asa, and B. K. Brandley, *J. Cell. Biol.*, **117,** 895 (1992).

28. S. Suzuki, Y. Toda, T. Tamatani, T. Watanabe, T. Suzuki, T. Nakao, K. Murase, M. Kiso, A. Hasegawa, K. Tadano-Aritomi, I. Ishizuka, and M. Miyasaka, *Biochem. Biophys. Res. Commun.*,**190,** 426 (1993).

29. H. M. Flowers, *Carbohydr. Res*, **2,** 371 (1966).

30. M. Patankar, S. Ohehninger, T. Barnett, R. Williams, and G. Clark, *J. Biol Chem.*, **268,** 21770 (1993).

31. (a) R. K. Jain and K.L. Matta, *Carbohydr. Res.*, **208,** 280 (1990); (b) R. K. Jain and K. L. Matta, *Carbohydr. Res.*, **208,** 51 (1990).

32. E. V. Chandrasekaran, R. K. Jain, and K. L. Matta, *J. Biol. Chem.*, **267**, 23806 (1992).

33 R. M. Ippolito, W. Haque, C. Jiang, R. Hanna, and A. Venot, *P C T Application*, **WO92/00245** (1992).

34. M. Nilsson and T. Norberg, *Carbohydr. Res.* **183**, 71 (1988).

35. B. K. Brandley, M. Kiso, S. Abbas, P. Nikrad, O. Srivasatava, C. Foxall, Y. Oda, and A. Hasegawa, *Glycobiology* , **3**, 633 (1993).

36. M. Bevilacqua, *Personal Communication* (1991).

37. K. C. Nicolaou, N. J. Bockwich, and D. R. Carcanague, *J. Am. Chem. Soc.*, **115**, 8843 (1993).

38. (a) T. Mukaiyama, Y. Muria, and S. Shoda, *Chem. Lett.*, 431 (1981); (b) T. Matsumoto, H. Maeta, K. Suzuki, and G. Tsuchihashi, *Tetrahedron Lett.*, **29**, 3567 (1988).

39. (a) Y. Ichikawa, M. Sim, and C-H. Wong, *J. Org. Chem.*, **57**, 2943 (1992); (b) R. R. Schmidt, B. Wegmann, and K. H. Jung, *Leibigs Ann. Chem.* 191 (1991); (c) G. H. Veeneman, H. J. G. Broxterman, G. A. van der Marel, and J. H. van Boom, *Tetrahedron Lett.*, **32**, 6175 (1991).

40. (a) P. Westerduin, G. H. Veeneman, J. E. Marugg, G. A. van der Marel and J. H. van Boom, *Tetrahedron Lett.*, **27**, 6271 (1986); (b) A. Nikolaev, I. A. Ivanova, V. Shibaev and N. K. Kochetkov, *Carbohydr. Res.*, **204**, 65 (1994).

41. For recent studies on inositol analogues see: (a) A. P. Kozikowski, A. H. Faug, G. Powis, P. Kurian, and F. T. Crews, *J. Chem. Soc., Chem. Commun.*, 362 (1992); (b) A. P. Kozikowski, A. H. Faug, G. Powis, and D. C. Melder, *Med. Chem. Res.*, **1**, 227 (1991); (c) F. McPhee, D. P. Downes, G. Lowe, *Biochem. J.*, **277**, 407 (1991).

42. (a) K. K. Reddy, J. R. Falck, and J. Capdevila, *Tetrahedron Lett.*, **34**, 7869 (1993); (b) C. Jaramillo, J. L. Chiara, and M. Martin-Lomas, *J. Org. Chem.*, **59**, 3135 (1994).

43. J. Thiem and M. Franzkowiak, *J. Carbohydr. Chem.*, **8**, 28 (1989).

44. (a) R. K. Jain, S. A. Abbas, and K. L. Matta, *Carbohydr. Res.*, **161**, 318 (1987); (b) K. L.Matta, M. S. Chowdhary, R. K. Jain, and S. A. Abbas, *Carbohydr. Res*, **150**, C1 (1986).

45. H-N. Caro, M. Martin-Lomas, and M. Bernabe, *Carbohydr. Res,* **240**, 119
     (1993).

46. A. Franke, H. H. Scheit, and F. Eckstein, *Angew. Chem. Int. Ed. Engl.,* **6**,
     362 (1967).

47. P. Westerduin, G. H. Veeneman, J. E. Marugg, G.A. van der Marel, and J.
     H. van Boom, *Tetrahedron Lett.,* **27**, 1211 (1986).

Chapter 18

SYNTHETIC GLYCOSYLTRANSFERASE ACCEPTORS AND INHIBITORS; USEFUL TOOLS IN GLYCOBIOLOGY

KHUSHI L. MATTA

Department of Gynecologic Oncology, Roswell Park Cancer Institute, Elm & Carlton Streets, Buffalo, NY 14263, USA

Abstract   This chapter focuses upon the role of carbohydrate chemistry in the study of glycosyltransferases. The chemical synthesis of glycosyltransferase acceptors, inhibitors and related reference compounds are reviewed with emphasis being placed on those compounds required for the study of selected enzymes which are involved in the biosynthesis of O-linked glycoproteins containing Galα→Ser/Thr*. Also discussed are ß-N-acetylglucosaminyl-transferases involved with N-glycosidically linked glycoproteins, some selected α-L-fucosyltransferases and certain sialyltransferase activities.

1. INTRODUCTION

Glycoproteins and glycolipids found on cell surface membranes play an important role in various biological functions. The oligosaccharide chains of these glycoconjugates can be involved in intercellular recognition and interaction[1-5] or in the control of cell growth and differentiation.[6,7] These cell surface glycoconjugates also interact with various other biological agents such as lectins, hormones and enzymes[1,2,8,9] and are involved in

---

*Abbreviations: Fuc = L-fucose; Gal = galactose; GalNAc = N-acetylgalactosamine; Glc = Glucose; GlcNAc = N-acetylglucosamine; NeuNAc = N-acetyl neuraminic acid; Ser = Serine; Thr = Threonine; GlcNAc-T = N-acetylglucosaminyltransferase. In Structural Formulae: All = allyl; Bn = Benzyl; Me = Methyl; SE represents the sulfate ester (-OSO₃⁻). All sugars are of D-configuration and pyranose form unless specified otherwise.

the binding of bacteriotoxins and viruses to host cells. One of the landmark events of current research is the recent finding that both sialyl Le^x and sialyl Le^a type structures can act as ligands for selectins.[17-19] The discovery of mannose-6-phosphate as a marker for lysosomal enzyme targeting is another notable example which illustrates the role of carbohydrate moieties in biological functions.[9,20,21]

A number of studies carried out by many different investigators have shown that oncogenic transformations of cells are accompanied by dramatic changes in the chemical composition, metabolism, and organization of cell-surface glycoconjugates. Study of these changes has revealed that many of the cell surface antigens are carbohydrates incorporated in either glycolipids or glycoproteins, which have been defined by monoclonal antibodies (Mabs) raised against cancer cells.[22-27] It has also been well documented that the carbohydrate portion imparts the immunological specificity of these antigens. The aberrantly glycosylated glycosphingolipids and glycoprotein antigens can be divided into three major classes: (i) Glycosphingolipids that have ganglio and globo series structures. (ii) The lacto series type 1 [Galß(1→3)GlcNAc] and type 2 [Galß(1→4)GlcNAc] structures which also may have a terminal sialic acid and fucose branching as seen in the sialyl Le^a and sialyl Le sequences. These structures can be expressed on both tumor-associated glycolipid and glycoprotein antigens. (iii) Tumor-associated glycoprotein antigens, specifically of the O-linked mucin-type, which in recent years have gained a great deal of attention.[28-31]

2. ENHANCED INTEREST IN GLYCOSYLTRANSFERASES

Although there may be other contributing factors, it has become increasingly evident that glycosyltransferases play a vital role in the aberrant expression of cell surface carbohydrates found on cancer cells.[22,23,32,33] For example, Fukuda et al.,[34,35] described the variation of O-linked oligosaccharides and suggested that in K562 erythroleukemic cells a key ß6-GlcNAc-T may be missing which prevents further branching of Galß(1→3)GalNAcα→ by GlcNAc. A rival enzyme, sialyltransferase, competes for this same acceptor. On the other hand, in activated T-lymphocytes the α(2→6)-sialyltransferase activity is decreased and ß(1→6)GlcNAc-transferase activity is increased, resulting in the synthesis of O-linked saccharides.[34] L-Fucose is an essential part of many of the tumor-associated glycoconjugates.[22-30] This has led to an immense interest in the study of α-L-fucosyltransferases in various tissues and cell lines.[35-45]

There has been a concurrent surge of interest in the study of glycosyltransferases in both normal and tumor tissues.[46-48] In just the past five years, a series of glycosyltransferases have been purified, cloning strategies for the isolation of cDNA have been developed and glycosyltransferase genes discovered, as described in a recent review by Kleene and Berger.[49] Thanks to the hard work of numerous biochemists and molecular biologists, we have gained a great deal of knowledge about glycosyltransferases and the biosynthesis of glycoconjugates. The subject has been documented in many reviews.[49-53] In addition, the contribution of synthetic carbohydrate chemistry in facilitating the availability of acceptor-substrates, sugar nucleotide inhibitors and their modified analogs is starting to gain recognition.

3. GLYCOSYLTRANSFERASES AND SOME FUNDAMENTAL RULES OF GLYCOCONJUGATE BIOSYNTHESIS

The glycosyltransferases are a class of enzymes that catalyze the synthesis of specific glycosides by the transfer of a monosaccharide from a sugar nucleotide donor substrate to an acceptor.[32] One of the unique features of glycosyltransferases is their strict specificity requirements. Each enzyme is specific not only to a sugar nucleotide but also to an acceptor substrate. One gene encodes only one glycosyltransferase, whose activity is highly restricted to the generation of a single glycosidic linkage between a glycosyl nucleotide and its acceptor. This concept which has been called the "one enzyme-one linkage" hypothesis[54] has become a central fundamental principle in studying the biosynthesis of glycoconjugates. Studies of the specificities of purified glycosyltransferases show that this concept, in general, appears to be valid with few exceptions.[50] For example, α-L-fucosyltransferase from human milk,[50,55-57] intestine, gall bladder, kidney and Colo 205 cell line can act on both oligosaccharide structures containing Galß(1→3)GlcNAc (type 1, fucosylation at C-4 position), and Galß(1→4)GlcNAc (type 2 chain, fucosylation at C-3 position).[37,58] In 1970 Roseman proposed that a multiglycosyltransferase system is responsible for the building of a given specific glycoconjugate structure through their unique properties of cooperative sequential specificity.[32] This implies that the resulting product of an action of one glycosyltransferase and its acceptor becomes the acceptor for the next glycosyltransferase action and that this cooperation leads to lengthening of the carbohydrate chain. In his review article[51] Schachter formulated some general rules which attempt to explain the branching patterns and microheterogeneity of protein bound oligosaccharides. These rules address such headings as: (i) competition by two or more enzymes for a common substrate, (ii) controls at the level of the enzyme substrate specificity, e.g. critical sugar residues which turn enzyme activity on or off, and (iii) substrate availability. These rules are quite helpful in unraveling the biosynthetic routes of various glycoconjugates, including those of tumor-associated antigens.

4. ISOLATION OF THE PRODUCT OF GLYCOSYLTRANSFERASE ACTIVITY

The procedures for isolation and identification of product(s) formed by glycosyltransferase reactions are dependent on the nature of the acceptor employed.[50-53,59] Identification procedures for low molecular weight oligosaccharides have utilized thin-layer and paper chromatography as well as HPLC, although its development is not as advanced for this application. For example, the use of an acceptor which contains a chromophore such as o-NO₂Ph-ß-D-galactopyranoside for α(1→2)-L-fucosyl-transferase[60] allows rapid separation by TLC of the acceptor and product without having to resort to radiochromatograms. Provided that an acceptor with a chromophore and expected product are both available as reference compounds, a rapid enzyme assay can be developed. SEP-PAK C-18 cartridge chromatography is now being used frequently with acceptors containing a benzyl, aryl or hydrophobic chain.[62,63]

An array of well-defined oligosaccharide moieties O-linked to $(CH_2)_7COOMe$ or $(CH_2)_7CH_3$ chains have been synthesized and used as acceptors for the assay of various enzymes by the laboratory of Palcic and Hindsgaul. The use of $(CH_2)_7COOMe$[64] as a spacer in linkage to a protein was introduced by Lemieux. We have used acceptor moieties containing ß-nitrophenyl group for analogous reasons. Complete structural determination of enzymatic reaction product(s) can be performed by standard methods. Purified glycosidases capable of differentiating the interglycosidic linkage produced by a glycosyltransferase and its acceptor are excellent tools for such analyses. The availability of well-defined oligosaccharides as reference compounds can also be helpful. Some examples which warrant mention are: (i) The presence of an α-L-fucosyl residue in the trisaccharide Galß(1→4)[Fucα(1→3)]GlcNAc has an inhibitory effect on the removal of galactose by the ß-D-galactosidase of *D-pneumoniae* which is known to hydrolyze Galß(1→4)GlcNAc linkages.[65] (ii) α(1→3/4)-fucosidase does not cleave fucose from sialylated Le^x and sialylated Le^a structures. Thus, sialic acid must be removed prior to treatment with α-L-fucosidase.[66] (iii) α-L-fucosidase purified from human serum cleaves fucose from compounds such as Fucα(1→2)Gal and Fucα(1→2)Galß-(1→4)Glc, but fails to act significantly on Fucα(1→2)Galß(1→3)-GalNAcα1→OPh and Fucα(1→2)[Galß(1→3)]-Galß1→OPh.[67] (iv) Kornfeld and co-workers[68] observed that hexosaminidase was incapable of cleaving a ß-GlcNAc linked to C-4 of the inner α-mannose residue in high-mannose oligosaccharides without the presence of α-D-mannosidase. Thus, phenomena attributed to the steric hindrance of other sugars present in a molecule should be given serious consideration for such investigations. Nevertheless, glycosidases have been very helpful in structural investigations.

5. BIOSYNTHESIS OF *N*-ACETYLGALACTOSAMINE-SER/THR *O*-LINKED GLYCOPROTEINS

The simple sequence, GalNAcα-Ser/Thr, (Tn antigen) found in many glycoproteins, and disaccharide sequences such as Galß(1→3)GalNAcα1→ (T antigen) and NeuAcα(2→6)-GalNAcα→ (sialyl Tn) are very common in *O*-linked glycoproteins. The disaccharide moieties, GlcNAcß(1→3)GalNAcα→ and GalNAcα(1→3)GalNAcα→, have also been found in such glycoproteins. The structures and biosynthetic pathways of *O*-linked glycoproteins containing different complex carbohydrate chains have been reviewed.[69,70] Carbohydrate chains, which can vary in size from three to more than twenty carbohydrate units, have been classified according to designated core structures by Schachter and co-workers.[70]

Galβ(1→3)GalNAcα1→OR	**Core 1**	
GlcNAcβ(1→6)[Galβ(1→3)]GalNAcα1→OR	**Core 2**	
GlcNAcβ(1→3)GalNAcα1→OR	**Core 3**	R = Ser/Thr
GlcNAcβ(1→6)[GlcNAcβ(1→3)]GalNAcα1→OR	**Core 4**	
GalNAcα(1→3)GalNAcα1→OR	**Core 5**	
GlcNAcβ(1→6)GalNAcα1→OR	**Core 6**	

Scheme 1.

The complex chains of *O*-linked glycoprotein consist of three distinct regions, core, backbone and non-reducing terminus.

Non-reducing terminus → Backbone → Coreα→ Ser/Thr

Scheme 2.

N-acetylglucosamine and galactose are key ß-linked sugars found in the backbone region, while Galα→, Fucα→, NeuAcα2→, and GalNAcα→ are generally located at the non-reducing terminus of a glycoprotein. The backbone region may also be branched by α-L-fucose. The sulfate groups of glycoproteins are important functionalities that are now gaining much attention. Some *O*-linked glycoproteins can lack a backbone region, thus the core structure or followed by the galactosylation of GlcNAc present therein may contain the non-reducing terminus residue(s).

The core structures mentioned above are unique for *O*-linked glycoproteins, while the backbone and non-reducing terminus sugar residues can be found in glycolipids and also as part of *N*-glycans.

6. ACCEPTORS FOR THE STUDY OF GLYCOSYLTRANSFERASES OF SIX CORES OF *O*-LINKED GLYCOPROTEIN

The biosynthesis of *N*-glycans in its initial steps involves dolichol derivatives. In comparison, *O*-linked glycoprotein biosynthesis is based upon the stepwise addition of each monosaccharide. Thus, α-D-GalNAc, the very first sugar in each of the six core structures listed in Scheme 1 is transferred by the polypeptide α-D-GalNAc-T from UDP-GalNAc into the polypeptide chain.[69,70]

6.A. Assembly of Core 6-GlcNAcß1→6GalNAcα

The GalNAcα→Ser/Thr moiety, once generated by GalNAcα-transferase can, by branching, give any of the six different core structures, or it can be terminated by α(2→6)-sialyltransferase to give sialyl Tn, NeuAcα(2→6)GalNAcα1→O Ser/Thr. At each step the biosynthetic pattern depends upon the levels of different enzyme activities and various other rules, as cited by Schachter.[51]

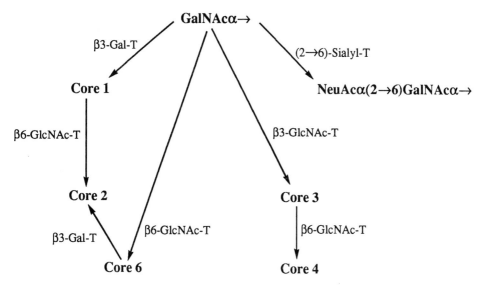

Figure 1.

A linear carbohydrate chain containing the core 6 moiety, GlcNAcß(1→6) GalNAcα-, was found as part of the mucous glycoproteins of human seminal plasma,[71] human meconium glycoprotein[72] and human K-casein[73]. This suggests the existence of ß6-GlcNAc-T capable of acting on the GalNAcα1→ linked structure. Human ovarian tissues have been reported to contain this enzyme activity.[74] Benzyl 2-acetamido-2-deoxy-α-D-galactopyranoside (GalNAcα1→OBn) can be used as an acceptor to give Core 6 GlcNAcß(1→6)GalNAcα1→. Formation of the GlcNAcß(1→3)GalNAcα1→OB Core 3 structure is also observed, due to the presence of ß3-GlcNAc-T.[74]

6.B. Synthesis of Galß1→3GalNAcα1→ - Core 1

ß3-Gal-T which is responsible for the generation of the Core 1 moiety has been examined in a number of different sources. This work has recently been reviewed.[52,69,70] Mucins containing the Tn (GalNAcα→) unit have been used as acceptors for this enzyme. However, synthetic GalNAcα1→OR (R = Bn or Ph) can also be used as an acceptor for this enzyme. The specificity of the partially purified rat liver Core 1 ß3-Gal-T has been studied in detail employing a series of analogs of GalNAcα1→OBn.[75] The enzyme recognizes the C-3 and C-4 hydroxy groups but not the C-6 group in GalNAcα1→OBn. The enzyme can also convert the Core 6 structure to a Core 2[75,76] type structure. A series of acceptors containing benzyl, p-NO$_2$Ph and fluoro-substituted moieties were used to determine specificity and levels of ß3-Gal-T activity present in normal and ovarian tumor tissues.[76] The contribution and importance of well-defined acceptors in providing glycosyltransferase assay procedures by which the biosynthetic pathways of glycoconjugates can be elucidated is evident. One advantage of employing Galß(1→4)GlcNAcß(1→6)GalNAcα1→OBn over GlcNAcß(1→6)GalNAcα1→OBn as an acceptor for ß3-Gal-T is that it can be used in the presence of commonly occurring ß4-Gal-T.

6.C.   Synthesis of GlcNAcß(1→6)[Galß(1→3)]GalNAcα1→ - Core 2

$$\text{Galß(1→3)GalNAcα1} \xrightarrow[\text{UDP-GlcNAc}]{\text{ß6-GlcNAc-T}} \text{GlcNAcβ(1→6)[Galβ(1→3)]GalNAc +UDP}$$

Scheme 3.

Williams and Schachter[77] were the first to demonstrate the existence of this enzyme using mucin and antifreeze glycoproteins containing Galß(1→3)GalNAcα1→Ser/Thr type structures. Synthetic Galß(1→3)GalNAcα1→OR [R = Ph, PhNO$_2$ (o or p) or Bn] type structures have been found to be effective acceptors for the enzyme.[78] This enzyme has also been extensively studied by other investigators.[34,35,49,70,79-81] Recently a series of modified analogs of the Galß(1→3)GalNAcα- acceptor moiety, including the 6-deoxy derivatives, have been prepared and employed to elucidate the specificity of this enzyme.[80] From these studies it was determined that the C-6 hydroxy group of galactose and the 4,6-hydroxy groups of GalNAc are important for action by the enzyme.[80,81] These specificity requirements often depend upon the source of the enzyme.[70,80] Sekine et $al.$[82] have found a ß6-GlcNAc-T capable of synthesizing the GlcNAcß(1→6)[Galß(1→3)]GalNAcß- branched structure in globoside.

6.D.   Acceptors for Enzymes Involved in Core 3 and Core 4 Synthesis

ß3-GlcNAc-T is responsible for the synthesis of Core 3 units which then become acceptors for ß6-GlcNAc-T to give Core 4 units. Synthetic GalNAcα1→OR [R = Bn, Ph, PhNO$_2$ (o or p)] can be used as acceptors in addition to Tn linked glycoprotein.[69,70]

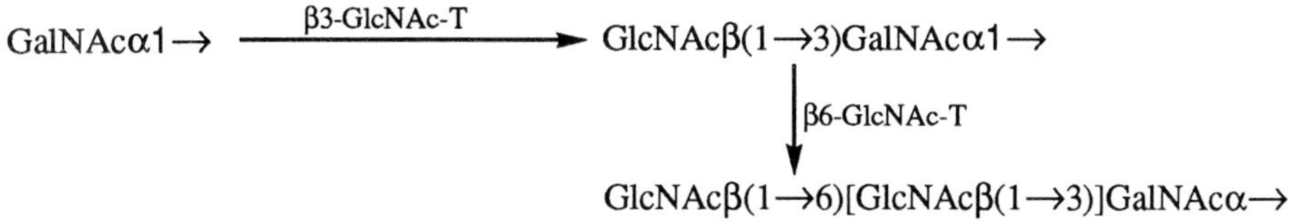

GalNAcα1→ $\xrightarrow{\text{ß3-GlcNAc-T}}$ GlcNAcβ(1→3)GalNAcα1→

$\downarrow$ ß6-GlcNAc-T

GlcNAcβ(1→6)[GlcNAcβ(1→3)]GalNAcα→

Scheme 4.

Modified analogs of GalNAcα1→OBn employed for the study of ß3-Gal-T can be helpful in examining the specificity of this enzyme. So far it has not been shown that either GlcNAcß(1→6)GalNAcα1-OBn or Galß(1→4)GlcNAcß(1→6) GalNAcα1→OBn type compounds can be used as acceptors for this enzyme. This ß3-GlcNAc-T is quite different from the enzyme which incorporates GlcNAc at the C-3 position of galactose in Core 1 and Core 2 structures. On the other hand, the ß6-GlcNAc-T capable of forming Core 4 structures may be the same enzyme which is responsible for the conversion of Core 1 to Core 2. As shown in the above equation, GlcNAcß(1→3)GalNAcα1→OR [R = Bn, Ph, PhNO$_2$ (*o* or *p*)] or the polypeptide of mucin glycoprotein containing this Core 3 sequence are acceptors for this enzyme.

6.E. Synthesis of Core 5 GalNAcα1→3GalNAcα1→O Ser/Thr

Kuroska *et al.*[83] reported the presence of UDP-GalNAc:GalNAc-mucin α-GalNAc-T activity in cancerous human intestinal tissues which are capable of synthesizing this Core 5 structure. The specificity of this enzyme remains to be established with the availability of synthetic GalNAcα(1→3)GalNAcα1→ type structures enabling investigators to study elongation reactions of this disaccharide moiety with different enzymes.

7. ENZYMES INVOLVED IN CORE STRUCTURE ELONGATION

It has become apparent that in the elongation of Core 1 and 2 structures, the C-3 position of galactose in these structures is generally the site of ß3-GlcNAc-T action as shown:

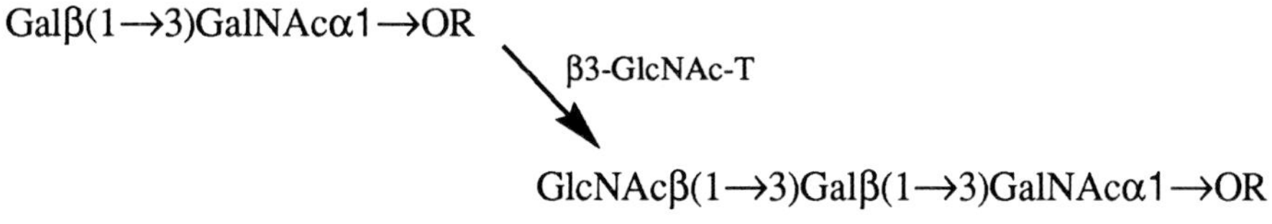

Galβ(1→3)GalNAcα1→OR

$\searrow$ ß3-GlcNAc-T

GlcNAcβ(1→3)Galβ(1→3)GalNAcα1→OR

Scheme 5.

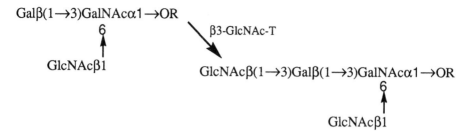

Scheme 5 (continued).

One of the more striking observations from this system is that simple disaccharide core structures (where R = Bn or Ph) are not acceptors for ß3-GlcNAc-T but the glycopolypeptide from mucin which contains this sequence is,[70] and interestingly, synthetic Core 2 type structures [where R = Bn or $PhNO_2$ (*o* or *p*)] can also be efficient acceptors for this enzyme. This enzyme is further distinguished from the ß3-GlcNAc-T which acts on both Galß(1→4)GlcNAc or Galß(1→3)GlcNAc type structures and from ß3-GlcNAc-T involved in the synthesis of Core 3, discussed earlier. A review by Schachter *et al.*,[70] stated that there are at least three types of ß3-GlcNAc transferases. Also, the ß6-GlcNAc transferase which converts Core 1 to the Core 2 sequence, Core 3 to Core 4, and the ßGlcNAc-transferase capable of linking GlcNAc at the C-6 position of the inner galactose, shown in the equation below, may be the same in some tissues (e.g. pig gastric mucosa) but in other tissues they are separate enzymes. Nevertheless, synthetic compounds [e.g. where R = Bn, $PhNO_2$ (*o* or *p*)] can be used as effective acceptors for these enzyme activities. Modified analogs of these acceptor moieties may also be useful in examining the specificity of ß6-GlcNAc-T.

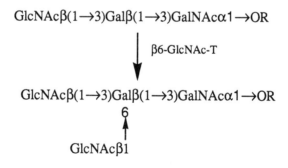

Scheme 6.

20. S. Kornfeld, *FASEB J.*, **1**, 462 (1987).

21. K. von Figura and A. Hasilik, *Annu. Rev. Biochem.*, **55**, 167 (1986).

22. S. Hakomori, *Cancer Cells*, **3**, 461 (1991).

23. S. Hakomori, *Advances in Cancer Res.*, **52**, 257 (1989).

24. M. Fukuda, *Biochim. Biophys. Acta.*, **780**, 119 (1985).

25. K.O. Lloyd and L.J. Old, *Cancer Res.*, **49**, 3445 (1989).

26. C.L.M. Stults, C.C. Sweeley, and B.A. Macher, *Methods Enzymol.*, **179**, 167 (1989).

27. S. Sell, *Human Pathology* **21**, 1003 (1990).

28. F.G. Hanisch, G. Uhlenbruck, J. Peter-Katalinic, and H. Egge, *Carbohydr. Res.*, **178**, 29 (1988).

29. T. Feizi, *Nature*, **314**, 53 (1985).

30. S.B. Ho and Y.S. Kim, *Cancer Biol.*, **2**, 389 (1991).

31. V.P. Bhavanandan, *Glycobiology*, **1**, 493 (1991).

32. S. Roseman, *Chem. Phys. Lipids*, **5**, 270 (1970).

33. R.O. Brady and P.H. Fishman, *Biochem. Biophys. Acta.*, **335**, 121 (1974).

34. M. Fukuda, S.R. Carlsson, J.C. Klock, and A. Dell, *J. Biol. Chem.*, **261**, 12796 (1986).

35. F. Piller, V. Piller, R.I. Fox, and M. Fukuda, *J. Biol. Chem.*, **263**, 15146 (1988).

36. G.C. Hansen and D. Zopf, *J. Biol. Chem.*, **260**, 9388 (1985).

37. R. Mollicone, J.J. Candelier, B. Mennesson, P. Couillin, A.P. Venot, and R. Oriol, *Carbohydr. Res.*, **228**, 265 (1992).

38. M. Basu, J.W. Hawes, Z. Li, S. Ghosh, F.A. Khan, B. Zhang, and S. Basu, *Glycobiology*, **1**, 527 (1991).

39. F.G. Hanisch, A. Mitsakos, H. Schroten, and G. Uhlenbruck, *Carbohydr. Res.*, **178**, 23 (1988).

40. P.H. Johnson and W.M. Watkin, *Biochem. Soc. Trans.*, **15**, 396 (1987).

41. D.R. Howard, M. Fukuda, M.N. Fukuda, and P. Stanley, *J. Biol. Chem.*, **262**, 16830 (1987).

42. C.S. Foster, D.R.B. Gillies, and M.C. Glick, *J. Biol. Chem.*, **266**, 3526 (1991).

43. S.E. Goetz, C. Hession, D. Goff, B. Griffiths, R. Tizard, B. Newman, G. Chi-Rosso, and R. Lobb, *Cell*, **63**, 1349 (1990).

44.   (a) K. Sasaki, K. Kurata, K. Funayama, F. Kikuko, M. Nagata, E. Watanabe, S. Ohta, N. Hanai, and T. Nishi, *J. Biol. Chem.*, **269**, 14730 (1994); (b) S. Natsuka, K.M. Gersten, K. Zenita, R. Kannagi, and J.B. Lowe, *J. Biol. Chem.*, **269**, 16789 (1994).

45.   E.V. Chandrasekaran, R.K. Jain, and K.L. Matta, *J. Biol. Chem.*, **267**, 23806 (1992).

46.   A.L. Pohl, *Cancer Detection and Prevention*, **7**, 299 (1984).

47.   M.M. Weiser, W.D. Klohs, D.K. Podolsky, and J.R. Wilson, In *The Glycoconjugates*, Vol. IV, Academic Press, New York, p. 301 (1982).

48.   D. Kessel, In *Progress in Clinical and Biological Research; Tumor Markers and Their Significance in the Management of Breast Cancer*, (T. Dao, A. Brodie, and C. Ip, Eds.), Alan R. Liss Inc., Vol. 204, p. 21 (1986).

49.   R. Kleene and E.G. Berger, *Biochim. Biophys. Acta.*, **1154**, 283 (1993).

50.   T.A. Beyer, J.E. Sadler, J.R. Rearick, J.C. Paulson, and R. L. Hill, *Adv. Enzymol.* **52**, 23 (1981).

51.   H. Schachter, *Biochem. Cell Biol.*, **64**, 163 (1986).

52.   H. Schachter, I. Brockhausen, and E. Hull, *Methods Enzymol.*, **179**, 351 (1989).

53.   W.M. Watkins, *Carbohydr. Res.*, **149**, 1 (1986).

54.   A. Hagopian and E.H. Eylar, *Arch. Biochem. Biophys.*, **128**, 422 (1968).

55.   P.H. Johnson and W.M. Watkins, *Glycoconjugate J.*, **9**, 241 (1992).

56.   J.P. Prieels, D. Monnom, M. Dolmans, T.A. Beyer, and R.L. Hill, *J. Biol. Chem.*, **256**, 10456 (1981).

57.   E.V. Chandrasekaran, J.M. Rhodes, R.K. Jain, and K.L. Matta, *Biochem. Biophys. Res. Commun.*, **198**, 350 (1994).

58.   E.H. Holmes and S.B. Levery, *Archives of Biochem. Biophys.*, **274**, 633 (1989).

59.   M. Basu, D.E. Tripiti, K.K. Das, J.W. Kyle, H.C. Chon, R. Schaeper, and S. Basu, *Methods Enzymol.*, **138**, 575 (1987).

60.   M.A. Chester, A.D. Yates, and W.M. Watkins, *Eur. J. Biochem.*, **69**, 583 (1976).

61.   R.A. DiCioccio, J.J. Barlow, and K.L. Matta, *Clinica. Chim. Acta.*, **108**, 41 (1980).

62.   M.M. Palcic, L.D. Heerze, M. Pierce, and O. Hindsgaul, *Glycoconjugate J.*, **5**, 49 (1988).

63.   S. Yazawa, R. Madiyalakan, R.K. Jain, M. Shimoda, and K.L. Matta, *Anal. Biochem.*, **187**, 374 (1990).

64. R.U. Lemieux, D.R. Bundle, and D.A. Baker, *J. Am. Chem. Soc.*, **97**, 4076 (1975).

65. J.C. Paulsen, J.P. Prieels, L.R. Glasgow, and R.L. Hill, *J. Biol. Chem.*, **253**, 5617 (1978).

66. K. Maemura and M. Fukuda, *J. Biol. Chem.*, **267**, 24379 (1992).

67. R.A. DiCioccio, J.J. Barlow, and K.L. Matta, *J. Biol. Chem.*, **257**, 714 (1982).

68. R. Couso, H. van Halbeek, V. Reinhold, and S. Kornfeld, *J. Biol. Chem.*, **262**, 4521 (1987).

69. I. Brockhausen, *Crit. Rev. Clin. Lab. Sci.*, **30**, 65 (1993).

70. H. Schachter and I. Brockhausen, In *Glycoconjugates, Composition, Structure, and Function*, (H.J. Allen and E.C. Kisailus, Eds.), Marcel Dekker, Inc., New York, p. 263 (1992).

71. F.G. Hanisch, H. Egge, J.P. Katalinic, and G. Uhlenbruck, *Eur. J. Biochem.*, **155**, 239 (1986).

72. E.F. Hounsell, A.M. Lawson, J. Feeney, H.C. Gooi, N.J. Pickering, M.S. Stoll, S.C. Lui, and T. Feizi, *Eur. J. Biochem.*, **148**, 367 (1985).

73. H. van Halbeek, J.F.G. Vliegenthart, A.M. Fiat, and P. Jolles, *FEBS Lett.*, **187**, 81 (1985).

74. S. Yazawa, S.A. Abbas, R. Madiyalakan, J.J. Barlow, and K.L. Matta, *Carbohydr. Res.*, **149**, 241 (1986).

75. I. Brockhausen, G. Moller, A. Pollex-Kruger, V. Rutz, H. Paulsen, and K.L. Matta, *Biochem. Cell Biol.*, **70**, 99 (1992).

76. E.V. Chandrasekaran, R.K. Jain, and K.L. Matta, *J. Biol. Chem.*, **267**, 19929 (1992).

77. D. Williams and H. Schachter, *J. Biol. Chem.*, **255**, 11247 (1980).

78. D. Williams, G.D. Longmore, K.L. Matta, and H. Schachter, *J. Biol. Chem.*, **255**, 11253 (1980).

79. W.E. Wingert and P. Cheng, *Biochemistry*, **23**, 690 (1984).

80. W. Kuhns, V. Rutz, H. Paulsen, K.L. Matta, M.A. Baker, M. Barner, M. Granovsky, and I. Brockhausen, *Glycoconjugate J.*, **10**, 381 (1993).

81. O. Hindsgaul, K.J. Kaur, G. Srivastava, M. Blasczyk-Thurin, S.C. Crawley, L.D. Heerze, and M.M. Palcic, *J. Biol. Chem.*, **266**, 17858 (1991).

82. M. Sekine, Y. Hashimoto, F. Inagaki, T. Yamakawa, and A. Suzuki, *J. Biochem.*, **108**, 103 (1990).

83.   A. Kurosaka, I. Funakoshi, M. Matsuyama, T. Nagayo, and I. Yamashina, *FEBS Lett.*, **190**, 259 (1985).

84.   W.M. Blanken, G.J.M. Hooghwinkel, and D.H. van den Eijnden, *Eur. J. Biochem.*, **127**, 547 (1982).

85.   R.K. Jain, S.A. Abbas, and K.L. Matta, *J. Carbohydr. Chem.*, **7**, 377 (1988).

86.   M.M. Palcic, O.P. Srivastava, and O. Hindsgaul, *Carbohydr. Res.*, **159**, 315 (1987).

87.   B.T. Sheares and D.M. Carlson, *J. Biol. Chem.*, **258**, 9893 (1983).

88.   B.T. Sheares, J.Y.T. Jau, and D.M. Carlson, *J. Biol. Chem.*, **257**, 599 (1982).

89.   R. Madiyalakan, S.A. Abbas, and K.L. Matta, *Proceedings of AACR*, **27**, p.3 (1986).

90.   W.D. Klohs, K.L. Matta, J.J. Barlow, and R.J. Bernacki, *Carbohydr. Res.*, **89**, 350 (1981).

91.   T. Henion, D. Handley, F. Vargas, K.L. Matta, M.S. Piver, and K.R. Diakun, *Immunol. Invest.*, **20**, 1 (1991).

92.   (a) E.V. Chandrasekaran, R.K. Jain, and K.L. Matta, *Abstracts of the XVIIth Int. Carbohyd. Symp.*, Ottawa, Canada, p. 475 (1994).

       (b) E.V. Chandrasekaran, R.K. Jain, R.D. Larsen, K. Wlasichuk, and K.L. Matta, *Biochemistry*, **34**, 2925 (1995).

93.   T.A. Beyer, J.I. Rearick, J.C. Paulson, J.P. Prieels, J.E. Sadler, and R.L. Hill, *J. Biol. Chem.*, **254**, 12531 (1979).

94.   M.L.E. Bergh, G.J.M. Hooghwinkel, and D.H. van den Eijnden, *Biochim. Biophys. Acta*, **660**, 161 (1981).

95.   D.H. van den Eijnden, M.L.E. Bergh, B. Dieleman, and W.E.C.M. Schiphorst, *Hoppe-Seyler Z. Physiol. Chem.*, **362**, 113 (1981).

96.   H. Baubichon-Cortray, P. Broquet, P. George, and P. Louisot, *Glycoconjugate J.*, **6**, 115 (1989).

97.   J.T.Y. Lau and T.P. O'Hanlon, In *Glycoconjugates, Composition, Structure, and Function*, (H.J. Allen and E.C. Kisailus, Eds.), Marcel Dekker, Inc., New York, p. 499 (1992).

98.   K. Sasaki, E. Watanabe, K. Kawashima, S. Sekine, T. Dohi, M. Oshima, N. Hanai, T. Nishi, and M. Hasegawa, *J. Biol. Chem.*, **268**, 22782 (1993).

99. J. Weinstein, E.U. Lee, K. McEntee, P. Lai, and J.C. Paulson, *J. Biol. Chem.*, **262**, 17735 (1987).

100. K. Yamashita, K. Umetsu, T. Suzuki, and T. Ohkura, *Biochemistry*, **31**, 11647 (1992).

101. D.X. Wen, B.D. Livingston, K.F. Medzihradszky, S. Kelm, A.L. Burlingame, and J.C. Paulson, *J. Biol. Chem.*, **267**, 21011 (1992).

102. A. Lubineau, C. Auge, and P. Francois, *Carbohydr. Res.*, **228**, 137 (1992).

103. Y. Kajihara, H. Kodama, T. Wakabayashi, and H. Hashimoto, *Abstracts of the XVIth Int. Carbohydr. Symp.*, Paris, France, p. 168 (1992).

104. E.V. Chandrasekaran, R.K. Jain, and K.L. Matta, *Abstracts of the 23rd Annual Meeting of Society for Glycobiology*, Notre Dame, USA, p. 742 (1994).

105. G. Pfeiffer, K.H. Strube, and R. Geyer, *Biochem. Biophys. Res. Commun.*, **189**, 1681 (1992).

106. A. Shilatifard, R.K. Merkle, D.E. Helland, J.L. Welles, W.A. Haseltine, and R.D. Cummings, *J. Virol.*, **67**, 943 (1993).

107. T.P. Mawhinney, D.C. Landrum, D.A. Gayer, and G.J. Barbero, *Carbohydr. Res.*, **235**, 179 (1992).

108. C. Capon, C.L. Laboisse, J.M. Wieruszeski, J.J. Maoret, C. Augeron, and B. Fournet, *J. Biol. Chem.*, **267**, 19248 (1992).

109. T. Kumazaki and A. Yoshida, *Proc. Natl. Acad. Sci., U.S.A.*, **81**, 4193 (1984).

110. A. Sarnesto, T. Kohlin, J. Thurin, J., and M. Blaszczyk-Thurin, *J. Biol. Chem.*, **265**, 15067 (1990).

111. T.L. Lowary, S.J. Swiedler, and O. Hindsgaul, *Carbohydr. Res.*, **256**, 257 (1994).

112. R. Madiyalakan, S. Yazawa, S.A. Abbas, J.J. Barlow, and K.L. Matta, *Anal. Biochem.*, **152**, 22 (1986).

113. R. Madiyalakan, S. Yazawa, J.J. Barlow, and K.L. Matta, *Cancer Letters*, **30**, 201 (1986).

114. S. Yazawa, T. Asao, Y. Nagamachi, and K.L. Matta, *J. Tumor Marker Oncol.*, **4**, 355 (1989).

115. A.A. Bergwerff, J.A. van Kuik, W.E.C.M. Schiphorst, C.A.M. Koeleman, D.H. van den Eijnden, J.P. Kamerling, and J.F.G. Vliegenthart, *FEBS Lett.*, **334**, 133 (1993).

116.   E.V. Chandrasekaran, R.K. Jain, S.A. Abbas, and K.L. Matta, *Abstracts of the 11th Int. Symp. on Glycoconj.*, Toronto, Canada, p. 272 (1991).

117.   K.L. Matta, R.K. Jain, and E.V. Chandrasekaran, U.S. Patent Application Submitted (1992).

118.   (a) C. Yuen, A.M. Lawson, W. Chai, M. Larkin, M.S. Stoll, A.C. Stuart, F.X. Sullivan, T.J. Ahern, and T. Feizi, *Biochemistry, 31*, 9126 (1992); (b) Steven Rosen, *Personal Communication* (1994).

119.   K.C. Nicolaou, N.J. Bockovich, and D.R. Carcanague, *J. Am. Chem. Soc., 115*, 8843 (1993).

120.   K.L. Matta, C.F. Piskorz, G.V. Reddy, E.V. Chandrasekaran, and R.K. Jain, *ACS Symp. Ser., 560*, 120 (1994).

121.   R.K. Jain, R.D. Locke, E.V. Chandrasekaran, and K.L. Matta, *Bioorgan. & Med. Chem. Lett., 2*, 63 (1992).

122.   M. Blaszczyk-Thurin, A. Sarnesto, J. Thurin, O. Hindsgaul, and H. Koprowski, *Biochem. Biophys. Res. Commun., 151*, 100 (1988).

123.   R.K. Jain, S.M. Pawar, E.V. Chandrasekaran, C.F. Piskorz, and K.L. Matta, *Bioorgan. Med. Chem. Lett., 3*, 1333 (1993).

124.   S. Yazawa, J. Nakamura, T. Asao, Y. Nagamachi, M. Sagi, K.L. Matta, T. Tachikawa, and M. Akamatsu, *Jpn. J. Cancer Res., 84*, 989 (1993).

125.   B. Wang, K. Akiyama, and H. Kimura, *Vox Sang., 66*, 280 (1994).

126.   Y. Kato and R.G. Spiro, *J. Biol. Chem., 264*, 3364 (1989).

127.   F. Vavasseur, K. Dole, J. Yang, K.L. Matta, A. Corfield, C. Paraskeva, and I. Brockhausen, *Eur. J. Biochem.*, In Press (1994).

128.   R.K. Jain and K.L. Matta, *Abstract of the XVIIth Int. Carbohyd. Symp.*, Ottawa, Canada, p. 443 (1994).

129.   B.L. Slomiany, J. Piotrowski, H. Nishikawa, and A. Slomiany, *Biochem. Biophys. Res. Commun., 157*, 61 (1988).

130.   R.K. Jain, X.-G. Liu, and K.L. Matta, *Carbohydr. Res.*, In Press (1995).

131.   Y. Nishikawa, W. Pegg, H. Paulsen, and H. Schachter, *J. Biol. Chem., 263*, 8270 (1988).

132.   G. Alton, G. Srivastava, K.J. Kaur, and O. Hindsgaul, *Bioorgan. and Med. Chem., 2*, 675 (1994).

133.   K.J. Kaur and O. Hindsgaul, *Carbohydr. Res., 226*, 219 (1992).

134. S.H. Khan, C.A. Compston, M.M. Palcic, and O. Hindsgaul, *Carbohydr. Res.*, **262**, 283 (1994).

135. I. Brockhausen, G. Moller, J.M. Yang, S.H. Khan, K.L. Matta, H. Paulsen, A.A. Grey, R.N. Shah, and H. Schachter, *Carbohydr. Res.*, **236**, 281 (1992).

136. J.W. Dennis, *Cancer Surveys* **7**, 573 (1988).

137. J.W. Dennis, and S. Laferte, *Cancer Res.*, **49**, 945 (1989).

138. M. Pierce, J. Arango, S.H. Tahir, and O. Hinsgaul, *Biochem. Biophys. Res. Commun.*, **146**, 679 (1987).

139. S.H. Khan, S.A. Abbas, and K.L. Matta, *Carbohydr. Res.*, **193**, 125 (1989).

140. S.H. Tahir and O. Hindsgaul, *Can. J. Chem.*, **64**, 1771 (1986).

141. O. Hindsgaul, S.H. Tahir, O. Srivastava, and M. Pierce, *Carbohydr. Res.*, **173**, 263 (1988).

142. S.H. Khan, S.C. Crawley, O. Kanie, and O. Hindsgaul, *J. Biol. Chem.*, **268**, 2468 (1993).

143. S.H. Khan, S.A. Abbas, and, and K.L. Matta, *Carbohydr. Res.*, **205**, 385 (1990).

144. O. Kanie, S.C. Crawley, M.M. Palcic, and O. Hindsgaul, *Carbohydr. Res.*, **243**, 139 (1993).

145. S.H. Khan and K.L. Matta, *Carbohydr. Res.*, **243**, 29 (1993).

146. O. Kanie, S.C. Crawley, M.M. Palcic, and O. Hindsgaul, *Bioorg. and Med. Chem.*, **2**, 1231 (1994).

147. I. Brockhausen, F. Reck, W. Kuhns, S.H. Khan, K.L. Matta, E. Meinjohanns, H. Paulsen, R.N. Shah, M.A. Baker, and H. Schachter, *Glycoconjugate J.*, In Press (1995).

148. (a) T.L. Lowary and O. Hindsgaul, *Carbohydr. Res.*, **249**, 163 (1994); (b) *ibid.* **251**, 33 (1994).

149. H. Paulsen, *Angew. Chem. Int. Ed. Engl.*, **21**, 155 (1982).

150. R.R. Schmidt, *Angew. Chem. Int. Ed. Engl.*, **25**, 212 (1986).

151. Y.C. Lee and R.T. Lee, In *Glycoconjugates: Composition, Structure and Function* (H.J. Allen and E.C. Kisailus, Eds.), Marcel Dekker, Inc., p. 121 (1992).

152. O. Hindsgaul, *Sem. Cell Biol.*, **2**, 319 (1991).

153. M.P. DeNinno, *Synthesis*, 583 (1991).

154. D.G. Drueckhammer, W.J. Hennen, R.L. Pederson, C.F. Barbas, C.M. Gautheron, T. Krach, and C. Wong, *Synthesis*, 499 (1991).

155. Y. Ichikawa, J.L. Liu, G.J. Shen, and C.H. Wong, *J. Am. Chem. Soc.*, **113**, 6300 (1991).

156. R. Zeitler, A. Giannis, S. Danneschewski, E. Henk, T. Henk, C. Bauer, W. Reutter, and K. Sandhoff, *Eur. J. Biochem.*, **204**, 1165 (1992).

157. S.F. Kuan, J.C. Byrd, C. Basbaum, and Y.S. Kim, *J. Biol. Chem.*, **264**, 19271 (1989).

158. J.D. Esko, T.A. Fritz, A. Sarkar, and F. Lugenwa, *Glycobiology*, **3**, 534 (1993).

Chapter 19

PRACTICAL SYNTHESIS OF OLIGOSACCHARIDES BASED ON GLYCOSYLTRANSFERASES AND GLYCOSYLPHOSPHITES

CHI-HUEY WONG

Department of Chemistry, The Scripps Research Institute,
10666 North Torrey Pines Road, La Jolla, CA 92307, USA

Abstract   This chapter describes the substrate specificity and practical synthetic application of glycosyltransferases. Regeneration of sugar nucleotides in glycosyltransferase-catalyzed synthesis of oligosaccharides is a key to large-scale processes.   In addition, glycosylation with glycosylphosphites and application of glycosyltransferases to solid-phase glycopeptide synthesis are described.

1. INTRODUCTION

Complex carbohydrates are often part of cell-surface structures. They are involved in various types of biochemical recognitions. These carbohydrates often exist in minute quantities and are difficult to isolate, characterize, and synthesize in quantities large enough for biological study and therapeutic evaluation. A limited number (7-8) of monosaccharides are found as monomers in the complex carbohydrates of mammalian systems and biologically active, complex carbohydrates often contain only 4-5 different monosaccharides. However, due to their multifunctionality, even this limited number of monomers can give rise to millions of possible oligosaccharide structures. The synthesis of various oligosaccharide structures in a practical manner for structural and biological study therefore still represents a significant challenge in synthetic chemistry. This chapter describes two of the most recently developed methods for the synthesis of

467

oligosaccharides: the use of glycosyltransferases and the other is based on glycosylphosphites as glycosylation reagents.

2. ENZYMATIC FORMATION OF GLYCOSIDIC BONDS

Nature employs two groups of enzymes in the biosynthesis of oligosaccharides: the enzymes of the Leloir pathway[1] and those of non-Leloir pathways. The Leloir pathway enzymes are responsible for the synthesis of most N- and O-linked glycoproteins and other glycoconjugates in mammalian systems. The N-linked glycoproteins are characterized by a β-glycosidic linkage between a GlcNAc residue and the δ-amide nitrogen of an asparagine. The less common O-linked glycoproteins contain an α-glycosidic linkage between a GalNAc (or xylose) and the hydroxyl group of a serine or threonine. The addition of oligosaccharide chains to glycoproteins occurs co-translationally for both O-linked and N-linked types, and takes place in the endoplasmic reticulum and the Golgi apparatus.[2,3] The biosynthesis of the N-linked type involves an initial synthesis of a dolichol pyrophosphoryl oligosaccharide intermediate in the endoplasmic reticulum by the action of GlcNAc-transferases and mannosyltransferases. This structure is further glucosylated, and then the entire oligosaccharide moiety is transferred to an Asn residue of the growing peptide chain by the enzyme oligosaccharyltransferase.[3] Before transport into the Golgi apparatus, the glucose residues and some mannose residues are removed by the action of glucosidase I and II and a mannosidase, in a process called trimming, to reveal a core pentasaccharide (peptide-Asn-$(GlcNAc)_2$-$(Man)_3$). The resulting core structure is further processed by mannosidases and glycosyltransferases present in the Golgi apparatus to produce either the high-mannose, complex, or the hybrid-type oligosaccharides. Monosaccharides are then added sequentially to this core structure to provide the fully-elaborated oligosaccharide chain.

In contrast to the dolichol pyrophosphate mediated synthesis of N-linked oligosaccharides, for the synthesis of O-linked oligosaccharides in the Golgi apparatus, monosaccharide residues are added sequentially to the growing oligosaccharide chain.[3]

The glycosyltransferases of the Leloir pathway in mammalian systems utilize as glycosyl donors monosaccharides which are activated as glycosyl esters of nucleoside mono- or diphosphates.[3] Non-Leloir transferases typically utilize glycosyl phosphates as activated donors. The Leloir glycosyltransferases utilize primarily eight nucleoside mono- or diphosphate sugars as monosaccharide donors for the synthesis of most oligosaccharides: UDP-Glc, UDP-GlcNAc, UDP-Gal, UDP-GalNAc, GDP-Man, GDP-Fuc, UDP-GlcUA, and CMP-NeuAc. Many other monosaccharides, such as the anionic or sulfated sugars of heparin and chondroitin sulfate, are also found in mammalian systems, but they usually are a result of modification of a particular sugar after it is incorporated into an oligosaccharide structure. A very diverse array of monosaccharides (e.g. xylose, arabinose, KDO, deoxysugars) and oligosaccharides are also present in microorganisms, plants, and invertebrates.[4,5] The enzymes responsible for their biosynthesis, however, have not been extensively exploited for synthesis, though they follow the same principles as do those in mammalian systems.

The glycosyltransferases from the Leloir and non-Leloir pathways have been used for the synthesis of oligosaccharides and glycoconjugates.[6,7] Glycosidases have also been exploited for synthesis.[8] Each group of enzymes has certain advantages and disadvantages for synthesis. Glycosyltransferases are highly specific in the formation of glycosides, however the availability of many of the necessary transferases is a limiting factor. Fortunately, the recent advances in genetic engineering and recombinant techniques have rapidly alleviated this drawback. In contrast to the transferases, glycosidases have the advantage of wider availability and lower cost but they are not as specific or high-yielding in synthetic reactions.

3. GLYCOSYLTRANSFERASES OF THE LELOIR PATHWAY

Leloir pathway glycosyltransferases utilize glycosyl esters of nucleoside phosphates as activated monosaccharide donors. Most of these sugar nucleoside phosphates are synthesized *in vivo* from the corresponding monosaccharides. The initial step is a kinase-mediated phosphorylation to produce a glycosyl phosphate. This glycosyl phosphate then reacts with a nucleoside triphosphate (NTP), catalyzed by a nucleoside diphosphosugar pyrophosphorylase, to afford an activated nucleoside diphosphosugar (eq. 1). Other sugar nucleoside phosphates, such as GDP-Fuc and UDP-GlcUA, are produced by further enzymatic modification of these key sugar nucleotide phosphates. Another exception is CMP-NeuAc, which is formed by the direct reaction of NeuAc with CTP (eq. 2). Some of the enzymes involved in the biosynthesis of sugar nucleotides also accept unnatural sugars as substrates. In general, however, the rates are quite slow, thus limiting the usefulness of this approach.

| Sugar-1-P | + | NTP | → | NDP-Sugar | + | PPi | (Eq. 1) |
| NeuAc | + | CTP | → | CMPNeuAc | + | PPi | (Eq. 2) |

The appropriate nucleoside triphosphates are utilized as substrates for the biosynthesis of sugar nucleoside phosphates. Therefore, enzymatic preparation of these donors for use in glycosylations requires a preparative scale synthesis of NTPs.

Most preparative-scale enzymatic syntheses of NTPs use commercially available nucleoside monophosphates (NMPs) as starting materials. Alternatively, all of the NMPs can be obtained from yeast RNA digests at low cost, or can be readily prepared chemically.[9] In general, these methods involve the sequential use of two kinases to transform NMPs to NTPs, via the corresponding NDPs. There are three kinases which may be used to synthesize NTPs from the corresponding NDPs, each uses a different phosphoryl donor: pyruvate kinase (EC 2.7.1.40) uses phosphoenolpyruvate (PEP)[10] as a phosphoryl donor, acetate kinase (EC 2.7.2.1) uses acetyl phosphate, and nucleoside diphosphate kinase (EC 2.7.4.6) uses ATP. Pyruvate kinase is generally the enzyme of choice because it is less expensive than nucleoside diphosphate kinase,[9] and because PEP is more stable and provides a more thermodynamically favorable driving force for phosphorylation than does acetyl phosphate (Figure 1).

The preparation of NDPs from NMPs is more complicated, and requires different enzymes for each NMP. Adenylate kinase (EC 2.7.4.3) phosphorylates AMP and CMP, and also slowly phosphorylates UMP.[9] Guanylate kinase (EC 2.7.4.8) catalyzes the phosphorylation of GMP. Nucleoside monophosphate kinase

(EC 2.7.4.4) uses ATP to phosphorylate AMP, CMP, GMP, and UMP; however this enzyme is relatively expensive and unstable. Both CMP and UMP kinases exist but are not commercially available.   For those kinases requiring ATP as a phosphorylating agent, ATP is usually used in a catalytic amount and recycled from ADP using pyruvate kinase/PEP or acetate kinase/acetylphosphate.[9] Phosphoenolpyruvate may be prepared chemically from pyruvate or generated enzymatically from D-3-phosphoglyceric acid[11] (Figure 2).

Comparisons of chemical and enzymatic methods for the synthesis of NTPs led to the conclusion that enzymatic methods provide the most convenient route to CTP and GTP.[9] Chemical deamination of CTP is the best method for preparing UTP. ATP is relatively inexpensive from commercial sources, although it has been synthesized enzymatically from AMP on a 50 mmol scale. Mixtures of NTPs can be prepared from RNA by sequential nuclease P1, polynucleotide phosphorylase, and pyruvate kinase-catalyzed reactions.[12]   This mixture can be selectively converted to a sugar nucleotide using a particular sugar nucleoside diphosphate pyrophosphorylase.

4. SUBSTRATE SPECIFICITY AND SYNTHETIC APPLICATIONS OF GLYCOSYLTRANSFERASES

In general, there are many glycosyltransferases available for each sugar nucleotide glycosyl donor, each transfering the particular donor to different acceptors. These enzymes are usually considered to be specific for a given glycosyl donor and acceptor, as well as for the stereochemistry and the linkage position of the newly-formed glycoside bond. Though systematic investigations of the *in vitro* substrate specificity of most glycosyltransferases have not been carried out, some deviations from this picture of absolute specificity have already emerged, both in the glycosyl donors and acceptors.

4.A. Galactosyltransferase

Because of its availability, β-(1→4)-galactosyltransferase (UDP-Gal:*N*-acetyl-glucosamine β-(1→4)-galactosyltransferase, EC 2.4.1.22)[2,7] is one of the most extensively studied mammalian glycosyltransferases with regard to synthesis and substrate specificity. This enzyme catalyzes the transfer of galactose from UDP-

Gal to the 4-position of β-linked GlcNAc residues to produce the Galβ-(1→4)-GlcNAc substructure. In the presence of lactalbumin, however, glucose is the preferred acceptor, resulting in the formation of lactose, Galβ-(1→4)-Glc. The enzyme has been employed in the *in vitro* synthesis of *N*-acetyllactosamine and its glycosides, as well as other galactosides.

Galactosyltransferase utilizes as acceptor substrates: *N*-acetylglucosamine and glucose and its β-glycosides, 2-deoxyglucose, D-xylose, 5-thioglucose, *N*-acetylmuramic acid, and myo-inositol.[13] Modifications at the 3- or 6-position of the acceptor GlcNAc are also tolerated. For example, Fucα-(1→6)-GlcNAc and NeuAcα-(2→6)-GlcNAc are substrates.[14,15] Acceptor substrates which are derivatized at the 3-position include: 3-O-methyl-GlcNAc, 3-deoxy-GlcNAc, 3-O-allyl-GlcNAcβOBn, and 3-oxo-GlcNAc.[16] All glycosides of GlcNAc which are reported to be substrates for the galactosyltransferase have β-glycosidic linkages. Both α- and β-glycosides of glucose are acceptable, but the presence of lactalbumin is required for galactosyl transfer onto α-glycosides. D-Mannose, D-allose, D-galactose, D-ribose, and D-xylose are not substrates, nor are monosaccharides which have a negative charge, such as glucuronic acid and α-glucose-1-phosphate. Figure 3 illustrates several disaccharides which have been synthesized with galactosyltransferase.[16-21] A particularly interesting example is the β,β-(1→1)-linked disaccharide, in which the anomeric hydroxyl of 3-acetamido-3-deoxyglucose serves as the acceptor moiety. The acetamido function apparently controls the position of glycosylation.

β-(1→4)-Galactosyltransferase has also been employed in solid-phase oligosaccharide synthesis, and has been used to galactosylate *gluco* or *cellobio* subunits of polymer-supported oligosaccharides and polysaccharides.[22,23] The resulting oligosaccharides can then be removed from the support by either a photochemical cleavage or a chymotrypsin-mediated hydrolysis. The types of polymer supports employed include polyacrylamide and a water-soluble poly(vinyl alcohol). *N*-Acetylglucosaminyl amino acids and peptides have also been used as substrates for galactosyltransferase to afford galactosylated glycopeptides.[21,24,25] The carbohydrate chain can then be extended further with other transferases, e.g., sialyltransferase. Similarly, the synthesis of a ceramide glycoside which was subsequently sialylated enzymatically provided a GM$_3$ analog.[26] A recent advance

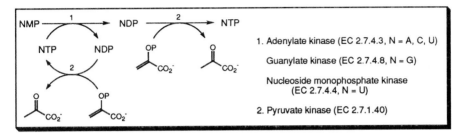

Figure 1. Enzymatic synthesis of nucleoside triphosphates.

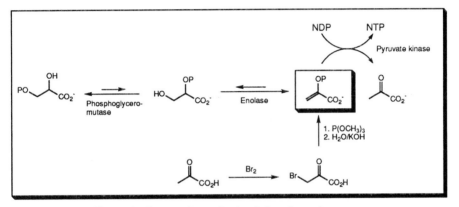

Figure 2. Enzymatic synthesis of phosphoenol pyruvate.

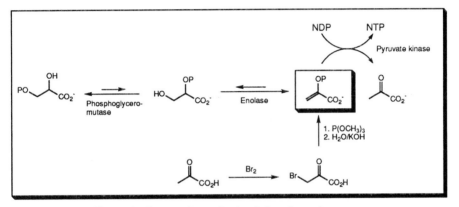

Figure 3. Some disaccharides synthesized using β-(1→4)-galactosyltransferase.

in the solid-phase synthesis is illustrated in the chemo-enzymatic synthesis of sialyl Lewis[x] glycopeptide.[27]

With regard to the donor substrate, the β-galactosyltransferase also transfers the D-isomers of glucose, 4- and 6-deoxygalactose, arabinose, glucosamine, galactosamine, N-acetylgalactosamine, 2-deoxygalactose, and 2-deoxyglucose via their respective UDP-derivatives, providing an enzymatic route to oligosaccharides which terminate in β-(1→4)-linked residues other than galactose (Figure 4). A noteworthy example example is the transfer of 5-thiogalactose to an acceptor. Although the rate of transfer of many of these unnatural donor substrates is quite slow, this method is useful for milligram-scale synthesis. The α-(1→3)-galactosyltransferase and α-(1→3)-GalNAc transferase involved in the synthesis of the B and A blood-group antigens have also been studied.[28]

4.B. Sialyltransferase

Several sialyltransferases, classified as either α-(2→6)- or α-(2→3)-sialyltransferases, have been used for oligosaccharide synthesis.[29-31] These sialyltransferases generally transfer N-acetylneuraminic acid to either the 3- or 6-position of terminal Gal or GalNAc residues. There is also a α-(2→8)-sialyltransferase which is involved in the synthesis of α-(2→8)-linked polysialic acids.[32] Some sialyltransferases have been shown to accept CMP-NeuAc analogs which are derivatized at the 9-position of the sialic acid side chain.[33-35] These include those analogs in which the hydroxyl group at C-9 is replaced with: amino, fluoro, azido, acetamido, or the benzamido group. Analogs of the acceptors Galβ-(1→4)-GlcNAc and Galβ-(1→3)-GalNAc, in which the acetamido function is replaced by an azide, phthalimide, carbamate, or pivaloyl functionality are also substrates for the enzymes.[36] A recent synthesis of a GM3 analog started with a disaccharyl ceramide derivative where the fatty acid amide group was replaced with the azide group.[37] Incorporation of sialic acid analogs into sialosides is, however, problematic because sialyltransferases are very specific for their natural substrates.

4.C. Fucosyltransferase

Fucosyltransferases are involved in the biosynthesis of many oligosaccharide structures such as the blood-group determinants and cell-surface and tumor-

Donor Substrate		Relative Rate	References
UDP-Gal	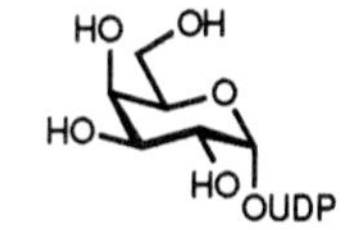	100	38
UDP-Glc	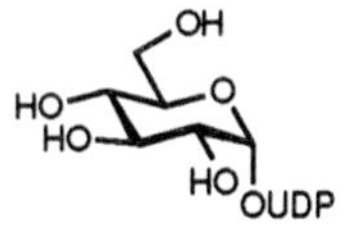	0.3	39
UDP-4-deoxy-Glc	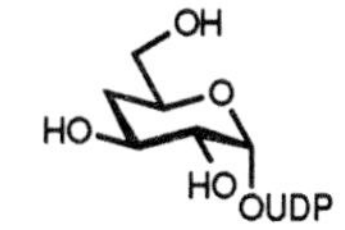	5.5	38
UDP-Ara	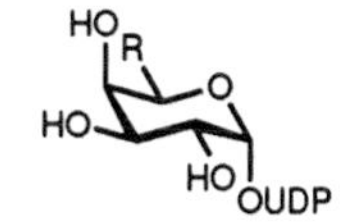	4.0 R = H 1.3 R = CH₃ 0.2 R = CH₂F	38
UDP-GalNAc	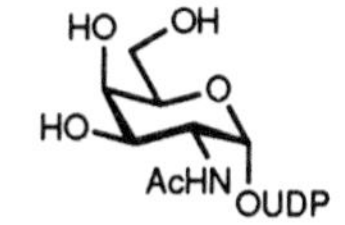	4.0	39
UDP-GlcNAc	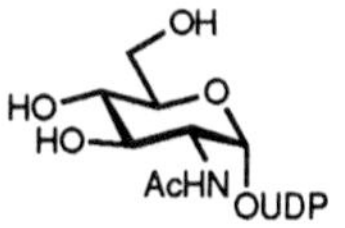	0.00	39
UDP-GlcN	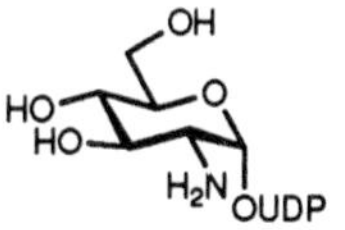	0.09	39
UDP-5-thio-Gal	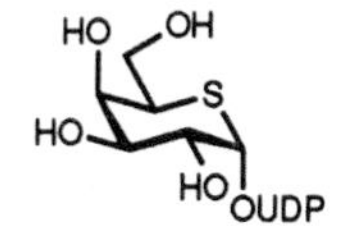	5.0	40
UDP-2-deoxy-Gal	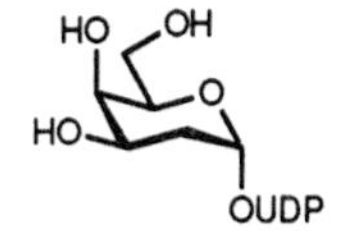	90	20

Figure 4. Relative rates of β-(1→4)-galactosyltransferase catalyzed transfer of donor substrates.

associated antigens.   Fucosylation is one of the last modifications of oligosaccharides *in vivo*. Several fucosyltransferases have been isolated and have been used for *in vitro* synthesis.[41-43] For example, α-(1→3)-fucosyltransferase has been used to L-fucosylate the 3-position of the GlcNAc of *N*-acetyllactosamine and of sialyl α-(2→3)-*N*-acetyllactosamine to provide the Lewis[x] and sialyl Lewis[x] structural motifs, respectively.[42,43]   Several other acceptor substrates with modifications in the GlcNAc residue can also be fucosylated (Figure 5).[42] Galβ-(1 →4)-Glc, Galβ-(1→4)-Glucal, and Galβ-(1→4/5)-thio-Glc are all substrates. A similar enzyme, α-(1→3/4)-fucosyltransferase, has also been used for synthesis. This enzyme fucosylates either the 3-position of the GlcNAc moiety in Galβ-(1→ 4)-GlcNAc or the 4-position of GlcNAc in Galβ-(1→3)-GlcNAc to afford Lewis X or Lewis[a], respectively.[42] The corresponding sialylated substrates have also been employed as acceptors.

The Lewis a α-(1→4)-fucosyltransferase has been shown to transfer unnatural fucose derivatives from their GDP esters. 3-Deoxyfucose and L-arabinose are transferred to Galβ-(1→4)-GlcNAcβ-O(CH$_2$)$_8$CO$_2$CH$_3$ at a rate of 2.3% and 5.9%, respectively, relative to L-fucose.[41] Furthermore, this enzyme will transfer a fucose residue which is substituted on C-6 by a very large, sterically crowded substituent.[44] This approach has been used to alter the antigenic properties of cell-surface glycoproteins.

4.D.   *N*-Acetylglucosaminyltransferase

The *N*-acetylglucosaminyltransferases *in vivo* control the branching pattern of *N*-linked glycoproteins.[45] Each of these enzymes transfers a β-GlcNAc residue from the donor UDP-GlcNAc to a mannose or other acceptor. The GlcNAc transferases I-VI, which catalyze the addition of the GlcNAc residues to the core pentasaccharide of *N*-linked asparagine glycoproteins as outlined in Figure 6 have been identified and characterized.[45-47]   These, as well as other GlcNAc transferases, have been exploited for purposes of oligosaccharide synthesis.[48,49]

GlcNAc transferases have also been utilized to transfer non-natural residues onto oligosaccharides. In addition to GlcNAc, *N*-acetylglucosaminyltransferase I from human milk catalyzes the transfer[49] of 3-, 4-, or 6-deoxy-GlcNAc from its respective   UDP   derivative   to   Manα-(1→3)-[Manα-(1→6)]-Manβ-

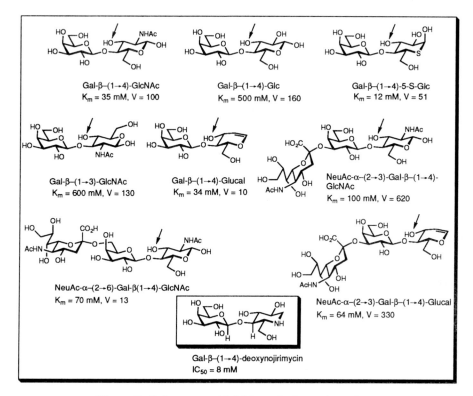

Figure 5. Substrates and inhibitors for fucosyltransferase.

Gal-β–(1→4)-GlcNAc
K_m = 35 mM, V = 100

Gal-β–(1→4)-Glc
K_m = 500 mM, V = 160

Gal-β–(1→4)-5-S-Glc
K_m = 12 mM, V = 51

Gal-β–(1→3)-GlcNAc
K_m = 600 mM, V = 130

Gal-β–(1→4)-Glucal
K_m = 34 mM, V = 10

NeuAc-α–(2→3)-Gal-β–(1→4)-GlcNAc
K_m = 100 mM, V = 620

NeuAc-α–(2→6)-Gal-β(1→4)-GlcNAc
K_m = 70 mM, V = 13

NeuAc-α–(2→3)-Gal-β–(1→4)-Glucal
K_m = 64 mM, V = 330

Gal-β–(1→4)-deoxynojirimycin
IC_{50} = 8 mM

Figure 6. Specificity of GlcNAc transferases I-VI

$O(CH_2)_8CO_2CH_3$. The 3-, 4-, and and 6-deoxy-GlcNAc analogs can also be transferred[49] by GlcNAc transferase II to tetrasaccharide acceptor GlcNAc-β(1→2)-Manα-(1→3)-[Manα-(1→6)]-Manβ-$O(CH_2)_8CO_2CH_3$.   In addition to the synthetic applications of GlcNAc tranferases, a GlcNAc transferase has been used to   attach   the   terminal   GlcNAc   of   GlcNAcβ-(1→4)-GlcNAcα dolichol pyrophosphate, a substance employed in the study of oligosaccharyl transferase.[50] A mouse kidney GlcNAc transferase was used in the synthesis of a hexasaccharide containing sialyl Lewis[x] structure. This enzyme catalyzes[51] the transfer of GlcNAc to 6-OH of Galβ-(1→3)-GlcNAc.

4.E. Mannosyltransferase

Various mannosyltransferases have been shown to transfer mannose and 4-deoxymannose from their respective GDP adducts to acceptors.[52] The α-(1→2)-mannosyltransferase from yeast has been cloned and overexpressed in E. coli (~ 1U/L) and was employed to transfer mannose to the 2-position of various derivatized α-mannosides and α-mannosyl peptides to produce the Manα-(1→2)-Man structural unit.[53] A recent report indicates that mannosyltransferases from pig liver accept GlcNAcβ-(1→4)-GlcNAc phytanyl pyrophosphate, an analog of the natural substrate in which the phytanyl moiety replaces dolichol.[54]

5. IN SITU CO-FACTOR REGENERATION IN GLYCOSYL-
TRANSFERASE REACTIONS

Though analytical- and small-scale synthesis using glycosyltransferases is extremely useful, the high cost of sugar nucleotides and the product inhibition caused by the released mono- or dinucleotides present major obstacles to their use in large-scale synthesis. A simple solution to both of these problems is to use a system in which the sugar nucleotide is regenerated in situ from the released nucleoside diphosphate.   The first example of the use of such a strategy is the galactosyltransferase-catalyzed synthesis of N-acetyllactosamine (Figure 7).[55]   A catalytic amount of UDP-Gal is used initially to glycosylate GlcNAc; UDP-Gal is regenerated from the product UDP and galactose using an enzyme-catalyzed reaction sequence which requires stoichiometric amounts of a phosphorylating agent.   A second regeneration system for UDP-Gal, which utilizes galactose-1-

phosphate uridyltransferase, has also been developed,[56] and has been used in the preparation of analogs such as 2'-deoxy-LacNAc and 2'-amino-2'-deoxy-LacNAc (Figure 8). A third regeneration method for UDP-Gal is based on sucrose synthetase, which catalyzes the formation of UDP-Glc from sucrose and UDP.[57]

In situ co-factor regeneration offers several advantages. First, a catalytic amount of nucleoside diphosphate and a stoichiometric amount of monosaccharide can be used as starting materials rather than a stoichiometric quantity of sugar nucleotide, thus tremendously reducing costs. Second, feed-back inhibition by the NDP is minimized due to its low concentration in solution. And third, isolation of the product is greatly facilitated.

A regeneration system for CMP-NeuAc has also been developed, and is illustrated in Figure 9.[58,59] The UDP-Gal and CMP-NeuAc regeneration schemes have been combined in a one-pot reaction (Figure 10) and applied to the synthesis[42] of sialyl Lewis[x]. N-Acetyllactosamine can also be generated in situ from the β-galactosidase catalyzed reaction[60] as a substrate for the sialyltransferase reaction (Figure 10).[61] The development of these regeneration systems, as well as the more recent development of regeneration schemes for UDP-GlcNAc,[62] GDP-Man,[53] GDP-Fuc[42] (Figure 11), and UDP-GlcUA[63] should facilitate the more widespread use of glycosyltransferases for oligosaccharide synthesis.

6. CLONING AND EXPRESSION OF GLYCOSYLTRANSFERASES AND SUGAR NUCLEOTIDE SYNTHETASES

The amount of a glycosyltransferase that can be isolated from a natural sources is often limited by the low concentrations of these enzymes present in most tissues and body fluids. The purification of glycosyltransferases is further complicated by the relative instability of this group of enzymes.[2] For this reason, the cloning of the glycosyltransferase genes into convenient expression systems has been of great interest.

To date, very few glycosyltransferases have been cloned, expressed, and produced in quantities sufficient for enzymatic.[64,65] However, given the advantages of the enzymatic synthesis of oligosaccharides over traditional schemes, research into the overexpression of glycosyltransferases will continue to flourish.

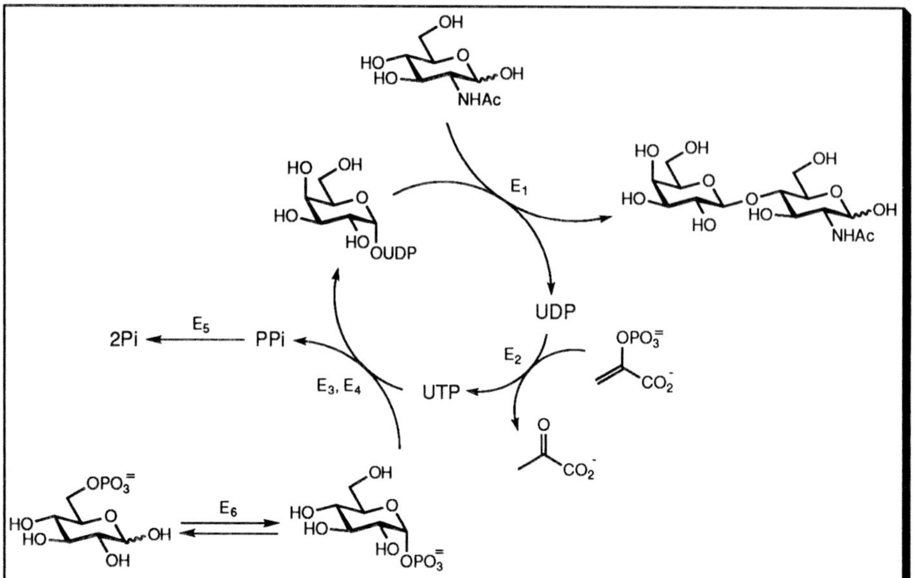

E$_1$: β–(1→4)-Galactosyltransferase; E$_2$: Pyruvate kinase; E$_3$: UDP-Glc pyrophosphorylase
E$_4$: UDP-Glc epimerase; E$_5$: Pyrophosphorylase; E$_6$: Phosphoglucomutase

Figure 7. Enzymatic transfer of galactose with galactosyltransferase.

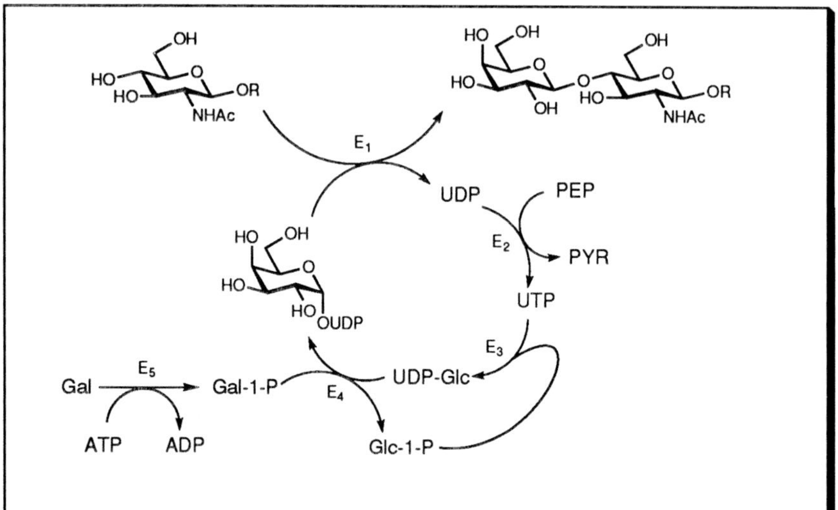

E$_1$: β–(1→4)-Galactosyltransferase; E$_2$: Pyruvate kinase; E$_3$: UDP-Glc pyrophosphorylase
E$_4$: Galactose-1-phosphate uridyltransferase; E$_5$: Galactokinase

Figure 8. Galactosyltransferase-catalyzed glycosylation with *in situ* regeneration
of UDP-Gal.

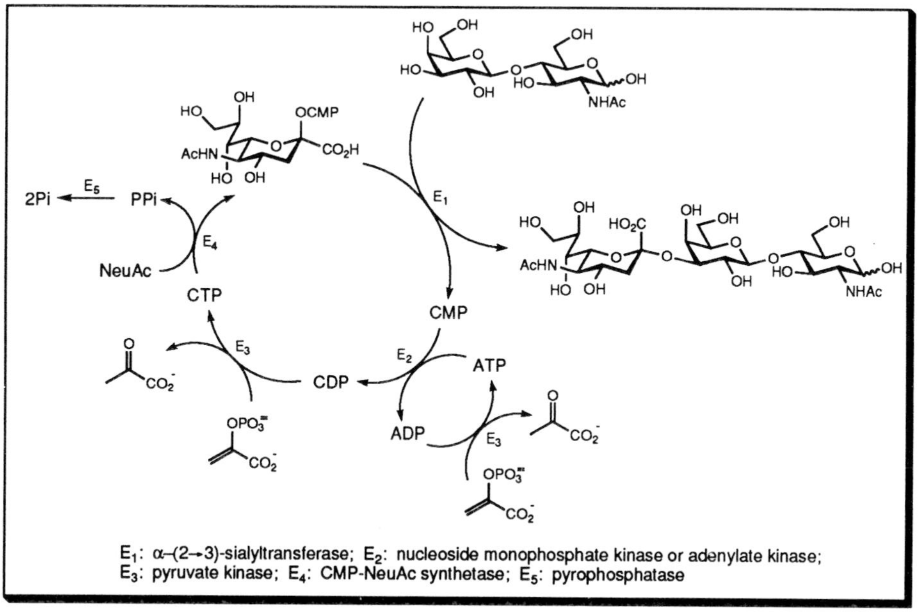

E_1: α–(2→3)-sialyltransferase; E_2: nucleoside monophosphate kinase or adenylate kinase;
E_3: pyruvate kinase; E_4: CMP-NeuAc synthetase; E_5: pyrophosphatase

Figure 9. Enzymatic sialylation with *in situ* regeneration of CMP-NeuAc.

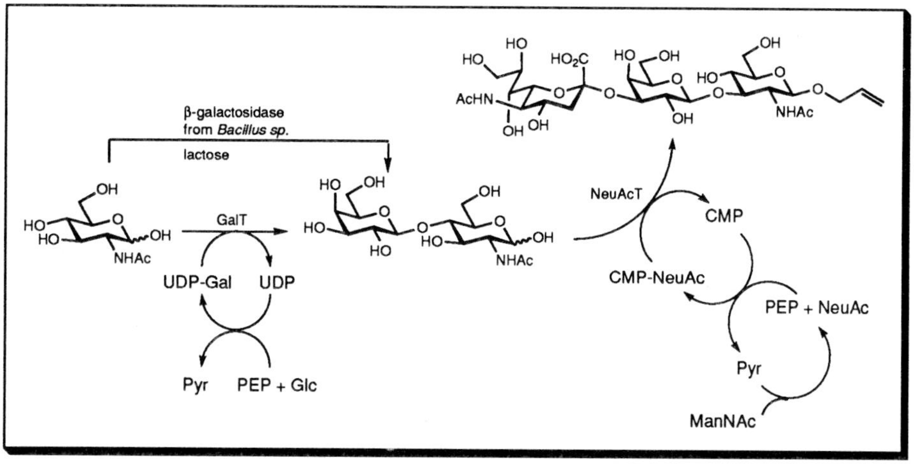

Figure 10. Synthesis of a trisaccharide using a galactosidase/sialyltransferase enzyme system.

The most practical expression systems are those based on the baculovirus[42] and yeast.[66]

7. GLYCOPEPTIDE SYNTHESIS AND GLYCOPROTEIN REMODELING

There is substantial interest in developing methods that will permit modification of oligosaccharide structures on glycoproteins by removing and adding sugar units ("remodeling") and in making new types of protein-oligosaccharide conjugates. The motivation for these efforts is the hope that modification of the sugar components of naturally-occurring or unnatural glycoproteins might increase serum lifetime and solubility, decrease antigenicity, and promote uptake by target cells and tissues.

Enzymes are plausible catalysts for manipulating the oligosaccharide content and structure of glycoproteins. The delicacy and polyfunctional character of proteins, and the requirement for high selectivity in their modification, indicate that classical synthetic methods will be of limited use. A major problem in the widespread use of enzymes in glycoprotein remodeling and generation is that many of the glycosyl transferases that are plausible candidates for this area are not available. Furthermore, glycosyltransferases most likely act on unfolded or partially folded proteins *in vivo* and might not be active at the surface of a completely folded protein.

A useful method for glycopeptide synthesis involves the incorporation of glycosylamino acids into oligopeptides, either chemically[67,68] or enzymatically,[69] followed by introduction of additional sugars by glycosyltransferases. Enzymatic formation of peptide and glycoside bonds in certain cases is quite effective as both procedures can be carried out in aqueous solution, thus minimizing the protection/deprotection steps needed. Glycosyl amino acids can be used as the P_2, P_3, P_2' and P_3' residues in subtilisin-catalyzed glycopeptide segment condensation. Using a thermostable variant (developed by site-directed mutagenesis) with the active-site Ser is converted to Cys, the enzymatic coupling of glycopeptide segments can be carried out effectively at 60°C in aqueous solution (Figure 12)[69] The enzyme prefers aminolysis to hydrolysis by a factor of ca.10,000, and kinetic

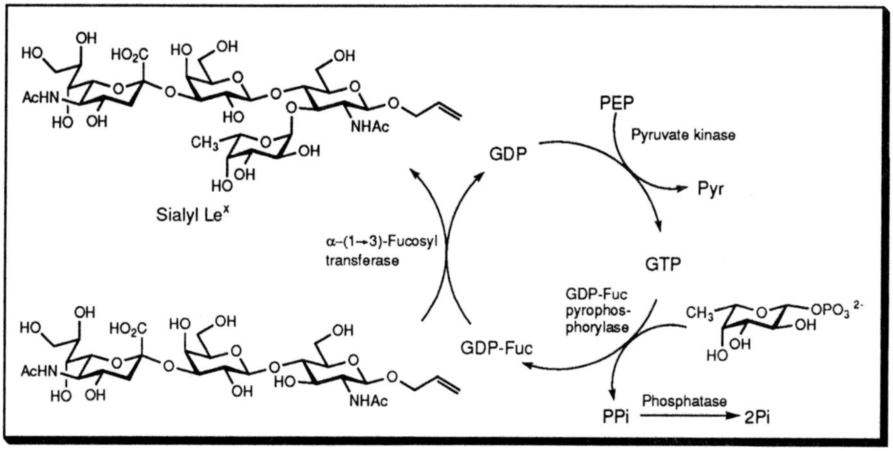

Figure 11. Enzymatic fucosylation with *in situ* regeneration of GDP-fucose for the synthesis of sialyl Lewis[x].

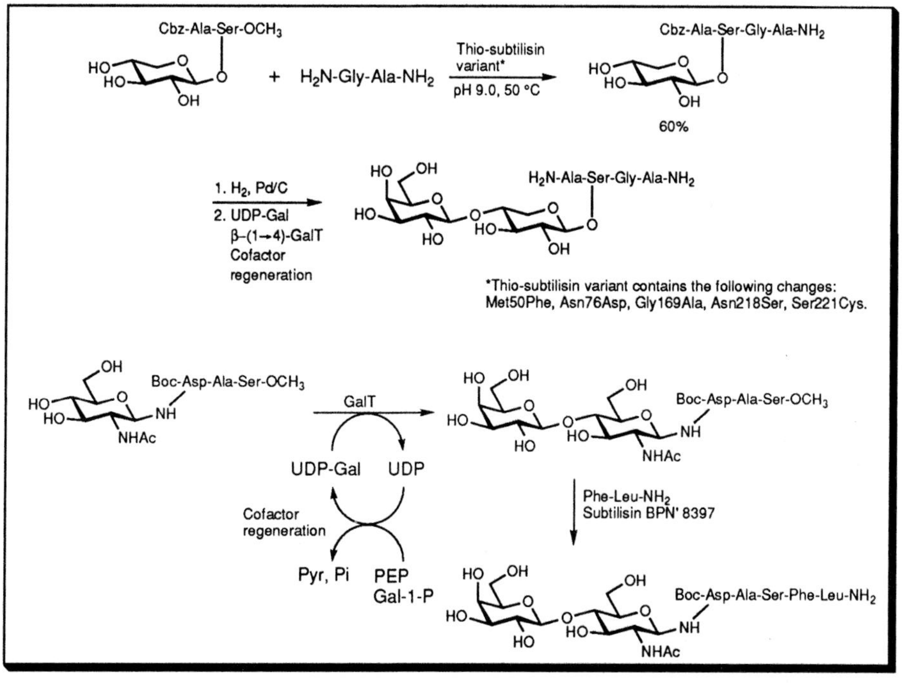

Figure 12 . Synthesis of glycopeptides using subtilisin and galactosyltransferase.

studies indicate that the selectivity comes from the acyl-enzyme intermediate reacting more selectively with the amine nucleophile than that from the wild-type enzyme (Figure 13). Alternatively, the solid-phase chemical synthesis of glycopeptides followed by enzymatic glycosylation, works well on aminopropyl-silica support.[27] A key element is the attachment of a proper acceptor-spacer group with a selectively cleavable bond. This strategy allows a rapid, iterative formation of peptide and glycosidic bonds in organic and aqueous solvents, respectively, and enables the release of glycopeptides from the support by enzymatic means (Figure 14).

8. GLYCOSYLATION USING GLYCOSYLPHOSPHITES

Although enzymatic synthesis of oligosaccharides is useful for large-scale processes, the method is quite limited, as many oligosaccharide analogs which contain modified monosaccharide moieties cannot be prepared enzymatically. Synthesis of these unnatural carbohydrates thus requires the use of chemical glycosylation methods. The use of the recently developed glycosylphosphites in glycosylation[70-72] has been examined in detail regarding the scope and mechanism.[73] It appears that the method can be applied to almost all monosaccharides, and the preparation of glycosylphosphites is quite straightforward (Figure 15). The activation reagent can be either TMSOTf (as a catalyst)[70,71] or $ZnCl_2$ (a stoichiometric activation reagent)[74,75] and the reaction can be carried out from -20 to -78°C in solvents such as CH_3CN, CH_2Cl_2 or ether. Mechanistic study of the glycosylation reaction indicates that triflic acid is involved in the activation when TMSOTf is used as catalyst[73] (Figure 15).

9. FUTURE OPPORTUNITIES

The pace of development of carbohydrate-derived pharmaceutical agents has, in general, been slower than with more convenient classes of materials. The difficulties in the synthesis and analysis of carbohydrates have undoubtedly contributed to this slow pace, but at least three areas of biology and medicinal chemistry have redirected attention to carbohydrates. First, interfering with the assembly of bacterial cell walls remains one of the most successful strategies for the

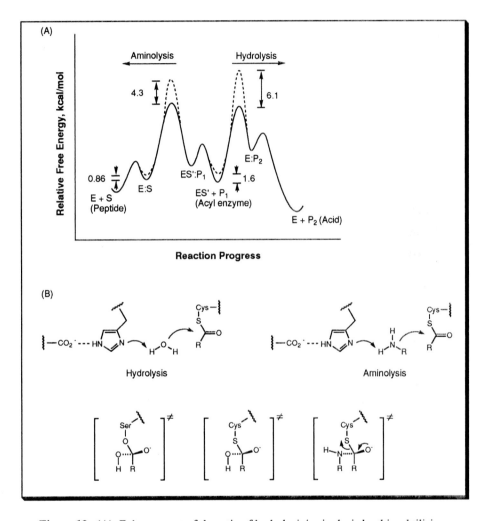

Figure 13. (A) Enhancement of the ratio of hydrolysis/aminolysis by thiosubtilisin.
(B) Mechanism of subtilisin and thiosubtilisin.

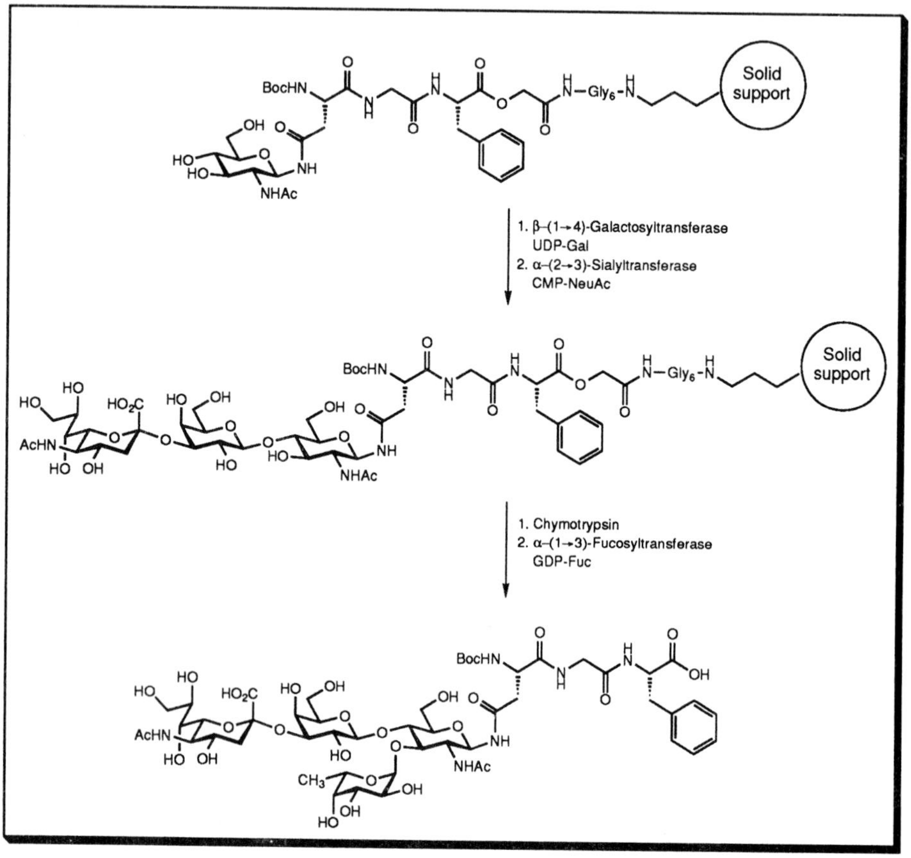

Figure 14 . Solid-phase chemo-enzymatic synthesis of a glycopeptide containing sialyl Lewis^x.

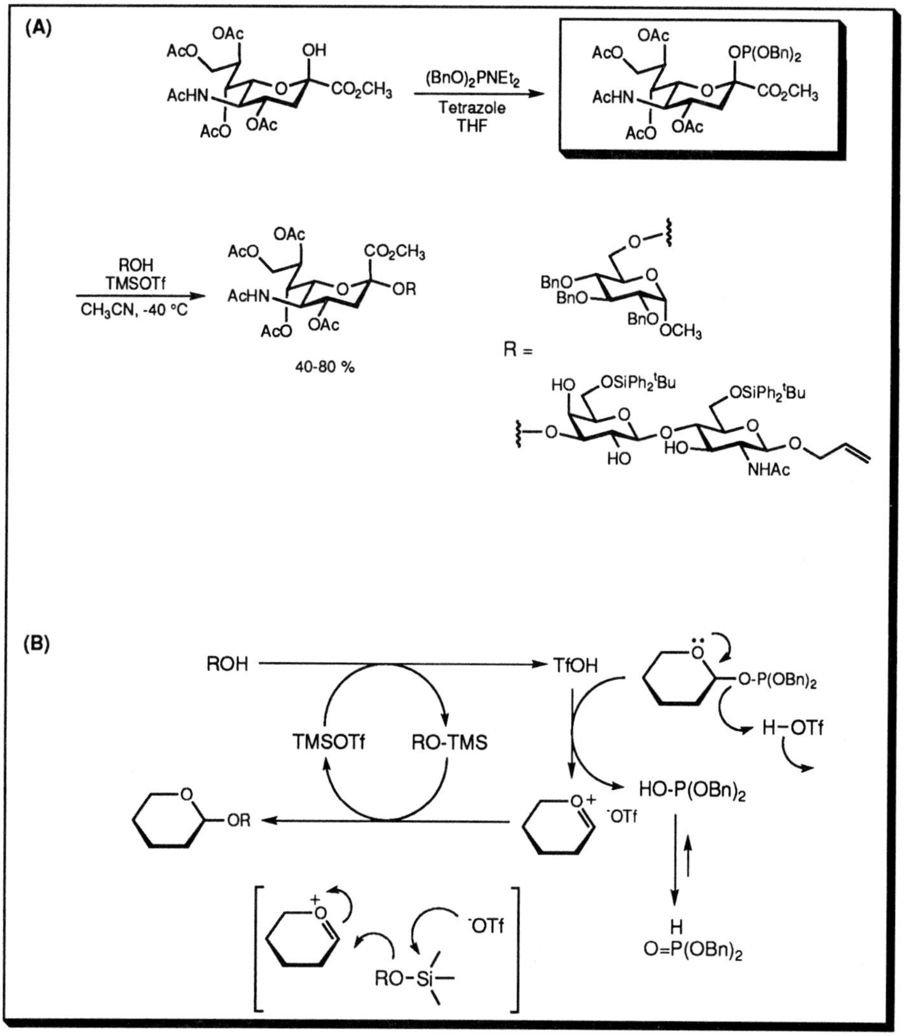

Figure 15. (A) Chemical sialylation using sialyl phosphites.
(B) Proposed mechanism of the reaction.

development of antimicrobials. As bacterial resistance to the commonly used penams and cephams becomes more widespread, there is increasing interest in interfering with the biosythesis of the characteristic carbohydrate components of the cell wall, especially KDO, heptulose, lipid A and related materials. Secondly, cell-surface oligosacchardes play a pivotal role in cell communication, cell adhesion, infection, differentiation, and development, and may be relevant to abnormal states of differentiation, such as those characterizing some malignant tumors. Third, the broad interest in diagnostics has finally begun to generate interest in carbohydrates as markers of human health. Enzymatic methods of synthesis, by rendering carbohydrates more accessible, will contribute to further research in all of these areas.

REFERENCES

1. L.F. Leloir, *Science,* **172**, 1299 (1971).
2. T.A. Beyer, J.E. Sadler, J.I. Rearick, J.C. Paulson, and R.L. Hill, *Adv. Enzymol.,* **52**, 23 (1981).
3. R. Kornfeld and S. Kornfeld, *Ann. Rev. Biochem.,* **54**, 631 (1985).
4. P.J. Oths, R.M. Hayer, and H.G. Floss, *Carbohydr. Res.,* **198**, 91 (1990).
5. R.N. Russell and H.-W. Liu, *J. Am. Chem. Soc.,* **113**, 7777 (1991).
6. O. Hindsgaul, *Seminars in Cell Biology,* **2**, 319 (1991).
7. Y. Ichikawa, G.C. Look, and C.-H. Wong, *Anal. Biochem.,* **202**, 215 (1992).
8. K.G.I. Nilsson, *TIBTECH,* **6**, 256 (1988).
9. J.E. Heidlas, K.W. Williams, and G.M. Whitesides, *Acc. Chem. Res.,* **25**, 307 (1992).
10. B.L. Hirschbein, F.P. Mazenod, and G.M. Whitesides, *J. Org. Chem.,* **47**, 3765 (1982).
11. E.S. Simon, S. Grabowski, and G.M. Whitesides, *J. Am. Chem. Soc.,* **111**, 8920 (1989).
12. C.-H. Wong, S.L. Haynie, and G.M. Whitesides, *J. Am. Chem. Soc.,* **105**, 115 (1983a).
13. L.J. Berliner, M.E. Davis, K.E. Ebner, T.A. Beyer, and J.E. Bell, *Mol. Cell Biochem.,* **62**, 37 (1984).

14. C. Auge, S. David, C. Mathieu, and C. Gautheron, *Tetrahedron Lett.*, **25**, 1467, (1984).

15. M.M. Palcic, O.P. Srivastava, and O. Hindsgaul, *Carbohydr. Res.*, **159**, 315 (1987).

16. C.-H. Wong, Y. Ichikawa, T. Krach, C. Gautheron-Le Narvor, D.P. Dumas, and G.C. Look, *J. Am. Chem. Soc.*, **113**, 8137 (1991).

17. Y. Nishida, T. Wiemann, and J. Thiem, *Tetrahedron Lett.*, **33**, 8043 (1992).

18. Y. Nishida, T. Wiemann, V. Sinwell, and J. Thiem, *J. Am. Chem. Soc.*, **115**, 2536 (1993).

19. Y. Nishida, T. Wiemann, and J. Thiem, *Tetrahedron Lett.*, **34**, 2905 (1993).

20. G. Srivastava, O. Hindsgaul, and M.M. Palcic, *Carbohydr. Res.*, **245**, 137 (1993).

21. J. Thiem and T. Wiemann, *Angew. Chem. Int. Ed. Engl.*, **29**, 80 (1990).

22. U. Zehavi and M. Herchman, *Carbohydr. Res.* **133**, 339 (1984).

23. U. Zehavi, S. Sadeh, and M. Herchman, *Carbohydr. Res.*, **124**, 23 (1983).

24. C. Auge, C. Gautheron, and H. Pora, *Carbohydr. Res.*, **193**, 288 (1989).

25. C. Unverzagt, H. Kunz, and J.C. Paulson, *J. Am. Chem. Soc.*, **112**, 9308 (1990).

26. B. Guilbert, T.H. Khan, and S.L. Flitsch, *J. Chem. Soc. Chem., Commun.* 1526 (1992).

27. M. Schuster, P. Wang, J.C. Paulson, and C.-H. Wong, *J. Am. Chem. Soc.*, **116**, 1135 (1994).

28. (a) T.L. Lowary and O. Hindsgaul, *Carbohydr. Res.*, **249**, 163 (1993); (b) *ibid.*, **251**, 33 (1994).

29. C. Auge and C. Gautheron, *Tetrahedron Lett.*, **19**, 789 (1988).

30. S. Sabesan and J.C. Paulson, *J. Am. Chem. Soc.*, **108**, 2068 (1986).

31. J. Thiem and W. Treder, *Angew. Chem. Int. Ed. Engl.*, **25**, 1096 (1986).

32. R.D. McCoy, E.R. Vimr, and F.A. Troy, *J. Biol. Chem.*, **260**, 12695 (1985).

33. H.S. Conradt, A. Bunsch, and R. Browmer, *FEBS Lett.*, **170**, 295 (1984).

34. H.J. Gross and R. Brossner, *Eur. J. Biochem.*, **177**, 583 (1988).

35. C.R. Petrie, M. Sharma, O.D. Simmons, and W. Korytnyk, *Carbohydr. Res.*, **186**, 326 (1989).

36. Y. Ito, J.J. Gaudino, and J.C. Paulson, *Pure & Appl. Chem.*, **65**, 753 (1993).

37. K.K.-C. Liu and S.J. Danishefsky, *J. Am. Chem. Soc.*, **115**, 4933 (1993).

38. L.J. Berliner and R.D. Robinson, *Biochemistry* **21**, 6340 (1982).

39. M.M. Palcic and O. Hindsgaul, *Glycobiology,* **1**, 205 (1991).

40. H. Yuasa, O. Hindsgaul, and M.M. Palcic, *J. Am. Chem. Soc.,* **114**, 5891 (1992).

41. U.B. Gokhale, O. Hindsgaul, and M.M. Palcic, *Can. J. Chem.,* **68**, 1063 (1990).

42. Y. Ichikawa, Y.-C. Lin, D.P. Dumas, G.-J. Shen, E. Garcia-Junceda, M.A. Williams, R. Bayer, C. Ketcham, L.E. Walker, J.C. Paulson, and C.-H. Wong, *J. Am. Chem. Soc.,* **114**, 9283 (1992).

43. M.M. Palcic, A.P. Venot, R.M. Ratcliffe, and O. Hindsgaul, *Carbohydr. Res.,* **190**, 1 (1989).

44. G. Srivastava, K.J. Kaur, O. Hindsgaul, and M.M. Palcic, *J. Biol. Chem.,* **267**, 22356 (1992).

45. H. Schachter, *Biochem. Cell. Biol.,* **64**, 163 (1986).

46. I. Brockhausen, J. Carver, and H. Schachter, *Biochem. Cell. Biol.,* **66**, 1134 (1988).

47. I. Brockhausen, E. Hull, O. Hindsgaul, H. Schachter, R.N. Shah, S.W. Michnick, and J.P. Carver, *J. Biol. Chem.,* **264**, 11211 (1989).

48. K.J. Kaur, G. Alton, and O. Hindsgaul, *Carbohydr. Res.,* **210**, 145 (1990).

49. O. Hindsgaul, K.J. Kaur, U.B. Gokhale, G. Srivastava, G. Alton, and M.M. Palcic, *ACS Symp. Ser.,* **466**, 38 (1991).

50. B. Imperiali and J.W. Zimmerman, *Tetrahedron Lett.,* **31**, 6485 (1990).

51. R. Oehrlein, O. Hindsgaul, and M.M. Palcic, *Carbohydr. Res.,* **244**, 149 (1993).

52. W. McDowell, T.J. Grier, J.R. Rasmussen, and R.T. Schwarz, *Biochem. J.,* **248**, 523 (1987).

53. P. Wang, G.-J. Shen, Y.-F. Wang, Y. Ichikawa, and C.-H. Wong, *J. Org. Chem.,* **58**, 3985 (1993).

54. S.L. Flitsch, H.L. Pinches, J.P. Taylor, and N.J. Turner, *J. Chem. Soc. Perkin Trans. 1,* **2087** (1992).

55. C.-H. Wong, S.L. Haynie, and G.M. Whitesides, *J. Org. Chem.,* **47**, 5416 (1982).

56. C.-H. Wong, R. Wang, and Y. Ichikawa, *J. Org. Chem.*, **57**, 4343 (1992).

57. L. Elling, M. Grothus, and M.-R. Kula, *Glycobiology,* **3**, 349 (1993).

58. Y. Ichikawa, J.L.-C. Liu, G.-J. Shen, and C.-H. Wong, *J. Am. Chem. Soc.,* **113**, 6300 (1991).

59. Y. Ichikawa, G.-J. Shen, and C.-H. Wong, *J. Am. Chem. Soc.,* **113**, 4698 (1991).

60. G.F. Herrmann, U. Kragl, and C. Wandrey, *Angew. Chem. Int. Ed. Engl.,* **32**, 1342 (1993).

61. G.F. Herrmann, Y. Ichikawa, C. Wandrey, F.C.A. Gaeta, J.C. Paulson, and C.-H. Wong, *Tetrahedron Lett.,* **34**, 3091 (1993).

62. G.C. Look, Y. Ichikawa, G.-J. Shen, P.-W. Cheng, and C.-H. Wong, *J. Org. Chem.,* **58**, 4326 (1993).

63. D. Gygax, P. Spies, T. Winkler, and U. Pfarr, *Tetrahedron,* **28**, 5119 (1991).

64. J.B. Lowe, *Seminars in Cell Biology,* **2**, 289 (1991).

65. J.C. Paulson and K.J. Colley, *J. Biol. Chem.,* **264**, 17615 (1989).

66. C.H. Krezdorn, G. Watgele, R.B. Kleene, S.X. Ivanov, and E.G. Berger, *Eur. J. Biochem.,* **212**, 113 (1993).

67. T. Bielfeldt, S. Peters, M. Meldal, K. Bock, and H. Paulsen, *Angew. Chem. Int. Ed. Engl.,* **31**, 857 (1992).

68. H. Kunz, *Pure & Appl. Chem.,* **65**, 1223 (1993).

69. C.-H. Wong, M. Schuster, P. Wang, and P. Sears, *J. Am. Chem. Soc.,* **115**, 5893 (1993).

70. H. Kondo, Y. Ichikawa, and C.-H. Wong, *J. Am. Chem. Soc.,* **114**, 8748 (1992).

71. T.J. Martin and R.R. Schmidt, *Tetrahedron Lett.,* **33**, 6123 (1992).

72. M.M. Sim, H. Kondo, and C.-H. Wong, *J. Am. Chem. Soc.,* **115**, 2260 (1993).

73. H. Kondo, S. Aoki, Y. Ichikawa, R.L. Halcomb, H. Ritzen, and C.-H. Wong, *J. Org. Chem.,* **59**, 864 (1994).

74. Y. Watanabe, C. Nakamoto, and S. Ozaki, *Synlett,* **115** (1993).

75. E.J. Corey and Y.-J. Wu, *J. Am. Chem. Soc.,* **115**, 8871 (1993).

Chapter 20

USE OF GLYCOSYLTRANSFERASES IN THE SYNTHESIS OF UNNATURAL OLIGOSACCHARIDE ANALOGUES

SUZANNE C. CRAWLEY* AND MONICA M. PALCIC[†]

*Amgen Inc., 1840 DeHavilland Drive, Thousand Oaks, CA 91320, USA
[†]Department of Chemistry, University of Alberta, Edmonton, Alberta, Canada TG6 2G2

Abstract Glycosyltransferase-mediated synthesis of unnatural oligo-saccharides is reviewed, with emphasis on the use of modified sugar-nucleotide donor precursors. General considerations regarding use of the glycosyltransferases as catalysts are discussed. Specific examples involving the use of β-$(1\rightarrow4)$-galactosyltransferase, sialyltransferases, fucosyl-transferases and N-acetylglucosaminyltransferases are described, as well as provision of their respective donor analogues.

1. INTRODUCTION

Recent interest in the preparation of natural and modified oligosaccharide structures stems from demonstrations of their role in a large number of cellular biological recognition events.[1] This is notable in the areas of inflammation,[2-4] infectious diseases[5] and cancer,[6,7] where complex oligosaccharides represent drug targets for the inhibition of carbohydrate dependent binding of cells. While the chemical synthesis of oligosaccharides is well established,[8-10] it requires experienced personnel and remains time consuming. Enzymatic synthesis provides an alternative approach for the preparation of naturally occurring oligosaccharides, and latitudes in donor and acceptor specificities can be exploited for the preparation of analogues.

The advantages of using glycosyltransferases for oligosaccharide synthesis have been well reviewed[11-14] and arise mostly from the mild reaction conditions, the lack of requirement for protection and de-protection steps (i.e. addition of a single sugar residue is accomplished in a single step) and the reaction regio- and stereoselectivity. Quantitative yields are often achieved and are mostly dependent upon the activity, stability and purity of the transferase. All of the naturally-occurring sugar-nucleotides utilized by these enzymes are commercially available and enzymatic methods, many pioneered by Whitesides *et al.*,[15] have been devised for their synthesis.[11, 14] Other new applications of glycosyltransferases have been in the areas of glycopeptide synthesis,[16] for addition of monosaccharides as protecting groups,[17] for product resolution and conversion ("capture") in glycosidase-mediated synthesis,[18] in the elaboration of multivalent oligosaccharide ligands,[19,20] and for incorporation of ^{13}C-labelled monosaccharide residues.[21] Since several excellent reviews on the area of general glycosyltransferase-mediated oligosaccharide synthesis have been published recently,[11,13,14,22] this review will focus on synthesis of unnatural oligosaccharides, with particular emphasis on the use of modified sugar-nucleotides.

Oligosaccharides with altered functional groups have applications in several areas of glycobiology: Modified oligosaccharides have been used to map, in atomic detail, the combining sites of carbohydrate-binding proteins such as lectins and antibodies,[23,24] glycosyltransferases[25-37] and, most recently, selectins.[38,39] Incorporation of unnatural sugar residues into synthetic oligosaccharides offers the potential for increasing their biochemical stability *in vivo*, by reducing reactivity with glycosidases and glycosyltransferases.[40] Introduction of modified sugar residues into oligosaccharides can also provide inhibitors for glycosyltransferases,[21, 41-47] as well as provides ligands to block other protein-carbohydrate interactions,[38,39] and offer a strategy for covalent affinity labeling of proteins *via* reactive functional groups.[43] Enzyme-mediated transfer of unnatural sugar residues from modified sugar-nucleotide donors also permits remodeling of glycoproteins and glycolipids in solution and on viable cells.[48-50]

2. PREPARATION OF GLYCOSYLTRANSFERASES

Glycosyltransferases are enzymes which can transfer an activated monosaccharide from a sugar-nucleotide (donor) to the hydroxyl oxygen of an acceptor sugar, usually on a glycoprotein or glycolipid. These transferases are commonly classified according to their sugar-nucleotide mono- or diphosphate donors, and are specific for a single α or ß configuration and, usually, position (2, 3, 4, 6 or 8)[51] (Figure 1). Although some glycosyltransferases can recognize and transfer to simple monosaccharide derivatives, sugar recognition by these and other carbohydrate binding proteins seems to involve direct and indirect contacts with key polar groups and with the complementary surfaces of at least two or three sugar residues.[25-37] Recognition of donors, on the other hand, seems to be driven by the nucleotide portions of these molecules, as evidenced by the ability of nucleotide mono- and diphosphates to inhibit many glycosyltransferases and to serve as affinity ligands for transferase purifications.[52]

Glycosyltransferases are type II membrane-bound proteins residing in the endoplasmic reticulum and Golgi compartments[53,54]. Each is comprised of a short cytoplasmic N-terminal sequence, followed by a hydrophobic transmembrane domain, and then lumenally-oriented stem and C-terminal catalytic domains[55]. Soluble versions of different glycosyltransferase activities have been isolated from and/or detected in various mammalian fluids such as serum and milk, and are thought to arise from proteolysis of the labile stem domain.[56] The full-length, Golgi-resident forms of these enzymes require detergent solubilization, which can complicate both enzyme purification and product isolation from preparative synthetic reactions.

Most glycosyltransferase purifications involve affinity chromatography using immobilized sugar-nucleotide donor or acceptor analogues.[52] Obtaining substrates and substrate analogues for synthesis of affinity matrices and for enzyme assays, and the relatively low abundance and yields of these enzymes, are still serious obstacles to their common usage for synthesis. On the other hand, many of the original descriptions of individual glycosyltransferases provide procedures for substrate preparations and for small-scale synthesis of enzyme products, often using very crude enzyme preparations. A few recent examples where amounts of

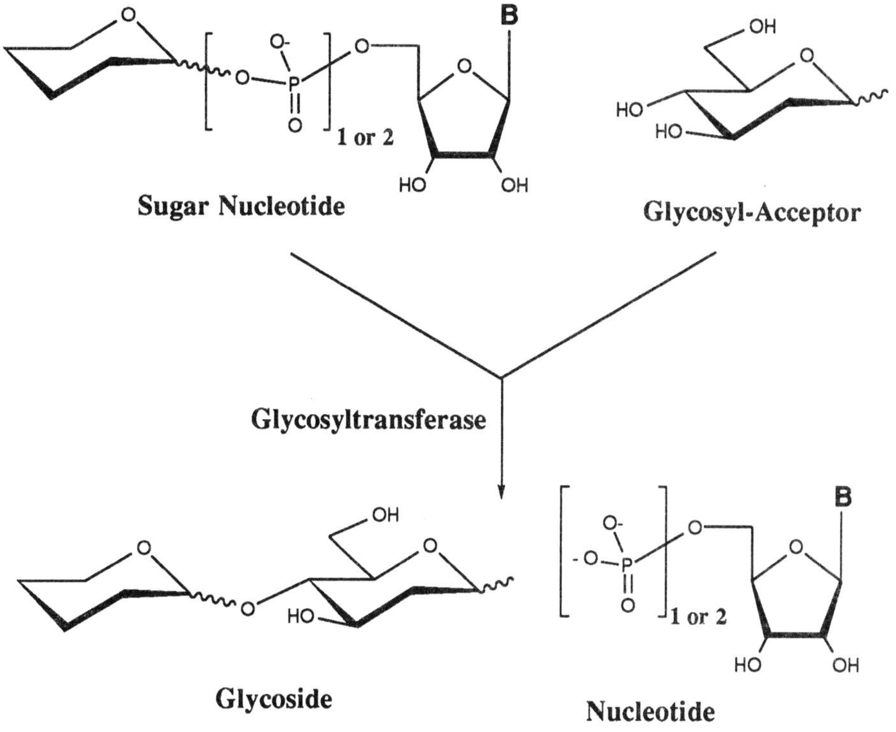

Figure 1. A generalized glycosyltransferase reaction.

material sufficient for ^{1}H-NMR have been synthesized using crude glycosyltransferase preparations include use of MDAY-D2 lysate for preparation of core 2 enzyme products,[57] use of rat kidney homogenate as a source of N-acetylglucosaminyltransferase-III (GlcNAcT-III) for incorporation of a bisecting GlcNAc into a synthetic biantennary N-linked type pentasaccharide,[58] use of porcine stomach microsomes as a source of blood-group A transferase for synthesis of a panel of tri- and tetrasaccharides,[59] and use of hen oviduct microsomes for addition of various GlcNAc branches on N-linked oligosaccharides.[60-63] Although yields can be low and product purification may be challenging, this type of synthesis has the advantage of being within the capacity of most biochemical laboratories.

For most synthetic purposes, glycosyltransferases are purified to the extent that formation of a unique product, and isolation of this product, are not impaired. This involves limiting the activities of competing glycosyltransferases, glycosidases, proteases and phosphatases, which is accomplished either by purification or by manipulation of the synthetic reaction conditions with addition of enzyme inhibitors, alteration of buffer, pH and divalent cations. Other innovations to increase yields include *in situ* generation of sugar-nucleotide donors,[14] addition of an ATP/theophylline/dimercaptopropanol cocktail to prevent sugar-nucleotide hydrolysis,[64] and addition of alkaline phosphatase for the decomposition of inhibitory nucleotide diphosphate by-products, which otherwise accumulate as the glycosyltransferase reaction progresses.[65]

Although milliunit amounts of crude glycosyltransferases are often adequate for synthesis of small amounts of natural-type oligosaccharides, larger scale synthesis using labile, precious, and somewhat unreactive substrate analogues requires unit amounts of well-defined glycosyltransferase preparations. Cloning has increased the availability of glycosyltransferases within the past ten years; genes and/or cDNA's for over 22 catalytically-different transferases have been sequenced, mostly within the last five years.[54] Despite similarities in the structures of both donor and acceptor substrates, sequence homologies have only been found within four different classes of glycosyltransferases. A sialyl motif was used to determine the primary structure of 4 out of 7 $\alpha-(2\rightarrow3)$- and $\alpha-(2\rightarrow6)$- sialyltransferases [$\alpha-(2\rightarrow3)$- and $\alpha-(2\rightarrow6)$-STs].[67,68] Expression cloning of the $\alpha-(1\rightarrow3/4)$-fucosyltransferases [$\alpha-(1\rightarrow3/4)$-FTs] has shown that this is a class of homologous transferases which differ in their substrate specificities and in the inter-relatedness of their respective amino acid sequences.[69] The two mucin-type $\beta-(1\rightarrow6)$-N-acetylglucosaminyltransferases (GlcNAcTs) which have been cloned show some homology,[70] on the other hand, these enzymes show no sequence homology to the only other $\beta-(1\rightarrow6)$-GlcNAc transferase (V) which has been cloned;[71] nor are they related to the three other branching GlcNAc transferases cloned to date (I-III). The $\alpha-(1\rightarrow3)$-galactosyltransferases ($\alpha-(1\rightarrow3)$-GalTs) constitute a fourth class of homologous glycosyltransferases. Transferases derived from different mammalian species but catalyzing the same reactions typically display greater than 85% homology at the amino acid level, suggesting that characterization of an activity from one source may be useful for predicting some properties of the same activity from different sources.

Stable expression of soluble glycosyltransferases in mammalian cell culture has been achieved by replacing N-terminal cytoplasmic and transmembrane domains with a cleavable signal peptide sequence.[72] Further modification by addition of an N-terminal sequence encoding a polypeptide "handle", such as the IgG-binding domain of protein A, was originally introduced to permit isolation, measurement and characterization of the putative glycosyltransferase gene product transiently expressed in COS cells.[54,73] Besides providing useful amounts of pure, well-defined transferase activities, this type of strategy also makes it possible to circumvent purification problems such as the requirement for precious affinity matrices, providing immobilized enzyme that can be used directly in oligosaccharide synthesis.[74]

Aoki et al.[26,75] have reported production of mutant and native forms of the catalytic domain of $\beta-(1\rightarrow4)$-galactosyltransferase $\beta-(1\rightarrow4)$-GalT) in Escherichia coli. Similarly, the production and refolding of a $\beta-(1\rightarrow4)$-T-glutathione-S-transferase (GST) fusion protein in E. coli has also been described.[76] N-Acetylglucosaminyltransferase I (GlcNAcT-I) has also been expressed as a GST fusion protein, which was detergent extractable at levels of 1-5 mg/ml (1 milliunit/mg) from the bacterial pellet,[77] and at levels of 2 milliunits/mg of bacterial cell lysate.[78] The catalytic domain of a GDP-Man: Man$\alpha-(1\rightarrow2)$-ManαThr/Ser $\alpha-(1\rightarrow2)$-mannosyltransferase has been expressed in the periplasmic space of E. coli at levels of 0.7 to 0.8 U per liter of bacterial culture.[63] Krezdorn et al.[79] have expressed $\beta-(1\rightarrow4)$-GalT in Saccharomyces cerevisiae, achieving 4.5 mU/ml of crude detergent extract and specific activities comparable to the native enzyme. Also, glycosyltransferases have been produced in Sf-9 insect cell culture: Gal$\beta-(1\rightarrow3/4)$-GlcNAc $\alpha-(2\rightarrow3)$-ST was secreted at levels of 2-3 U/10^8 cells,[21] GlcNAcT-I produced at levels of 1-5 mg/ml,[80] and 100 mU per 2 X 10^6 cells has been achieved for $\alpha-(1\rightarrow3)$-GalT.[81] Some of the vectors and expression systems used in this work are commercially available.

3.   GALACTOSYLTRANSFERASES

At present, bovine milk $\beta(1\rightarrow4)$-GalT is the only glycosyltransferase that is commercially available in multi-unit quantities; the human milk enzyme is also commercially available in milliunit quantities. Bovine milk galactosyltransferase is one of the most extensively mapped in terms of its donor and acceptor specificities, and has been widely utilized for the synthesis of oligosaccharides. In the absence of α-lactalbumin, the enzyme (designated as E.C. 2.4.1.90) catalyzes the transfer of galactose from uridine-5'-diphosphogalactose (UDP-Gal) to the 4-position of terminal β-linked GlcNAc, giving N-acetyllactosamine structures.[51] In the presence of α-lactalbumin, the acceptor specificity is modified such that $\beta-(1\rightarrow4)$-GalT transfers to the reducing sugar glucose; this activity is designated as E.C. 2.4.1.22.

Pioneering studies from the Berliner group[82,83] demonstrated that bovine milk galactosyltransferase could be used to transfer glucose, 4-deoxyglucose or arabinose from the corresponding UDP derivatives to glucose (Glc) or GlcNAc acceptors. The K_m values for these donors were in the same range as for UDP-Gal, however, the relative rates of transfer (V_{max}) were considerably reduced (Table 1). Modifications at each position of galactose were tolerated. Uridine-5'-diphospho-2-acetamido-2-deoxy-galactosamine (UDP-GalNAc) was transferred at only 0.2% of the rate of the parent donor; however, this rate of transfer was sufficient to produce milligram quantities of GalNAc$\beta-(1\rightarrow4)$-GlcNAc$\beta-(1\rightarrow2)$-Manα-O(CH$_2$)$_8$COOCH$_3$, analogous to the sequence found on oligosaccharide chains of glycoprotein hormones such as lutropin.[84] UDP-2NH$_2$-Glc was also transferred to GlcNAcβ-O(CH$_2$)$_8$COOCH$_3$, to give an amino-disaccharide which was acetylated to the di-N-acetyl-chitobiose derivative.[84] UDP-3-deoxy-Gal[85] and UDP-2NH$_2$.Gal[86] have also been used to prepare milligram quantities of the corresponding N-acetyllactosamine analogues. To date, UDP-2-deoxy-Gal is the only modified donor with a relative rate of transfer comparable to that of the parent UDP-Gal.[87] When this donor was generated in a recycling system using either free cofactor regeneration enzymes[86] or immobilized enzymes,[88,89] greater than 50 mg quantities of 2'-deoxy-N-acetyllactosamine were produced. Chemically synthesized UDP-6-deoxy- and UDP-6-fluoro-galactose have been used for the glycosylation of monosaccharides and biantennary carbohydrate chains.[90] Bovine milk galactosyltransferase also catalyzes the transfer of 5'-thiogalactose from its UDP derivative to give a disaccharide in which the ring oxygen in the non-reducing monosaccharide is replaced with sulfur.[40] A summary of all modifications on the donor that have been mapped is shown in Figure 2.

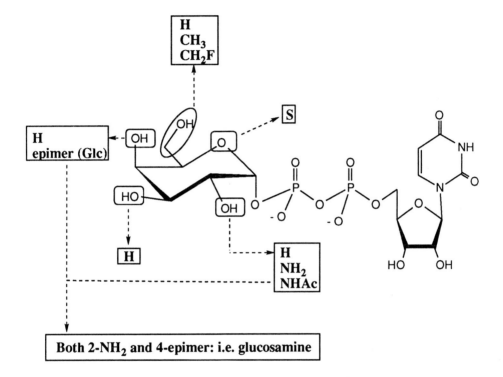

Figure 2. UDP-Gal analogs which have been used as donors for bovine β-(1→4)-galactosyltransferase (see Table 1).

In all of the previous examples the formation of the glycosidic linkage by bovine milk galactosyltransferase was both stereospecific and regiospecific, giving only β–(1→4)- linkages in the reaction products. The acceptor substrates utilized were all glucose or GlcNAc derivatives. The only exception reported to date is the formation of a β–(1→1)- linkage when 3-acetamido-3-deoxy-glucosamine was employed as the acceptor substrate.[91]

Table 1. Modified oligosaccharide structures which have been synthesized using bovine milk galactosyltransferase (added sugar residue is shown in bold).

Oligosaccharide Products	% Rel. Rate	% Rel. V_{max}	K_m (μM)	Ref.
Gal-β-(1→4)-GlcNAc	100	100	14	83
Gal-β-(1→4)-GlcNAcβ-OR$_1$[a]	100	100	44	87
4-Deoxy-Gal-β-(1→4)-Glc				83
4-Deoxy-Gal-β-(1→4)-GlcNAc		5.5	26	83
Glc-β-(1→4)-Glc		0.3	23	82,83
Ara-β-(1→4)-GlcNAc		4.0	14	83
Glc-β-(1→4)-GlcNAcβ-OR$_1$	0.4			84
GalNAc-β-(1→4)-GlcNAc-β-(1→2)-Manα-OR$_1$	0.19			84
GalNAc-β-(1→4)-GlcNAc-β-(1→2)-Man-α-(1→6)-\ Manβ-OR$_1$ / **GalNAc**-β-(1→4)-GlcNAc-β-(1→2)-Man-α-(1→3)-				84
2-NH$_2$-Glc-β-(1→4)-GlcNAcβ-OR$_1$	0.9			84
2-NH$_2$-Gal-β-(1→4)-GlcNAc				86
3-Deoxy-Gal-β-(1→4)-GlcNAcβ-OR$_1$	0.16			85
3-Deoxy-Gal-β-(1→4)-GlcNAc-β-(1→6)-Man-α-(1→6)-Manβ-OR$_1$				85
2-Deoxy-Gal-β-(1→4)-Glc				86,88,89
2-Deoxy-Gal-β-(1→4)-GlcNAcβ-OR$_1$	82	51		87
5-ThioGal-β-(1→4)-GlcNAc-OR$_1$	5			40
6-Deoxy-Gal-β-(1→4)-GlcNAc	1.3			90
6-Fluoro-Gal-β-(1→4)-GlcNAcβ-OR$_2$	0.2			90
Gal-β-(1→1)-Glc3NAcβ				91
6-Deoxy-Gal-β-(1→4)-GlcNac-β-(1→2)-\ Manβ-OR$_3$ / **6-Deoxy-Gal**-β-(1→4)-GlcNAc-β-(1→6)-				90

[a] R$_1$=(CH$_2$)$_8$COOCH$_3$, R$_2$=CH$_3$, R$_3$=(CH$_2$)$_2$CH$_3$

4. SIALYLTRANSFERASES

Most sialyltransferases (ST) transfer to terminal galactose residues and are classified according to the glycosidic linkages which they form ($\alpha-(2\rightarrow3)$- or $\alpha-(2\rightarrow6)$- or $\alpha-(2\rightarrow8)$-, and the structures of their acceptors: N-linked type I and type II disaccharides (Gal$\beta-(1\rightarrow3)$-GlcNAc and Gal$\beta-(1\rightarrow4)$-GlcNAc, respectively); Gal$\beta-(1\rightarrow3)$-GalNAcα and GalNAcα-O-Ser/Thr O-linked type structures; lipid-linked Gal$\beta-(1\rightarrow4)$-Glc and Gal$\beta-(1\rightarrow3)$-GalNAc structures; and terminal sialic acid structures which serve as acceptors for $\alpha-(2\rightarrow8)$-ST.[51,66] Sialylated structures have been observed for which the corresponding transferases have not been found. Since sialyltransferases often recognize only the terminal disaccharide of an oligosaccharide, a given enzyme may transfer to more than one class of glycoconjugates. Typically, asialo-α_1-acid glycoprotein is used as a type II (Gal$\beta-(1\rightarrow4)$-GlcNAc) acceptor; antifreeze glycoprotein is used to assay O-type sialyltransferases which recognize Gal$\beta-(1\rightarrow3)$-GalNAcα-O-Ser/Thr structures; sheep submaxillary mucin provides acceptor sites for GalNAcα-O-Ser/Thr α-$(2\rightarrow6)$-ST; and asialofetuin is typically used as a source of both N- and O- linked acceptors.[51] The rat liver Gal$\beta-(1\rightarrow4)$-GlcNAc $\alpha-(2\rightarrow6)$-ST and porcine submaxillary Gal$\beta-(1\rightarrow3)$-GalNAc $\alpha-(2\rightarrow3)$-ST are commercially available. Addition of acetyl, lactoyl, methyl, sulfate and phosphate groups to sialic acid can occur after the transferase reaction, due to the action of other Golgi-resident enzymes.[92]

More work has been done to define the donor flexibility of sialyltransferases than almost any other glycosyltransferase, aided in large part by early advances in their purification and characterization,[93-96] and by the utilization of various CMP-sialic acid synthetase preparations to activate NeuAc analogues.[97-99] Most of these reactions were done under initial rate conditions on a nanomole scale, and they employed glycoprotein rather than oligosaccharide acceptors. As a consequence, characterization of products relied upon the use of different sialidase-mediated hydrolysis steps, with chromatographic identification and quantitation of released sialic acids.

Beau and Schauer used equine submandibular gland CMP-sialic acid synthetase to activate tritiated and unlabelled 4-O-Me-NeuAc, which was subsequently transferred to asialo-fetuin using a horse liver extract as a source of unspecified sialyltransferase(s).[98] Later, Conradt, Bünsch and Brossmer[100] reported the use of CMP-sialic acid synthetase in the preparation of CMP-9-fluoro-9-deoxy-N-[2-^{14}C]NeuAc; this was employed as the donor for a crude rat liver sialyltransferase, which transferred the labelled sugar residue to asialo-α_1-acid glycoprotein. Higa and Paulson[99] investigated the abilities of six mammalian sialyltransferases to transfer NeuAc, N-glycolyl neuraminic acid (NeuGc), 4-O-Ac-NeuAc and 9-O-Ac-NeuAc from their corresponding CMP-glycosides to different glycoproteins. The transferases which were evaluated included Galβ–(1→4)-GlcNAc α–(2→6)-ST from rat liver and bovine colostrum, rat liver Galβ–(1→3/4)-GlcNAc α–(2→3)-ST, porcine submaxillary Galβ–(1→3)-GalNAcα–O–Ser/Thr α–(2→3)-ST and GalNAcα–O–Ser/Thr α–(2→6)-ST from bovine and porcine submaxillary. Initial rate measurements showed that the 4-O-acetylated derivative was not a substrate. Sialyltransferases acting on N-linked oligosaccharides were more efficient at transferring the NeuGc and 9-O-Ac-NeuAc analogues (see Table 2). GalNAcα–Ser/Thr α–(2→6)-ST and Galβ–(1→3)-GalNAcα–O–Ser/Thr α–(2→3)-ST, on the other hand, showed less than 20% efficiency with the 9-O-Ac derivative, but their rates of transfer of NeuGc were essentially the same as for NeuAc. Donor kinetics for bovine and porcine GalNAcα–Ser/Thr α–(2→6)-STs showed that K_m's for the 9-O-Ac analogues were actually slightly lower than for CMP-NeuAc and CMP-NeuGc, and that reduced V_{max} values were responsible for the 10% relative transfer rate. Kinetic parameters and sugar structures synthesized in these studies are listed in Table 2.

Gross et al.[101] described the CMP-sialic acid synthetase-catalyzed activation of five different 9-substituted NeuAc derivatives (azido, amino, acetamido, hexanoylamido, and benzamido) (Figure 3), and transfer of these NeuAc analogues by rat liver Galβ–(1→4)-GlcNAc α–(2→6)-ST to asialo-α_1-acid glycoprotein (Table 2). This work was subsequently amplified[102] by a kinetic analysis of the five different CMP-glycosides with four different sialyltransferases: rat liver Galβ–(1→4)-GlcNAc α–(2→6)-ST and Galβ–(1→3/4)-GlcNAc α–(2→3)-ST, and, from porcine submaxillary, GalNAc α–(2→6)-ST and Galβ–(1→3)-GalNAc α–(2→3)-ST. These enzymes varied in their abilities to transfer the modified NeuAc to the N-linked Galβ–(1→4)-GlcNAc acceptors of asialo-α_1-acid

503

Table 2.   Structures which have been synthesized on glycoproteins using sialyltransferases (added sugar residue in bold).

Oligosaccharide Products	E. C. #	Source	% Rel. rate	% Rel. Vmax	Km^a (µM)	Ref.
4-O-Me-NeuAc-α-(2→3/6)- N- and/or O-linked structures of asialo-fetuin		Equine Liver		6	3400 (400)	98
9-F-[2-14C]NeuAc-α-(2→3)-Gal-β-(1→3/4)-GlcNAc- and/or	2.4.99.6	Rat Liver	< 60			100
9-F-[2-14C]NeuAc-α-(2→6)-Gal-β-(1→4)-GlcNAc	2.4.99.1					
9-O-AcNeuAc-α-(2→6)-Gal-β-(1→4)-GlcNAc	2.4.99.1	Rat Liver Bov. Colostrum	75 50			99
9-O-AcNeuAc-α-(2→3)-Gal-β-(1→3/4)-GlcNAc	2.4.99.6	Rat Liver	60			99
9-O-AcNeuAc-α-(2→3)-Gal-β-(1→3)-GalNAcα	2.4.99.4	Porc.Submax.	25			99
9-O-AcNeuAc-α-(2→6)-GalNAcα	2.4.99.3	Porc. Submax. Bov. Submax.	20 10	10 9	80 (150) 100 (270)	99
9-R^b-NeuAc-α-(2→6)-Gal-β-(1→4)-GlcNAc	2.4.99.1	Rat Liver	36-55			101
9-R-NeuAc-α-(2→6)-Gal-β-(1→4)-GlcNAc	2.4.99.1	Rat Liver	71-148	90-110	30-720 (50)	102
9-R-NeuAc-α-(2→3)-Gal-β-(1→3/4)-GlcNAc	2.4.99.6	Rat Liver	10-62	60	220-245 (70)	102
9-R-NeuAc-α-(2→3)-Gal-β-(1→3)-GalNAcα	2.4.99.4	Porc. Submax.	6-143			102
9-R-NeuAc-α-(2→6)-GalNAcα	2.4.99.3	Porc. Submax.	9-188			102

Table 2. (continued) Structures which have been synthesized on glycoproteins using sialyltransferases (added sugar residue in bold).

Oligosaccharide Products	E. C. #	Source	% Rel. rate	% Rel. Vmax	Km^a (μM)	Ref.
5-Formyl-Neu-α-(2→6)-Gal-β-(1→4)-GlcNAc	2.4.99.1	Rat Liver	114			106
		Human Liver	113			
5-Aminoacetyl-Neu-α-(2→6)-Gal-β-(1→4)-GlcNAc	2.4.99.1	Rat Liver	48			106
		Human Liver	65			
4-Deoxy-NeuAc-α-(2→6)-Gal-β-(1→4)-GlcNAc	2.4.99.1	Rat Liver		≥100		103
9-Fluoresceinyl-NeuAc-α-(2→6)-Gal-β-(1→4)-GlcNAc	2.4.99.1	Rat Liver		100	7 (45)	105
		Human Liver		80	2 (15)	
9-Fluoresceinyl-NeuAc-α-(2→3)-Gal-β-(1→3/4)-GlcNAc	2.4.99.6	Rat Liver		25	8 (65)	105
9-Fluoresceinyl-NeuAc-α-(2→3)-Gal-β-(1→3)-GalNAcα	2.4.99.4	Porcine		15	2.5 (6)	105
9-Fluoresceinyl-NeuAc-α-(2→6)-GalNAcα	2.4.99.3	Porcine		20	30 (500)	105

[a]Values in parenthese are K_m values for CMP-NeuAc, obtained under the same conditions. [b]**R**=azido, amino, acetamido, benzamido, hexanoylamid.

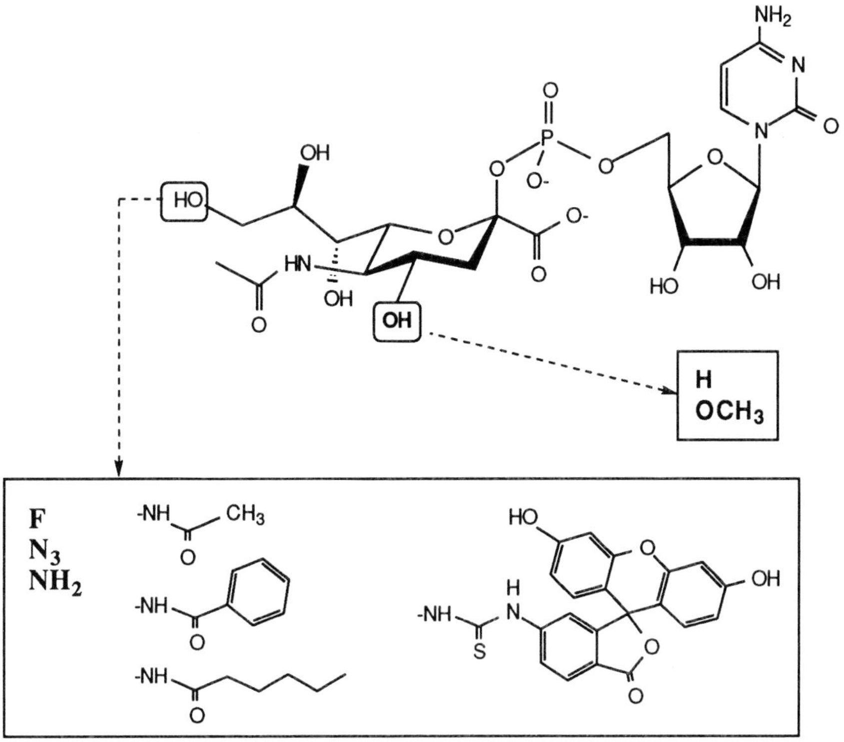

Figure 3. CMP-sialic acid analogs which have been used as donors for sialyltransferase reactions described in Table 2 and the text.

glycoprotein and to the O-linked Galβ–(1→3)-GalNAcα sugars of antifreeze glycoprotein. Initial rates of the sialyl transfer reaction relative to CMP-NeuAc (2 mM) ranged from 70-150% for the rat α–(2→6)-ST, from 10 to 150% for rat α–(2→3)-ST, from 5 to 140% for porcine α–(2→3)-ST, and from negligible to 165% for porcine GalNAcα–Ser/Thr α–(2→6)-ST. The slowest analogue in all cases was the 9-amino derivative, whereas the most reactive donor tended to be CMP-9-azido-NeuAc. Since glycoprotein acceptors were used for this work, preparative sialylations were not carried out, but these descriptions of sialyltransferase reaction kinetics with different CMP-glycosides indicate that oligosaccharides terminating with the trisaccharides 9-R-NeuAcα–(2→3)-Galβ–(1→3/4)-GlcNAcβ, 9-R-NeuAcα–(2→6)-Galβ–(1→4)-GlcNAcβ and 9-R-NeuAcα–(2→3)-Galβ–(1→3)-

GalNAcα (where R=azido, amino, acetamido, benzamido, and hexanoylamido groups) may be as amenable to enzymatic synthesis as the natural analogues. Even the reactions which seemed to proceed at a very slow rate could be made synthetically useful by optimizing substrates and/or conditions and, especially, by using larger amounts of enzymes available through cloning and over-expression.

Gross and Brossmer[103] also used the rat liver Galβ–(1→4)-GlcNAc α–(2→6)-ST to transfer the 4-deoxy derivative of NeuAc from CMP-4-deoxy-NeuAc to asialo-α₁-acid glycoprotein, which has N-linked oligosaccharides terminating in the type II disaccharide Galβ–(1→4)-GlcNAcβ (Table 2). They observed that this CMP-glycoside is particularly labile but displays kinetic parameters similar to the natural donor for the rat liver α–(2→6)-ST.

It was also found[104] that a fluorescent analogue of CMP-NeuAc could be synthesized, albeit at less than 10% the rate of the natural donor, by taking advantage of the ability of the bovine brain CMP-sialic acid synthetase to tolerate bulky substitutions at C9 of NeuAc. 9-Fluoresceinyl-NeuAc (Figure 3) could also be transferred onto glycoprotein by six sialyltransferases, yielding fluorescent products and forming the basis for a sensitive fluorometric sialyltransferase assay.[105] Kinetic evaluations of human and rat liver Galβ–(1→4)-GlcNAc α–(2→6)-ST, and of rat liver Galβ–(1→3/4)-GlcNAc α–(2→3)-ST showed that the K_m for CMP-9-fluoresceinyl-NeuAc was consistently lower than the K_m for CMP-NeuAc, whereas the relative V_{max} was appreciably lower only for the α–(2→3)-ST. The donor K_m for the porcine submaxillary Galβ–(1→3)-GalNAc α–(2→3)-ST was essentially the same for natural and fluoresceinated sugar-nucleotide donors, but the relative V_{max} was only 15% with the latter substrate. Similarly GalNAc α–(2→6)-ST had a lower V_{max} with the fluorescent donor (20%), but the K_m was also substantially lower (30 mM instead of 500 μM).[105] CMP-9-fluoresceinyl-NeuAc is commercially available.

CMP-NeuAc analogues modified at the C5 N-acetyl, to give CMP-5-formyl-Neu and CMP-5-aminoacetyl-Neu, have also been enzymatically synthesized; they were found to be good donors for human and rat liver α–(2→6)-STs (see Table 2).[106] These investigators also evaluated the kinetics for these two enzymes with four C9 derivatives of CMP-NeuAc.

Some other CMP-sialic acid analogues which have been enzymatically synthesized include CMP-4-*O*-Me-NeuAc (17 micromoles),[97] CMP-9-iodo-NeuAc,[100] and CMP-9-fluoro-NeuAc.[100,107,109] *N*-carbomethoxy- and *N*-carbobenzyloxy derivatives of NeuAc have also been coupled to CMP, using recombinant CMP-sialic acid synthetase over-expressed in *E. coli* .[108] CMP-deaminoneuraminic acid has been synthesized with *in situ* generation of CTP from CMP using immobilized nucleoside monophosphokinase and pyruvate kinase, plus immobilized CMP-sialic acid synthetase and sialyltransferase.[110]

The similar reactivity of CMP-NeuAc analogues versus the natural donor, and the recent successful uses of sialyltransferases in preparation of gram amounts of oligosaccharides[21] both suggest that syntheses using unnatural CMP-glycosides would not require substantial modification of existing protocols. CMP-sialic acid synthetase required for activation of donor analogues is readily available: Numerous simple extraction procedures have been published,[97-99] the *E. coli* enzyme has been over-expressed in bacterial culture,[108, 109, 111] and recombinant enzyme is commercially available. Using the recombinant CMP-sialic acid synthetase, a one-pot synthesis of sialyl oligosaccharide with *in situ* generation of CTP, ATP and CMP-NeuAc has been described;[111] 3 mmol (2 g) of trisaccharide product NeuAcα–(2→6)-Galβ–(1→4)-GlcNAc in 97% yield was achieved.[111]

Modified acceptors can be used to achieve additional, unnatural variations in sialylated and other oligosaccharides. Important features of the acceptor binding sites have recently been described for rat liver Galβ–(1→3/4)-GlcNAc α–(2→3)-STase and Galβ–(1→4)-GlcNAc α–(2→6)-STase[28] and for Galβ–(1→3)-GalNAcα α–(2→3)–ST from human placenta and acute myeloid leukemia cells.[33]

5.   FUCOSYLTRANSFERASES

Three major classes of fucosyltransferase have been described: α–(1→6)-FT, which is responsible for fucosylation of the chitobiose core of *N*-linked sugars, α–(1→2)-FT which recognizes terminal Gal residues, and the α–(1→3)-FTs which form α–(1→3)- and α–(1→4)- linkages to the GlcNAc residues of type I and type II disaccharides.[51] There are only a few examples of the use of fucosyltransferases for the synthesis of oligosaccharide analogues using modified donors. This stems in part from the difficulties associated with the chemical synthesis of guanosine-5'-

diphosphofucose (GDP-Fuc) and its derivatives, and from the lack of reliable recycling systems to prepare GDP-Fuc enzymatically.[14,112] The examples listed in Table 3 have all used human milk α–$(1\rightarrow3/4)$-FT with modified GDP-Fuc in milligram scale syntheses.

GDP-3-deoxy-Fuc and GDP-arabinose showed relative transfer rates of only 2.3% and 5.9% , respectively, compared to GDP-Fuc. These reactions used Galβ–$(1\rightarrow3)$-GlcNAcβO(CH$_2$)$_8$COOCH$_3$ as well as this same disaccharide conjugated to bovine serum albumin[112] (Table 3). Extensive modifications at the C-6 of fucose were tolerated by the milk enzyme. These included the 6-O-propyl substituted fucose, which was transferred to soluble acceptor at about 50% of the rate of the parent GDP-Fuc,[113] and even included blood group B trisaccharide covalently linked to C-6 of fucose through an amino group.[113] This reaction proceeded at approximately 10% of the rate obtained with GDP-Fuc. Product was detected using enzyme-linked immunosorbent assays and also by agglutination of erythrocytes rendered "B" blood-group type by the transfer of a fucose-"B"-trisaccharide moiety. Recent reports of improved recycling systems[14,114] should facilitate the synthesis of GDP-Fuc and analogues thereof, provided that modifications to fucose are tolerated by the regeneration enzymes.

6.   OTHER GLYCOSYLTRANSFERASES

A comparatively limited amount of synthetic work has been done using mannosyltransferases and N-acetylglucosaminyltransferases, in part because the focus has been more on structures at the non-reducing terminals of complex oligosaccharides.

Some information about the donor specificity of mannosyltransferases has been contributed by the work of Schwartz's group: McDowell et al.[115] studied effects of GDP-4-deoxy-Man and GDP-4-fluoro-Man on BHK and chick-embryo mannosyltransferases. GDP-4-deoxy-Man was determined to be a substrate for the mannosyltransferases involved in the biosynthesis of Dol-PP-(GlcNAc)$_2$Man$_9$ and of Dol-P-Man, whereas GDP-4-F-Man appeared to be an inhibitor rather than a substrate for these latter mannosyltransferases. Dolichol phosphate mannosyl-transferase donor recognition was evaluated[116] by measuring apparent inhibition

Table 3.   Modified oligosaccharide structures that have been synthesized with fucosyltransferases (added sugar residue is shown in bold).

Oligosaccharide Product	%Rel. Rate	Ref.
Galβ–(1→3)-[**Fucα**–(1→4)-]GlcNAcβ-OR$_1$	100	112
Galβ–(1→3)-[**Fucα**–(1→4)-]GlcNAcβ-OR$_2$		
Galβ–(1→3)-[**3-deoxy-Fucα**–(1→4)-]GlcNAcβ-OR$_1$	2.3	112
Galβ–(1→3)-[**3-deoxy-Fucα**–(1→4)-]GlcNAcβ-OR$_2$		
Galβ–(1→3)-[**Araα**–(1→4)-]GlcNAcβ-OR$_1$	5.9	112
Galβ–(1→3)-[**Araα**–(1→4)-]GlcNAcβ-OR$_2$		
Galβ–(1→3)-[**6-O-propyl-Fucα**–(1→4)-]GlcNAcβ-OR$_1$	50	113
Galβ–(1→3)-[**Galα**–(1→3)-[**Fucα**–(1→2)-]GalβO(CH$_2$)$_8$CO-NH(6)-**Fuc-α**–(1→4)-]GlcNAcβ-OR$_2$		113

R$_1$=(CH$_2$)$_8$COOCH$_3$, R$_2$=(CH$_2$)$_8$CO)$_{15-20}$NH-bovine serum albumin

constants (K$_i$'s) for 2-, 3-, 4- and 6-deoxy and 2-fluoro versions of GDP-Man ; K$_i$ values were comparable to the K$_m$ for GDP-Man (0.4 to 3.1 mM versus K$_m$ = 0.52 mM), with the fluorine substitution most disfavored (K$_i$ = 15 mM).

GlcNAcTs are involved in the internal branching and elongation of *N*- and *O*-linked structures (recently reviewed in 117). Partially purified human milk GlcNAcT-I has been used to synthesize a series of deoxygenated tetrasaccharides from Manα–(1→3)-(Man-α–(1→6)-Manβ-*O*-(CH$_2$)$_8$COOCH$_3$.[118] 3-deoxy, 4-deoxy and 6-deoxy-GlcNAc analogues of UDP-GlcNAc were chemically prepared (13-25mg) in 5-9 steps from commercially available precursors, and enzymatically transferred to the above synthetic trimannoside acceptor. Each reaction contained less than 1 mU of enzyme (4.2 mU/mg), but, in common with many preparative glycosyltransferase reactions, were driven by an excess of donor. Yields relative to the natural UDP-GlcNAc donor, deduced by ^{1}H-NMR analysis of the final reaction mixtures, were 70, 10 and 37% for the 3-, 4- and 6-deoxy donor derivatives, respectively, indicating that removal of any one of these hydroxyl groups is tolerated by the GlcNAc transferase. Bulkier substitutions may not be as well-tolerated. Even better yields were obtained by using the same deoxygenated UDP-

GlcNAc analogues and GlcNAcT-II from the same source to synthesize 3-, 4-, and 6-deoxy GlcNAcβ–(1→2)-Manα–(1→6)-(GlcNAcβ–(1→2)-Manα–(1→3)-Manβ-O-(CH$_2$)$_8$COOCH$_3$ from a tetrasaccharide precursor.[85] As was observed for GlcNAcT-I, the 4-deoxygenated version of this donor was not as readily utilized. GlcNAcT-I and -II, co-purified from rabbit liver, have also been used in larger scale synthetic reactions to produce natural-type tetra- and pentasaccharides in 60 to 100 micromole yields.[119] The acceptor specificities of rat liver GlcNAcT-I[32] and -II[31] have been recently described.

Recombinant GlcNAcT-I, expressed in insect Sf-9 cells, has also been used to synthesize labile tetrasaccharides GlcNAcβ–(1→2)-3R-Manα–(1→6)-(Manα–(1→3)-Manβ1-O-octyl (where R = (4,4-azo)pentyl and 5-iodoacetamide) from a synthetic trimannoside precursor.[43] These compounds were irreversible inhibitors of GlcNAcT-II. Two groups have recently described the use of two different isolated β–(1→6)-GlcNAc transferases in early steps of the *in vitro* synthesis of natural versions of O-linked oligosaccharides related to the sLe^x epitope.[120,121] As mentioned previously, lysates of MDAY-D2 tumor cells have been used as a source of core 2 β–(1→6)-GlcNAc transferase, along with bovine milk β–(1→4)-GalTase and rat liver Galβ–(1→3/4)-GlcNAc α–(2→3)-STase, to synthesize 0.01 to 1 mmol of O-linked type tri-, tetra-, penta- and hexasaccharides with the structures ±NeuAcα–(2→3)-Galβ–(1→3)-(±NeuAcα–(2→3)±Galβ–(1→4)-GlcNAc-β–(1→6)-GalNAcα-C$_6$H$_4$-NO$_2$(p). The acceptor specificity of two core 2 β–(1→6)-GlcNAcTs was recently analyzed.[33] A crude extract of rat kidney GlcNAcT III has been used to add a bisecting GlcNAcβ–(1→4)- to the Manβ of synthetic pentasaccharides GlcNAcβ–(1→2)-6R-Manα–(1→6)-(GlcNAcβ–(1→2)-Manα–(1→3)-Manβ1-O-octyl, where R=H[122] or OCH$_3$.[123] These reactions were done on a micromole scale, and all products were characterized by ^{1}H-NMR. Such work demonstrates the utility of GlcNAc transferases in general.

7.   CONCLUSION

It has been determined that glycosyltransferases display considerable flexibility in their donor specificities, and this may be used to expand the synthetic utility of these enzymes for the synthesis of unnatural, substituted oligosaccharides. Availability of modified sugar-nucleotide donors, which depends largely upon the availability of systems for their enzymatic synthesis, continues to be a major limitation in this area. This seems to be particularly true for enzymes utilizing UDP-GlcNAc, UDP-GalNAc and GDP-fucose, but systems *have* been described for their enzymatic generation,[14,15] so it simply remains to be seen how well these systems tolerate substituted monosaccharides. On the other hand, an extensive literature exists demonstrating that CMP-sialic acid synthetase from different sources will tolerate a variety of modifications to NeuAc (see section 4 above). Enzymes such as hexokinase,[88,89] glucokinase and galactose-1-phosphate uridyltransferase,[86] and UDP-Gal-4-epimerase,[40] used for *in situ* generation and recycling of UDP-Gal, appear to be similarly accommodating. Since a wide variety of glycosyltransferases display flexibility in their donor recognition, a limited number of donor analogues could be used for the synthesis of a large number of different oligosaccharides.[85]

ACKNOWLEDGMENTS

This work was supported in part by a grant from the Medical Research Council of Canada.

REFERENCES

1. A. Varki, *Glycobiology, 3*, 97 (1993).
2. M. L. Phillips, E. Nudelman, F. C. A. Gaeta, M. Perez, A. K. Singhal, S. Hakomori, and J. C. Paulson, *Science,* **250**, 1130 (1990).
3. G. Walz, A. Aruffo, W. Kolanus, M. Bevilacqua, and B. Seed, *Science,* **250**, 1332 (1990).
4. B. K. Brandley, *Seminars Cell Biol.,* **2**, 281 (1991).
5. C. A. Lingwood, *Curr. Opin. Struct. Biol.,* **2**, 693 (1991).
6. S. Hakomori, *Adv. Cancer Res.,* **52**, 257 (1989).
7. J. W. Dennis, *Cancer Surveys,* **7**, 753 (1988).
8. H. Paulsen, *Angew. Chem. Int. Ed. Eng.,* **29**, 823 (1990).
9. P. Sinaÿ, *Pure Appl. Chem.,* **63**, 519 (1991).
10. O. Kanie and O. Hindsgaul, *Curr. Opin. Struct., Biol.* **2**, 674 (1992).
11. E. J. Toone, E. S. Simon, M. D. Bednarski, and G. M. Whitesides, *Tetrahedron,* **45**, 5365 (1989).

12. S. David, C. Augé, and C. Gautheron, *Adv. Carbohydr. Chem. Biochem.*, **49**, 175 (1991).

13. D. G. Drueckhammer, W. J. Hennen, R. L. Pederson, C. F. Barbas, C. M. Gautheron, T. Krach, and C.-H. Wong, *Synthesis*, **7**, 499 (1991).

14. Y. Ichikawa, G. C. Look, and C.-H. Wong, *Anal. Biochem.*, **202**, 215 (1992).

15. J. E. Heidlas, K. W. Williams, and G. M. Whitesides, *Acc. of Chem. Res.*, **25**, 307 (1992).

16. C.-H. Wong, M. Schuster, P. Wang, and P. Sears, *J. Am. Chem. Soc.*, **115**, 5893 (1993).

17. M. A. Kashem, K. B. Wlasichuk, J. M. Gregson, and A. P. Venot, *Carbohydr. Res.*, **250**, 129 (1993).

18. G. F. Herrmann, Y. Ichikawa, C. Wandrey, F. C. A. Gaeta, J. C. Paulson, and C.-H. Wong, *Tetrahedron Lett.*, **34**, 3091 (1993).

19. S. A. DeFrees, F. C. A. Gaeta, Y.-C. Lin, Y. Ichikawa, and C.-H. Wong, *J. Am. Chem. Soc.*, **115**, 7549 (1993).

20. C. Unverzagt, S. Kelm, and J. C. Paulson, *Carbohydr. Res.*, **251**, 285 (1994).

21. Y. Ichikawa, Y.-C. Lin, D. P. Dumas, G.-J. Shen, E. Garcia-Junceda, M. A. Williams, R. Bayer, C. Ketcham, L. E. Walker, J. C. Paulson, and C.-H. Wong, *J. Am. Chem. Soc.*, **114**, 9283 (1992).

22. M. M. Palcic, *Methods in Enzymol.*, **230**, 300 (1994).

23. R. U. Lemieux, *Chem. Soc. Rev.*, **18**, 347 (1989).

24. D. R. Bundle and N. M. Young, *Curr. Opin. Struct. Biol.*, **2**, 666 (1992).

25. C. Gautheron-Le Narvor and C.-H. Wong, *J. Chem. Soc., Chem. Commun.*, 1130 (1991).

26. C.-H. Wong, Y. Ichikawa, T. Krach, C. Gautheron-Le Narvor, D. P. Dumas, and G. C. Look, *J. Am. Chem. Soc.*, **113**, 8137 (1991).

27. C.-H. Wong, T. Krach, C. Gautheron-Le Narvor, Y. Ichikawa, G. C. Look, F. Gaeta, D. Thompson, and K. C. Nicolaou, *Tetrahedron Lett.*, **32**, 4867 (1991).

28. K. Wlasichuk, M. A. Kashem, P. V. Nikrad, P. Bird, C. Jiang, and A. P. Venot, *J. Biol. Chem.*, **268**, 13971 (1993).

29. T. L. Lowary and O. Hindsgaul, *Carbohydr. Res.,* **249**, 163 (1993).

30. T. L. Lowary and O. Hindsgaul, *Carbohydr. Res.,* **251**, 33 (1994).

31. H. Paulsen, E. Meinjohanns, F. Reck, and I. Brockhausen, *Liebigs Ann. Chem.*, 721 (1993).

32. G. Möller, F. Reck, H. Paulsen, K. J. Kaur, M. Sarkar, H. Schachter, and I. Brockhausen, *Glycoconj. J.,* **9**, 180 (1992).

33. W. Kuhns, V. Rutz, H. Paulsen, K. L. Matta, M. A. Baker, M. Barner, M. Granovsky, and I. Brockhausen, *Glycoconj. J.*, **10**, 381 (1993).

34. O. Kanie, S. C. Crawley, M. M. Palcic, and O. Hindsgaul, *Carbohydr. Res.*, **243**, 139 (1993).

35. T. Linker, S. C. Crawley, and O. Hindsgaul, *Carbohydr. Res.*, **245**, 323 (1993).

36. S.C. Crawley, *Ph.D. Dissertation*, University of Alberta, Edmonton, Canada (1993).

37. O. Kanie, S. C. Crawley, M. M. Palcic, and O. Hindsgaul, *Bioorg. and Med. Chem.*, **2**, 1231 (1994).

38. B. K. Brandley, M. Kiso, S. Abbas, P. Nikrad, O. Srivastava, C. Foxall, Y. Oda, and A. Hasegawa, *Glycobiology,* **3**, 633 (1993).

39. R. M. Nelson, S. Dolich, A. Aruffo, O. Cecconi, and M. P. Bevilacqua, *J. Clin. Invest.,* **91**, 1157 (1993).

40. H. Yuasa, O. Hindsgaul, and M. M. Palcic, *J. Am. Chem. Soc.,* **114**, 5891 (1992).

41. O. Hindsgaul, K. J. Kaur, G. Srivastava, M. Blaszczyk-Thurin, S.C. Crawley, L. D. Heerze, and M. M. Palcic, *J. Biol. Chem.,* **266**, 17858 (1991).

42. S. H. Khan, S. C. Crawley, O. Kanie, and O. Hindsgaul, *J. Biol. Chem.,* **268**, 2468 (1993).

43. F. Reck, H. Paulsen, I. Brockhausen, M. Sarkar, and H. Schachter, *Glycobiology*, **2**, 483 (1992).

44. S.-C. Ats, J. Lehmann, and S. Petry, *Carbohydr. Res.,* **233**, 125 (1992).

45. S.-C. Ats, J. Lehmann, and S. Petry, *Carbohydr. Res.,* **233**, 141 (1992).

46. C.-H. Wong, D. P. Dumas, Y. Ichikawa, K. Koseki, S. J. Danishefsky, B. W. Weston, and J. B. Lowe, *J. Am. Chem. Soc.,* **114**, 7321 (1992).

47. Y. Kajihara, H. Hashimoto, and H. Kodsma, *Carbohydr. Res.*, **229**, C5 (1992).

48. G. Herrler, H.-J. Gross, A. Imhof, R. Brossmer, G. Milks, and J. C. Paulson, *J. Biol. Chem.*, **267**, 12501 (1992).

49. R. E. Kosa, R. Brossmer, and H.-J. Gross, *Biochem. Biophys. Res. Comm.*, **190**, 914 (1993).

50. J. C. Paulson and G. N. Rogers, *Methods in Enzymol.*, **138**, 162 (1987).

51. J. A. Beyer, J. E. Sadler, J. I. Rearick, J. C. Paulson, and R. L. Hill, *Adv. Enzymol. Relat. Areas Mol. Biol.*, **52**, 23 (1981).

52. J. E. Sadler, T. A. Beyer, C. L. Oppenheimer, J. C. Paulson, J.-P. Priels, J. I. Rearick, and R. L. Hill, *Methods in Enzymol.*, **83**, 458 (1982).

53. R. Kornfeld and S. Kornfeld, *Ann. Rev. Biochem..*, **54**, 631 (1985).

54. R. Kleene and E. G. Berger, *Biochim. Biophys. Acta*, **1154**, 283 (1993).

55. J. C. Paulson and K. J. Colley, *J. Biol. Chem.*, **264**, 17615 (1989).

56. J. Weinstein, E. U. Lee, K. McEntee, P.-H. Lai, and J. C. Paulson, *J. Biol. Chem.*, **262**, 17735 (1987).

57. D. Zhuang, A. Grey, M. Harris-Brandts, E. Higgins, M. A. Kashem, and J. W. Dennis, *Glycobiology*, **1**, 425 (1991).

58. N. Taniguchi, A. Nishikawa, S. Fujii, and J. Gu, *Methods in Enzymol.*, **179**, 397 (1989).

59. C. A. Compston, C. Condon, H. R. Hanna, and M. A. Mazid, *Carbohydr. Res.*, **239**, 167 (1993).

60. S. Narasimhan, *J. Biol. Chem.*, **257**, 10235 (1982).

61. P. Gleeson and H. Schachter, *J. Biol. Chem.*, **258**, 6162 (1983).

62. I. Brockhausen, E. Hull, O. Hindsgaul, H. Schachter, R. N. Shah, S. W. Michnick, and J. P. Carver, *J. Biol. Chem.*, **264**, 11211 (1989).

63. I. Brockhausen, G. Möller, J.-M. Yang, S. H. Khan, K. L. Matta, H. Paulsen, A. A. Grey, R. N. Shah, and H. Schachter, *Carbohydr. Res.*, **236**, 281 (1992).

64. P. Wang, G.-J. Shen, Y.-F. Wang, Y. Ichikawa, and C.-H. Wong, *J. Org. Chem.*, **58**, 3985 (1993).

65. C. Unverzagt, H. Kunz, and J. C. Paulson, *J. Am. Chem. Soc.*, **112**, 9308 (1990).

66. D. H. van den Eijnden, and D. H. Joziasse, *Curr. Op. Struct. Biol.*, **3**, 711 (1993).

67. D. X. Wen, B. D. Livingston, K. F. Medzihradszky, S. Kelm, A. L. Burlingame, and J. C. Paulson, *J. Biol. Chem.*, **267**, 21011 (1992).

68. H. Kitagawa and J. C. Paulson, *J. Biol. Chem.*, **269**, 1394 (1994).

69. B. W. Weston, P. L. Smith, R. J. Kelley, and J. B. Lowe, *J. Biol. Chem.*, **267**, 24575 (1992).

70. M. F. A. Bierhuizen, M.-G. Mattei, and M. Fukuda, *Genes and Dev.*, **7**, 468 (1993).

71. M. F. A. Bierhuizen and M. Fukuda, *Trends Glycosci. Glycotech.*, **6**, 17 (1994).

72. K. J. Colley, E. U. Lee, B. Adler, J. K. Browne, and J. C. Paulson, *J. Biol. Chem.*, **264**, 17619 (1989).

73. J. F. Kukowska-Latallo, R. D. Larsen, R. P. Nair, and J. B. Lowe, *Genes Dev.*, **4**, 1288 (1990).

74. T. de Vries, D. H. van den Eijnden, J. Schultz, and R. A. O'Neill, *FEBS Lett.*, **330**, 243 (1993).

75. D. Aoki, H. E. Appert, D. Johnson, S. S. Wong, and M. N. Fukuda, *EMBO J.*, **9**, 3171 (1990).

76. E. E. Boeggeman, R. V. Balaji, N. Sethi, A. S. Masibay, and P. K. Qasba, *Protein Engineering*, **6**, 779 (1993).

77. Burke, J. M. Pettit, H. Schachter, M. Sarkar, and P. A. Gleeson, *J. Biol. Chem.*, **267**, 24433 (1992).

78. R. Kumar, J. Yang, R. L. Eddy, M. G. Byers, T. B. Shows, and P. Stanley, *Glycobiology*, **2**, 383 (1992).

79. C. H. Krezdorn, G. Watzele, R. B. Kleene, S. X. Ivanov, and E. G. Berger, *Eur. J. Biochem.*, **212**, 113 (1993).

80. M. Sarkar, and H. Schachter, *Glycobiology*, **2**, 483 (1992).

81. D. H. Joziasse, N. L. Shaper, L. S. Salyer, D. H. van den Eijnden, A. C. van der Spoel, and J. H. Shaper, *Eur. J. Biochem.*, **191**, 75 (1990).

82. P. J. Andree and L. J. Berliner, *Biochim. Biophys. Acta*, **544**, 489 (1978).

83. L. J. Berliner and R. D. Robinson, *Biochemistry*, **21**, 6340 (1982).

84. M. M. Palcic and O. Hindsgaul, *Glycobiology*, **1,** 205 (1991).

85. O. Hindsgaul, K. J. Kaur, U. B. Gokhale, G. Srivastava, G. Alton, and M. M. Palcic, *ACS Symp. Ser.,* **466,** 38 (1991).

86. C.-H. Wong, R. Wang, and Y. Ichikawa, *J. Org. Chem.*, **57,** 4343 (1992).

87. G. Srivastava, O. Hindsgaul, and M. M. Palcic, *Carbohydr. Res.*, **245,** 137 (1993).

88. J. Thiem and T. Wiemann, *Angew. Chem. Int. Ed. Eng.*, **30,** 1163 (1991).

89. J. Theim and T. Wiemann, *Synthesis*, **1992,** 141 (1992).

90. H. Kodama, Y. Kajihara, T. Endo, and H. Hashimoto, *Tetrahedron Lett.*, **34,** 6419 (1993).

91. Y. Nishida, T. Wiemann, V. Sinnwell, and J. Thiem, *J. Am. Chem. Soc.* **115,** 2536 (1993).

92. A. Varki, *Glycobiology,* **2,** 25 (1992).

93. J. Weinstein, U. de Souza-e-Silva, and J. C. Paulson, *J. Biol. Chem.,* **257,** 13835 (1982).

94. J. E. Sadler, J. I. Rearick, J. C. Paulson, and R. L. Hill, *J. Biol. Chem.,* **254,** 4434 (1979).

95. J. E. Sadler, J. I. Rearick, and R. L. Hill, *J. Biol. Chem.,* **254,** 5934 (1979).

96. J. I. Rearick, J. E. Sadler, and R. L. Hill, *J. Biol. Chem.,* **254,** 4444 (1979).

97. J. Haverkamp, J.-M. Beau, and R. Schauer, *Hoppe-Seyler's Z. Physiol. Chem.,* **360,** 159 (1979).

98. J.-M. Beau and R. Schauer, *Eur. J. Biochem.,* **106,** 531 (1980).

99. H. Higa and J. C. Paulson, *J. Biol. Chem.,* **260,** 8838 (1985).

100. H. S. Conradt, A. Bünsch, and R. Brossmer, *FEBS Lett.,* **170,** 295 (1984).

101. H. J. Gross, A. Bünsch, J. C. Paulson, and R. Brossmer, *Eur. J. Biochem.,* **168,** 595 (1987).

102. H. J. Gross, U. Rose, J. M. Krause, J. C. Paulson, K. Schmid, R. E. Feeney, and R. Brossmer, *Biochemistry,* **28,** 7386 (1989).

103. H. J. Gross and R. Brossmer, *Glycoconj. J.,* **4,** 145 (1987).

104. H. J. Gross and R. Brossmer, *Eur. J. Biochem.*, **177**, 583 (1988).

105. H. J. Gross, U. Sticher, and R. Brossmer, *Anal. Biochem.*, **186**, 127 (1990).

106. U. Sticher, H. J. Gross, and R. Brossmer. *Glycoconj. J.*, **8**, 45 (1991).

107. C. R. Petrie and W. Korytnyk, *Anal. Biochem.*, **131**, 153 (1983).

108. S. L. Shames, E. S. Simon, C. W. Christopher, W. Schmid, and G. M. Whitesides, *Glycobiol.*, **1**, 187 (1991).

109. J. L.-C. Liu, G. J. Shen, Y. Ichikawa, J. F. Rutan, G. Zapata, W. F. Vann, and C.-H. Wong, *J. Am. Chem. Soc.*, **114**, 3901 (1992).

110. C. Augé and C. Gautheron, *Tetrahedron Lett.*, **29**, 789 (1988).

111. Y. Ichikawa, G.-J. Shen, and C.-H. Wong, *J. Am. Chem. Soc.*, **113**, 4698 (1991).

112. U. B. Gokhale, O. Hindsgaul, and M. M. Palcic, *Can. J. Chem.*, **68**, 1063 (1990).

113. G. Srivastava, K. J. Kaur, O. Hindsgaul and M. M. Palcic, *J. Biol. Chem.*, **267**, 22356 (1992).

114. R. Stiller and J. Thiem, *Liebigs Ann. Chem.*, 467, (1992).

115. W. McDowell, T. J. Grier, J. R. Rasmussen, and R. T. Schwartz, *Biochem. J.*, **248**, 523 (1987).

116. W. McDowell and R. T. Schwartz, *FEBS Lett.*, **243**, 413 (1989).

117. H. Schachter, *Curr. Opin. Struct. Biol.*, **1**, 755 (1991).

118. G. Srivastava, G. Alton, and O. Hindsgaul, *Carbohydr. Res.*, **207**, 259 (1990).

119. K. J. Kaur, G. Alton, and O. Hindsgaul, *Carbohydr. Res.*, **210**, 145 (1991).

120. G. C. Look, Y. Ichikawa, G.-J. Shen, P.-W. Cheng, and C.-H. Wong, *J. Org. Chem.*, **58**, 4326 (1993).

121. R. Oehrlein, O. Hindsgaul, and M. M. Palcic, *Carbohydr. Res.*, **244**, 149 (1993).

122. G. Alton, Y. Kanie, and O. Hindsgaul, *Carbohydr. Res.*, **238**, 339 (1993).

123. S. H. Khan, C. A. Compston, M. M. Palcic, and O. Hindsgaul, *Carbohydr. Res.*, **262**, 283 (1994).

Chapter 21

SYNTHESIS WITH GLYCOSIDASES

KURT G. I. NILSSON

Glycorex AB, Sölveg. 41, S-223 70 Lund, Sweden

Abstract The glycosidase-catalyzed synthesis of glycosides and oligosaccharides is reviewed with emphasis on structures related to glycoconjugates. The general enzymatic properties of glycosidases such as substrate and reaction specificity, availability and stability, are briefly discussed, followed by a review of the three main techniques available for synthesis with these enzymes, hydrolysis, reversed hydrolysis and transglycosylation. Manipulation of the regioselectivity of the glycosylation reactions is described and examples of structures synthesized on a preparative scale are given. Finally, the use of glycosidases for glycosylation of various organic compounds, synthesis of partially protected carbohydrates/analogs and the combined use of glycosidases and glycosyltransferases for oligosaccharide synthesis are discussed.

1. INTRODUCTION

Historically, glycosidases were among the first enzymes to be investigated and studies on glycosidases have been important for the formulation of some of the basic principles of enzyme action. Glycosidases (EC 3.2) are ubiquitous in Nature and have a vital catabolic function in living organisms. They are also important for the trimming of N-linked glycoproteins *in vivo* and several inherited metabolic disorders are associated with the lack of specific glycosidases.[1] Today glycosidases are used in a number of different applications:

- Analysis, e.g. in diagnostic reagents and in the analysis of carbohydrate

The type of synthesis determines the degree of purity required. Thus, whereas pure glycosidases are required for synthesis by reversed hydrolysis to obtain a stereospecific synthesis, crude enzyme preparations and even intact bacteria or yeast cells can be used successfully in the synthesis of glycosidic bonds by transglycosylation (see below).

Thus, the general properties of glycosidases can be summarized as follows:
- Abundant, easily available catalysts which can be used on an industrial scale without prior cloning.
- Stereo- and reaction specific, but with a wider flexibility towards different aglycons.
- Allows for the use of abundant, easily available substrates of several different types.

These properties of glycosidases are of great value in the selective synthesis of glycosides, glycoconjugate glycans and derivatives.

3. SYNTHESIS OF GLYCOCONJUGATE GLYCAN SEQUENCES WITH GLYCOSIDASES

Three types of glycosidase reactions have been used for the preparation of oligo-saccharides/derivatives:
- Hydrolysis
- Reversed hydrolysis and
- Transglycosylation

Synthesis via transglycosylation has been the major route used. Glycosidases of the exo-type have been used mainly for the synthesis of the shorter carbohydrate sequences of glycoconjugates (di- and trisaccharides) but it is to be expected that endoglycosidases will increasingly be applied to the synthesis of higher structures.

3.A. Use of Hydrolysis for Preparation of Glycoconjugate Saccharides

Endoglycosidases are useful for preparative scale synthesis of di-, tri- and higher oligosaccharide fragments by selective hydrolysis of larger oligosaccharides or polysaccharides. Thus, the chemical synthesis of Galα(1→4)Gal (a proposed receptor structure for the common uropathogen P-fimbriated *E. coli*)[19] was facilitated by the preparation of large quantities of digalacturonic acid (kg-scale) by the controlled hydrolysis of pectin by pectinase.[20] The disaccharide receptor structure and its glycosides, which were suitable for inhibition studies of bacterial adhesion or for attachment to proteins and chromatography material, were conveniently obtained by chemical modification of the digalacturonic acid.

$$\text{Pectin} \longrightarrow \text{Digalacturonic acid} \longrightarrow \longrightarrow \text{Gal}\alpha(1{\rightarrow}4)\text{Gal-OR} \longrightarrow \begin{array}{l} \text{Inhibition Studies} \\ \text{Neoglycoproteins} \\ \text{Affinity Beads} \\ \text{Biosensors} \end{array}$$

Pectinase

Recently, the above galabiose structure was used in a biosensor application for the specific determination of pathogenic *E. coli*.[21]

Hydrolysis with endoglycosidases is widely used for the release of oligosaccharide chains from glycolipids and glycoproteins (cf. Table 1).[1,22] Examples are the release of asparagine linked carbohydrate chains with recombinant N-glycanase (EC 3.5.1.52) and the use of *Diplococcus pneumoniae* endo-ß-galactosidase, which selectively releases blood group A and B trisaccharides from the nonreducing end of glycoconjugates.[23]

In a third type of approach, the hydrolytic specificity of glycosidases towards

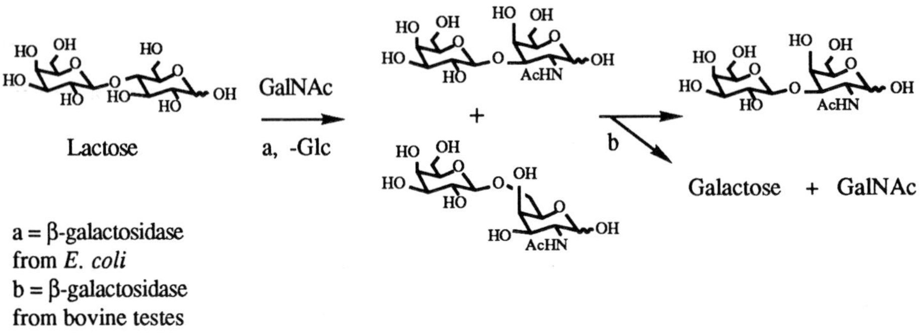

a = β-galactosidase
from *E. coli*
b = β-galactosidase
from bovine testes

Figure 1. Sequential use of two glycosidases for preparation of Galβ(1→3)GalNAc isomers.

positional isomers is exploited (cf. Table 2 and Figure 1). Thus, when a mixture of products is obtained in a glycosidase-catalyzed transglycosylation reaction it can be resolved by the use of another glycosidase with a different regioselectivity in a second (hydrolysis) reaction for selective hydrolysis of the contaminating oligosaccharide. This method has been used to produce pure Galß(1→3)GalNAc synthesised from lactose and GalNAc, using bovine testes and *E.coli* ß-galactosidase in the transglycosylation and hydrolytic reactions, respectively (Figure 1).[24] The *E. coli* enzyme hydrolyzes the ß(1→6)-linked isomer with high selectivity (cf. Table 2). A similar system was used for production of Galß(1→3)GlcNAc.[24,25]

3.B. Equilibrium Controlled Synthesis

Even though the hydrolytic reaction of glycosidases is favored under physiological conditions, synthesis of glycosidic linkages via reversed hydrolysis can be achieved at high substrate concentrations as illustrated in the scheme below where DOH is a monosaccharide and HOA is a hydroxyl group containing compound:

$$DOH \; + \; HOA \; \rightleftharpoons \; DOA \; + \; H_2O$$

$$Keq = 0.3 - 4 \text{ (including the water concentration)}$$
$$\Delta S = 30\text{-}50 \text{ J/Mol/K}; \quad \Delta H = \text{often} > 0$$

The synthesis of maltose by reversed hydrolysis from glucose used at high concentration (40 %), was one of the first examples of enzymatic synthesis in vitro.[26] Despite this early success, few examples of the preparative synthesis of the oligosaccharide structures found in glycoconjugates have been reported.

The equilibrium constants (Keq; including the water concentration) for the formation of maltose and isomaltose from glucose with glucoamylase are 0.28 and 3.10, respectively.[27] Similar differences between the equilibrium constants for the formation of linkage isomers were reported for α-linked dimannose formation catalyzed by jack bean α-mannosidase.[28]

To achieve reasonable yields high substrate concentrations or systems of decreased water activity have to be used to obtain a reasonable yield of product. The method has been mainly applied in the synthesis of various glycosides, e.g. in the synthesis of alkyl glucosides which are useful as detergents, such as octyl glucoside,[29,30] and also for the preparation of amino acid glycosides.[31,32] The equilibrium controlled reaction between mannose and different ergot alkaloids was used for the preparation of different ergot alkaloid α-mannosides.[33]

Isomers not formed to an appreciable extent in the kinetically controlled transglycosylation reactions, where rate constants may be unfavorable, may be formed in equilibrium controlled reactions. This has been demonstrated in the formation of maltose and isomaltose with glucoamylase.[27] Moreover, while no synthesis of amino acid sugars was observed with glycosides as donors and amino acids (serine, threonine, tyrosine) as acceptors (transglycosylation conditions), synthesis by reversed hydrolysis was observed when high concentrations of either N-acetyl-galactosamine, galactose or mannose were used as glycosyl donors (Figure 2).[31]

β-Glucosidase

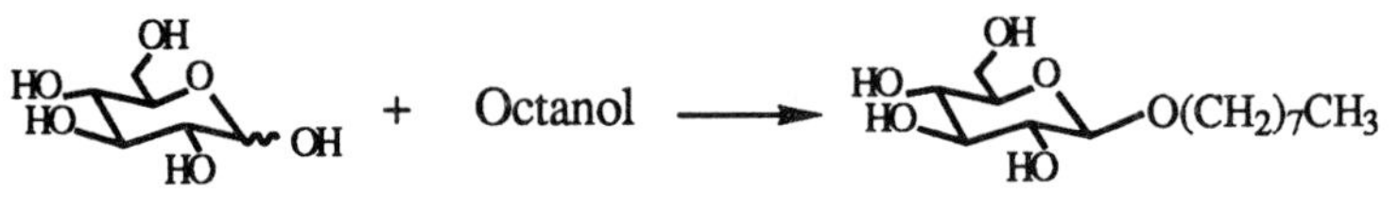

α-N-Acetyl-D-Galactosaminidase

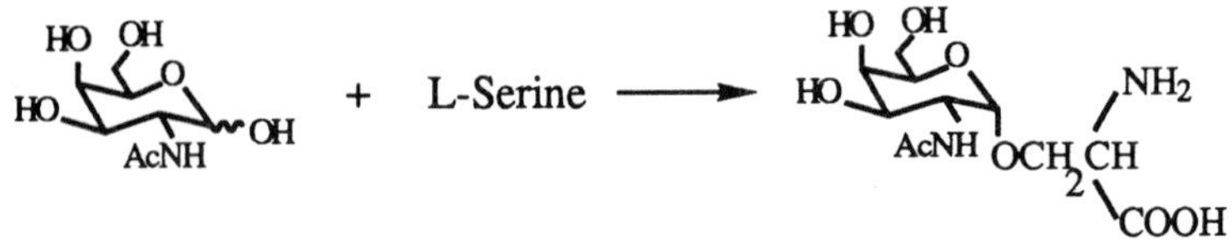

Figure 2. Stereospecific synthesis of octyl β-glucoside and GalNAcα-L-Serine employing equilibrium-controlled synthesis.

An obvious advantage of the equilibrium approach is that cheap, unmodified donor substrates are used. In transglycosylation reactions, the donor glycoside is consumed and thus can not be reused. However, glycosides are much better substrates for glycosidases than monosaccharides with a free reducing end and, therefore, much higher amounts of enzyme and/or longer reaction times are required in the equilibrium controlled reactions. In addition, pure enzyme preparations have to be used to avoid the simultaneous formation of α- and ß-glycosides.

3.C. Synthesis by Transglycosylation

The preparative syntheses of the common disaccharides Galß(1→4)GlcNAc and Galß(1→3)GlcNAc employing transglycosylation reaction conditions with *Lactobacillus bifidus* or mammalian ß-galactosidases as catalysts were reported in 1955[34] and in 1956,[35] respectively. Several different types of glycosides and glycoconjugate oligosaccharide structures and derivatives have now been synthesized with glycosidase-catalyzed transglycosylation reactions.[5,24,25,34-42]

3.C.i. General

In this approach a glycoside (e.g. an oligosaccharide or an alkyl or an aryl glycoside) is used as glycosyl donor and the glycon part of the glycoside is transferred to a hydroxyl group containing acceptor (HOA in the scheme below):

$$\text{D-R} + \text{EH} \xrightarrow{\quad \text{-RH} \quad} \text{E:D} \xrightarrow{\quad \text{+HOA} \quad} \text{D-OA} + \text{EH}$$

D-R = Donor glycoside or oligosaccharide; R = Saccharide unit, aromatic or aliphatic group, fluorine; HOA = hydroxyl group containing compound; EH = enzyme; E:D = glycosyl-enzyme intermediate.

Transglycosylations are stereospecific and occur with retention of the donor anomeric configuration. In the initial phase of the reaction the glycosyl-enzyme intermediate (E:D in the scheme above) will be preferentially formed from the donor glycoside.[1,3,6] The intermediate can react with a wide range of hydroxyl group containing acceptors other than water, including monosacharides, disaccharides and their analogs as well as aliphatic or aromatic alcohols, and a number of different compounds can thus be prepared from the donor glycoside (Table 3).

Compared with reversed hydrolysis, higher yields may be obtained (usually in the range 20-65 % depending on conditions and reaction) and with considerably

Table 3. Examples of substrates and products in glycosidase-catalyzed trans-glycosylation reactions.

Donors	Acceptors	Products
Di- , tri- and higher oligosaccharides Aliphatic-, aromatic- or F-glycosides	Carbohydrates Modified Carbohydrates Other hydroxyl-group containing aliphatic, aromatic compounds, including amino acids, peptide derivatives, drugs, vitamins	Oligosaccharides Modified Oligosaccharides Glycosides

shorter reaction times (minutes to hours at ambient temperature) due to the higher reactivity of the donor glycoside. The rate of transglycosylation relative to hydrolysis depend on the structure of the acceptor and this selectivity is different from enzyme to enzyme. Secondary hydrolyis of product is minimized by the use of a highly reactive donor glycoside having a high kcat/Km ratio compared with that of the product.[40-42] The reactions can be carried out under mild conditions near neutral pH or in aqueous-organic solvent mixtures and at different temperatures depending on the stability of the particular glycosidase used.

Crude or partially purified enzyme preparations may be used, whereas the equilibrium approach requires pure glycosidase preparations. In the case of bacteria or yeast, the enzyme may be used *in situ* and thus, no isolation of enzyme is required. Contaminating glycosidases are of less importance since transglycosylation with the active donor substrate is specific (e.g. β-galactosidase, but not e.g. α-galactosidase or β-glucosidase, is active on lactose) and considerably faster than reversed hydrolysis catalyzed by contaminating glycosidases. Several examples of successful syntheses of oligosaccharides with crude glycosidase preparations can be found in the literature.[34,35,37,42]

Several glycosidases useful for transglycosylation reactions, including α-L-fucosidases, α-D-galactosidases, β-D-galactosidases, α-D-hexosaminidases, β-D-hexosaminidases, α-D-mannosidases and α-sialidases, are commercially available in quantities sufficient for preparative synthesis, or can be isolated within a few days from their natural environment in a form suitable for transglycosylation reactions. This constitutes an important advantage compared with the less abundant glycosyltransferases.[43]

3.C.ii. Glycosyl donors

A number of glycosides suitable as glycosyl donors in transglycosylation reactions are commercially available, and are easy to prepare enzymatically[40] or by means of standard chemical procedures. Cheap oligosaccharides (lactose, oligomannosides, chitobiose, raffinose) have been used successfully as donors.[5,36,40] Nitrophenyl glycosides are often used as donors due to their high reactivity, and the ease of estimating the amount of consumed donor during the reaction by spectrophotometric measurement of the amount of nitrophenol liberated.[38] Such glycosides of interest for synthesis on a larger scale have been produced in kg-quantities in our laboratory.

As indicated above, the source of the enzyme is important when choosing a suitable donor, e.g. o-nitrophenyl ß-galactoside is an excellent substrate for ß-galactosidase from *E. coli*, whereas the calf intestine ß-galactosidase hydrolyzes lactose more rapidly than the nitrophenyl glycoside.[44]

It should be observed that the glycosyl donor can also act as acceptor and this has been used for the facile synthesis of various disaccharide glycosides such as Galα(1→3)Galα-OMe, Galα(1→3)Galα-OPhNO$_2$-p, Manα(1→2)Manα-OPhNO$_2$-p[38,41] starting from Galα-OMe, Galα-OPhNO$_2$-p, and Manα-OPhNO$_2$-p, respectively. Thus, if this reaction is not desired a high ratio of glycosyl acceptor to glycosyl donor should be used.

3.C.iii. Acceptors, regioselectivity

A broad range of structures can be used as acceptors in transglycosylation reactions and thus a broad range of glycosylated products have been prepared with glycosidases (cf. Table 3). The demands on the catalyst in oligosaccharide synthesis are higher than in, for instance, peptide synthesis.[45] Both types of reactions involve condensation between an activated donor electrophile and an acceptor nucleophile, but hydroxyl groups are less nucleophilic than amino groups and, in addition to the possibility of alternative configuration (α- or ß-), the multiple hydroxyl groups of the acceptor sugar put high demands on regioselectivity in carbohydrate synthesis. Low or wrong regioselectivity, such as predominant formation of undesired 1-6-linkages, was previously a problem in glycosidase-catalyzed synthesis of oligosaccharides, but developments during the last decade have solved this problem at least for the preparation of disaccharides. Thus, it is now possible to prepare such compounds and analogs stereospecifically and with high regioselectivity in one reaction step with glycosidases (see below).

3.C.iv. Influence of enzyme source

The source of the enzyme is important for regioselectivity (Figure 3). Thus, either of ß(1→3),[35] ß(1→4)[34] or ß(1→6)-linked[46] Gal-GlcNAc can be prepared provided the right source of enzyme is used. The enzymes and substrates are abundant and these type of reactions can be used for preparation of lactosamine and derivatives thereof on a kg-scale and with a yield of over 20 %.

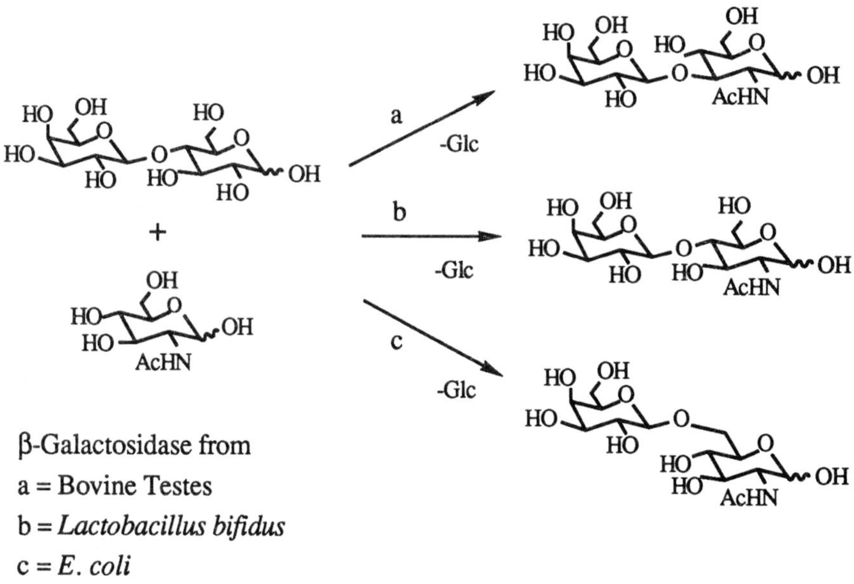

β-Galactosidase from
a = Bovine Testes
b = *Lactobacillus bifidus*
c = *E. coli*

Figure 3. Influence of enzyme source on the regioselectivity of glycosidase-catalyzed synthesis: Synthesis of β(1→3)-, β(1→4)- and β(1→6)-linked Gal-GlcNAc with β-galactosidase from different sources.

The rate of hydrolysis of isomeric oligosaccharides can be useful for predicting the regioselectivity of transglycosylation. Thus, in the case of E. coli ß-galactosidase the rates of hydrolysis of Gal-Glc were ß(1→6) > ß(1→4) > ß(1→3) and the synthetic rates decreased in the same order.[44] With the ß-galactosidases from calf intestine and bovine testes the reversed order for both hydrolysis and transglycosylation was observed.[44]

The rate of hydrolysis may have more direct effects on the regioselectivity of transglycosylations. For example, a more rapidly formed product isomer may be more susceptible to secondary hydrolysis than a more slowly formed product, and in such a case, the ratio of product isomers will be dependent on when the reaction is terminated.[41] Provided that an excess of glycosyl donor is used and the difference in secondary hydrolysis between isomers is high, the same enzyme can be used for the preferential synthesis of different product isomers (unpublished results). Furthermore, one glycosidase can be used for synthesis and another glycosidase with a different linkage specificity can be used for hydrolysis of unwanted isomers (cf. Figure 1).

3.C.v. Influence of the acceptor structure on the regioselectivity

The regioselectivity of glycosidase-catalysed transglycosylations can be changed by using different α- or ß-glycosides as acceptors (Figure 4).[38,42] This finding facilitated the application of glycosidases for preparative synthesis of a broad range of biologically active carbohydrate sequences. Both the anomeric configuration and the structure of the aglycon of the acceptor are important.

The effect of changing the anomeric configuration is further illustrated in Table 4 for α-galactosidase, ß-galactosidase, ß-hexosaminidase and α-sialidase. With α-galactosidase, methyl ß-D-galactopyranoside gave mainly the α(1→6)-linked digalactoside, whereas the α-anomer gave almost exclusively the α(1→3)-linked disaccharide. The reverse result was found with the ß-glycosidases. The ratios between product isomers formed (i.e. 1→3/1→6) were drastically changed, i.e. with a factor of about 30 and 100 for the α- and ß-galactosidase, respectively.

The nature of the acceptor aglycon also may have a pronounced influence on regioselectivity (Table 4). Thus, with α-galactosidase, the α(1→3)-linked digalactoside was almost exclusively formed when p-nitrophenyl α-D-galactopyranoside was the acceptor whereas, with the corresponding o-nitrophenyl glycoside, the 1→2-linked product predominated.[38] In this way the abundant coffee bean α-galactosidase can be used for preparative scale synthesis of either α(1→2)-α(1→3)- or α(1→6)-linked digalactosides. As mentioned above, the α(1→4)-linked digalactoside can be prepared via pectinase-catalyzed hydrolysis of pectin.

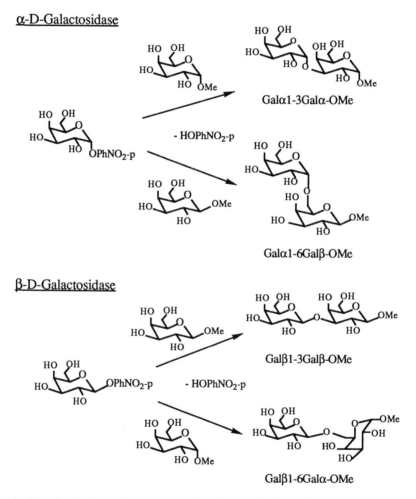

Figure 4. Manipulation of the regioselectivity of glycosidase-catalyzed trans-glycosylation.

The size or hydrophobicity of the acceptor aglycon also seems to be important. Thus, with the bovine testes ß-galactosidase, ß(1→3) and ß(1→4)-linked Gal-

Table 4.   Influence on regioselectivity of glycosidase-catalyzed transglycosylation reactions of acceptor aglycon structure and anomeric configuration.[38,42]

Enzyme	Acceptor	Main glycoside formed
α-D-galactosidase	Galα-OPhNO$_2$-o	Galα(1→2)Galα-OPhNO$_2$-o
	Galα-OPhNO$_2$-p	Galα(1→3)Galα-OPhNO$_2$-p
	Galα-OMe	Galα(1→3)Galα-OMe
	Galβ-OMe	Galα(1→6)Galβ-OMe
β-D-galactosidase	Galβ-OMe	Galβ(1→3)Galβ-OMe
	Galα-OMe	Galβ(1→6)Galα-OMe
	GlcNAcβ-OMe	Galβ(1→3(6))GlcNAcβ-OMe
	GlcNAcβ-OEtSiMe$_3$	Galβ(1→3)GlcNAcβ-OEtSiMe$_3$
β-D-glucosaminidase	Galβ-OMe	GlcNAcβ(1→3)Galβ-OMe
	Galα-OMe	GlcNAcβ(1→6)Galα-OMe

GlcNAc-OMe is formed in about equal amounts when GlcNAcß-OMe is used as acceptor while ß(1→3)-linked Gal-GlcNAcß-OEtSiMe$_3$ is formed with high regio-selectivity (ca 90 %) when the trimethylsilylethyl glycoside is the acceptor.[37] Similarly, with jack bean α-mannosidase, the ratio of α(1→2) and α(1→6)-linked products was 5:1 with methyl α-D-mannopyranoside as acceptor, whereas the ratio was 19:1 with the corresponding p-nitrophenyl glycoside.

In addition to the structures mentioned above, several other disaccharide glycosides related to glycoconjugate saccharides have been synthesized with the procedure[24,42,47-51] as exemplified in Table 5. Controlling the regioselectivity by this method thus allows the use of one glycosidase for the preferential synthesis of several different linkages. Glycosidases known to prefer primary hydroxyl groups, i.e. formation of 1-6-linkages, such as the α-galactosidase from green coffee beans and the ß-galactosidase from E. coli, can be used for preparative synthesis of other linkages, thus extending the number of structures that can be synthesized with readily available glycosidases. Moreover, a range of "unnatural" linkages can be produced.

Table 5.   Examples of disaccharide glycosides prepared by glycosidases (see also Table 4).

Enzyme	Acceptor	Main glycoside formed
α-L-Fucosidase	Galβ-OMe	Fucα(1→6)Galβ-OMe[47-49]
α-D-Galactosidase	GalNAcα-OEt	Galα(1→3)GalNacα-OEt[37]
β-D-Galactosidase	GlcNAcβ-SEt	Galβ(1→3)GlcNAcβ-SEt[24]
	GlcNAcβ-SEt	Galβ(1→6)GlcNAcβ-SEt[24]
β-D-Glucosaminidase	Manα-OMe	GlcNAcβ(1→6)Manα-OMe[37]
α-D-Galactosaminidase	Galα-OMe	GalNAcα(1→3)Galα-OMe[42]
β-D-Galactosaminidase	Galβ-OMe	GalNAcβ(1→3)Galβ-OMe[42]
β-D-Galactosaminidase	GlcNAcβ-OMe	GalNAcβ(1→4)GlcNAcα-OMe[50]
α-Sialidase	Galβ-OMe	NeuAcα(2→3)Galβ-OMe[51]

3.C.vi. Yields

Isolation of products in pure form is simplified when acceptor glycosides are used, since no anomerisation of the product glycosides occurs. Even if some of the reactions are not completely regioselective, product glycosides with hydrophobic aglycons are separated on Sephadex (e.g. nitrophenyl glycosides)[38,52] or reverse-phase C-18 resins. Recovery of excess of reagents by extraction, precipitation and/or column chromatography is usually straightforward.

The method has been used for the synthesis of disaccharide glycosides on a g - kg scale and with molar yields usually in the range 20 - 65 % depending on the conditions. Because of the simplicity of these reactions, the yields are sufficiently high to make the method attractive (for a discussion of yields under various conditions of substrate concentration, pH, temperature and solvents, see references 52 and 53).

3.C.vii. Application of products, preparation of acceptor glycosides

The use of acceptor glycosides provides an approach to glycosides commonly used for inhibition studies and to glycoside intermediates suitable for various applications. For example, in addition to methyl or ethyl glycosides, allyl-, benzyl-, 2-bromoethyl-, nitrophenyl-, thioethyl- and trimethylsilylethyl glycosides have been prepared.[37-42]   In this way glycosides used for temporary anomeric protection in chemical synthesis of oligosaccharides as well as glycosides suitable as "blocks" in convergent block synthesis of higher oligosaccharides, e.g. thioethyl glycosides,[54] or glycosides suitable for affinity labelling or for preparation of neo-glycolipids, neoglycoproteins or affinity adsorbents can be obtained. The use of amino acid or peptide glycosides as acceptors is also possible[55,56] and in this way various glycosylated amino acids and glycopeptides can be produced.

Moreover, some disaccharide glycosides have been prepared in "one-pot" reactions from a suitable saccharide donor and alcohol (Figure 5).[40]

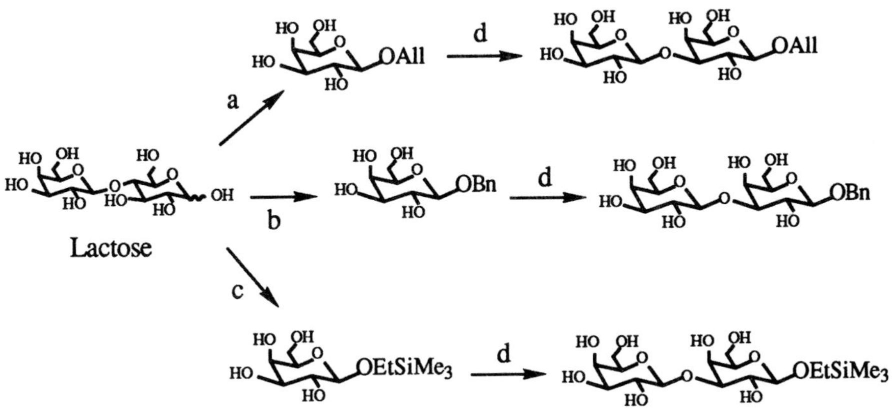

a = Allyl alcohol; b = Benzyl alcohol
c = Trimethylsilylethanol; d = Lactose
a - d: β-D-Galactosidase

Figure 5. One-pot synthesis of β-linked mono- and digalactosides with β-D-galac-tosidase (*E. coli*) as catalyst.

As an example of preparative scale synthesis, about 20 g of allyl ß-galactoside and 1 g of Galβ(1→3)Galß-OCH$_2$CH=CH$_2$ were prepared from lactose and allyl alcohol, employing a small amount of ß-galactosidase (2.5 mg)[40] although this reaction has not been optimized. Similar results were obtained with several other alcohols.

3. D.   Glycosylation of Various Organic Compounds

As mentioned above, transglycosylation can also be used to glycosylate amino acids, peptides and various organic compounds including drugs. This is of interest since the glycosidase-catalyzed reactions are carried out rapidly (a few minutes to hours) and under mild conditions (near neutral pH and often at ambient temperature) which minimize chemical side-reactions such as racemization. A few examples of such reactions are shown in Figures 6 and 7. As mentioned above (see 3.B), transglycosylations do not seem to be efficient with non-protected amino acids. However, successful galactosylation with lactose as the glycosyl donor can

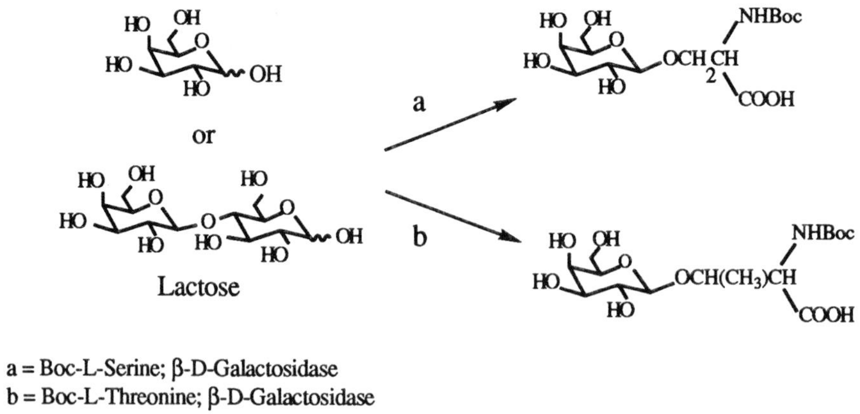

a = Boc-L-Serine; β-D-Galactosidase
b = Boc-L-Threonine; β-D-Galactosidase

Figure 6.   Synthesis of galactosylated serine and threonine.

be achieved if the hydroxyl-containing amino acids and dipeptides are protected at their amino- and carboxyl terminal groups.[56,57] It was also found that N-protected serine with an unprotected carboxyl group could be mannosylated or galactosylated by using either the transglycosylation or reversed hydrolysis approaches (Figure 6).[32]

Glycosylation of compounds of pharmaceutical interest is likely to alter their properties *in vivo*. There are several reports of successful glycosylation of such compounds which are sensitive to acids or bases employing glycosidase-catalyzed transglycosylations. Thus, as an example, the preparative galactosylation of the steroid gitoxigenin and other sensitive genins by *Aspergillus oryzae* β-galactosidase was achieved in 50 % water-acetonitrile, which allowed isolation of 3-O-cardiac glycosides in 25-74 % yield.[58] We found that the ergot alkaloids a, b and c in Figure 7, elymoclavine, chanoclavine and ergometrine, respectively, could be successfully mannosylated by employing jack bean α-mannosidase with either mannose in high concentration (reversed hydrolysis) or Manα-PhNO$_2$-p (trans-glycosylation conditions).[33] Important in these reactions are the mild conditions (near neutral pH, ambient temperature).

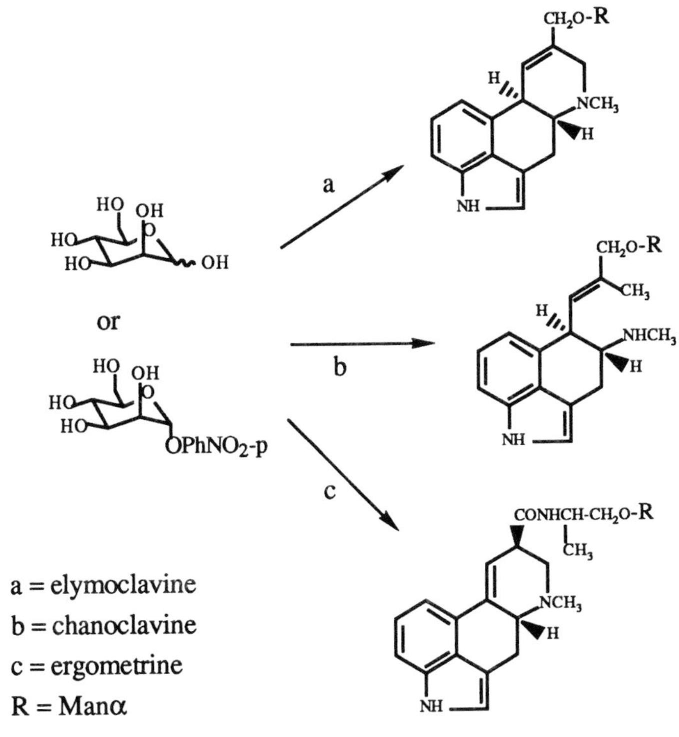

a = elymoclavine
b = chanoclavine
c = ergometrine
R = Manα

Figure 7.   Synthesis of ergot alkaloid α-mannosides with α-mannosidase.

3.E. Synthesis of Partially Modified Carbohydrates

Specifically modified carbohydrates are required for the block-wise chemical synthesis of larger saccharides or for the preparation of analogs. We found that glycosidases are well suited for preparation of carbohydrates which are specifically modified in one or more positions. This can be achieved by using acceptors which are protected/modified in one or more of the ring hydroxyl positions in the following type of reaction where HOA(R') represents the acceptor (DR represents the glycosyl donor):

$$\text{D-R} \; + \; \text{HOA(R')} \; \longrightarrow \; \text{D-OA(R')} \; + \; \text{RH}$$

$$\text{Glycosidase}$$

Saccharide acceptors modified with the common protection groups (R' in the scheme above) used in carbohydrate synthesis for production of larger saccharides or for the preparation of analogs/derivatives can be used. These include acetyl, allyl or benzyl groups for hydroxyl group modification and the phthalimido-group for amino group protection (e.g. on glucosamine). A few examples of reactions catalyzed by the coffee bean α-galactosidase are shown in Figure 8.[59]

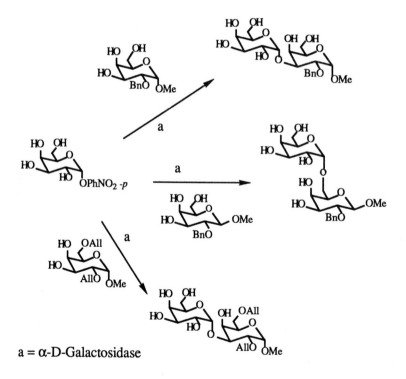

a = α-D-Galactosidase

Figure 8.   Synthesis of partially protected digalactosides with α-D-galactosidase.

As shown the galactoside acceptor could be modified in three positions and still act efficiently as an acceptor. In fact, the rate of reaction and yield of product are comparable and in some cases higher with modified acceptors than with the corresponding unprotected acceptor glycosides. This was the case with for example the acceptors (2-O-Bn)Galα-OMe and (2-OAll,6-O-All)Galα-OMe in the α-galactosidase catalyzed reactions shown in Figure 8. We have found similar results with other glycosidases, including α-fucosidase, α-mannosidase, β-hexosaminidase and β-galactosidases, of which the latter allows for preparation of lactosamine derivatives and regioisomers modified in the amino group and in one or more of the hydroxyl groups.

The above method allows derivatives suitable for use as inhibitors or for further synthesis of carbohydrate analogs or oligosaccharides to be synthesized, as exemplified in the scheme below:

Galα(1->3)(6-All)Galα-OMe
Galα(1->3)(2,6-All)Galα-OMe } ——————→ Analogs, Oligosaccharides
Galα(1->3)(2-Bn)Galα-OMe   Galα(1->3)(Fucα1->2)Galα-OMe
(Blood group B)

Chemical synthesis at present allows for preparation of a broader range of derivatives than the above method, but remains a complex task for preparation of specifically modified saccharides.[43] The efficiency of glycosidases to produce these types of compounds are attractive and are in contrast with glycosyltransferase catalyzed reactions, which usually are highly sensitive to modifications of the acceptor structure.

As mentioned above (see 2.A.i.) glycosidases also show some activity towards donor substrates modified in the glycon part (D in the scheme above). Even if transglycosylation reactions with such modified donor substrates are relatively slow they can be used to produce disaccharides modified in the non-reducing end. Thus, the use of glucal and galactal as glycosyl donors resulted in the formation of β-2-deoxy-D-glycosides upon prolonged reaction with various saccharide acceptors employing β-glucosidase or β-galactosidase as catalysts.[60]

3. F. Synthesis of Larger Structures

The above examples relate to syntheses with exoglycosidases of glycosides and disaccharide structures present in glycoconjugates. The synthesis of some trisaccharides has been achieved as well. Manα(1→2)Manα(1→2)Manα-OMe is formed in a "one-pot" reaction from Manα-OPhNO$_2$-p and Manα-OMe with jack bean α-mannosidase as catalyst.[38] Similarly, with bovine testes ß-galactosidase, Galß(1→3)Galß(1→4)GlcNAcß-OEtSiMe$_3$ is formed from Galß-OPhNO$_2$-o (glycosyl donor) and GlcNAcß-OEtSiMe$_3$ (acceptor).[37] Interestingly, the ß-galactosidase seems to be selective for the acceptor disaccharide Galß(1→4)GlcNAcß-OEtSiMe$_3$, since the ß(1→3) isomer which was the major disaccharide glycoside formed in the reaction, was a poor acceptor. The reaction was quantitative (about 50 % yield) and this enzyme seems to be useful for the preparative synthesis of various trisaccharides and even higher oligosaccharide structures.[37] Recently, the sequential use of α-sialidase and β-galactosidase for the regiospecific synthesis of NeuAcα(2→3)Galβ-(1→4)GlcNAc was reported.[61] Although the yield of the final sialidase-reaction was rather low, it might be possible to improve the reaction to allow for preparative scale synthesis.

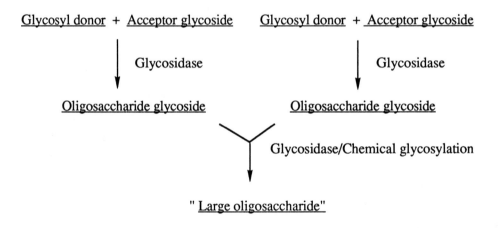

Figure 9. A general strategy for convergent block synthesis of oligosaccharides employing glycosidases.

Endoglycosidases such as lysozyme[62] and endo-ß-N-acetylglucosaminidase F[63] have transfer activity and endo-α-N-acetyl-galactosaminidase from *Diplococcus pneumoniae* was reported to catalyze the formation of trisaccharides of the type Galß(1→3)GalNAcα-OR.[64]

The use of exoglycosidases for synthesis of shorter fragments, e.g. partially modified disaccharide glycosides such as thioethyl glycosides which are suitable for convergent block synthesis, and the use of endoglycosidases or chemical synthesis for condensation of these fragments is a useful strategy for synthesis of higher oligosaccharides with glycosidases (Figure 9).

3.G. Sequential Use of Glycosidase and Glycosyltransferase.

Glycosidases are suitable for use alone or in combination with other methods for achieving synthetic goals. The broader application of glycosyltransferases for the synthesis of oligosaccharides is at present limited by several factors including the low availability, cofactor dependence and linkage specificity (one enzyme - one linkage).

In one approach for synthesis of oligosaccharides, the more available glycosidases are used to produce acceptor oligosaccharides and the use of glycosyltransferases is limited to the final glycosylation steps when the demand for regiospecificity is higher.[5,37] In this way the glycosyltransferase reaction can be used like a molecular trap for the product formed in the glycosidase-catalyzed reaction. Two examples are shown in Figure 10. Here, ß-galactosidase and ß-galactoside α(2→3)sialyltransferase were used for the synthesis of the methyl glycoside of the sialylated sequence Neu5Acα(2→3)Galß(1→3)GlcNAcß- and the bromoethyl glycoside of Neu5Acα(2→3)Galß(1→3)GalNAcß-.[39] The latter trisaccharide sequence is a common terminal sequence of glycolipids and O-linked glycoproteins. The use of immobilized enzymes facilitated reuse with high retention of activity.[65] The 2-bromoethyl glycoside was obtained in a reaction catalysed by a ß-hexosaminidase. The nucleotide sugar CMP-NeuAc can be produced by straightforward enzymatic synthesis[65-67] or by a multi-enzyme system for cofactor recycling.[68]

This type of approach is expected to be important for large scale synthesis when more efficient expression systems have been developed for large-scale production of glycosyltransferases. The method is an attractive complement to e.g. the chemoenzymatic approach in which chemical methods are used to produce acceptor saccharides for subsequent glycosylation with glycosyltransferase.[43]

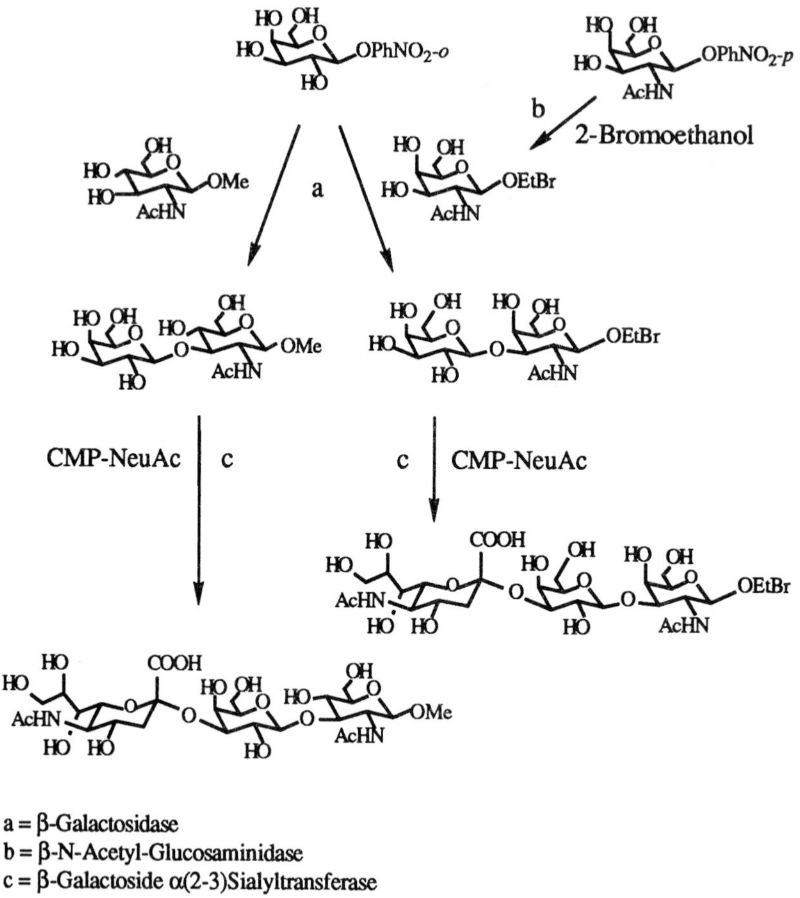

a = β-Galactosidase
b = β-N-Acetyl-Glucosaminidase
c = β-Galactoside α(2-3)Sialyltransferase

Figure 10.   Synthesis of sialylated trisaccharides by the sequential use of β-galactosidase and sialyltransferase.

4. CONCLUSION

The use of glycosidases for glycosylation reactions to prepare glycosides and glycoconjugate glycans have been briefly reviewed. Previously, glycosidases have been mainly associated with hydrolysis reactions both on an analytical and industrial scale. The previous obstacles of employing glycosidases in synthetic applications have been largely solved at least for the synthesis of shorter sequences and the last decade has witnessed an increasing number of successful synthetic applications.

The general characteristics of the glycosidases in the described reactions can be summarized as follows:

- Glycosidases are easily available. An abundance of enzymes which are relatively easy to prepare in sufficient quantities for either small scale or large scale synthesis are available.

- Broad range of reaction conditions. The glycosidases are often quite stable and can be used as soluble or as immobilized, reusable catalysts either under mild conditions (neutral buffered water, ambient temperature) or under more harsh conditions, such as aqueous-organic solvent mixtures or increased temperatures with high substrate concentrations. This allows for glycosylation of both sensitive molecules and molecules with lower solubility in water.

- Broad range of substrates and products. A broad range of substrates from simple sugars, glycosides or alcohols to more complex molecules, can be used with approximately equal efficiency as acceptors which allows for preparative, stereospecific and regioselective synthesis of a relatively broad range of products, even of linkages not yet identified in living organisms. Thus, a wide range of glycosides, saccharides and modified carbohydrates have been selectively prepared with glycosidases.

It is to be expected that these catalysts will continue to be important for the scientific and clinical development of biologically active carbohydrates and glycosides in the future.

REFERENCES

1. H. M. Flowers and N. Sharon, *Adv. Enzymol.,* **48**, 29 (1979).

2. *Enzyme Nomenclature,* Recommendations of the International Union of Biochemistry, Academic Press, New York, p. 306 (1984).

3. K. Wallenfels and R. Weil, *The Enzymes,* **7**, 617 (1972).

4. P. M. Dey and E. Campillo, *Adv. Enzymol.,* **56**, 141 (1984).

5. K. G. I. Nilsson, *Trends Biotechnol.,* **6**, 256 (1988).

6. P. M. Dey and J. B. Pridham, *Adv. Enzymol.,* **36**, 91 (1972).

7. S.-C. Li and Y.-T. Li, *J. Biol. Chem.,* **245**, 5153 (1970).

8. A. D. Elbein, S. Adya, and Y. C. Lee, *J. Biol. Chem.,* **252**, 2026 (1977).

9. S. Yasawa, R. Madiyalakan, R. P. Chawda, and K. L. Matta, *Biochem.*

Biophys. Res. Commun., **136**, 563 (1986).

10.   K. Nisizawa and Y. Hashimoto, In *The Carbohydrates, Chemistry and Biochemistry,* (W. Pigman and D. Horton, Eds.) Academic Press, New York, p. 241 (1975).

11.   R. A. DiCioccio, P. J. Klock, J. J. Barlow and K. L. Matta, *Carbohydr. Res.,* **81**, 315 (1980).

12.   J. J. Distler and G. W. Jourdian, *J. Biol. Chem.,* **248**, 6772 (1973).

13.   R. A. DiCioccio, J. J. Barlow, and K. L. Matta, *J. Biol. Chem.,* **257**, 714 (1982).

14.   D. H. Van den Eijnden, W. M. Blanken, and A. van Vliet, *Carbohydr. Res.,* **151**, 329 (1986).

15.   P. M. Dey, E. M. Diel Campillo, and R. Pont Lezica, *J. Biol Chem.,* **258**, 923 (1983).

16.   E. Steers Jr., P. Cuatrecasas, and H. B. Pollard, *J. Biol. Chem.,* **246**, 196 (1971).

17.   P. Cuatrecasas, *Methods Enzymol.,* **28**, 897 (1972).

18.   R. Barker, C. K. Chiang, I. P. Trayer, and R. L. Hill, *Methods Enzymol.,* **34**, 317 (1974).

19.   C. Svanborg-Edén, R. Freter, L. Hagberg, R. Hull, H. Leffler, and G. Schoolnik, *Nature,* **298**, 560 (1982).

20.   J. Dahmen, T. Frejd, F. Lave, F. Lindh, G. Magnusson, G. Noori, and K. Pålsson, *Carbohydr. Res.,* **113**, 219 (1983).

21.   K. G. I. Nilsson and C.-F. Mandenius, *Bio/Technology,* In Press (1994).

22.   A. Kobata, *Anal. Biochem.,* **100**, 1 (1979).

23.   A. Kobata and S. Takasaki, *Methods Enzymol.,* **50**, 560 (1978).

24.   (a) L. Hedbys, E. Johansson, K. Mosbach, P.-O. Larsson, A. Gunnarsson, and S. Svensson, *Carbohydr. Res.,* **186**, 217 (1989); (b) L. Hedbys, E. Johansson, K. Mosbach, P.-O. Larsson, A. Gunnarsson, and S. Svensson, *Glyconjugate J.,* **6**, 161 (1989).

25.   K. Sakai, R. Katsumi, H. Ohi, T. Usui, and Y. Ishido, *J. Carbohydr. Chem.,* **11**, 553 (1992).

26.   A. C. Hill, *J. Chem. Soc., Trans.,* **73**, 634 (1898).

27.   S. Adachi, Y. Ueda, and K. Hashimoto, *Biotechnol. Bioeng.,* **26**, 121

(1984).

28. E. Johansson, L. Hedbys, K. Mosbach, P.-O. Larsson, A. Gunnarsson, and S. Svensson, *Enzyme Microb. Technol.*, **11**, 349 (1989).

29. E.N. Vulfsson, R. Patel, and B. A. Law, *Biotechnol. Lett.*, **12**, 397 (1990).

30. G. Ljunger, P. Adlercreutz, and B. Mattiasson, *Enzyme Microb. Technol.*, **16**, 751 (1994).

31. E. Johansson, L. Hedbys, and P.-O. Larsson, *Enzyme Microb. Technol.*, **13**, 781 (1991).

32. K. G. I. Nilsson and M. Scigelová, *Biotechnol. Lett.*, **16**, 671 (1994).

33. M. Scigelová, V. Krén, and K. G. I. Nilsson, *Biotechnol. Lett.*, **16**, 683 (1994).

34. F. Zilliken, P. N. Smith, C. S. Rose, and P. György, *J. Biol. Chem.*, **217**, 79 (1955).

35. A. Allesandrini, E. Schmidt, F. Zilliken, and P. György, *J. Biol. Chem.*, **220**, 71 (1956).

36. E. J. Toone, E. S. Simon, M. D. Bednarski, and G. M. Whitesides, *Tetrahedron*, **45**, 5365 (1989).

37. K. G. I. Nilsson, *Carbohydr. Res.*, **188**, 9 (1989).

38. K. G. I. Nilsson, *Carbohydr. Res.*, **167**, 95 (1987).

39. K. G. I. Nilsson, In *Sialic Acids*, (R. Schauer and T. Yamakawa, Eds.) Kieler Verlag, Wissenschaft und Bildung, p. 28 (1988).

40. K. G. I. Nilsson, *Carbohydr. Res.*, **180**, 53 (1988).

41. K. G. I. Nilsson, *Ann. N. Y. Acad. Sci.*, **542**, 383 (1988).

42. K .G. I. Nilsson, *Carbohydr. Res.*, **204**, 79 (1990).

43. S. H. Khan and O. Hindsgaul, In *Molecular Glycobiology*, (M. Fukuda and O. Hindsgaul, Eds.) IRL Press, Oxford, p. 206 (1994).

44. K. Wallenfels and O. P. Malhotra, *Adv. Carbohydr. Chem. Biochem.*, **16**, 239 (1961).

45. K.G.I. Nilsson, In *Studies in Organic Chemistry*, (C. Laane, J. Tramper and M.D. Lilly, Eds.) Elsevier, Amsterdam, p. 369 (1987).

46. R. Kuhn, H. H. Baer, and A. Gauhe, *Chem. Ber.*, **188**, 1713 (1955)

47. K. G. I. Nilsson, In *Opportunities in Biotransformation*, (I. G. Copping, R.E. Martin, J. A. Pickett, C. Bucke, and A.W. Bunch, Eds.) Elsevier

Applied Science, London, p. 131 (1990).

48. S. C. T. Svensson and J. Thiem, *Carbohydr. Res.*, **200**, 391 (1990).

49. K. Ajisaka and M. Shirakabe, *Carbohydr. Res.*, **224**, 291 (1992).

50. D. H. G. Crout, S. Sing, B. E. P. Swoboda, P. Critchley, and W. T. Gibson, *J. Chem. Soc., Chem. Commun.*, 704 (1992).

51. J. Thiem and B. Sauerbrei, *Angew. Chem. Int. Ed. Engl.*, **30**, 1503 (1991).

52. K. G. I. Nilsson, In *Applied Biocatalysis*, (H. W. Blanch and D. S. Clark, Eds.) Marcel Dekker, Inc. New York, p. 117 (1991).

53. K. G. I. Nilsson, *ACS Symp. Ser.*, **466**, 51 (1991).

54. P. Fügedi, P. J. Garegg, H. Lönn, and T. Norberg, *Glycoconjugate J.*, **4**, 97 (1987).

55. K. G. I. Nilsson, US Patent No. **4,918,009** (1990).

56. D. Cantacuzene and S. Attal, *Carbohydr. Res.*, **211**, 327 (1991).

57. N. J. Turner and M. C. Webberley, *J. Chem. Soc., Chem. Commun.*, 1349 (1991).

58. Y. Ooi, T. Hashimoto, N. Mitsuo, and T. Satoh, *Chem. Pharm. Bull.*, **33**, 1808 (1985).

59. K. G. I. Nilsson and A. Fernandez-Mayoralas, *Biotechnol. Lett.*, **13**, 914 (1991).

60. J.-M. Petit, F. Paquet, and J.-M. Beau, *Tetrahedron Lett.*, **32**, 6125 (1991).

61. K. Ajisaka, H. Fujimoto, and M. Isomura, *Carbohydr. Res.*, **259**, 103 (1994).

62. E. Deya, K. Nojiri, and S. Igarashi, *Nippon Nogei Kagaku Kaishi*, **58**, 273 (1984)

63. R. B. Trimble, P. H. Atkinson, A. L. Tarentino, T. Plummer, F. Maley, and K. B. Tomer, *J. Biol Chem.*, **261**, 12000 (1986).

64. R. M. Bardales and V. P. Bhavanandan, *J. Biol. Chem.*, **264**, 19893 (1989).

65. K. G. I. Nilsson and B.-M. E. Gudmundsson, *Meth. Mol. Cellul. Biol.*, **1**, 195 (1990).

66. E. S. Simon, M. D. Bednarski, and G. M. Whitesides, *J. Am. Chem. Soc.*, **110**, 7159 (1988).

67. C. Augé and C. Gautheron, *Tetrahedron Lett.*, **29**, 789 (1988).

68. G. F. Herrmann, Y. Ichikawa, C. Wandrey, F. C. A. Gaeta, J. C. Paulson,

and C.-H. Wong, *Tetrahedron Lett.*, **34**, 3091 (1993).

INDEX